Solid State
Physical Electronics

SOLID STATE PHYSICAL ELECTRONICS SERIES
Nick Holonyak, Jr., *Editor*

ANKRUM   Semiconductor Electronics

BURGER and DONOVAN, Editors   Fundamentals of Silicon Integrated
Device Technology:
Vol. I: Oxidation, Diffusion and Epitaxy.
Vol. II: Bipolar and Unipolar Transistors

GENTRY, et al.   Semiconductor Controlled Rectifiers:
Principles and Applications of p-n-p-n Devices

LAUDISE   The Growth of Single Crystals

NUSSBAUM   Applied Group Theory for Chemists, Physicists and Engineers

NUSSBAUM   Electromagnetic and Quantum Properties of Materials

NUSSBAUM   Semiconductor Device Physics

PANKOVE   Optical Processes in Semiconductors

ROBERTS and VANDERSLICE   Ultrahigh Vacuum and Its Applications

STREETMAN   Solid State Electronic Devices

UMAN   Introduction to the Physics of Electronics

VAN DER ZIEL   Solid State Physical Electronics, 3rd ed.

WALLMARK and JOHNSON, Editors   Field-Effect Transistors:
Physics, Technology and Applications

WESTINGHOUSE ELECTRIC CORPORATION   Integrated Electronic Systems

Solid State
Physical Electronics

Third Edition

Aldert van der Ziel

Department of Electrical Engineering
University of Minnesota

Department of Electrical Engineering
University of Florida

PRENTICE-HALL INC., *Englewood Cliffs, New Jersey*

Library of Congress Cataloging in Publication Data

VAN DER ZIEL, ALDERT
  Solid state physical electronics.
  (Solid state physical electronics series)

  Includes bibliographies.
  1. Electronics.   2. Semiconductors.   3. Solids.
  I. Title.
TK7835.V35 1976   621.381   75-5661
ISBN 0-13-821603-7

© 1976 by Prentice-Hall, Inc.,
Englewood Cliffs, New Jersey

10 9 8 7 6 5 4 3 2 1

Printed in the United States of America

PRENTICE-HALL INTERNATIONAL Inc., *London*
PRENTICE-HALL OF AUSTRALIA, PTY. LTD., *Sydney*
PRENTICE-HALL OF CANADA, LTD., *Toronto*
PRENTICE-HALL OF INDIA PRIVATE LTD., *New Delhi*
PRENTICE-HALL OF JAPAN, INC., *Tokyo*
PRENTICE-HALL OF SOUTHEAST ASIA (PTE.) LTD., *Singapore*

To my children

JAN P. (*Bell Telephone Laboratories*)
CORNELIA H. J., M.D.
JOANNA C. (*University of Minnesota*)

who are pursuing a career in science

Contents

19 MISCELLANEOUS SEMICONDUCTOR DEVICES   464

20 MISCELLANEOUS SEMICONDUCTOR PROBLEMS   485

APPENDICES   507

INDEX   519

Preface

For the third edition of *Solid State Physical Electronics* to keep up with the rapidly changing field, a major revision and extension of the second edition was necessary. As a consequence the part of dielectric, piezoelectric and magnetic devices had to be eliminated.

This edition consists of three parts: I. Introductory chapters (1-6), dealing with wave mechanics, atomic physics, statistics, the theory of solid state materials in general, and of semiconductors in particular; II. Electron emission and conduction devices (Chapters 7-13), dealing with thermionic emission, field emission, photoemission, secondary electron emission, photoconduction, luminescence, and the application of these principles to electron devices; III. Semiconductor devices (Chapters 14–20), dealing with metal-semiconductor diodes, *p-n* junction diodes, bipolar transistors, field effect transistors, junction luminescence, *p-n* junction lasers, avalanche oscillators, Gunn oscillators, space-charge-limited devices, Hall devices, thermoelectric devices, amorphous semiconductor devices, and their applications.

The author is indebted to Drs. A. Nussbaum, W. T. Peria and J. P. van der Ziel for discussing sections of the book, and to his graduate students at the Universities of Minnesota and Florida for trying out parts of the new manuscript. He is indebted to Dr. K. M. van Vliet for his work on high-level injection effects in *p-n* junctions and bipolar transistors, and to Mrs. van der Ziel for her help in preparing the manuscript.

A. VAN DER ZIEL

Minneapolis, Minnesota

Solid State
Physical Electronics

1

Structure of
the Solid State

In some solids, such as glass, the atoms or atom groups are arranged in an irregular, more or less random fashion like the atoms or molecules in a liquid; such solids are called *noncrystalline* or *amorphous.* In other solids the atoms or atom groups are arranged in a regular order; such solids are called *crystalline.* If the regular order extends over a whole piece of material, the material is said to be a *single crystal;* if the regularity extends over only a small part of the material, so that it consists of an agglomerate of smaller or larger crystallites, the material is said to be *polycrystalline.* If the crystallites are very small, the crystalline character of the solid may manifest itself only in x-ray or electron-diffraction experiments; such techniques may in fact be used to determine the size of the crystallites.

1.1. CRYSTAL SYSTEMS AND NOMENCLATURE

Since the atoms or atom groups of a crystal are arranged in a regular order, it is possible to move the crystal in certain directions over certain distances such that each atom again coincides with an atom of the same kind. Thus we can define *three fundamental translation vectors* **a, b,** and **c,** such that the most general movement that makes each atom coincide with one of the same kind is a movement in the direction of and over the length of a vector **v,**

$$\mathbf{v} = n_1\mathbf{a} + n_2\mathbf{b} + n_3\mathbf{c} \tag{1.1}$$

where n_1, n_2, and n_3 are arbitrary integers. The points in space defined by Eq. (1.1) are called *lattice points;* the assembly of lattice points defined by (1.1) is called a *space lattice;* the parallelepiped defined by the vectors **a, b,** and **c** is called a *primitive cell.* An infinite number of primitive cells can be defined for a given crystal; usually the one with the highest symmetry is chosen. The space lattice is a geometrical structure that has the same symmetry as the crystal, but that has nothing to do with the crystal otherwise. In particularly simple cases we can assign an atom or atom group to each lattice point.

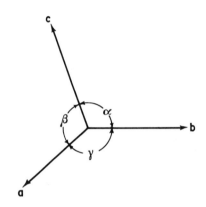

Fig. 1.1. Arbitrary primitive cell $a \neq b \neq c$, $\alpha \neq \beta \neq \gamma \neq 90°$.

A primitive cell has lattice points only at the corners. It is sometimes more convenient to draw *unit cells,* which are somewhat larger than the primitive cells and which have lattice points not only at the corners but also at the centers of certain faces or at the center of volume of the unit cell. This is done so that the symmetry of the unit cell comes closer to the symmetry of the crystal itself.

The sides of the unit cell define *fundamental translation vectors* **a, b,** and **c.** We obtain *seven crystal systems* characterized by these vectors—that is, by their lengths a, b, and c, and their angles α, β, and γ (Fig. 1.1):

Triclinic:	$\alpha \neq 90°, \beta \neq 90°, \gamma \neq 90°$	$c \leqq a \leqq b$
Monoclinic:	$\alpha = \gamma = 90°, \beta \neq 90°$	$c \leqq a; b$ arbitrary
Orthorhombic:	$\alpha = \beta = \gamma = 90°$	$c < a < b$
Tetragonal:	$\alpha = \beta = \gamma = 90°$	$a = b \neq c$
Hexagonal:	$\alpha = \beta = 90°, \gamma = 120°$	$a = b; c$ arbitrary
Cubic:	$\alpha = \beta = \gamma = 90°$	$a = b = c$
Rhombohedral:	$\alpha = \beta = \gamma \neq 90°$	$a = b = c$

The unit cells form together the *space lattice.* A space lattice is called *simple* if the unit cell has lattice points only in the corners, *base-centered* if it also has lattice points at the center of two opposite faces, *face-centered* if it also has lattice points at the center of all faces, and *body-centered* if it has a lattice point at the center of volume (Fig. 1.2). This does not yield 4×7 possible arrangements, since not all these possibilities are actually needed. In some cases the unit cell can be replaced by an elementary cell of the same symmetry, in other cases the extra lattice points would destroy the symmetry

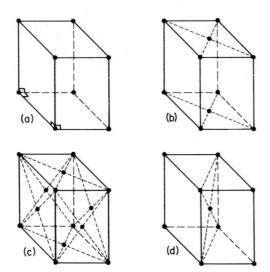

Fig. 1.2. Orthorhombic unit cells. (a) Simple; (b) base-centered; (c) face-centered; (d) body-centered.

of the unit cell, and so on. It can be shown by geometrical arguments that only the following *fourteen space lattices occur:*

1. Triclinic: simple
2. Monoclinic: simple, body-centered
3. Orthorhombic: simple, base-centered, body-centered, face-centered
4. Tetragonal: simple, body-centered
5. Hexagonal: simple
6. Cubic: simple, face-centered, body-centered
7. Rhombohedral: simple

It is common practice to specify the orientation of a crystal plane by its *Miller indices*. These are defined as follows:

1. Let the plane intercept the basis axis of the crystal at the points n_1a, n_2b, and n_3c, where a, b, and c are the side lengths of the unit cell. The plane can thus be characterized by the set of numbers (n_1, n_2, n_3).

2. The Miller indices (h, k, l) are now defined as the smallest set of integers m/n_1, m/n_2, m/n_3, with m being the lowest integer for which all three numbers m/n_1, m/n_2, and m/n_3 are integral. Examples are given in Fig. 1.3.

If the atoms that constitute the crystal can be considered hard spheres, there are three possible ways of packing the spheres densely.

(a)100 plane   (b)110 plane   (c)111 plane

Fig. 1.3. Location of the crystal planes described by the Miller indices (100), (110), and (111).

1. If the layers of spheres are arranged so that each sphere has *four nearest neighbors*, the spheres of the second layer fit into the holes of the first layer, the spheres of the third layer fit into the holes of the second layer, and so on. It is easy to see that the spheres of the third layer are right above the spheres of the first layer and that the arrangement forms a *body-centered cubic lattice*. In this case the atoms *touch along the body diagonals* of the unit cell.

2. If the layers of spheres are arranged so that each sphere has *six closest neighbors*, the spheres of the second layer fit into the holes of the first layer and the spheres of the third layer fit into the holes of the second layer. The latter can happen in two ways:

(a) The spheres of the third layer are placed directly over the spheres of the first layer, so that the arrangement has a hexagonal symmetry; this is the *hexagonal close-packed structure* (Fig. 1.4a).

(b) The spheres of the third layer are placed directly over those holes of the first layer that are not occupied by the second layer. We can see by inspection that this leads to a face-centered cubic structure (*cubic close-packed structure*)(Fig. 1.4b). Both close-packed structures have equal density; their density is slightly greater than for the body-centered cubic structure, since the packing is closer.

In both structures there are two types of holes between the spheres. Between four spheres packed in a tetrahedron there are *tetrahedral* holes, and between six spheres packed in an octahedron there are *octahedral* holes. The octahedral holes are somewhat larger than the tetrahedral ones. By drawing a unit cell and inspecting it closely (Fig. 1.5), the reader can verify for himself that the eight tetrahedral holes are in the eight corners of the unit cell. One octahedral hole is wholly inside the cell. The cell also contains parts of twelve octahedral holes. Each of these octahedral holes is shared by four neighboring cells; we thus have in total $1 + \frac{12}{4} = 4$ octahedral holes per unit cell.

This arrangement has an important bearing on the crystal structure of ionic crystals (Sec. 1.2a). Often the large negative ions form a close-packed structure and the small positive ions are located in the holes. Small ions are

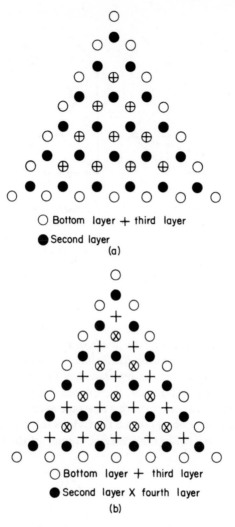

○ Bottom layer + third layer
● Second layer
(a)

○ Bottom layer + third layer
● Second layer X fourth layer
(b)

Fig. 1.4. Dense packing of hard spheres (hexagonal and cubic). A common feature of both structures is that one can designate crystal planes in which each atom is surrounded symmetrically by six neighbors so that the atoms of the next plane fit into the holes of the previous one. In the hexagonal structure these are the 100 planes, in the cubic structure the 111 planes.

The figure shows a projection of the subsequent close-packed atomic layers upon the bottom layer. (a) Hexagonal close-packed structure. The third atomic layer is above the first layer, giving the structure a hexagonal symmetry; the hexagonal axis is perpendicular to the bottom layer. (b) Cubic close-packed structure. The fourth atom layer is above the first layer. The body diagonal of the cubic structure is perpendicular to the bottom layer.

5

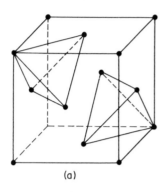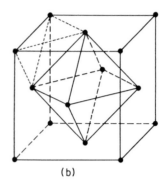

(a)   (b)

Fig. 1.5. Location of the tetrahedral and octahedral holes in a close-packed cubic structure. (a) The tetrahedral holes are formed between one corner sphere and its three adjacent spheres in the face centers. Two of these tetrahedrons are drawn. (b) One full octahedral hole is formed by the six spheres in the face centers. Twelve additional octahedral holes, having one of the sides of the cube as body diagonal, are shared by four neighboring unit cells. One of them is partly shown by two full-drawn lines and four dotted lines.

preferably located in the tetrahedral holes; if they are too big for the tetrahedral holes, they are located in the octahedral ones.

For example, the Cl ions in the *NaCl structure* form a cubic close-packed structure, and the Na ions are so big that they fit only into the octahedral holes (Fig. 1.6a). In CsCl, however, the Cs ions are too big to fit even into the octahedral holes; as a consequence a new type of crystal structure appears, the *CsCl structure* (Fig. 1.6b), in which the Cl ions form a simple cubic lattice and the Cs ions are located in the center of volume of the unit cells.

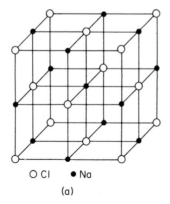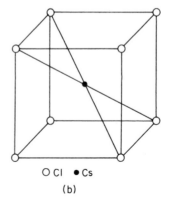

○ Cl   ● Na   ○ Cl   ● Cs

(a)   (b)

Fig. 1.6. (a) NaCl structure. The Cl ions form a cubic close-packed structure, the Na ions occupy the octahedral holes and are located in the middle of the body diagonals of the octahedron. (b) CsCl structure.

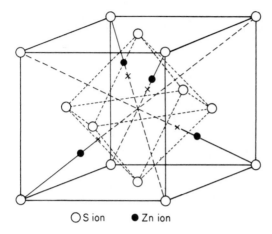

○ S ion   ● Zn ion

Fig. 1.7. Cubic ZnS (zincblende structure). The S ions form a close-packed structure; the Zn ions are located in the center of the tetrahedral holes, half of which are filled.

In ZnS, the *zincblende structure* occurs; the S ions form a cubic, close-packed structure and the very small Zn ions again fit into the tetrahedral holes (Fig. 1.7). Another modification of ZnS, the *wurtzite structure*, differs in that the S ions form a hexagonal close-packed structure, the Zn ions again being located in the tetrahedral holes.

1.2. TYPES OF CRYSTAL BINDING

The most important examples of crystal binding that concern us here are the ionic, the metallic, the covalent, and the hydrogen bonds.

1.2a. Ionic Bond

In the ionic bond the atoms are ionized so that the electron structure of each ion consists of closed shells. Examples are Na^+Cl^- and $Mg^{2+}O^{2-}$. Since each ion has a closed shell, and therefore has quantum numbers $L = 0$ and $S = 0$, the charge distribution has spherical symmetry (see Chapter 2). One would therefore expect crystals of high symmetry—for example, crystals of the cubic and the hexagonal type. In crystals in which the negative ions are atom complexes, the crystal structure may be determined by the structure of the negative ions.

The binding forces at larger distances are strong electrostatic forces; at very small distances the space-charge clouds start to penetrate, and a very strong repulsion occurs. As a consequence, the ions are drawn so closely together that the space-charge clouds just start to penetrate. This is the explanation for the close-packed structure of the negative ions in many ionic crystals and for the large binding energy of those crystals.

1.2b. Metallic Bond

The metallic bond structure is formed by positively charged metal ions embedded in a sea of mobile free electrons. The negative charges are not fixed to atoms bound in fixed positions, as in the ionic bond, but are associated with the mobile free electrons that are distributed more or less uniformly through the crystal. This explains the large differences in the properties of metals and ionic crystals. The attractive force is electrostatic, and usually the binding is so strong that the metal ions form a close-packed structure of either the hexagonal or the cubic type. In the case of the alkali metals the atomic distance is relatively large and the body-centered cubic lattice is energetically more favorable; this explains the relatively small binding energy for these metals.

1.2c. Covalent Bond

With the covalent bond a strong bond between the atoms is obtained because of the mutual sharing (exchange) of electrons by adjacent atoms; bonds in which two atoms mutually share an *electron pair* are particularly stable and are called *covalent bonds*. The binding is electrostatic but of a form that cannot be explained on a classical basis; the interaction is a consequence of the wave character of the electron. It is known as *exchange interaction*.

A good example of such a bond is the structure of diamond, silicon, and germanium. Each of these atoms has four outer electrons and therefore can share an electron pair with four neighbors, which are arranged around it in a symmetrical fashion; hence the four neighbors form a regular tetrahedron with the atom itself in its center (Fig. 1.8). This gives rise to a cubic structure known as the *diamond* structure.

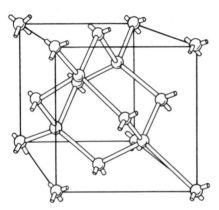

Fig. 1.8. Diamond structure (cubic). Each carbon atom is tetrahedrally surrounded by four neighbors. (Courtesy *Bell System Technical Journal*.)

Often the binding in a crystal is a mixture of two types of bonds. In some oxides the bond is a mixture of ionic and covalent bonds. In metals having an incomplete d shell ($l = 2$; see Chapter 2), the d electrons, which are relatively deep inside the atom, form a kind of covalent bond; the outer electrons are free and give the element its metallic character. This case is thus a mixture of the metallic and covalent bonds; it explains why such metals have a very large binding energy.

The difference in these three structures is manifested in the electron density between the ions making up the lattice. This electron density is small in ionic crystals, large in valence-bond crystals, and more or less uniform in metallic crystals. Such differences can be detected with the help of accurate intensity measurements in the x-ray diffraction pattern of the crystal.

1.2d. Hydrogen Bond

In the hydrogen bond two negative ions are linked together by means of a hydrogen ion (the hydrogen having lost its electron to one of its neighbors or to both neighbors, which share it). This bond occurs in ice.

REFERENCES

BLAKEMORE, J. S., *Solid State Physics*. W. B. Saunders Company, Philadelphia, 1969.

DEKKER, A. J., *Solid State Physics*. Prentice-Hall, Inc., Englewood Cliffs, N.J., 1957.

HERSHBERGER, W. D., ed., *Topics in Solid State and Quantum Electronics*. John Wiley & Sons, Inc., New York, 1972.

KANO, K. A., *Physical and Solid State Electronics*. Addison-Wesley Publishing Company, Inc., Reading, Mass., 1972.

KITTEL, C., *Introduction to Solid State Physics*, 4th ed. John Wiley & Sons, Inc., New York, 1971.

MARTON, L., ed., *Methods of Experimental Physics*, vol. 6, parts A and B, *Solid State Physics*. Academic Press, Inc., New York, 1959.

MCKELVEY, J. P., *Solid State and Semiconductor Physics*. Harper & Row, Inc., New York, 1966.

NUSSBAUM, A., *Electronic and Magnetic Behavior of Materials*, Prentice-Hall, Inc., Englewood Cliffs, N.J., 1967.

SACHS, M., *Solid State Theory*. McGraw-Hill Book Company, New York, 1963.

SEITZ, F., *The Modern Theory of Solids*. McGraw-Hill Book Company, New York, 1940.

STREETMAN, B. G., *Solid State Electronic Devices*, Prentice-Hall, Inc., Englewood Cliffs, N.J., 1972.

WANG, S., *Solid State Electronics*. McGraw-Hill Book Company, New York, 1966.

ZIMAN, J. M., *Principles of the Theory of Solids*. Cambridge University Press, New York, 1965.

PROBLEMS

1.   Show that a simple orthorhombic lattice, a body-centered orthorhombic lattice, and a face-centered orthorhombic lattice have one, two, and four lattice points per unit cell, respectively.

2.   Show that a base-centered tetragonal lattice can be replaced by a simple tetragonal lattice. *Hint:* Draw two unit cells with the bases in the same plane and then draw new simple unit cells with half the volume of the original unit cell.

3.   Show that a face-centered tetragonal lattice can be replaced by a body-centered tetragonal lattice. *Hint:* Draw two unit cells with the bases in the same plane and then draw a new tetragonal unit cell that has half the volume and is body-centered.

4.   Show from symmetry considerations that a base-centered cubic lattice does not exist.

5.   Show that in a cubic crystal the normal to the (h, k, l) planes has direction cosines

$$l_1 = \frac{h}{(h^2 + k^2 + l^2)^{1/2}}, \qquad l_2 = \frac{k}{(h^2 + k^2 + l^2)^{1/2}}, \qquad l_3 = \frac{l}{(h^2 + k^2 + l^2)^{1/2}}$$

6.   Show for a cubic crystal with a side length a of the elementary cell that the first (h, k, l) plane not passing through the origin meets the X axis, Y axis, and Z axis at a/h, a/k, and a/l, respectively.

7.   Show that the plane of problem 6 has a distance $a/(h^2 + k^2 + l^2)^{1/2}$ to the origin.

Hint: For the last few problems remember that the normalized equation of a plane with direction cosines (l_1, l_2, l_3) of its normal and with a distance d to the origin is

$$l_1 x + l_2 y + l_3 z = d$$

2

Wave Mechanics

2.1. THE TIME-INDEPENDENT WAVE EQUATION AND ITS PROPERTIES

We shall here derive the time-independent wave equation of atomic particles and discuss its properties. To that end, we first deal with the classical wave equation and then establish the atomic wave equation by analogy.

2.1a. Classical Wave Equation

Any classical wave phenomenon (sound, light, and so on) can be described by the classical time-dependent wave equation

$$\nabla^2 \psi - \frac{1}{v^2} \frac{\partial^2 \psi}{\partial t^2} = 0 \qquad (2.1)$$

Here $\psi(x, y, z, t)$ is the time-dependent wave function describing the wave phenomenon, v is the velocity with which the wave propagates, and ∇^2 is the operator

$$\nabla^2 = \frac{\partial^2}{\partial x^2} + \frac{\partial^2}{\partial y^2} + \frac{\partial^2}{\partial z^2} \qquad (2.1a)$$

operating on the wave function ψ.

To illustrate, consider the one-dimensional case

$$\frac{\partial^2 \psi}{\partial x^2} - \frac{1}{v^2} \frac{\partial^2 \psi}{\partial t^2} = 0 \qquad (2.1b)$$

As is seen by substitution, $g(x - vt)$ and $g(x + vt)$, where $g(u)$ is an arbitrary decent function of u, are solutions of the wave equation. The first solution represents a wave propagating in the positive X direction with a velocity v and the second a wave propagating in the negative X direction with a velocity v. The fact that $g(u)$ is an arbitrary function of u means that arbitrary wave pulses can propagate.

It is not difficult to extend these considerations to three dimensions. One then obtains as solutions arbitrary wave pulses propagating in arbitrary but fixed direction.

We now look for sinusoidal solutions, that is, solutions in which the time dependence of $\psi(x, y, z, t)$ is as $\exp(-j\omega t)$ and $\omega = 2\pi f$, where f is the frequency of the sinusoidal wave. We thus substitute

$$\psi(x, y, z, t) = \varphi(x, y, z) \exp(-j\omega t) \tag{2.2}$$

and obtain the time-independent wave equation

$$\nabla^2 \varphi + \left(\frac{2\pi}{\lambda}\right)^2 \varphi = 0 \tag{2.3}$$

with

$$\frac{2\pi}{\lambda} = \frac{\omega}{v} = \frac{2\pi f}{v} \quad \text{or} \quad \lambda = \frac{v}{f} \tag{2.3a}$$

Here λ is the *wavelength* of the wave and $\varphi(x, y, z)$ is the time-independent wave function, corresponding to the complex amplitude of the wave. Consequently, $\varphi^* \varphi \, dV$ is a measure for the wave energy contained in the volume element dV, so that

$$\int_V \varphi^* \varphi \, dV \tag{2.3b}$$

taken over the volume V, measures the energy in that volume.

If boundary conditions are applied to Eq. (2.3), one obtains solutions in particular cases only. These solutions represent the *standing-wave patterns* that are possible for the system under consideration. The most general solution is then a linear superposition of these standing-wave patterns.

In the case of wave propagation in inhomogeneous media, the wavelength λ is a function of position. For homogeneous media, λ is a constant.

We illustrate this for a one-dimensional wave, that is, a wave in which $\varphi(x, y, z)$ depends only on x. Equation (2.3) then becomes

$$\frac{d^2\varphi}{dx^2} + \left(\frac{2\pi}{\lambda}\right)^2 \varphi = 0 \tag{2.4}$$

with solutions $\varphi(x) = \exp(\pm 2\pi jx/\lambda)$, as is seen by substitution. The general solution of this equation is therefore

$$\varphi(x) = A \exp\left(2\pi j \frac{x}{\lambda}\right) + B \exp\left(-2\pi j \frac{x}{\lambda}\right) \tag{2.4a}$$

where A and B are arbitrary constants. We thus see that the wave form repeats itself when x is changed by an amount λ; that is,

$$\varphi(x + n\lambda) = \varphi(x), \qquad n = \pm 1, \pm 2, \ldots$$

This illustrates the meaning of the wavelength concept.

Multiplying by $\exp(-j\omega t)$, we see that the solution (2.4a) represents free-running plane waves; the first term in Eq. (2.4a) represents a plane wave propagating in the positive X direction, whereas the second wave propagates in the negative X direction.

If we apply boundary conditions such as

$$\varphi(x) = 0 \qquad \text{at } x = 0 \text{ and } x = L$$

we write the general solution as

$$\varphi(x) = C \sin\left(2\pi \frac{x}{\lambda}\right) + D \cos\left(2\pi \frac{x}{\lambda}\right) \tag{2.4b}$$

where C and D are arbitrary constants; it is seen by substitution that $\sin(2\pi x/\lambda)$ and $\cos(2\pi x/\lambda)$ are also solutions of Eq. (2.4). Applying the boundary condition at $x = 0$ yields $D = 0$, or

$$\varphi(x) = C \sin\left(2\pi \frac{x}{\lambda}\right) \tag{2.4c}$$

If we apply the boundary condition at $x = L$, then Eq. (2.4c) is only a solution when the wavelength λ satisfies the condition

$$\frac{2\pi L}{\lambda} = n\pi \quad \text{or} \quad \lambda = \frac{2L}{n}, \qquad n = 1, 2, \ldots \tag{2.4d}$$

Substituting this into Eq. (2.4c) and multiplying by $\exp(-j\omega t)$ shows that Eq. (2.4c) represents a standing-wave pattern. This verifies the statement made earlier.

2.1b. Wave Equation for Atomic Particles

Electron diffraction phenomena show that a wavelength

$$\lambda = \frac{h}{mv} \tag{2.5}$$

must be ascribed to a particle of mass m moving with a speed v; here h is Planck's constant. While the experiments prove the validity of Eq. (2.5) for constant v only, it is sensible to assume that it will also be valid if $v = v(x, y, z)$ is an arbitrary function of position.

If the particle has a total energy E and a potential energy $V(x, y, z)$, then the energy law requires that the kinetic energy must be written

$$\tfrac{1}{2}mv^2 = E - V(x, y, z) \tag{2.5a}$$

Solving for mv and substituting into Eq. (2.5) yields

$$\lambda = \frac{h}{[2m(E - V)]^{1/2}} \tag{2.5b}$$

Because of Eq. (2.5), the motion of an atomic particle becomes a problem of wave propagation to which a time-independent wave function $\varphi(x, y, z)$ can be assigned. In analogy with the classical case, one would thus expect that

$$\nabla^2\varphi + \left(\frac{2\pi}{\lambda}\right)^2 \varphi = 0 \tag{2.6}$$

where λ is given by Eq. (2.5b). Substituting Eq. (2.5b) into (2.6) yields

$$\nabla^2\varphi + \frac{2m}{\hbar^2}[E - V(x, y, z)]\varphi = 0 \tag{2.6a}$$

where $\hbar = h/2\pi$. This may be written in an alternative form, called *Schrödinger's time-independent wave equation:*

$$-\frac{\hbar^2}{2m}\nabla^2\varphi + V(x, y, z)\varphi = E\varphi \tag{2.7}$$

As a first example, consider the hydrogen atom. The potential energy is

$$V(x, y, z) = V(r) = -\frac{e^2}{4\pi\epsilon_0 r} \tag{2.8}$$

Here $r = (x^2 + y^2 + z^2)^{1/2}$, e is the charge of the electron, and $\epsilon_0 = 10^7/(4\pi c^2)$, where c is the velocity of light, is the MKS conversion factor. Hence the wave equation for the hydrogen atom is

$$-\frac{\hbar^2}{2m}\nabla^2\varphi - \frac{e^2}{4\pi\epsilon_0 r}\varphi = E\varphi \tag{2.9}$$

As a second example, consider a one-dimensional linear harmonic oscillator. Let the atomic particle be bound to an equilibrium position by a force $F_x = -bx$, where b is a constant; then the potential energy $V(x)$ is

$$V(x) = -\int_0^x F(u)\, du = \tfrac{1}{2}bx^2 \tag{2.10}$$

The particle now vibrates with a frequency $\omega_0 = (b/m)^{1/2}$, where m is the mass of the particle; so we may write

$$V(x) = \tfrac{1}{2}m\omega_0^2 x^2 \tag{2.10a}$$

The wave equation for the linear harmonic oscillator thus becomes

$$-\frac{\hbar^2}{2m}\frac{d^2\varphi}{dx^2} + \frac{1}{2}m\omega_0^2 x^2\varphi = E\varphi \tag{2.11}$$

It is sometimes convenient to rewrite the Schrödinger equation in operational form. Starting with Eq. (2.5a) and introducing the momenta $p_x = mv_x$,

$p_y = mv_y$, and $p_z = mv_z$, we may write

$$E = \frac{1}{2m}(p_x^2 + p_y^2 + p_z^2) + V(x, y, z) = H(p_x, p_y, p_z, x, y, z) \qquad (2.12)$$

The function H, which expresses the energy E in terms of the coordinates and the components of momentum, is called the *Hamilton function* of the particle or the *Hamiltonian*. We now arrive at Eq. (2.7) by the following prescription: Change the H function into an operator H_{op} by putting

$$p_x = \frac{\hbar}{j}\frac{\partial}{\partial x}, \qquad p_y = \frac{\hbar}{j}\frac{\partial}{\partial y}, \qquad p_z = \frac{\hbar}{j}\frac{\partial}{\partial z} \qquad (2.13)$$

and let it operate on the wave function $\varphi(x, y, z)$ so that

$$H_{op}\varphi = E\varphi \qquad (2.14)$$

Substituting Eq. (2.13) immediately yields Eq. (2.7).

The advantage of this new formulation is twofold. First, one can now extend the Schrödinger equation to the case of several particles as soon as the Hamilton function is known. Second, it makes the equation simpler so that it can be handled more easily in formal manipulations. It should be noted, however, that the H function must here be expressed in terms of the *Cartesian* coordinates of the individual particles.

Having derived the wave equation, we must now see what restrictions should be placed on the wave functions. In some cases one can apply boundary conditions, for example, when an electron is contained in a metal cube of dimensions $L \times L \times L$. In other cases no such boundary conditions can be applied. Then $\varphi(x, y, z)$ must be single valued and well behaved at all points in space where the wave equation can be defined. One can now distinguish between two cases:

1. The atomic particle cannot reach infinity; we call this a *bound* particle. In that case a well-behaved wave function exists *only* if the energy E has certain discrete values E_i $(i = 1, \ldots)$ called *eigenvalues* or *energy levels*. The wave functions φ_i associated with these energies E_i are called *eigenfunctions*. The energy is said to be *quantized* in this case.

2. The atomic particle can reach infinity; we call this a *nonbound* particle. In this case the energy E is not quantized.

We need a further distinction. If the energy E is quantized, then it may be that for each energy E_i there is one and only one wave function φ_i. The problem is then said to be *nondegenerate*. Or it may be that for some energies E_i there are several wave functions φ_i. The problem is then said to be *degenerate*.

2.1c. Additional Properties of the Wave Function

1. For light waves the expression $\varphi\varphi^* dV$ is a measure for the energy in the volume element dV. Or, what amounts to the same thing, it is a measure for the probability of finding a photon in the volume element dV. For that reason, in the case of atomic particles, it is sensible to call $\varphi_i\varphi_i^*\, dV$ the probability of finding a particle of energy E_i in the volume element dV. Since the particle must be somewhere in space with probability unity, we write

$$\int_V \varphi_i^*\varphi_i \, dV = 1 \qquad (2.15)$$

where the asterisk-denotes the complex conjugate, and the integration is performed over the volume V for which the wave equation is defined. This is called the condition of *normalization*. Only if the particle is a nonbound particle, for example, a particle described by the plane-wave approach discussed in Sec. 2.2a, does this normalization procedure need modification.

2. Since a bound particle cannot reach infinity, we must require that $\varphi_i^*\varphi_i$ go to zero at infinity sufficiently fast, so that the outlying regions of space give a negligible contribution to the integral (2.15). Let us investigate for a particular example what this means. Let φ_i be a function $\varphi_i(r)$ for large r; then we must require that

$$\int_{r>R} \varphi_i^*(r)\varphi_i(r)\cdot 4\pi r^2 \, dr \ll 1 \qquad (2.15a)$$

if R is sufficiently large. This means that $\varphi_i(r)$ must go to zero faster than $1/r^{3/2}$ for large r.

3. The eigenfunctions are *orthogonal*; that is, if φ_i and φ_j are two wave functions, corresponding to energies E_i and E_j, then

$$\int_V \varphi_i^*\varphi_j \, dV = 0 \qquad \text{for } E_i \neq E_j \qquad (2.16)$$

The proof is as follows. We observe that φ_i and φ_j are solutions of the equations

$$\frac{h^2}{2m}\nabla^2\varphi_i^* + (E_i - V)\varphi_i^* = 0$$

$$\frac{h^2}{2m}\nabla^2\varphi_j + (E_j - V)\varphi_j = 0$$

Multiplying the first equation by φ_j and the second by φ_i^*, and subtracting the results, yields

$$(E_i - E_j)\varphi_i^*\varphi_j + \frac{h^2}{2m}(\varphi_j\nabla^2\varphi_i^* - \varphi_i^*\nabla^2\varphi_j) = 0$$

Integrating over the volume V yields

$$(E_i - E_j)\int_V \varphi_i^*\varphi_j \, dV + \frac{h^2}{2m}\int_V (\varphi_j\nabla^2\varphi_i^* - \varphi_i^*\nabla^2\varphi_j) \, dV = 0 \qquad (2.16a)$$

We now apply Green's theorem to the second half. If A is the area of the surface enclosing the volume V, Green's theorem states

$$\int_V (\varphi_j \nabla^2 \varphi_i^* - \varphi_i^* \nabla^2 \varphi_j)\, dV = \int_A (\varphi_j \nabla \varphi_i^* - \varphi_i^* \nabla \varphi_j)_n\, dA \qquad (2.16b)$$

where ∇ is the gradient vector with components $\partial/\partial x$, $\partial/\partial y$, $\partial/\partial z$, and n represents the normal component of the vector between parentheses, that is, the component perpendicular to the surface element dA. Since φ_i and φ_j must go to zero at infinity at least as fast as $1/r^{3/2}$, the second half of Eq. (2.16b) goes to zero if V goes to infinity. Hence Eq. (2.16) is valid.

If the system is degenerate, that is, if φ_i and φ_j correspond to the same energy E, the preceding proof does not hold, so that one cannot be sure that the orthogonality condition is satisfied.

4. The particle current density $\mathbf{J}$ can be defined as

$$\mathbf{J} = \frac{h}{2mj}(\varphi^* \nabla \varphi - \varphi \nabla \varphi^*) \qquad (2.17)$$

where $\nabla \varphi$ is the gradient of φ with components $\partial\varphi/\partial x$, $\partial\varphi/\partial y$, and $\partial\varphi/\partial z$. We shall prove this in Sec. 2.3a. This definition is important when dealing with streams of particles.

2.2. APPLICATIONS

2.2a. Free Electron Model of a Conductor

The free electron model of a conductor is a model in which the electrons moving in the material have a potential energy $V(x, y, z) = -V_0$ with a kinetic energy $\frac{1}{2}mv^2$. We shall consider this model for a one-dimensional as well as for a three-dimensional lattice. The one-dimensional lattice is a fictitious crystal in which the electrons move in only one direction, say the (1 0 0) direction. The three-dimensional lattice is a lattice in which the electrons can move in all directions.

The wave equation for the free electron model is

$$-\frac{\hbar^2}{2m}\nabla^2\varphi = (E + V_0)\varphi \qquad (2.18)$$

Introducing

$$E + V_0 = \frac{\hbar^2}{2m}k^2 \quad \text{or} \quad k = \frac{[2m(E + V_0)]^{1/2}}{\hbar} = \frac{mv}{\hbar} \qquad (2.18a)$$

or

$$mv = \hbar k \quad \text{and} \quad E = -V_0 + \frac{\hbar^2}{2m}k^2 \qquad (2.18b)$$

Eq. (2.18) becomes

$$\nabla^2\varphi = -k^2\varphi \qquad (2.19)$$

For the one-dimensional case the solution is

$$\varphi(x) = \exp(jkx) \qquad (2.20)$$

without any restrictions for k.

For the three-dimensional case we use the method of separation of variables; that is, we substitute into Eq. (2.19)

$$\varphi(x, y, z) = \varphi_x(x)\varphi_y(y)\varphi_z(z) \qquad (2.21)$$

Since

$$\nabla^2\varphi = \frac{d^2\varphi_x}{dx^2}\varphi_y\varphi_z + \frac{d^2\varphi_y}{dy^2}\varphi_x\varphi_z + \frac{d^2\varphi_z}{dz^2}\varphi_x\varphi_y$$

we find, if we substitute into Eq. (2.19) and divide by $\varphi_x\varphi_y\varphi_z$, that

$$\frac{1}{\varphi_x}\frac{d^2\varphi_x}{dx^2} + \frac{1}{\varphi_y}\frac{d^2\varphi_y}{dy^2} + \frac{1}{\varphi_z}\frac{d^2\varphi_z}{dz^2} = -k^2 = -k_1^2 - k_2^2 - k_3^2 \qquad (2.22)$$

This can only be true if the individual terms of the left side are constants; that is,

$$\frac{1}{\varphi_x}\frac{d^2\varphi_x}{dx^2} = -k_1^2, \qquad \frac{1}{\varphi_y}\frac{d^2\varphi_y}{dy^2} = -k_2^2, \qquad \frac{1}{\varphi_z}\frac{d^2\varphi_z}{dz^2} = -k_3^2 \qquad (2.22a)$$

In analogy with Eq. (2.20), the solutions are of the form

$$\varphi_x = \exp(jk_1x), \qquad \varphi_y = \exp(jk_2y), \qquad \varphi_z = \exp(jk_3z) \qquad (2.22b)$$

Substituting back into Eq. (2.21) yields

$$\varphi(x, y, z) = \exp[j(k_1x + k_2y + k_3z)] = \exp[j(\mathbf{k}\cdot\mathbf{r})] \qquad (2.23)$$

where $\mathbf{k}\cdot\mathbf{r} = k_1x + k_2y + k_3z$ is the dot product of the vectors $\mathbf{k}$ and $\mathbf{r}$, and $k^2 = k_1^2 + k_2^2 + k_3^2$. It is easily shown that the waves propagate in the direction perpendicular to the planes $\mathbf{k}\cdot\mathbf{r} = $ constant, and that the direction cosines l, m, n of the direction of propagation are

$$l = \frac{k_1}{(k_1^2 + k_2^2 + k_3^2)^{1/2}}, \qquad m = \frac{k_2}{(k_1^2 + k_2^2 + k_3^2)^{1/2}},$$

$$n = \frac{k_3}{(k_1^2 + k_2^2 + k_3^2)^{1/2}} \qquad (2.23a)$$

We thus have as the most general solution a superposition of free-running planar waves without any restrictions on the values of k. The vector $\mathbf{k}$ is called the *wave vector*.

Since, according to Eq. (2.23a),

$$\exp[j(k_1x + k_2y + k_3z)] = \exp[jk(lx + my + nz)]$$

$$= \exp\left[j\frac{2\pi}{\lambda}(lx + my + nz)\right]$$

the time-dependent wave function of an electron traveling in a direction identified by the direction cosines l, m, n is

$$A \exp \left[j\frac{2\pi}{\lambda}(lx + my + nz) - j\omega t \right] \qquad (2.23b)$$

We shall use this result in Chapter 3.

We see that the normalization condition

$$\int \varphi^*\varphi \, dV = 1 \qquad (2.24)$$

is not satisfied here, because the integral diverges. To remedy the situation, boundary conditions must be applied. We shall see that this quantizes the components k_1, k_2, and k_3 of the wave vector **k**.

As a first example of a boundary condition, we require that the wave function periodically repeat itself after a distance L in the X, Y, or Z direction. The solution is

$$\varphi(x, y, z) = C \exp \left[j(k_1 x + k_2 y + k_3 z) \right] \qquad (2.25)$$

If we now apply periodic boundary conditions, we obtain a solution only if $k_1 L = 2\pi n_1, k_2 L = 2\pi n_2, k_3 L = 2\pi n_3$, or

$$k_1 = \frac{2\pi n_1}{L}, \qquad k_2 = \frac{2\pi n_2}{L}, \qquad k_3 = \frac{2\pi n_3}{L} \qquad (2.25a)$$

where $n_1 = 0, \pm 1, \ldots$, $n_2 = 0, \pm 1, \ldots$, and $n_3 = 0, \pm 1, \ldots$, so that

$$E = -V_0 + \frac{\hbar^2}{2m}k^2 = -V_0 + \frac{\hbar^2}{2m}\left(\frac{2\pi}{L}\right)^2(n_1^2 + n_2^2 + n_3^2) \qquad (2.26)$$

That is, the energy E is quantized and the electron is characterized by its three *quantum numbers* n_1, n_2, and n_3. Since

$$n_1 = \frac{k_1 L}{2\pi} = \frac{k}{2\pi}\frac{k_1 L}{k} = \frac{lL}{\lambda}$$

condition (2.25a) may be written

$$n_1 = \frac{lL}{\lambda}, \qquad n_2 = \frac{mL}{\lambda}, \qquad n_3 = \frac{nL}{\lambda} \qquad (2.25b)$$

We shall use this result in Chapter 3.

As a second example, we enclose the particle in a cubic box of side length L and require that $\varphi(x, y, z) = 0$ at the walls. We must use another set of solutions of Eq. (2.22a). It is easily seen that $\sin k_1 x$, $\cos k_1 x$; $\sin k_2 y$, $\cos k_2 y$; and $\sin k_3 z$, $\cos k_3 z$ are solutions. Since the cosine function does not satisfy the boundary conditions at $x = 0$, $y = 0$, or $z = 0$, the solution becomes

$$\varphi(x, y, z) = C \sin k_1 x \sin k_2 y \sin k_3 z \qquad (2.27)$$

which satisfies the conditions at $x = 0$, $y = 0$, and $z = 0$ automatically. The conditions at $x = L$, $y = L$, and $z = L$ are only satisfied if $k_1 L = n_1 \pi$,

$k_2 L = n_2 \pi$, and $k_3 L = n_3 \pi$, or

$$k_1 = \frac{\pi n_1}{L}, \qquad k_2 = \frac{\pi n_2}{L}, \qquad k_3 = \frac{\pi n_3}{L} \qquad (2.27a)$$

with $n_1 = 1, 2, \ldots$; $n_2 = 1, 2, \ldots$; $n_3 = 1, 2, \ldots$, so that

$$E = -V_0 + \frac{\hbar^2}{2m} k^2 = -V_0 + \frac{\hbar^2}{2m}\left(\frac{\pi}{L}\right)^2 (n_1^2 + n_2^2 + n_3^2) \qquad (2.28)$$

Again the energy is quantized and the electron is characterized by the three quantum numbers n_1, n_2, and n_3.

The energy steps in Eq. (2.26) and (2.28) are quite small. For example, the coefficient in front of $(n_1^2 + n_2^2 + n_3^2)$ in Eq. (2.26) is $\hbar^2/(2mL^2)$. Putting $h = 6.62 \times 10^{-34}$ J, $m = 9.11 \times 10^{-31}$ kg, and $L = 1$ cm yields $h^2/2mL^2 = 2.4 \times 10^{-33}$ J $= 1.5 \times 10^{-14}$ eV. The increments are thus very small indeed, and for all practical purposes the energy E can be considered a continuous function of k.

We now normalize the wave function in each case by integrating over the volume $V = L^3$. In the first case this yields

$$\int_L \varphi^* \varphi \, dV = C^2 L^3 = 1 \quad \text{or} \quad C = \frac{1}{L^{3/2}} \qquad (2.29)$$

and in the second case one obtains

$$\int_V \varphi^* \varphi \, dV = C^2\left(\frac{1}{2}L\right)^3 \quad \text{or} \quad C = \left(\frac{2}{L}\right)^{3/2} \qquad (2.29a)$$

In the first case one can also determine C from the particle current. According to Eqs. (2.17) and (2.25), for the X component J_x of the particle current density $\mathbf{J}$, we have

$$J_x = \frac{\hbar C^2}{2mj}\left(\varphi_x^* \frac{d\varphi_x}{dx} - \varphi_x \frac{d\varphi_x^*}{dx}\right)\varphi_y^* \varphi_y, \varphi_z^* \varphi_z = \frac{\hbar C^2}{2mj}(jk_1 + jk_1)$$

$$= \frac{\hbar}{m} k_1 C^2 = C^2 v_x \qquad (2.30)$$

where we have written $\varphi = \varphi_x(x)\varphi_y(y)\varphi_z(z)$; Eq. (2.30) follows, since $\varphi_x = \exp(jk_1 x)$, $\varphi_y^* \varphi_y = \varphi_z^* \varphi_z = 1$, and $v_x = \hbar k_1/m$. We thus see that

$$C^2 = \frac{J_x}{v_x} \qquad (2.30a)$$

which is nothing but the particle density in the stream of particles.

2.2b. Linear Harmonic Oscillator

According to Eq. (2.11), the wave equation of the one-dimensional linear harmonic oscillator is

$$\frac{\hbar^2}{2m}\frac{d^2\varphi}{dx^2} + \left(E - \frac{1}{2}m\omega_0^2 x^2\right)\varphi = 0 \qquad (2.31)$$

where ω_0 is the frequency of the oscillator. We shall give here an outline of the way this equation is solved. A few of the mathematical steps will be omitted, but enough information is given for the reader to understand how the quantization of the harmonic oscillator is arrived at.

Substituting $x = (\hbar/m\omega_0)^{1/2}u$ gives the equation

$$\frac{d^2\varphi}{du^2} + \left(\frac{2E}{\hbar\omega_0} - u^2\right)\varphi = 0 \tag{2.32}$$

Next substitute

$$\varphi = \exp\left(-\frac{u^2}{2}\right)H(u) \tag{2.33}$$

Then the differential equation for $H(u)$ becomes

$$\frac{d^2H}{du^2} - 2u\frac{dH}{du} + \left(\frac{2E}{\hbar\omega_0} - 1\right)H = 0 \tag{2.34}$$

This resembles Hermite's differential equation

$$\frac{d^2H}{dz^2} - 2z\frac{dH}{dz} + 2vH = 0 \tag{2.34a}$$

which has only a univalued and well-behaved solution if v is integral ($v = 0, 1, 2, \ldots$). Consequently,

$$\frac{2E}{\hbar\omega_0} - 1 = 2v \quad \text{or} \quad E = \hbar\omega_0\left(v + \frac{1}{2}\right) \tag{2.35}$$

The energy can thus only change in steps $\hbar\omega_0$ and there is a ground-state energy $\frac{1}{2}\hbar\omega_0$.

The solutions of Hermite's differential equation are

$$H_v(u) = (-1)^v \exp(u^2)\frac{d^v[\exp(-u^2)]}{du^v}, \quad v = 0, 1, 2, \ldots \tag{2.36}$$

These functions are known as Hermite polynomials; the first ones are $H_0(u) = 1$, $H_1(u) = 2u$, $H_2(u) = 4u^2 - 2$, $H_3(u) = 8u^3 - 12u$, and so on. Since

$$\int_{-\infty}^{\infty} \exp(-u^2)H_v^2(u)\,dx = \left(\frac{\hbar}{m\omega_0}\right)^{1/2} \cdot 2^v v!\,\pi^{1/2}$$

the normalized wave function is

$$\varphi_v(x) = \left(\frac{m\omega_0}{\pi\hbar}\right)^{1/4}\frac{1}{2^{(1/2)v}\sqrt{v!}}\exp\left(-\frac{1}{2}u^2\right)H_v(u) \quad \text{with } u = \left(\frac{m\omega_0}{\hbar}\right)^{1/2}x \tag{2.36a}$$

These wave functions are orthogonal.

2.2c. The Hydrogen and Alkali Atoms

In Eq. (2.9) we gave the wave equation of the hydrogen atom. It has a potential energy $V(r)$ depending on the distance r only. An alkali atom can be considered as a hydrogen-like atom in which the one outer electron moves in the field of force of the nucleus and the other electrons, which are much closer to the nucleus. This force field has spherical symmetry, and for that reason the potential energy due to the outer electron in that force field can also be written as $V(r)$. The wave equation for the hydrogen atom as well as for the alkali atom may thus be written as

$$\frac{h^2}{2m} \nabla^2 \varphi + [E - V(r)]\varphi = 0 \qquad (2.37)$$

where $V(r) = -e^2/(4\pi\epsilon_0 r)$ for the hydrogen atom. For the alkali atom this expression for $V(r)$ is only true for larger distances; closer to the nucleus important deviations from the $1/r$ law occur.

We shall give here a short outline of the way in which this equation is solved. Some mathematical details will be omitted; the aim is to demonstrate how the quantum numbers n, l, and m_l are introduced and what they quantize.

Since the potential energy depends only on r, it is convenient to introduce polar coordinates r, ϑ, and ϕ. Equation (2.37) then becomes

$$\frac{1}{r^2}\frac{\partial}{\partial r}\left(r^2\frac{\partial\varphi}{\partial r}\right) + \frac{1}{r^2 \sin\vartheta}\frac{\partial}{\partial\vartheta}\left(\sin\vartheta\frac{\partial\varphi}{\partial\vartheta}\right) + \frac{1}{r^2 \sin^2\vartheta}\frac{\partial^2\varphi}{\partial\phi^2} + \frac{2m}{h^2}[E - V(r)]\varphi = 0$$

$$(2.38)$$

as is found by expressing $\nabla^2\varphi$ in polar coordinates.

Applying the method of separation of variables, we write

$$\varphi(r, \vartheta, \phi) = R(r)\Theta(\vartheta)\Phi(\phi) \qquad (2.39)$$

substitute into Eq. (2.38), and divide by $R\Theta\Phi$. This gives the equation

$$\frac{1}{R}\left[\frac{1}{r^2}\frac{d}{dr}\left(r^2\frac{dR}{dr}\right)\right] + \frac{1}{r^2}\left[\frac{1}{\Theta \sin\vartheta}\frac{d}{d\vartheta}\left(\sin\vartheta\frac{d\Theta}{d\vartheta}\right) + \frac{1}{\sin^2\vartheta}\left(\frac{1}{\Phi}\frac{d^2\Phi}{d\phi^2}\right)\right]$$

$$+ \frac{2m}{h^2}[E - V(r)] = 0 \qquad (2.40)$$

Equation (2.40) thus requires that Φ satisfy the relation

$$\frac{1}{\Phi}\frac{d^2\Phi}{d\phi^2} = C_1 = -m_l^2 \quad \text{or} \quad \Phi = \exp(jm_l\phi) \qquad (2.41)$$

where C_1 is a constant. This is only single valued if Φ does not change when ϕ increases by 2π. Consequently, $m_l = 0, \pm 1, \pm 2, \ldots$.

Equation (2.40) requires that Θ satisfy the relation

$$\frac{1}{\Theta \sin\vartheta}\frac{d}{d\vartheta}\left(\sin\vartheta\frac{d\Theta}{d\vartheta}\right) - \frac{m_l^2}{\sin^2\vartheta} = C_2 \qquad (2.42)$$

where C_2 is a constant. This differential equation has a well-behaved solution only if $C_2 = -l(l + 1)$, where $l \geq |m_l|$ is integral. The solutions are the *associated Legendre functions*

$$\Theta(\vartheta) = P_l^{|m_l|}(\cos \vartheta) = \sin^{|m_l|}\vartheta \frac{d^{|m_l|}P_l(\cos \vartheta)}{d(\cos \vartheta)^{|m_l|}}$$

$$P_l(\cos \vartheta) = \frac{1}{2^l l!} \frac{d^l(\cos^2 \vartheta - 1)^l}{d(\cos \vartheta)^l}$$

(2.43)

The Legendre polynomials $P_l(\cos \vartheta)$ are polynomials in $\cos \vartheta$. For all other values of C_2 the function does not behave properly in the interval $-\pi/2 \leqq \vartheta \leqq \pi/2$, and hence these solutions must be discarded.

The first few associated Legendre polynomials are

$P_0^0(\cos \vartheta) = 1$
$P_1^0(\cos \vartheta) = \cos \vartheta$, $P_1^1(\cos \vartheta) = \sin \vartheta$
$P_2^0(\cos \vartheta) = \frac{1}{2}(3 \cos^2 \vartheta - 1)$, $P_2^1(\cos \vartheta) = 3 \sin \vartheta \cos \vartheta$, $P_2^2(\cos \vartheta) = 3 \sin^2 \vartheta$, and so on.

Substituting back into Eq. (2.40), we obtain the following equation for R:

$$\frac{d^2R}{dr^2} + \frac{2}{r}\frac{dR}{dr} + \left[\frac{2mE}{\hbar^2} - \frac{2m}{\hbar^2}V(r) - \frac{l(l+1)}{r^2}\right]R = 0 \qquad (2.44)$$

It may now be shown that for $E > 0$ there is no restriction on the energy; every energy value gives a well-behaved function $R(r)$. For $E < 0$ this is not the case, as we shall see.

Up to this point the discussion applies for both the hydrogen and alkali atoms. To treat the case of the hydrogen atom, we assume $E < 0$ and substitute $V(r) = -e^2/(4\pi\epsilon_0 r)$. We introduce the parameters A and B,

$$A = -\frac{2m}{\hbar^2}E = \frac{1}{r_0^2}, \qquad B = \frac{e^2m}{4\pi\epsilon_0\hbar^2} \qquad (2.45)$$

and take $u = 2r/r_0$ as an independent variable. We thus obtain

$$\frac{d^2R}{du^2} + \frac{2}{u}\frac{dR}{du} + \left[-\frac{1}{4} + \frac{B}{\sqrt{A}}\frac{1}{u} - \frac{l(l+1)}{u^2}\right]R = 0 \qquad (2.46)$$

Substituting

$$R = \exp\left(-\frac{u}{2}\right)u^l w(u) \qquad (2.47)$$

yields the following equation for $w(u)$:

$$u\frac{d^2w}{du^2} + [2(l+1) - u]\frac{dw}{du} + \left[\frac{B}{\sqrt{A}} - (l+1)\right]w = 0 \qquad (2.48)$$

This equation has a proper solution only if $B/\sqrt{A} = n$, where n is integral and $\geqq l + 1$; the solution is a polynomial in u. For all other values of $B/\sqrt{A}$, the function is an infinite series in u that does not behave properly at infinity. The solution of Eq. (2.48) can be expressed in terms of the *asso-*

ciated Laguerre polynomials

$$w_{n,l}(u) = L_{n+l}^{2l+1}(u) \tag{2.48a}$$

where
$$L_k(u) = \exp(u)\frac{d^k[u^k \exp(-u)]}{du^k} \tag{2.48b}$$

is the kth Laguerre polynomial and the associated Laguerre polynomials are

$$L_k^p(u) = \frac{d^p L_k(u)}{du^p} \tag{2.48c}$$

The first few of these functions are

$$w_{1,0}(u) = L_1^1(u) = -1$$
$$w_{2,0}(u) = L_2^1(u) = -4 + 2u, \ w_{2,1}(u) = L_3^3(u) = -6$$
$$w_{3,0}(u) = L_3^1(u) = -18 + 18u - 3u^2, \ w_{3,1}(u) = L_4^3(u) = -96 + 24u$$
$$w_{3,2}(u) = L_5^5(u) = -120, \text{ and so on}$$

The condition $B/\sqrt{A} = n$ yields for the energy

$$E = -\frac{\hbar^2}{2m}\frac{B^2}{n^2} = -\frac{e^4 m}{8\epsilon_0^2 h^2}\frac{1}{n^2} = -\frac{13.60}{n^2}\text{ eV} \tag{2.49}$$

which is the well-known Bohr formula. We notice that the energy does not depend on l and m_l, so that the problem is degenerate for $n > 1$. For a given $n > 1$, the parameter l has the values $0, 1, \ldots, n-1$, and for a given l the parameter m_l can have the values $-l, -l+1, \ldots, +l$.

A calculation shows that there is no selection rule for the quantum number n. The wavelengths of the emitted radiation thus follow from

$$\frac{1}{\lambda} = \frac{E_m - E_n}{hc} = \frac{e^4 m}{8\epsilon_0^2 h^3 c}\left(\frac{1}{n^2} - \frac{1}{m^2}\right) = R\left(\frac{1}{n^2} - \frac{1}{m^2}\right) \tag{2.49a}$$

where $m = n+1, \ n+2, \ldots, n = 1, 2, \ldots$. The case $n = 2$ corresponds to the well-known *Balmer series*. The parameter

$$R = \frac{e^4 m}{8\epsilon_0^2 h^3 c} \tag{2.49b}$$

is called the *Rydberg constant*.

We must now normalize the wave function by evaluating

$$\int |\varphi(r, \vartheta, \phi)|^2 \, dV = \int_0^\infty R^2(r)r^2 \, dr \int_{-\pi/2}^{\pi/2} \Theta^2(\vartheta) \sin\vartheta \, d\vartheta \int_0^{2\pi} \Phi\Phi^* \, d\phi \tag{2.50}$$

Now

$$\int_0^{2\pi} \Phi\Phi^* \, d\phi = 2\pi \tag{2.50a}$$

$$\int_{-\pi/2}^{\pi/2} [P_l^{|m_l|}(\cos\vartheta)]^2 \sin\vartheta \, d\vartheta = \frac{2}{2l+1}\frac{(l+|m_l|)!}{(l-|m_l|)!} \tag{2.50b}$$

$$\int_0^\infty R^2(r)r^2\,dr = \left(\frac{r_0}{2}\right)^3 \int_0^\infty \exp{(-u)}u^{2l}[w_{n,l}(u)]^2 u^2\,du$$
$$= \left(\frac{r_0}{2}\right)^3 \frac{2n[(n+l)!]^3}{(n-l-1)!} \tag{2.50c}$$

Here

$$r_n = nr_0 = nA^{-1/2} = \frac{n^2}{B} = \frac{4\pi\epsilon_0\hbar^2 n^2}{e^2 m} \tag{2.51}$$

is the radius of the orbit of the Bohr atom for the quantum number n.

From this the normalized wave function is found to be

$$\left(\frac{2}{r_0}\right)^{3/2}\left\{\frac{(n-l-1)!}{2n[(n+l)!]^3}\right\}^{1/2}\left[\frac{2l+1}{2}\frac{(l-|m_l|)!}{(l+|m_l|)!}\right]^{1/2}$$
$$\left(\frac{1}{2\pi}\right)^{1/2}\cdot\exp\left(-\frac{1}{2}u\right)u^l w_{n,l}(u)P_l^{|m_l|}(\cos\vartheta)\exp{(jm_l\phi)} \tag{2.52}$$

These wave functions are orthogonal for all n, l, and m_l. The parameter

$$r_1 = \frac{4\pi\epsilon_0\hbar^2}{e^2 m} \tag{2.52a}$$

is the radius of the orbit of the Bohr atom for the lowest energy state ($n = 1$); it has a value of 0.53×10^{-8} cm.

Having evaluated the energy levels of the hydrogen atom, we now turn to the alkali atom. If we introduce approximate values for $V(r)$, we find that the energy E depends on l and is more negative as l decreases (see Fig. 2.1).

It should be noted that the two problems have the quantum numbers l and m_l in common. We shall see in Sec. 2.3c that m_l measures the component of the angular momentum along the Z axis and that l measures the total angular momentum. Both are therefore quantized.

In neither of the two problems does the energy depend on m_l, since Eq. (2.46) does not contain m_l. The energy does depend on m_l, however, if a magnetic field is applied.

2.3. THE TIME-DEPENDENT WAVE EQUATION; MATRIX REPRESENTATION

2.3a. Derivation of the Time-Dependent Wave Equation

To describe wave pulses, we need a time-dependent wave equation. We shall also see that we need it for the derivation of other useful relationships.

We shall now derive this equation. Classically, the time-dependent wave function for a given frequency ω may be written

$$\psi(x, y, z, t) = \varphi(x, y, z)\exp{(-j\omega t)}$$

But according to quantum theory, the energy E and the frequency ω are related by $\omega_i = E_i/\hbar$, and with E_i a wave function φ_i is associated. For the time-dependent wave function of an atomic particle, it is thus sensible to use

$$\psi_i(x, y, z, t) = \varphi_i(x, y, z) \exp\left[-j\left(\frac{E_i}{\hbar}\right)t\right] \qquad (2.53)$$

Multiplying both sides of the equation

$$H_{op}\varphi_i = E_i\varphi_i$$

by $\exp[-j(E_i/\hbar)t]$ yields the equation

$$H_{op}\psi_i = E_i\psi_i = -\frac{\hbar}{j}\frac{\partial\psi_i}{\partial t} \qquad (2.54)$$

We have already seen that the energy E is quantized—that is, E can only have discrete values E_i. For each energy E_i there is thus a wave function ψ_i.

If $\Psi(x, y, z, t)$ is an arbitrary atomic wave function, then we would expect Ψ to satisfy an equation similar to (2.54):

$$H_{op}\Psi = -\frac{\hbar}{j}\frac{\partial\Psi}{\partial t} \qquad (2.55)$$

We shall now see that this equation indeed has special solutions of the form (2.53).

To prove this, we apply the method of separation of variables to Eq. (2.55). Substituting

$$\Psi(x, y, z, t) = \varphi(x, y, z)\chi(t)$$

yields the following equation for $\chi(t)$:

$$\frac{1}{\varphi}H_{op}\varphi = -\frac{\hbar}{j}\frac{1}{\chi}\frac{d\chi}{dt} = \text{constant} = E \quad \text{or} \quad \chi = \exp\left[-j\left(\frac{E}{\hbar}\right)t\right]$$

and hence

$$H_{op}\varphi = E\varphi$$

which has the eigenvalues $E = E_i$ and the associated eigenfunctions $\varphi = \varphi_i$. We thus see that the constant E is the energy and that the solution is of the form (2.53), as expected.

Since Eq. (2.55) is linear, the most general solution is a linear superposition of all the possible wave functions found by the method of separation of variables. That is, the solution must be of the form

$$\Psi(x, y, z, t) = \sum_i c_i\psi_i(x, y, z, t) \qquad (2.56)$$

where the c_i's must be so chosen that the initial conditions are satisfied. *An equation of the form (2.56) describes an arbitrary atomic wave function.*

We shall use the symbol $\psi(x, y, z, t)$ to describe any particular wave function of the form (2.53) and reserve the symbol $\Psi(x, y, z, t)$ for an arbitrary wave function of the form (2.56). Both satisfy Eq. (2.55).

If $H_{op} = -(\hbar^2/2m)\,\nabla^2 + V(x, y, z)$, Eq. (2.55) may be written

$$-\frac{\hbar^2}{2m}\nabla^2\Psi + V(x, y, z)\Psi = -\frac{\hbar}{j}\frac{\partial\Psi}{\partial t} \tag{2.55a}$$

Unless one is dealing with free particles or with escaping particles, the wave function $\Psi(x, y, z, t)$ disappears sufficiently fast at infinity. This is important for the derivation of several relationships.

In the time-dependent case it is also reasonable to call $\Psi\Psi^*\,dV$ the *probability that the particle is in the volume element dV*. Thus we should normalize $\Psi(x, y, z, t)$ by requiring

$$\int_V \Psi\Psi^*\,dV = 1 \tag{2.57}$$

where the integration is extended over all space. Except for free or escaping particles, the integrand disappears sufficiently fast at infinity to make the integral converge. For a finite volume V, we can see that

$$\frac{\partial}{\partial t}\int_V \Psi\Psi^*\,dV = \int_V \left(\Psi\frac{\partial\Psi^*}{\partial t} + \Psi^*\frac{\partial\Psi}{\partial t}\right) dV \tag{2.57a}$$

We shall now derive a certain property of the wave function and an expression for the particle current density. According to Eq. (2.55a),

$$\frac{\hbar}{j}\frac{\partial\Psi^*}{\partial t} = -\frac{\hbar^2}{2m}\nabla^2\Psi^* + V(x, y, z)\Psi^*$$

$$-\frac{\hbar}{j}\frac{\partial\Psi}{\partial t} = -\frac{\hbar^2}{2m}\nabla^2\Psi + V(x, y, z)\Psi$$

Multiply the first equation by Ψ and the second by Ψ^* and subtract. This yields

$$\frac{\hbar}{j}\left(\Psi\frac{\partial\Psi^*}{\partial t} + \Psi^*\frac{\partial\Psi}{\partial t}\right) = \frac{\hbar^2}{2m}(\Psi^*\,\nabla^2\Psi - \Psi\,\nabla^2\Psi^*)$$

We now integrate over a finite volume V:

$$\int_V \left(\Psi\frac{\partial\Psi^*}{\partial t} + \Psi^*\frac{\partial\Psi}{\partial t}\right) dV = -\frac{\hbar}{2mj}\int_V (\Psi^*\,\nabla^2\Psi - \Psi\,\nabla^2\Psi^*)\,dV$$

$$= -\frac{\hbar}{2mj}\int_A (\Psi^*\,\nabla\Psi - \Psi\,\nabla\Psi^*)_n\,dA \tag{2.57b}$$

according to Green's theorem, where A is the area of the surface enclosing the volume V. Since the left side of Eq. (2.57b) equals

$$\frac{\partial}{\partial t}\int_V \Psi\Psi^*\,dV$$

this suggests that the vector

$$\mathbf{J} = \frac{\hbar}{2mj}(\Psi^*\,\nabla\Psi - \Psi\,\nabla\Psi^*) \tag{2.58}$$

represents the particle current density. Substituting

$$\Psi = \varphi_i(x, y, z) \exp\left[-j\left(\frac{E_i}{\hbar}\right)t\right]$$

gives Eq. (2.17).

2.3b. Position and Momentum of Wave Packet; Matrix Representation

Let an atomic particle be represented by a wave packet $\Psi(x, y, z, t)$ that disappears sufficiently fast at infinity. We assume Ψ to be normalized. Since $\Psi\Psi^* \, dV$ represents the probability of finding the particle, described by Ψ, in the volume element dV, the "position" of the particle must be defined as the average value of x:

$$\bar{x} = \int x\Psi\Psi^* \, dV = \int \Psi^* x\Psi \, dV \tag{2.59}$$

so that

$$\frac{\partial \bar{x}}{\partial t} = \int x\left(\Psi \frac{\partial \Psi^*}{\partial t} + \Psi^* \frac{\partial \Psi}{\partial t}\right) dV \tag{2.59a}$$

Next we calculate the momentum $\bar{p}_x = m\, \partial\bar{x}/\partial t$ associated with $\bar{x}$. To that end we start from the time-dependent wave equation

$$\frac{\hbar}{j}\frac{\partial \Psi^*}{\partial t} = -\frac{\hbar^2}{2m}\nabla^2\Psi^* + V(x, y, z)\Psi^*$$

$$-\frac{\hbar}{j}\frac{\partial \Psi}{\partial t} = -\frac{\hbar^2}{2m}\nabla^2\Psi + V(x, y, z)\Psi$$

We multiply the first equation by $x\Psi$ and the second by $x\Psi^*$ and subtract. This yields

$$\frac{\hbar}{j}x\left(\Psi \frac{\partial \Psi^*}{\partial t} + \Psi^* \frac{\partial \Psi}{\partial t}\right) = \frac{\hbar^2}{2m}(x\Psi^* \, \nabla^2\Psi - \Psi x \, \nabla^2\Psi^*)$$

Integrating over all space gives, since the left side yields $(\hbar/j)\partial\bar{x}/\partial t$ according to Eq. (2.59a),

$$\frac{\hbar}{j}\frac{\partial \bar{x}}{\partial t} = \frac{\hbar^2}{2m}\int (x\Psi^* \, \nabla^2\Psi - \Psi x \, \nabla^2\Psi^*) \, dV$$

$$= \frac{\hbar^2}{2m}\int (x\Psi^* \, \nabla^2\Psi - \Psi \, \nabla^2 x\Psi^*) \, dV + \frac{\hbar^2}{m}\int \Psi \frac{\partial \Psi^*}{\partial x} \, dV$$

since

$$\nabla^2 x\Psi^* = x \, \nabla^2\Psi^* + 2\frac{\partial \Psi^*}{\partial x} \quad \text{or} \quad x \, \nabla^2\Psi^* = \nabla^2 x\Psi^* - 2\frac{\partial \Psi^*}{\partial x}$$

But the first integral is zero, as we find by applying Green's theorem and bearing in mind that Ψ disappears sufficiently fast at infinity. Consequently,

$$\bar{p}_x = m\frac{\partial \bar{x}}{\partial t} = -\frac{\hbar}{j}\int \Psi \frac{\partial \Psi^*}{\partial x} dV = \int \Psi^*\left(\frac{\hbar}{j}\frac{\partial}{\partial x}\right)\Psi \, dV \tag{2.60}$$

for since $\bar{p}_x$ is real, it is equal to its complex conjugate. *We must thus make up* $\int \Psi^* p_x \Psi \, dV$ *and replace* p_x *by the operator* $(\hbar/j)\partial/\partial x$.

This can be generalized as follows. If the average must be taken of an arbitrary function $f(p_1, \ldots, p_n, x_i, \ldots, x_n)$, we turn the function f into an operator by replacing p_i by $p_i = (\hbar/j)\partial/\partial x_i$ and then defining

$$\overline{f(p_1, \ldots, p_n, x_1, \ldots, x_n)} = \int \Psi^* f\left(\frac{\hbar}{j}\frac{\partial}{\partial x_1}, \ldots, \frac{\hbar}{j}\frac{\partial}{\partial x_n}, x_1, \ldots, x_n\right)\Psi \, dV$$

(2.61)

This result is applied in two ways. If Ψ is associated with an eigenfunction, that is,

$$\Psi(x, y, z, t) = \psi_i(x, y, z, t) = \varphi_i(x, y, z) \exp\left[-j\left(\frac{E_i}{\hbar}\right)t\right]$$

then substituting into (2.61) yields

$$\bar{f} = f_{ii}^0 = \int \varphi_i^* f\left(\frac{\hbar}{j}\frac{\partial}{\partial x_1}, \ldots, \frac{\hbar}{j}\frac{\partial}{\partial x_n}, x_1, \ldots, x_n\right)\varphi_i \, dV \qquad (2.61a)$$

Next we apply the result to the most general solution (2.56) of the wave equation (2.55). For the sake of normalization (we assume all the φ_i's to be normalized) we require that $\sum_i |c_i|^2 = 1$. Applying Eq. (2.61) to this wave function Ψ yields

$$\bar{f} = \sum_i \sum_k c_i^* c_k f_{ik}(t) = \sum_i \sum_k c_i^* c_k f_{ik}^0 \exp(j\omega_{ik}t)$$

$$f_{ik} = f_{ik}^0 \exp(j\omega_{ik}t)$$

(2.61b)

where $\omega_{ik} = (E_i - E_k)/\hbar$ and

$$f_{ik}^0 = \int \varphi_i^* f\left(\frac{\hbar}{j}\frac{\partial}{\partial x_1}, \ldots, \frac{\hbar}{j}\frac{\partial}{\partial x_n}, x_1, \ldots, x_n\right)\varphi_k \, dV \qquad (2.61c)$$

Thus $\bar{f}$ contains time-independent terms involving the f_{ii}^0's and time-dependent terms involving the frequencies ω_{ik} and the f_{ik}^0's, so that $\bar{f}$ is a kind of Fourier series and the f_{ik}'s are a kind of Fourier term.

It is thus possible to represent the function $f(p_1, \ldots, p_n, x_1, \ldots, x_n)$ by its matrix

$$\mathbf{F} = \{f_{ik}\} \qquad (2.62)$$

There are two types of representation. It may happen that $\mathbf{F}$ is a diagonal matrix. In that case $\bar{f}$ contains only time-independent terms; then $\bar{f}$ is said to be conserved. Or it may happen that the matrix has off-diagonal elements; then $\bar{f}$ is time-dependent.

As an example of the first type, consider the Hamiltonian

$$H(p_1, \ldots, p_n, x_1, \ldots, x_n)$$

According to (2.61c),

$$H_{ik}^0 = \int \varphi_i^* H\varphi_k \, dV = \int \varphi_i^* E_k\varphi_k \, dV = E_k \int \varphi_i^* \varphi_k \, dV$$

But since the wave functions are orthogonal,

$$H_{ik} = 0 \quad \text{for} \quad i \neq k, \qquad H_{ii} = E_i$$

If this is applied to a general wave function of the form (2.56), $\bar{H}$ is independent of time, so that energy is conserved, and

$$\bar{H} = \sum_i |c_i|^2 E_i$$

We shall see in Sec. 2.3c that the angular momentum of the electron in the hydrogen atom is also a diagonal matrix, indicating that angular momentum is conserved.

2.3c. Quantization of the Angular Momentum

By applying the foregoing theory to the angular momentum **M** of the hydrogen or the alkali atom, we shall see that the Z component M_z of **M** and the absolute value $M^2 = \mathbf{M} \cdot \mathbf{M}$ are quantized and have the values

$$\bar{M}_z = \hbar m_l, \qquad \bar{M^2} = \hbar^2 l(l+1) \tag{2.63}$$

respectively. For the quantization of M_z, notice that the Z direction must be a direction of preferred orientation; that is, an electric or magnetic field must be applied in that direction. Observe also that the wave functions of the alkali and hydrogen atoms differ only in their radial parts, so that the quantization of the angular momentum is the same in both cases.

The proof of (2.63) is simple. Classically, the angular momentum $\mathbf{M} = \mathbf{r} \times \mathbf{p}$, where $\mathbf{p}$ is the linear momentum. Hence the components of **M** are

$$M_x = yp_z - zp_y, \qquad M_y = zp_x - xp_z, \qquad M_z = xp_y - yp_x$$

We now convert these expressions into operators in the usual manner and apply the previous theory. We can greatly simplify the resulting expressions by changing over to polar coordinates; we do so here without proof. For example, M_z corresponds to the operator

$$\frac{\hbar}{j}\left(x\frac{\partial}{\partial y} - y\frac{\partial}{\partial x}\right) = \frac{\hbar}{j}\frac{\partial}{\partial \phi}$$

and $M^2 = M_x^2 + M_y^2 + M_z^2$ corresponds to the operator

$$-\hbar^2\left[\left(y\frac{\partial}{\partial z} - z\frac{\partial}{\partial y}\right)^2 + \left(z\frac{\partial}{\partial x} - x\frac{\partial}{\partial z}\right)^2 + \left(x\frac{\partial}{\partial y} - y\frac{\partial}{\partial x}\right)^2\right]$$

$$= -\hbar^2\left[\frac{1}{\sin \vartheta}\frac{\partial}{\partial \vartheta}\left(\sin \vartheta \frac{\partial}{\partial \vartheta}\right) + \frac{1}{\sin^2 \vartheta}\frac{\partial^2}{\partial \phi^2}\right]$$

We now take into account that $\varphi_{n,l,m_l}(r, \vartheta, \phi) = CR(r)\Theta(\vartheta)\Phi(\phi)$, where C is the normalizing factor. According to the theory of Sec. 2.2c,

$$\frac{\hbar}{j}\frac{\partial}{\partial \phi}\varphi_{n,l,m_l} = \hbar m_l \varphi_{n,l,m_l}$$

and

$$-\hbar^2\left[\frac{1}{\sin\vartheta}\frac{\partial}{\partial\vartheta}\left(\sin\vartheta\frac{\partial}{\partial\vartheta}\right)+\frac{1}{\sin^2\vartheta}\frac{\partial^2}{\partial\phi^2}\right]\varphi_{n,l,m_l}=\hbar^2 l(l+1)\varphi_{n,l,m_l}$$

But since the wave functions are normalized and orthogonal,

$$\bar{M}_z=\hbar m_l\iiint\varphi^*_{n',l',m'_{l'}}\varphi_{n,l,m_l}r^2\,dr\,\sin\vartheta\,d\vartheta\,d\phi$$

is equal to $\hbar m_l$ if $n'=n$, $l'=l$, and $m'_{l'}=m_l$, and zero otherwise. And

$$\overline{M^2}=\hbar^2 l(l+1)\iiint\varphi_{n',l',m'_{l'}}\varphi_{n,l,m_l}r^2\,dr\,\sin\vartheta\,d\vartheta\,d\phi$$

is equal to $\hbar^2 l(l+1)$ if $n'=n$, $l'=l$, and $m'_{l'}=m_l$, and zero otherwise, again because of orthogonality and normalization.

We thus see that the momentum matrices are diagonal matrices. This is an indication of the validity of the law of conservation of angular momentum. We have also proved Eq. (2.63).

2.3d. Transition Probabilities

We now apply the theory of Sec. 2.3b to calculate transition probabilities. We shall see that these probabilities are associated with the matrix elements of the electric moment $\mathbf{m}=-e\mathbf{r}$, if we are dealing with a moving electron that radiates.

Classically, radiation is associated with the second time derivative of the electric moment $\mathbf{m}$ of the atom. In quantum theory terms, therefore, the emitted radiation is associated with the *matrix elements* of this electric moment.

Classically, an oscillating electric dipole with an electric moment

$$\mathbf{m}=\mathbf{m}_0+\sum_{k=1}\mathbf{m}_k\exp(j\omega_k t)$$

radiates the power

$$\bar{S}=C^2\langle|\ddot{\mathbf{m}}|^2\rangle=C^2\sum_k(|m_{xk}|^2+|m_{yk}|^2+|m_{zk}|^2)\omega_k^4$$

where C is a universal constant that is of no interest here, $\langle\ \rangle$ denotes a time average, $\ddot{\mathbf{m}}$ is a second-order time derivative, and m_{xk}, m_{yk}, and m_{zk} are the components of the vector $\mathbf{m}_k$.

If we translate this into the atomic domain, we must replace ω_k by ω_{ik}, and $\mathbf{m}_k$ by $\mathbf{m}_{ik}$. Consequently, the average power radiated at the frequency ω_{ik} is

$$S^0_{ik}=C^2\omega_{ik}^4(|m_{xik}|^2+|m_{yik}|^2+m_{zik}|^2)\tag{2.64}$$

where

$$m_{xik}=m^0_{xik}\exp(j\omega_{ik}t)\qquad\text{and so on}\tag{2.64a}$$

and, if $-e$ is the electric charge of the electron,

$$m_{xik}^0 = -e \int \varphi_i^* x \varphi_k \, dV$$

$$m_{yik}^0 = -e \int \varphi_i^* y \varphi_k \, dV \qquad (2.64\text{b})$$

$$m_{zik}^0 = -e \int \varphi_i^* z \varphi_k \, dV$$

The radiated frequencies thus correspond to $\omega_{ik} = (E_i - E_k)/h$; a positive frequency means emission, a negative frequency means absorption. If the matrix elements m_{xik}^0, m_{yik}^0, m_{zik}^0 are zero, the transition is *forbidden*. This establishes the *selection rules* for atomic and molecular spectra. We also note that $S_{ik}^0 = 0$ for $i = k$, since $\omega_{ik} = 0$ for $i = k$, indicating that only transitions *between* states emit radiation.

We now apply this to some of the examples discussed earlier. Since the calculations are tedious, we give only the final results.

For the harmonic oscillator of frequency f we find the selection rule $\Delta v = \pm 1$, so that the oscillator only emits and absorbs quanta hf. For the hydrogen and the alkali atoms we find no selection rule for n, but for l and m_l the following rules are obtained:

$$\Delta l = \pm 1, \qquad \Delta m_l = 0, \pm 1$$

At first sight the selection rules for l and m_l seem meaningless, since in our approximation the energy of the hydrogen atom is independent of l and m_l. We observe, however, that the alkali atom and hydrogen atom wave functions differ only in their radial parts, so that the same selection rules apply. Since in the alkali atom the energy depends on l, the selection rule $\Delta l = \pm 1$ is meaningful. The selection rule for m_l becomes meaningful if a magnetic field is applied.

The calculation of the matrix representation of the electric moment not only gives the selection rules, but also provides a means of calculating relative intensities. The details are beyond the scope of this book.

2.4. ATOMIC SPECTRA

2.4a. Alkali Atom; X-Ray Spectra

The degeneracy in l is removed if $V(r)$ is not given by Coulomb's law, as is the case, for example, for the outer electron in the alkali atom. Here different l values correspond to different energies. The smaller l is, the deeper the outer electron penetrates into the core formed by the inner electrons and the nucleus, and the more negative the energy becomes. The degeneracy in m_l can be removed by a magnetic field applied in the Z direction.

In the alkali atom it has been possible to assign quantum numbers l to the observed spectra. The most important series are the *sharp series*, the *principal series*, the *diffuse series*, and the *fundamental series*. Writing the emitted frequency ω as

$$\hbar\omega = E_{n_2, l_2} - E_{n_1, l_1}$$

physicists have made the following assignment (Fig. 2.1):

Sharp series: $l_2 = 0,\, l_1 = 1$ n_1 fixed, $n_2 > n_1$
Principal series: $l_2 = 1,\, l_1 = 0$ n_1 fixed, $n_2 > n_1$
Diffuse series: $l_2 = 2,\, l_1 = 1$ n_1 fixed, $n_2 > n_1$
Fundamental series: $l_2 = 3,\, l_1 = 2$ n_1 fixed, $n_2 > n_1$

We see that this assignment is compatible with the selection rule $\Delta l = \pm 1$.

The names of the alkali series of spectral lines are still reflected in atomic nomenclature. An electron with $l = 0$ is called an s electron, an electron with $l = 1$, a p electron, an electron with $l = 2$, a d electron, an electron with $l = 3$, an f electron, and so on. An electron is thus characterized by its quantum numbers n and l. Electrons with $n = 2$, $l = 0$ are called $2s$ electrons, electrons with $n = 3$, $l = 1$ are called $3p$ electrons, and so on. The number of electrons of a given kind is denoted by a superscript to the right; for example, $(2p)^3$ means three $2p$ electrons.

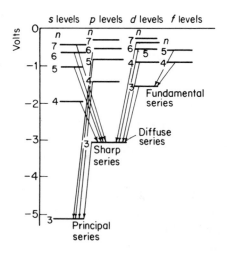

Fig. 2.1. Energy-level diagram of sodium, showing the sharp, principal, diffuse, and fundamental series. Note that the lowest energy levels have $n = 3$, except for the lowest f level, which has $n = 4$.

The introduction of the quantum number l did not explain all the fine details of the alkali spectra, however. The spectral lines were multiple lines rather than single lines. This problem was resolved by introduction of the electron spin.

Electrons with $l \neq 0$ have a magnetic moment and will orient themselves in a magnetic field. Electrons also spin around their axis, and this produces a magnetic moment that interacts with the magnetic moment caused by the orbital motion of the outer electron. The two magnetic moments will orient themselves in such a manner that the orbital angular momentum $\hbar\sqrt{l(l + 1)}$ and the spin angular momentum $\hbar\sqrt{s(s + 1)}$, where $s = \frac{1}{2}$, combine to give the

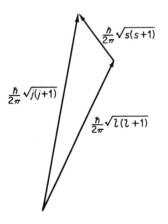

Fig. 2.2. Vector diagram of the atom, showing the orbital angular momentum $(\hbar/2\pi)\sqrt{l(l+1)}$ and the spin angular momentum $(\hbar/2\pi)\sqrt{s(s+1)}$ combining to the total angular momentum $(\hbar/2\pi)\sqrt{j(j+1)}$. In this example $s = \frac{1}{2}$, $l = 2$, $j = 2\frac{1}{2}$. The same vector diagram also holds if many electrons interact. In that case, one need only to replace l by L, s by S, and j by J (Sec. 2.4b).

total angular momentum $\hbar\sqrt{j(j+1)}$ with $j = l \pm \frac{1}{2}$ (Fig. 2.2). This is known as "spin–orbit" coupling. Since different j values give slightly different energies, all energy levels with $l \neq 0$ are double. The observed multiplicity of the lines can be explained with the help of the additional selection rules $\Delta j = 0, \pm 1$, which can be deduced from a more detailed wave-mechanical analysis of the problem. As a consequence, the sharp and the principal series have double lines, whereas the diffuse and the fundamental series have triple lines.

Whereas the alkali spectra give indications about the quantum numbers of the outer electron of an alkali atom, x-ray absorption and emission spectra give information about the inner electrons. Figure 2.3 shows the x-ray absorption coefficient of a moderately heavy element plotted against the wavelength. The explanation is simple. The inner electrons have well-defined binding energies; when the quantum $\hbar\omega$ does not have enough energy to bring an electron outside

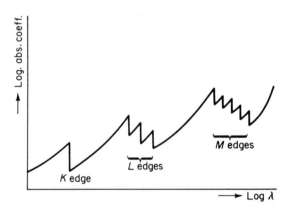

Fig. 2.3. The logarithm of the x-ray absorption coefficient for a moderately heavy element is plotted schematically against the wavelength λ. The plot shows one K edge, three L edges, and five M edges.

the atom, that electron can no longer absorb, and the absorption coefficient drops.

Apparently, then, there are electrons with different binding energies, and it is obvious to associate them with the differing quantum number n. The K edge is due to the fact that the electrons with $n = 1$ (K electrons) cease to absorb; the L edges are due to the fact that the electrons with $n = 2$ (L electrons) cease to absorb, and so on. This explains the "shell structure" of the atom.

The multiplicity of the L edges, M edges, and so forth, can be understood from the fact that for every $n > 1$ there are n possible l values. The orbital angular momentum $\hbar\sqrt{l(l+1)}$ and the spin angular momentum $\hbar\sqrt{s(s+1)}$ orient themselves in such a way that the total angular momentum $\hbar\sqrt{j(j+1)}$ is quantized and $j = l \pm \frac{1}{2}$ (spin–orbit coupling). For every n there are $2n - 1$ different l, j combinations with slightly different energies. This explains why there are three L edges, five M edges, and so on.

This agreed with the results obtained from the alkali spectrum. The lowest energy level of the outer electron (with $l = 0$) had $n = 2$ for Li, $n = 3$ for Na, $n = 4$ for K, $n = 5$ for Rb, and $n = 6$ for Cs. This, together with the x-ray evidence, showed that the various "shells" become gradually filled when one goes to heavier elements.

2.4b. More Complicated Spectra; Shell and Subshell Structure

The unraveling of more complicated spectra has generally proceeded with the help of the "vector model" of the atom, which is an extension of the considerations of the alkali atom that led to the new quantum number j. First, we assign orbital motion quantum numbers l_i and spin quantum numbers s_i (with $s_i = \frac{1}{2}$) to each individual electron. Then the various interactions are taken into account in order of importance.

1. The most important interaction is that between the orbital motions of the individual electrons. We can describe this by adding the orbital angular momenta $p_i = \hbar\sqrt{l_i(l_i+1)}$ together vectorially to form the total angular momentum

$$P = \hbar\sqrt{L(L+1)}$$

where L is integral, and its maximum value L_{max} is $\sum_i |l_i|$. Different L's usually correspond to widely different energies.

2. The next important interation is the interaction between the spins. We can describe it by adding together vectorially the electron spin momenta $\hbar\sqrt{s_i(s_i+1)}$, with $s_i = \frac{1}{2}$, to form the total spin momentum

$$\hbar\sqrt{S(S+1)}$$

S has integral values if the number of electrons is even and half-integral values if the number of electrons is odd; its maximum value $S_{max} = \sum_i s_i$.

3. The total orbital angular momentum and the total spin momentum now interact; we describe this by adding the two momenta together vectorially to form the total angular momentum

$$\hbar\sqrt{J(J + 1)}$$

where J can have the values $(L + S)$, $(L + S - 1)$, ..., $|L - S|$ (see Fig. 2.2).

For $S \leq L$ there are $(2S + 1)$ possible J values; this number is called the "multiplicity" of the energy level described by the quantum number L. Spectroscopic evidence suggests and theory confirms the following selection rules: $\Delta L = \pm 1$, $\Delta S = 0$; $\Delta J = 0, \pm 1$, $J = 0 \rightarrow J = 0$ forbidden.

The following nomenclature is in common use: energy levels with $L = 0, 1, 2, 3, \ldots$ are, respectively, called S, P, D, and F levels; the multiplicity $(2S + 1)$ is added as a superscript in front and the J values as a subscript behind the symbol. For instance, $^3D_1$ means a level with $L = 2$ (symbol D), $S = 1$ (superscript 3), and $J = 1$ (subscript 1).

It was observed that this method gave many more energy levels than were actually found. This difficulty was removed by the introduction of *Pauli's exclusion principle*, which is discussed in Sec. 2.4d.

With the help of this vector model, the spectra of more complicated atoms were gradually analyzed, the energy levels were determined, and quantum numbers were assigned to the individual electrons in the ground state. A great help was the following general rule: *Atoms of nuclear charge Ze, singly ionized atoms of nuclear charge $(Z + 1)e$, doubly ionized atoms of nuclear charge $(Z + 2)e$, ..., have similar spectra.*

Combining all information from atomic, ionic, and x-ray spectra, we can obtain the following electron structure for the ground states of the light elements:

H: $(1s)$
He: $(1s)^2$
Li: $(1s)^2(2s)$
Be: $(1s)^2(2s)^2$
B: $(1s)^2(2s)^2(2p)$
C: $(1s)^2(2s)^2(2p)^2$
N: $(1s)^2(2s)^2(2p)^3$

and so forth. When this was extended to the elements with larger nuclear charge Ze, the following general rule was obtained: *A subshell (n given, l given) is filled if it contains $2(2l + 1)$ electrons.* An s subshell can thus have

2 electrons, a p subshell 6 electrons, a d subshell 10 electrons, and an f subshell 14 electrons. We shall see in Sec. 2.4d how this leads to Pauli's exclusion principle.

2.4c. Absorption and Emission Spectra; Fluorescence Spectra

The spectral lines that can be *absorbed* by an atom can also be *emitted* by the atom upon excitation. In general the emission spectrum contains many more spectral lines than the absorption spectrum. The reason is that at room temperature all atoms are in their ground state; therefore, only allowed transitions from the ground state to excited states can contribute to the *absorption* spectrum. On the other hand, allowed transitions between *all* excited states (including the ground state, of course) can contribute to the *emission* spectrum.

If an atom has been brought from the ground state of energy E_0 to an excited state of energy E_i owing to the absorption of a quantum $hf = E_i - E_0$, it may then either reemit the quantum while returning to the ground state or it may emit smaller quanta hf', owing to transitions from the state of energy E_i to excited states of energy E_k such that $E_0 < E_k < E_i$. The spectrum thus emitted is called a *fluorescence spectrum;* the wavelengths of the emitted spectral lines are longer than the wavelength of the absorbed line (Stokes's law).

The number of spectral lines emitted by the excited atom depends upon the energy state E_i, that is, upon the absorbed quantum hf. In some cases only the transition back to the ground state is allowed; the emitted frequency is then equal to the absorbed frequency. The absorption line for which this is the case is called the *resonance line* of the atom; it represents the transition from the ground state to the lowest excited state for which the transition is allowed.

2.4d. Pauli's Exclusion Principle

Section 2.2c indicates that, for a hydrogen-like atom in which the Z direction is a preferred direction (for example, when applying a magnetic field), a new quantum number m_l must be introduced for an orbiting electron of quantum number l, and that m_l has $2l + 1$ values $-l, -l + 1, \ldots, l - 1, l$, corresponding to a component $m_l \hbar$ of angular momentum in the Z direction. The orbiting electron acts as a tiny magnet so that different orientations of the orbit in a magnetic field have slightly different energy. The spinning electron also behaves as a tiny magnet, which can have a component $m_s \hbar$ of angular momentum in the Z direction where $m_s = \pm \frac{1}{2}$. An electron can thus have $2(2l + 1)$ different quantum number combinations m_l, m_s for any given l.

Now a filled subshell of quantum number l has $2(2l + 1)$ electrons. We therefore conclude that *no two electrons in an atom can have an identical set of quantum numbers n, l, m_l, m_s*. This is called *Pauli's exclusion principle.*

If this is applied to finding the possible L, S combinations for a number of electrons with individual quantum numbers l_i, one comes to exactly the number found experimentally. In particular, one finds that closed subshells always have $L = S = 0$, so that only the outer electrons contribute to L and S.

We can apply this principle also to electrons in a solid. Here the electron is characterized by its quantum numbers n_1, n_2, and n_3 and its spin quantum number $m_s = \pm\frac{1}{2}$. According to Pauli's principle, each level characterized by n_1, n_2, and n_3 can at most be occupied by two electrons.

REFERENCES

BOHM, D., *Quantum Theory.* Prentice-Hall, Inc., Englewood Cliffs, N.J., 1951.

EISBERG, R. M., *Fundamentals of Modern Physics.* John Wiley & Sons, Inc., New York, 1961.

HERZBERG, G., *Atomic Spectra and Atomic Structure.* Dover Publications, Inc., New York, 1959.

SCHIFF, L. I., *Quantum Mechanics*, 2d ed. McGraw-Hill Book Company, New York, 1955.

SHERWIN, C. W., *Introduction to Quantum Mechanics.* Holt, Rinehart and Winston, Inc., New York, 1960.

More Advanced Reading

MESSIAH, A., *Quantum Mechanics*, vols. I and II. North-Holland Publishing Company, Amsterdam, 1961.

PROBLEMS

1.   Since a quantum has an energy $hf = hc/\lambda = eV$, the quantum energy may be measured in the same units as $1/\lambda$ (cm^{-1}) or in electron volts. Express the one unit in terms of the other.

Answer: $1 \text{ cm}^{-1} = 1.24 \times 10^{-4} \text{ eV}$; $1 \text{ eV} = 8066 \text{ cm}^{-1}$.

2.   Electrons with an energy of 10 eV are shot through a space containing a monatomic gas at low pressure. Some of the electrons collide inelastically with the atoms, thereby losing energy under simultaneous excitation of atoms. The excitation potentials of the atoms are 4.5, 8.0, and 11.0 eV; calculate the possible energies of the electrons that have traversed the region.

Answer: 5.5, 2.0, 1.0 eV.

3. The mercury resonance line, having a wavelength of 2537 Å, is due to a transition from the ground state to the next excited state. What is the excitation potential of that state?

Answer: 4.89 eV.

4. A monatomic gas has an absorption spectrum consisting of a number of wavelengths $\lambda_1, \lambda_2, \lambda_3, \ldots$ converging toward a low-wavelength limit λ_∞. Show that the ionization potential of the atom is $hc/e\lambda_\infty$).

5. Derive the Bohr formula for the hydrogen atom from the force equation, the angular momentum, and the energy equation. Assume circular orbits.

Force equation:
$$\frac{mv^2}{r} = \frac{e^2}{4\pi\epsilon_0 r^2}$$

Angular momentum: $mvr = n\hbar$

Energy equation:
$$E = -\frac{e^2}{4\pi\epsilon_0 r} + \frac{1}{2}mv^2$$

Answer: $E = -e^4 m/(8n^2 h^2 \epsilon_0^2); \ r = n^2 h^2 \epsilon_0/(\pi e^2 m)$.

6. Calculate the radius of the electron orbit in the ground state of hydrogen.

Answer: 0.53 Å.

7. Calculate the velocity of the electron in the ground state of hydrogen, assuming a circular orbit.

Answer: 2.19×10^8 cm/s.

8. Calculate the first three excitation potentials of the hydrogen atom.

Answer: 10.20, 12.09, 12.75 eV.

9. Calculate the angular momentum of an electron spinning around its axis.

Answer: 0.914×10^{-34} kg m²/s.

10. From the differential equation for the Hermite polynomials $H_v(u)$, prove the relation

$$H_{v+2}(u) - 2uH_{v+1}(u) + 2(v+1)H_v(u) = 0 \qquad (1)$$

Hint: Substitute the expression (2.36) for $H_v(u)$ into the differential equation (2.34a) and carry out the differentiations.

11. Making use of (1), show that the matrix elements of the electric moment *eu* are zero unless $v' = v \pm 1$.

Hint: Make use of the fact that the $H_v(u)$'s are orthogonal.

12. (a) Show from the orthogonality of the functions

$$P_l^{|m_l|}(\cos \vartheta) \exp (jm_l\phi)$$

that

$$\int_{-\pi/2}^{\pi/2} P_l^{|m_l|}(\cos \vartheta) P_{l'}^{|m_l|}(\cos \vartheta) \sin \vartheta \, d\vartheta = 0, \qquad \text{unless } l' = l$$

(b) With the help of examples for associated Legendre polynomials given in the text, show that

$$\int_{-\pi/2}^{\pi/2} P_l^{|m_l|}(\cos \vartheta) P_{l'}^{|m_{l'}|}(\cos \vartheta) \sin \vartheta \, d\vartheta$$

is not necessarily zero if $|m_{l'}| \neq |m_l|$.

13. In the hydrogen atom the electric moment of the electron is

$$\mathbf{m} = \mathbf{i}m_x + \mathbf{j}m_y + \mathbf{k}m_z = -er(\mathbf{i} \sin \vartheta \cos \phi + \mathbf{j} \sin \vartheta \sin \phi + \mathbf{k} \cos \vartheta)$$

Prove by calculating the matrix elements of the electric moments that the following selection rules hold for $0 < |m_l| < l$:

$$\Delta m_l = \pm 1, \qquad \Delta l = \pm 1 \qquad (m_x, m_y)$$
$$\Delta m_l = 0, \qquad \Delta l = \pm 1 \qquad (m_z)$$

Hint: Make use of the relations

$$\cos \vartheta \, P_l^u(\cos \vartheta) = \frac{l + u}{2l + 1} P_{l-1}^u(\cos \vartheta) + \frac{l - u + 1}{2l + 1} P_{l+1}^u(\cos \vartheta)$$

$$\sin \vartheta \, P_l^u(\cos \vartheta) = \frac{1}{2l + 1}[P_{l+1}^{u+1}(\cos \vartheta) - P_{l-1}^{u+1}(\cos \vartheta)]$$

and make use of the result of problem 12(a).

3

Statistics

In many physics problems one deals with large numbers of particles. Therefore, it is not possible to describe them on an individual basis; one has to be content with statistical considerations. Fortunately, these statistical considerations usually give all the needed information. For instance, the pressure of the gas refers to the average momentum transferred to a unit area of the wall per second; the current flowing through a conductor refers to the average number of electrons crossing a cross section of the conductor per second; and so on. We shall now develop the statistical methods needed in our discussion of solid-state devices.

3.1. ENERGY AND VELOCITY DISTRIBUTION FUNCTIONS

Consider a physical system consisting of a huge number of particles such as electrons, atoms, and molecules. These particles need not be identical. If they differ, we assume that large numbers of each kind exist. What can we learn in such a case from statistical considerations? We shall show this with the help of an example.

Let the system consist of a large number N of molecules kept at a temperature T, which interact with one another only weakly. The energy of a molecule can then be split into two parts; the *internal energy* of the molecule (rotation of the molecule, vibration of the atoms around equilibrium positions, excitation of electrons) and the *energy due to translatory motion*.

According to the quantum theory, the *internal energy* can assume only discrete values $E_1, E_2, \ldots, E_n$. Let N_m molecules have, on the average, an energy E_m. The probability $P(E_m)$ that the molecules have an internal energy E_m is then defined by

$$P(E_m) = \frac{N_m}{N} \tag{3.1}$$

$P(E_m)$ is called the internal *energy distribution function*. In view of these definitions,

$$\sum_{m=1}^{n} N_m = N \quad \text{and} \quad \sum_{m=1}^{n} P(E_m) = \frac{\sum_{m=1}^{n} N_m}{N} = 1 \tag{3.2}$$

The total average internal energy of the system is thus given by the relation

$$E_{\text{tot}} = \sum_{m=1}^{n} E_m N_m = N \sum_{m=1}^{n} E_m P(E_m) = N\bar{E}_i \tag{3.3}$$

where

$$\bar{E}_i = \sum_{m=1}^{n} E_m P(E_m) \tag{3.3a}$$

is called the *average internal energy* of a molecule.

Because the energy of translation of a molecule can be varied continuously between zero and infinity, our definitions have to be modified accordingly. If, on the average, ΔN molecules have an energy of translation between E and $(E + \Delta E)$, then ΔN is proportional to ΔE as long as ΔE is not too large.

$$\Delta P = \frac{\Delta N}{N} = g(E)\,\Delta E \tag{3.4}$$

is the probability that the energy has a value between E and $(E + \Delta E)$, and $g(E)$ is called the *energy distribution function* of the translatory motion. In view of these definitions,

$$\sum \Delta N = N, \qquad \sum \Delta P = \sum g(E)\,\Delta E = \int_0^\infty g(E)\,dE = 1 \tag{3.5}$$

The total average energy of translation is

$$E_{\text{tot}} = \sum E\,\Delta N = \sum N E g(E)\,\Delta E = N \int_0^\infty E g(E)\,dE = N\bar{E}_t \tag{3.6}$$

where the integration is carried out over all allowed energies ($E \geq 0$) and

$$\bar{E}_t = \int_0^\infty E g(E)\,dE \tag{3.6a}$$

is the average energy of translation of a molecule.

The total average energy W of the system is therefore

$$W = N(\bar{E}_i + \bar{E}_t) \tag{3.7}$$

notice whether distribution is normalized

We thus see that, if statistical considerations can give us $P(E_m)$ and $g(E)$ as functions of the temperature T, we can calculate W as a function of temperature and thus evaluate quantities such as the specific heat $\partial W/\partial T$ of the system.

Besides the energy distribution function $g(E)$, we may also need the related velocity distribution function $f(v)$. If ΔN molecules have on the average a velocity between v and $(v + \Delta v)$, we have instead

$$\Delta P = \frac{\Delta N}{N} = f(v)\,\Delta v \tag{3.8}$$

The average velocity would, of course, be

$$\bar{v} = \int_0^\infty vf(v)\,dv \tag{3.9}$$

To find the average energy from (3.8), we would have, in analogy with (3.9),

$$\bar{E} = \tfrac{1}{2}m\overline{v^2} = \tfrac{1}{2}m\int_0^\infty v^2 f(v)\,dv \tag{3.10}$$

According to the definition (3.8),

$$\int_0^\infty f(v)\,dv = 1 \tag{3.11}$$

A distribution function $f(v)$ having this property is said to be *normalized*. If the distribution function $f(v)$ is *not* normalized, we introduce a normalized distribution function $f_1(v) = Cf(v)$ and determine C from the condition

$$\int_0^\infty f_1(v)\,dv = C\int_0^\infty f(v)\,dv = 1 \tag{3.11a}$$

Substituting $f_1(v)$ into (3.9) and (3.10), we obtain

$$\bar{v} = \frac{\displaystyle\int_0^\infty vf(v)\,dv}{\displaystyle\int_0^\infty f(v)\,dv} \tag{3.9a}$$

$$\overline{v^2} = \frac{\displaystyle\int_0^\infty v^2 f(v)\,dv}{\displaystyle\int_0^\infty f(v)\,dv} \tag{3.10a}$$

holding for an unnormalized velocity distribution function $f(v)$. *Formulas (3.3a), (3.6a), (3.9), and (3.10) hold only for normalized distribution functions.*

In some cases the velocity distribution function $f(v_x, v_y, v_z)$ of the individual velocity components (v_x, v_y, v_z) of the molecule is needed. If ΔN molecules have, on the average, velocity components between v_x and $(v_x + \Delta v_x)$, between v_y and $(v_y + \Delta v_y)$, and between v_z and $(v_z + \Delta v_z)$, respectively, then we have, instead of (3.8),

$$\Delta P = \frac{\Delta N}{N} = f(v_x, v_y, v_z)\, \Delta v_x\, \Delta v_y\, \Delta v_z \tag{3.12}$$

This again satisfies the normalization condition

$$\int_{-\infty}^{+\infty} \int_{-\infty}^{+\infty} \int_{-\infty}^{+\infty} f(v_x, v_y, v_z)\, dv_x\, dv_y\, dv_z = 1 \tag{3.13}$$

By analogy with (3.9) and (3.10) we have, for example, for $\bar{v}_x$ and $\overline{v_x^2}$,

$$\bar{v}_x = \int_{-\infty}^{+\infty} \int_{-\infty}^{+\infty} \int_{-\infty}^{+\infty} v_x f(v_x, v_y, v_z)\, dv_x\, dv_y\, dv_z \tag{3.14}$$

$$\overline{v_x^2} = \int_{-\infty}^{+\infty} \int_{-\infty}^{+\infty} \int_{-\infty}^{+\infty} v_x^2 f(v_x, v_y, v_z)\, dv_x\, dv_y\, dv_z \tag{3.15}$$

and the average energy of the particle is

$$\bar{E} = \tfrac{1}{2}m(\overline{v_x^2} + \overline{v_y^2} + \overline{v_z^2}) \tag{3.16}$$

If the distribution function $f(v_x, v_y, v_z)$ is not normalized, we have, by analogy with (3.9a) and (3.10a),

$$\bar{v}_x = \frac{\displaystyle\int_{-\infty}^{+\infty} \int_{-\infty}^{+\infty} \int_{-\infty}^{+\infty} v_x f(v_x, v_y, v_z)\, dv_x\, dv_y\, dv_z}{\displaystyle\int_{-\infty}^{+\infty} \int_{-\infty}^{+\infty} \int_{-\infty}^{+\infty} f(v_x, v_y, v_z)\, dv_x\, dv_y\, dv_z} \tag{3.14a}$$

$$\overline{v_x^2} = \frac{\displaystyle\int_{-\infty}^{+\infty} \int_{-\infty}^{+\infty} \int_{-\infty}^{+\infty} v_x^2 f(v_x, v_y, v_z)\, dv_x\, dv_y\, dv_z}{\displaystyle\int_{-\infty}^{+\infty} \int_{-\infty}^{+\infty} \int_{-\infty}^{+\infty} f(v_x, v_y, v_z)\, dv_x\, dv_y\, dv_z} \tag{3.15a}$$

The proof is the same as for (3.9a) and (3.10a). Usually $\bar{v}_x = \bar{v}_y = \bar{v}_z = 0$.

3.2. CLASSICAL AND SEMICLASSICAL STATISTICS

In classical physics there is no restriction on the energy that a system can assume. In the atomic domain the energy is quantized. It is obvious that quantization can have some influence on the velocity and energy distributions of particles; this, however, turns out to be important at low temperatures only. In the case of electrons there is severe restriction in the form of Pauli's exclusion principle, according to which no two electrons can have the same set of quantum numbers. We shall discuss that problem in Sec. 3.3. In this section we deal with those systems for which the restriction due to Pauli's exclusion principle does not exist.

3.2a. Boltzmann Factor

If there is no restriction on the number of particles that can occupy a given energy level, the particles may be able to assume either an almost continuous range of energies or a discrete range of energies. There may be only one kind of particle or there may be several kinds. If several kinds exist, there must be a very large number of particles of each kind. We shall now show that the probability $a(E)$ that a particle has an energy E at a given temperature T under these conditions is

$$a(E) = A \exp\left(-\frac{E}{kT}\right) \qquad (3.17)$$

where A is a constant, $k = R/N$ is Boltzmann's constant (1.38×10^{-23} J/°K), R is the gas constant in joules per mole, and N is Avogadro's number. The factor $\exp(-E/kT)$ is known as the *Boltzmann factor;* it will show up in many further theoretical considerations.

The proof is in two steps. In the first step it is shown that $a(E) = A \exp(-\beta E)$; in the second step it is proved that $\beta = 1/kT$.

Consider two particles with energies E_1 and E_2 that collide and obtain energies E_1' and E_2'. According to the energy conservation law,

$$E_1 + E_2 = E_1' + E_2'$$

If, on the average, p collisions per second occur in which particles of energy E_1 and E_2 collide and change their energies into E_1' and E_2', then p obviously is proportional to the probability $a(E_1)$ that the one particle has an energy E_1 and to the probability $a(E_2)$ that the other particle has an energy E_2; that is,

$$p = Ca(E_1)a(E_2) \qquad (3.18)$$

where C is a constant. If, on the average, p' collisions occur per second in which the initial energies are E_1' and E_2' and the final energies are E_1 and E_2, then, for the same reasons,

$$p' = Ca(E_1')a(E_2') \qquad (3.18a)$$

But in equilibrium these numbers should be equal. Substituting $E_1' = E_1 - x$ and $E_2' = E_2 + x$, we thus have the equilibrium condition

$$p = p' \quad \text{or} \quad a(E_1)a(E_2) = a(E_1 - x)a(E_2 + x) \qquad (3.18b)$$

The only function that satisfies this condition is

$$a(E) = A \exp(-\beta E) \qquad (3.19)$$

as can be seen by substitution; the minus sign is used because $a(E)$ must be small if E is large.

We shall now show that $\beta = 1/kT$. It is easily seen that β is a universal factor. For if we had considered two kinds of particles for which the probabilities of having an energy E were $a(E)$ and $b(E)$, respectively, we would have obtained from collisions between the two kinds [compare (3.18b)]

$$a(E_1)b(E_2) = a(E_1 - x)b(E_2 + x) \qquad (3.18c)$$

and this is satisfied only if

$$a(E) = A \exp(-\beta E) \quad \text{and} \quad b(E) = B \exp(-\beta E) \qquad (3.19a)$$

as is again found by substitution. Hence the factor $\exp(-\beta E)$ is a universal factor.

It is therefore necessary to prove the relation $\beta = 1/kT$ for one simple case only in order to show its validity for all cases. We choose as our example 1 gram molecule of a perfect gas, consisting of N monatomic molecules that interact only during collisions; here N is Avogadro's number. The energy E of such a particle consists only of kinetic energy:

$$E = \tfrac{1}{2}mv^2 = \tfrac{1}{2}m(v_x^2 + v_y^2 + v_z^2) \tag{3.20}$$

and the normalized velocity distribution of the particles, satisfying the normalization condition (3.13), is

$$f(v_x, v_y, v_z)\,\Delta v_x\,\Delta v_y\,\Delta v_z = \left(\frac{m\beta}{2\pi}\right)^{3/2} \exp\left[-\frac{1}{2}m\beta(v_x^2 + v_y^2 + v_z^2)\right]\Delta v_x\,\Delta v_y\,\Delta v_z \tag{3.21}$$

The total energy W of the system is equal to $N\bar{E}$, where E is given by (3.20) and $\overline{v_x^2}$, $\overline{v_y^2}$, and $\overline{v_z^2}$ can be calculated from integrals of the type (3.15). Carrying out the integration, we obtain

$$W = \frac{3N}{2\beta} \tag{3.22}$$

The specific heat of the system at constant volume is equal to dW/dT. But we also know from gas theory that this value is equal to $\tfrac{3}{2}R$, where R is the gas constant. Hence, since $W = 0$ at $T = 0$,

$$\frac{3}{2}N\frac{d(1/\beta)}{dT} = \frac{3}{2}R \quad\text{or}\quad \frac{1}{\beta} = \frac{R}{N}T = kT \tag{3.23}$$

so that Eq. (3.17) is now proved.

The normalized velocity distribution of the particles thus becomes

$$f(v_x, v_y, v_z)\,\Delta v_x\,\Delta v_y\,\Delta v_z$$
$$= \left(\frac{m}{2\pi kT}\right)^{3/2} \exp\left[-\frac{\tfrac{1}{2}m(v_x^2 + v_y^2 + v_z^2)}{kT}\right]\Delta v_x\,\Delta v_y\,\Delta v_z \tag{3.24}$$

This is called a *Maxwellian velocity distribution.* The normalized distribution function for the speed v is[†]

$$f(v)\,\Delta v = \left(\frac{m}{2\pi kT}\right)^{3/2} \exp\left(-\frac{\tfrac{1}{2}mv^2}{kT}\right)4\pi v^2\,\Delta v \tag{3.25}$$

Substituting $\tfrac{1}{2}mv^2 = E$, we obtain the normalized energy distribution,

$$g(E)\,\Delta E = \frac{2}{\sqrt{\pi}}\exp\left(-\frac{E}{kT}\right)\left(\frac{E}{kT}\right)^{1/2}\Delta\left(\frac{E}{kT}\right) \tag{3.26}$$

---

† This follows most easily if we use a rectangular coordinate system with coordinates v_x, v_y, v_z; $\Delta v_x\,\Delta v_y\,\Delta v_z$ represents a small volume element in that coordinate system, and $4\pi v^2\,\Delta v$ represents the volume contained within a spherical shell of radii v and $(v + \Delta v)$.

If the energy levels are discrete, the energy distribution becomes

$$P(E_m) = CZ(E_m) \exp\left(-\frac{E_m}{kT}\right)$$ (3.27)

where $Z(E_m)$ represents the number of energy levels with energy E_m, and C is chosen so that the distribution is normalized.

3.2b. Equipartition Law

Substituting (3.24) back into (3.15), we obtain

$$\tfrac{1}{2}m\overline{v_x^2} = \tfrac{1}{2}m\overline{v_y^2} = \tfrac{1}{2}m\overline{v_z^2} = \tfrac{1}{2}kT$$ (3.28)

This result is known as the *equipartition law*. It holds for any particle that satisfies the following conditions:

1. The energy E of the particle depends upon a number of variables (coordinates, velocities, angles, angular velocities); the minimum number of variables needed to describe the system is called the number of *degrees of freedom*.
2. The energy E can be written as a sum of quadratic terms:

$$E = a_1 q_1^2 + a_2 q_2^2 + \ldots + a_n q_n^2$$ (3.29)

in each of the variables with none of the coefficients equal to zero.
3. Each variable is continuously variable from $-\infty$ to $+\infty$.
4. The system of particles is kept at the uniform temperature T.

Under these conditions the average energy per degree of freedom is $\tfrac{1}{2}kT$ (*equipartition law*).

We need prove it for one variable q_i only. The normalized distribution for the variable q_i is

$$f(q_i)\,\Delta q_i = \left(\frac{a}{\pi kT}\right)^{1/2} \exp\left(-\frac{aq_i^2}{kT}\right)\Delta q_i$$ (3.30)

and hence the average energy $\bar{E}_i$ corresponding to this degree of freedom is

$$\bar{E}_i = \int_{-\infty}^{+\infty} (aq_i^2)\left(\frac{a}{\pi kT}\right)^{1/2} \exp\left(-\frac{aq_i^2}{kT}\right) dq_i = \frac{1}{2}kT$$ (3.31)

3.2c. Deviations from the Equipartition Law

Equation (3.31) holds for energies that are continuously variable. For discrete energies $E_1, E_2, \ldots, E_n$, according to (3.27) and (3.3a) we would have

$$\bar{E} = \sum_{m=1}^{n} E_m CZ(E_m) \exp\left(-\frac{E_m}{kT}\right)$$ (3.32)

The average energy may then differ from the value expected from the equipartition law. It turns out that if $E_m - E_{m-1} \ll kT$ for the lowest energy levels (m small), the equipartition law is approximately satisfied. Large deviations can be expected, however, if $E_m - E_{m-1} > kT$ even for the lowest energy levels.

As an example, consider the vibration of a one-dimensional harmonic oscillator, such as an atom bound to an equilibrium position and free to vibrate in one direction only. The system has two degrees of freedom: the position x and the velocity v of the particle. If m is the mass and $-bx$ is the retarding force driving the system back to its equilibrium position, the energy E is

$$E = \tfrac{1}{2}bx^2 + \tfrac{1}{2}mv^2 = \tfrac{1}{2}m\omega_0^2 x^2 + \tfrac{1}{2}mv^2 \qquad (3.33)$$

so that the conditions for the equipartition law are satisfied. According to the quantum theory, Eq. (2.35), however, the energy of the system can have only the values

$$E_v = hf(v + \tfrac{1}{2}), \qquad v = 0, 1, 2, \ldots \qquad (3.34)$$

where

$$f = \frac{1}{2\pi}\sqrt{\frac{b}{m}} \equiv \frac{\omega_0}{2\pi} \qquad (3.34a)$$

is the frequency of vibration of the oscillator. The energy $\tfrac{1}{2}hf$ is called the *zero-point energy*. Since it means a shift only in the zero point of the energy scale, we can neglect its influence here. The energy distribution of the harmonic oscillator thus becomes

$$P(E_v) = C \exp\left(-\frac{vhf}{kT}\right), \qquad v = 0, 1, 2, \ldots \qquad (3.35)$$

where C is determined such that the distribution is normalized:

$$\sum_{v=0}^{\infty} P(E_v) = 1 \quad \text{or} \quad C^{-1} = \sum_{v=0}^{\infty} \exp\left(-\frac{vhf}{kT}\right) = \left[1 - \exp\left(-\frac{hf}{kT}\right)\right]^{-1}$$
$$(3.35a)$$

Consequently, the average energy of the harmonic oscillator is[†]

$$\bar{E} = \sum_{v=0}^{\infty} E_v P(E_v) = C \sum_{v=0}^{\infty} vhf \exp\left(-\frac{vhf}{kT}\right) = \frac{hf}{\exp(hf/kT) - 1} \qquad (3.36)$$

This reduces to kT as long as hf/kT is a small quantity, but large deviations occur if $hf/kT > 1$.

This problem is pertinent to the contribution of the vibrational energy of the atoms in gas molecules and in solids to the specific heat. In diatomic molecules the atoms can vibrate around the line joining the nuclei; since

---

[†]
$$\sum_{n=0}^{\infty} \exp(-nx) = \frac{1}{1 - \exp(-x)}$$

$$\sum_{n=0}^{\infty} nx \exp(-nx) = -x\frac{d}{dx}\sum_{n=0}^{\infty} \exp(-nx) = \frac{x\exp(-x)}{[1 - \exp(-x)]^2}$$

the center of gravity of the atom does not move in this vibration, this molecule has only two degrees of freedom of vibration. At sufficiently high temperatures the vibration of the atoms gives a contribution $R = kN$ to the specific heat per gram molecule at constant volume.

In crystals an atom can vibrate in three independent directions; hence it has six degrees of freedom. The specific heat per gram molecule of an element at sufficiently high temperatures is therefore $3kN = 3R$. This is known as *Dulong and Petit's law*.

3.3. QUANTUM STATISTICS

3.3a. Fermi–Dirac Statistics

In a conductor the outer electrons of the atoms can move freely through the material, except for collisions with the fixed positive ions. Neglecting the interaction between the electrons and the positive ions, we can treat the electrons as a "free-electron gas" imprisoned in a box the sides of which coincide with the walls of the cube. A fuller discussion of this "free-electron" model is given in Chapter 4; here we discuss only the statistics of such a free-electron gas.

The motion of electrons in the conductor is restricted in two respects:

1. The *wave character* of the electrons restricts the number of allowed energy levels between E and $(E + \Delta E)$.
2. According to *Pauli's exclusion principle*, the number of electrons with an allowed energy between E and $(E + \Delta E)$ is drastically restricted (Fig. 3.1).

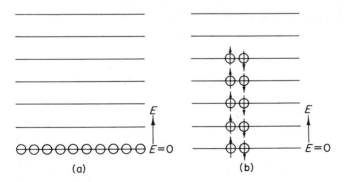

Fig. 3.1. Distribution of 10 particles over a number of energy levels at $T = 0$. (a) The number of particles that can occupy a given energy level is not restricted; hence at $T = 0$ all particles are in the lowest energy state. (b) The number of particles that can occupy a given energy level is restricted by Pauli's exclusion principle. Since only two particles with opposite spin can occupy the same energy level, some particles have $E > 0$, even at $T = 0$.

First we shall show that the number of energy levels per unit volume, corresponding to a speed between v and $(v + \Delta v)$, is

$$\Delta N = 4\pi v^2 \, \Delta v \frac{2m^3}{h^3} \tag{3.37}$$

because of these restrictions. Then we shall prove that the probability $a(E)$ that an energy level with energy E is occupied is

$$a(E) = \left[\exp\left(\frac{E - E_f}{kT}\right) + 1 \right]^{-1} \tag{3.38}$$

where E_f is a constant, characteristic for the material. This is called the *Fermi level*, the magnitude of which is calculated in Sec. 3.3b. This quantity plays an extremely important part in subsequent chapters.

The proof of (3.37) is in two steps. The first step takes into account the influence of the wave character of the electrons, whereas the second step deals with the influence of Pauli's exclusion principle. The proof of (3.38) then follows along similar lines as the proof of (3.17)—with modifications, of course, imposed by Pauli's exclusion principle.

We take into account the first restriction by observing that a particle of mass m moving with a speed v exhibits wave character and that the corresponding wavelength λ is

$$\lambda = \frac{h}{mv} \tag{3.39}$$

Consider now a conducting cube. It is known that the conduction of electricity in the conductor is due to the fact that some of the electrons (one or two or more per atom) can move more or less freely inside the conductor. Because of the wave character of the electron we can consider the motion of the electrons as a problem of wave propagation.

As is well known, the equation of a wave of angular frequency ω and wavelength λ traveling in free space is [see Eq. (2.23b)]

$$A \exp\left[-j\omega t + \frac{j2\pi(lx + my + nz)}{\lambda} \right] \tag{3.40}$$

where l, m, and n are the direction cosines of the direction of propagation. If the requirement is made that the wave periodically repeat itself if x, y, and z are increased by L, we have [see Eq. (2.25b)]

$$\frac{lL}{\lambda} = n_1, \quad \frac{mL}{\lambda} = n_2, \quad \frac{nL}{\lambda} = n_3 \tag{3.41}$$

or, since $l^2 + m^2 + n^2 = 1$,

$$\left(\frac{L}{\lambda}\right)^2 = n_1^2 + n_2^2 + n_3^2 \tag{3.41a}$$

where n_1, n_2, and n_3 are integral. This equation determines the allowed "wave numbers" $(1/\lambda)$.

How many numbers are allowed between $1/\lambda$ and $[1/\lambda + \Delta(1/\lambda)]$? To determine this, consider n_1, n_2, and n_3 as rectangular coordinates. Equation (3.41a) represents a sphere of radius $r = L/\lambda$. The number Δn of allowed wave numbers in the range between $1/\lambda$ and $[1/\lambda + \Delta(1/\lambda)]$ is therefore equal to the volume between two concentric spheres of radii r and $(r + \Delta r)$:

$$\Delta n = 4\pi r^2 \, \Delta r = 4\pi \left(\frac{L}{\lambda}\right)^2 \Delta \frac{L}{\lambda}$$
$$= \frac{4\pi V}{\lambda^2} \Delta \frac{1}{\lambda} \tag{3.42}$$

where $V = L^3$ is the volume of the cube. The total number of wave numbers per unit volume in the range between $1/\lambda$ and $[1/\lambda + \Delta(1/\lambda)]$ is therefore

$$\frac{\Delta n}{V} = \frac{4\pi}{\lambda^2} \Delta \frac{1}{\lambda} \tag{3.43}$$

Or, putting $\lambda = h/mv$, we find

$$\frac{\Delta n}{V} = 4\pi v^2 \, \Delta v \frac{m^3}{h^3} \tag{3.44}$$

This then, is the restriction imposed by the wave character of the electrons.

To take the second restriction into account, we apply Pauli's exclusion principle (Sec. 2.4d). We saw that because of the electron spin the allowed number of energy states characterized by the quantum numbers n_1, n_2, and n_3 was two, and hence the allowed number ΔN of electrons per unit volume with a velocity between v and $(v + \Delta v)$ is $2 \, \Delta n/V$.

$$\Delta N = \frac{2 \, \Delta n}{V} = 4\pi v^2 \, \Delta v \frac{2m^3}{h^3} \tag{3.44a}$$

Equation (3.37) has now been proved.

We now calculate the probability $a(E)$ that an energy level of energy E is occupied by considering collisions between an electron and the lattice. Since energy is conserved in the process,

$$E_1 + E_2 = E_1' + E_2' \tag{3.45}$$

where E_1 and E_1' refer to the electron and E_2 and E_2' to the lattice vibrations. Let $b(E)$ be the probability that the lattice vibrations have an energy E. The occupancy of an electron energy state E is restricted by Pauli's exclusion principle.

In equilibrium the average number p of transitions $(E_1, E_2) \rightarrow (E_1', E_2')$ equals the average number of transitions $(E_1', E_2') \rightarrow (E_1, E_2)$. Considering transitions of the first type, we see that p is proportional to $a(E_1)b(E_2)[1 - a(E_1')]$, where the last factor takes account of the exclusion principle; for transitions of the second type, p is proportional to $a(E_1')b(E_2')[1 - a(E_1)]$. Since the proportionality factors are equal,

$$a(E_1)b(E_2)[1 - a(E_1')] = a(E_1')b(E_2')[1 - a(E_1)] \tag{3.46}$$

for all sets of E_1, E_2, E'_1, and E'_2. But

$$b(E) = B \exp\left(-\frac{E}{kT}\right) \tag{3.47}$$

Substituting into (3.46), dividing by $a(E_1)b(E_2)a(E'_1)$, and using (3.45) gives

$$\frac{1}{a(E'_1)} - 1 = \left[\frac{1}{a(E_1)} - 1\right] \exp\left(\frac{E'_1 - E_1}{kT}\right) \tag{3.48}$$

which must hold for all E_1 and E'_1. This can be true only if

$$\frac{1}{a(E)} - 1 = A \exp\left(\frac{E}{kT}\right) \tag{3.49}$$

as we can see by substituting, so that

$$a(E) = \left[A \exp\left(\frac{E}{kT}\right) + 1\right]^{-1} \tag{3.49a}$$

Putting

$$A = \exp\left(-\frac{E_f}{kT}\right) \tag{3.50}$$

we may write Eq. (3.49a) in the form (3.38):

$$a(E) = \left[\exp\left(\frac{E - E_f}{kT}\right) + 1\right]^{-1} \tag{3.51}$$

3.3b. Fermi–Dirac Distribution

Multiplying (3.37) and (3.38), we obtain for the unnormalized distribution in the speed per unit volume

$$\Delta N = f(v)\, \Delta v = \frac{2(m^3/h^3)4\pi v^2\, \Delta v}{\exp\left[(E - E_f)/kT\right] + 1} \tag{3.52}$$

Substituting $E = \tfrac{1}{2}mv^2$, we obtain for the unnormalized distribution in energy per unit volume

$$\Delta N = g(E)\, \Delta E = \frac{8\pi\sqrt{2}\,(m^{3/2}/h^3)E^{1/2}\, \Delta E}{\exp\left[(E - E_f)/kT\right] + 1} \tag{3.53}$$

The velocity distribution (3.52) expressed in terms of the velocity components [compare also (3.24) and (3.25)] becomes

$$\Delta N = f(v_x, v_y, v_z)\, \Delta v_x\, \Delta v_y\, \Delta v_z = \frac{2(m^3/h^3)\, \Delta v_x\, \Delta v_y\, \Delta v_z}{\exp\left[(E - E_f)/kT\right] + 1} \tag{3.54}$$

The distributions (3.52) to (3.54) are known as the *Fermi–Dirac distribution*.

The factor $\{\exp\left[(E - E_f)/kT\right] + 1\}^{-1}$ depends upon the energy E in a peculiar way. At $T = 0$ the factor is unity for $E < E_f$, and zero if $E > E_f$. At higher temperature the factor is close to unity if $E_f - E \gg kT$, is equal to $\tfrac{1}{2}$ at $E = E_f$, and varies as $\exp\left[-(E - E_f)/kT\right]$ for $E - E_f \gg kT$ (Fig. 3.2).

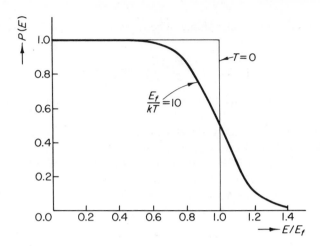

Fig. 3.2. Values of $P(E) = \{\exp[(E - E_f)/kT] + 1\}^{-1}$ as a function of E/E_f, for $T = 0$ and for $E_f/kT = 10$. The reference energy level is taken at the bottom of the conduction band.

The value of E_f follows from the normalization condition

$$\int_0^\infty g(E)\, dE = N$$

where N is the number of free electrons per unit volume. The calculation is often quite difficult. For metals at $T = 0$, $a(E) = 1$ for $E < E_f$ and $a(E) = 0$ for $E > E_f$. We then have

$$N = 8\pi\sqrt{2}\,\frac{m^{3/2}}{h^3} \int_0^{E_f} E^{1/2}\, dE = 8\pi\sqrt{2}\,\frac{m^{3/2}}{h^3}\,\frac{2}{3}\,E_f^{3/2} \tag{3.55}$$

so that

$$E_f = E_{f0} = \frac{h^2}{2m}\left(\frac{3N}{8\pi}\right)^{2/3} \tag{3.56}$$

For $T > 0$, the following expression has been obtained with the help of approximation methods:

$$E_f = E_{f0}\left[1 - \frac{\pi^2}{12}\left(\frac{kT}{E_{f0}}\right)^2\right] \tag{3.57}$$

The considerations should hold for electrons in conductors in those cases where the electrons can be considered as free. If this condition is not satisfied, the factor $a(E)$ is not altered, since its derivation does not depend upon the assumption. In that case (3.53) should be written

$$\Delta N = \frac{Z(E)\,\Delta E}{\exp[(E - E_f)/kT] + 1} \tag{3.58}$$

where $Z(E)\,\Delta E$ is the number of available energy states in the interval between

E and $(E + \Delta E)$ per unit volume. The quantity E_f then follows from the condition

$$\int \frac{Z(E)\, dE}{\exp\left[(E - E_f)/kT\right] + 1} = N \tag{3.59}$$

where the integration extends over all energy levels that the N electrons can occupy.

In many cases the energy-level density $Z(E)$ is of the form $CE^{1/2}$ as required by the free-electron theory, but C differs strongly from the free-electron value $8\pi\sqrt{2}\,(m^{3/2}/h^3)$. We can take this into account by introducing an "effective mass" m^* for the electrons. A large density $Z(E)$ corresponds to a large value of m^*, and a small density $Z(E)$ corresponds to a small value of m^*. In these cases we find, by substituting for m and h, that

$$\begin{aligned}
Z(E) &= 8\pi\sqrt{2}\,\frac{m^{3/2}}{h^3}\left(\frac{m^*}{m}\right)^{3/2} E^{1/2} \\
&= 6.814 \times 10^{27}\left(\frac{m^*}{m}\right)^{3/2} E^{1/2}, \qquad \text{in m}^{-3}
\end{aligned} \tag{3.60}$$

where E is measured in electron volts.

3.3c. Bose–Einstein Statistics

From Fermi–Dirac statistics it is only a small step to *Bose–Einstein* statistics. The same framework of discussion can be applied—with one exception.

Consider a black body of volume $V = L^3$. The expression (3.43) for the density of wave modes is still valid, but we must take into account that for photons $\lambda = c/f$, where c is the velocity of light and f the frequency. Moreover, it must be taken into account that for each possible wave there are two independent directions of polarization. Consequently, Eq. (3.43) for the density of states must be replaced by

$$\Delta N_s = \frac{2\,\Delta n}{V} = 8\pi\left[\frac{1}{\lambda^2}\,\Delta\!\left(\frac{1}{\lambda}\right)\right] = \frac{8\pi f^2\,\Delta f}{c^3} \tag{3.61}$$

This equation must be modified in media with a refractive index n, because in that case $\lambda = c/nf$, and n depends on frequency. Thus, as the first part of (3.61) indicates,

$$\Delta N_s = \frac{8\pi n^2 f^2\,\Delta(nf)}{c^3} \tag{3.61a}$$

Pauli's exclusion principle does not hold here. But we know that each black-body mode behaves like a harmonic oscillator, so that the average energy per mode is given by Eq. (3.36). Hence the density of photons of frequency between f and $f + \Delta f$ is

$$\Delta N_p = \frac{8\pi f^2 \, \Delta f}{c^3} \frac{1}{\exp(hf/kT) - 1} \tag{3.62}$$

which gives Planck's radiation law when multiplied by hf.

If we now introduce $E = hf$ as the energy of the photon, Eq. (3.62) may be written

$$\Delta N_p = \frac{Z(E) \, \Delta E}{\exp(E/kT) - 1} \tag{3.62a}$$

where

$$Z(E) \, \Delta E = \frac{8\pi}{h^3 c^3} E^2 \, \Delta E \tag{3.62b}$$

Equation (3.62a) is the basic equation of Bose–Einstein statistics. It is in a form that closely resembles Eq. (3.58), except for the minus sign.

3.4. THERMODYNAMICS AND EQUILIBRIUM

3.4a. Helmholtz Function

According to the second law of thermodynamics, for any reversible process one has

$$T \, dS = dU + p \, dV - \mu \, dN - \psi \, dQ \tag{3.63}$$

Here T is the temperature, dS the change in entropy, and dU the change in internal energy. Furthermore, p is the pressure, dV the change in volume, and $p \, dV$ the mechanical work done by the system. In addition, μ is the chemical potential, dN the change in the number of particles, and $-\mu \, dN$ the chemical work done by the system. Finally, ψ is the electrostatic potential, dQ the change in charge, and $-\psi \, dQ$ the electrical work done by the system.

To describe equilibrium situations it is customary to introduce certain state functions of the system. We need here the Helmholtz function F. The function F is defined as $U - TS$ and is considered to be a function of the temperature T, the volume V, the particle number N, and the charge Q:

$$F = F(T, V, N, Q) = U - TS \tag{3.64}$$

Taking total differentials yields

$$dF = dU - T \, dS - S \, dT = -S \, dT - p \, dV + \mu \, dN + \psi \, dQ \tag{3.65}$$

The Helmholtz function is especially useful for studying equilibrium conditions, where T and V are kept constant.

In most solid-state problems of interest to us, the charges are associated with charged particles. For *electrons*, $dQ = -e \, dN$, and hence Eq. (3.65) can be simplified as

$$dF = -S \, dT - p \, dV + \bar{\mu} \, dN \tag{3.65a}$$

μ- chemical potential

$\bar{\mu}$ - electrochemical potential · Ch. 3

where

$$\bar{\mu} = \mu - e\psi \qquad (3.66)$$

is called the *electrochemical potential* of the system.

3.4b. Equilibrium Conditions

We shall now apply the foregoing to an equilibrium situation. Suppose that two conductors (two metals, a metal and a semiconductor, or two semiconductors) are brought into contact. Then electrons may flow from the one material to the other until an equilibrium is established. What is this equilibrium situation?

Since T and V are kept constant and the Helmholtz functions of the two materials are additive, for the total change in Helmholtz potential upon transfer of electrons from one material to the other we have

$$dF_{\text{tot}} = dF_1 + dF_2 = \bar{\mu}_1 \, dN_1 + \bar{\mu}_2 \, dN_2 \qquad (3.67)$$

where $\bar{\mu}_1$ and $\bar{\mu}_2$ are the electrochemical potentials. But since the electrons flow from the one material to the other, we have $dN_2 = -dN_1$, so that

$$dF_{\text{tot}} = (\bar{\mu}_1 - \bar{\mu}_2) \, dN_1 \qquad (3.67a)$$

According to thermodynamical rules, there is equilibrium if F_{tot}, taken as a function of the only variable N_1, is a minimum. Hence

$$\frac{\partial F_{\text{tot}}}{\partial N_1} = 0 \quad \text{or} \quad \bar{\mu}_1 = \bar{\mu}_2 \qquad (3.68)$$

We therefore have the following rule: *At equilibrium for two arbitrary conductors in contact the electrochemical potentials are equal;* that is, they are at the same level or position.

We now prove that for conductors in equilibrium the Fermi level E_f and the electrochemical potential $\bar{\mu}$ are equal. We take an arbitrary conductor and add a small number δn of electrons while keeping the temperature T and the volume V constant. The fluctuation δF in $F = U - TS$ is then a function of δn only. Since $S = k \ln W$ (according to a well-known thermodynamical definition, where W is the number of particle arrangements that have the same distribution of electrons over the energy levels), under the stated conditions we have

$$\delta F = \delta U - T \, \delta S = \delta U - kT \, \delta \ln W \qquad (3.69)$$

We evaluate δF, put $\delta F = \bar{\mu} \, \delta n$ according to the definition of $\bar{\mu}$, and find $\bar{\mu}$.

To do this we split the distribution of the energy levels into small sections ΔE. Let there be ΔN_i levels with energies between E_i and $(E_i + \Delta E)$ and let Δn_i of them be occupied by an electron. If this particular distribution has a probability W_i, then

$$W = \prod_{i=1}^{\infty} W_i, \qquad \ln W = \sum_{i=1}^{\infty} \ln W_i, \qquad \delta \ln W = \sum_{i=1}^{\infty} \delta \ln W_i \qquad (3.70)$$

But W_i is equal to the possible number of particle arrangements that give Δn_i filled levels and $\Delta N_i - \Delta n_i$ empty energy levels, and this is equal to

$$W_i = \frac{\Delta N_i!}{\Delta n_i!(\Delta N_i - \Delta n_i)!} \tag{3.71}$$

We now add a small number δn of electrons to the sample. Let Δn_i change by δn_i. If Δn_i increases by unity, $\Delta N_i - \Delta n_i$ decreases by unity, and the corresponding change in $\ln W_i$ is

$$\delta \ln W_i = \ln \frac{\Delta N_i - \Delta n_i}{\Delta n_i + 1} \simeq \ln \left[\frac{\Delta N_i}{\Delta n_i} - 1 \right] = \frac{E_i - E_f}{kT} \tag{3.72}$$

since $\Delta n_i \gg 1$ and, according to Fermi–Dirac statistics [Eq. (3.38)]

$$\frac{\Delta N}{\Delta n_i} = \exp \left(\frac{E_i - E_f}{kT} \right) + 1$$

Hence for small fluctuations δn_i the corresponding changes in $\ln W_i$ are δn_i times as large as (3.72) predicts, since all the contributions $\delta \ln W_i$ add, so that

$$\delta \ln W_i = \frac{E_i - E_f}{kT} \delta n_i, \qquad \delta \ln W = \sum_{i=1}^{\infty} \left(\frac{E_i - E_f}{kT} \right) \delta n_i \tag{3.73}$$

Since $\delta U_i = E_i \, \delta n_i$, substituting into (3.69) yields

$$\begin{aligned}
\delta F = \bar{\mu} \, \delta n &= \sum_{i=1}^{\infty} E_i \, \delta n_i - \sum_{i=1}^{\infty} (E_i - E_f) \, \delta n_i \\
&= \sum_{i=1}^{\infty} E_f \, \delta n_i = E_f \, \delta n
\end{aligned} \tag{3.74}$$

for $\sum_{i=1}^{\infty} \delta n_i = \delta n$, the total fluctuation in n. Hence

$$\bar{\mu} = E_f \tag{3.75}$$

In other words, *in equilibrium situations the electrochemical potential $\bar{\mu}$ is equal to the Fermi level.* The equilibrium condition for arbitrary conductors in contact is thus as follows: *When two arbitrary conductors are brought into contact, the Fermi levels are at equal height.*

REFERENCES

BLAKEMORE, J. S., *Semiconductor Statistics*. Pergamon Press, Inc., Elmsford, N.Y., 1962.

MORSE, P. M., *Thermal Physics*. W. A. Benjamin, Inc., Menlo Park, Calif., 1964.

REIF, F., *Fundamentals of Statistical and Thermal Physics*. McGraw-Hill Book Company, New York, 1965.

TER HAAR, D., *Elements of Thermostatistics*, 2d ed. Holt, Rinehart and Winston, Inc., New York, 1966.

PROBLEMS

1. The distribution function for the speed v of gas molecules of mass m at a temperature T may be written

$$\frac{\Delta N}{N} = f(v)\,\Delta v = C_1 \exp\left(-\frac{\frac{1}{2}mv^2}{kT}\right)v^2\,\Delta v$$

where v can assume all values between 0 and ∞.
(a) Normalize this distribution by determining C_1 such that

$$\int_0^\infty f(v)\,dv = 1$$

(b) Calculate the speed v_0 for which the function $f(v)$ has a maximum ($=$ the most probable velocity).
(c) Calculate the average value of v.
(d) Calculate the average value of v^2.

Answer:

$$C = 4\pi\left(\frac{m}{2\pi kT}\right)^{3/2}, \quad v_0 = \left(\frac{2kT}{m}\right)^{1/2}, \quad \bar{v} = \frac{2}{\sqrt{\pi}}\left(\frac{2kT}{m}\right)^{1/2}, \quad \overline{v^2} = \frac{3kT}{m}$$

2. Show that the foregoing velocity distribution may be written as an energy distribution:

$$\frac{\Delta N}{N} = g(E)\,\Delta E = C_2 \exp\left(-\frac{E}{kT}\right)E^{1/2}\,\Delta E$$

where $E = \frac{1}{2}mv^2$.
(a) Normalize this distribution by determining the required value of C_2.
(b) Find the most probable energy E_0.
(c) Find the average energy $\bar{E}$.

Answer:

$$C_2 = \frac{2}{\pi^{1/2}(kT)^{3/2}}, \quad E_0 = \frac{1}{2}kT, \quad \bar{E} = \frac{3}{2}kT$$

3. At $T = 0$ the energy levels in a metal are all occupied for $E < E_f$ and all empty for $E > E_f$. Treating the electrons as free, we have an energy distribution

$$\frac{\Delta N}{N} = h(E)\,\Delta E = CE^{1/2}\,\Delta E, \qquad \text{for } E < E_f$$

$$\frac{\Delta N}{N} = 0, \qquad\qquad\qquad \text{for } E > E_f$$

Show that the average energy of the electron is $\frac{3}{5}E_f$.

4. A metal has 3×10^{22} atoms/cm³, and each atom gives one electron to the conduction band. Find the position of the Fermi level above the bottom of the conduction band at $T = 0$. You may assume $m^* = m$.

Answer: 3.5 eV.

4

Band Theory
of Solids

4.1. EXPERIMENTAL EVIDENCE FOR ENERGY BANDS IN SOLIDS

There is considerable experimental evidence that the energy levels of electrons in solids are not discrete but that they form "bands" of allowed energies separated by gaps in which ordinarily no energy levels occur.

Clear-cut evidence of the band structure of *occupied* energy levels of electrons in solids comes from *x-ray emission spectra*. If a crystal is bombarded by high-energy electrons, electrons are knocked out of the innermost part of the atom (K shell, L shell, and so on) and transitions can then take place from the outer to the inner shells. These give rise to narrow discrete lines, except for one or two lines that are broadened; this broadening corresponds to an energy difference of several electron volts. The most sensible explanation is to attribute these lines to a transition of one of the outer electrons to the inner shells. But this means that the energy levels of the outer electrons are not discrete, but form a band. This can be attributed to the fact that the outer electrons of a given atom strongly interact with the outer electrons of neighboring atoms and, through this interaction, with all the other outer electrons in the crystal.

In ionic crystals, such as NaCl, there are two energy bands occupied by electrons, and it is necessary to attribute one to the outer electrons of the positive ion and the other to the outer electrons of the negative ion. In valence-bond crystals, such as diamond, there is only one energy band occupied by electrons, and it is obvious to attribute it to the electrons in the valence bond.

Experimental evidence for the band structure of *unoccupied* energy levels comes from *x-ray absorption spectra.* If a crystal is irradiated by continuous x rays, and one determines the absorption spectrum of the crystal, at certain frequencies one finds more or less abrupt jumps in the absorption. This can be attributed to the fact that the x-ray quanta bring electrons from one of the inner shells to unoccupied levels of the crystal. The minimum energy needed is to bring the electron from the given shell (K shell, L shell) to the *lowest unoccupied* energy level of the crystal. Since for higher frequencies the absorption shows only slight fluctuations with frequency, it may be assumed that all higher energies are allowed.

It is found that in ionic crystals and in valence-bond crystals the highest occupied energy level is several electron volts lower than the highest unoccupied energy level, whereas for metals there is no such gap. This explains why the first are insulators and the second are conductors. For if an electric field is applied to the crystal, the electrons in an occupied energy band, separated from the lowest unoccupied energy band by a gap of several electron volts, cannot gain energy from the field because there are no unoccupied energy levels to which they can go. In metals, however, there are unoccupied energy levels in the immediate vicinity of the occupied levels, and hence electrons can gain energy from the electric field and move through the crystal more or less freely.

It is found that optical transitions from occupied energy levels to unoccupied energy levels are governed by selection rules. For example, it seems that the condition $\Delta l = \pm 1$ still holds. Apparently, then, one must distinguish between s bands, p bands, d bands, f bands, and so on, even in cases when the electrons seem to move more or less freely through the crystal.

If the optical absorption spectrum of an insulator is measured, it is usually found that the absorption rises very sharply beyond a certain frequency that is characteristic of the crystal. The main absorption is associated with photoconductivity. Apparently, the photon causes the electron to transfer from an occupied energy level to an unoccupied energy level at which it can move more or less freely through the crystal. Optical absorption thus allows one to determine the energy gap between the occupied band and the unoccupied band (compare, however, Sec. 4.3e).

If this energy gap E_g is less than 2 eV, the material is a good insulator at low temperatures but becomes a conductor at elevated temperatures. The reason is that at elevated temperatures some of the electrons can gain enough energy from the crystal lattice vibrations to reach the unoccupied energy levels of the crystal, which allow more or less free movement through the crystal. The number that do so varies as $\exp(-eE_g/2kT)$, so that the conductivity increases very rapidly with increasing temperature. Such materials are called *semiconductors.*

We shall now introduce a few names that are commonly used. The band of unoccupied energy levels is called the *conduction band*. The band of occupied energy levels in valence-bond crystals is called the *valence band*, whereas the bands associated with ionic crystals are named after the ions responsible for them. In insulators the conduction band is empty at temperatures below several hundred degrees centigrade, in semi-

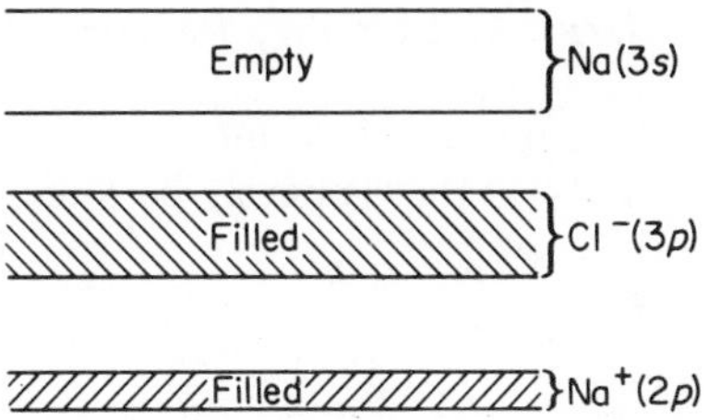

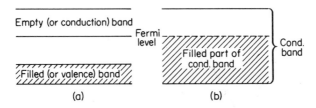

Fig. 4.1. Energy level diagram (schematic) of NaCl, showing the filled Na^+ $2p$ band and Cl^- $3p$ band and the empty Na $3s$ band, indicating that NaCl is an insulator.

conductors the conduction band is empty at low temperatures, and in metals the conduction band is partly occupied permanently (Figs. 4.1 and 4.2).

Empty (or conduction) band

Fermi level

/Filled (or valence) band/

(a)

Filled part of cond. band

Cond. band

(b)

Fig. 4.2. (a) Energy level diagram of an insulator at $T = 0$. (b) Energy level diagram of a metal at $T = 0$.

4.2. THE ONE-DIMENSIONAL LATTICE

A crystal lattice in which the electrons are constrained to move in only one direction, say the (1 0 0) direction, is called a *one-dimensional lattice*. We shall investigate the motion of electrons in such a lattice first, since it is simpler than the motion in the three-dimensional lattice.

We assume that the crystal has electrons that can move more or less freely through the crystal. This leaves positively charged ions behind that are supposed to be located in atomic planes perpendicular to the direction of motion of the electrons. Because of the interaction of a free electron with these ions and all the other electrons, a single electron can be considered as moving in a periodic potential.† The wave equation for a single electron is thus

$$\frac{\hbar^2}{2m}\frac{d^2\varphi}{dx^2} + [E - V(x)]\varphi = 0 \qquad (4.1)$$

where $V(x)$ is periodic with period a, and a is the lattice spacing.

---

† This periodic potential is called the Hartree–Fock potential.

4.2a. Solution by Perturbation Methods

We now split $V(x)$ into a (negative) part V_0 that is independent of x and a part $V'(x)$ that is purely periodic in x with

$$\int_0^a V'(x)\,dx = 0$$

Equation (4.1) may then be written

$$\frac{\hbar^2}{2m}\frac{d^2\varphi}{dx^2} + [E - V_0 - V'(x)]\varphi = 0 \qquad (4.1a)$$

We now neglect $V'(x)$ in zero-order approximation. Then the equation becomes

$$\frac{\hbar^2}{2m}\frac{d^2\varphi}{dx^2} + (E - V_0)\varphi = 0 \qquad (4.1b)$$

which has the solution (see Sec. 2.2a)

$$\varphi_k^0(x) = \exp(jkx) \qquad (4.2)$$

and the energy E_k^0 associated with it is

$$E_k^0 = V_0 + \frac{\hbar^2}{2m}k^2 \qquad (4.3)$$

The quantity k is called the *wave vector*.

If we now apply periodic boundary conditions, by requiring that the wave function periodically repeat itself every distance L, the wave function can be normalized and perturbation theory can be applied. The normalized wave function is then (compare Sec. 2.2a)

$$\varphi_k^0(x) = \frac{1}{L^{1/2}}\exp(jkx) \qquad (4.2a)$$

What happens to the wave function with wave number k if the potential energy $V'(x)$ is taken into account? The wave equation now becomes

$$\frac{\hbar^2}{2m}\frac{d^2\varphi_k}{dx^2} + [E_k - V_0 - V'(x)]\varphi_k = 0 \qquad (4.4)$$

This wave equation can be solved with perturbation methods. As shown in the Appendix, the first-order wave function is

$$\varphi_k = \varphi_k^0 + \sum_{k'}{}' \frac{V'_{k'k}}{E_k^0 - E_{k'}^0}\varphi_{k'}^0 \qquad (4.5)$$

where the symbol $\sum'$ indicates that the term $k' = k$ must be excluded, and the second-order energy becomes

$$E_k = E_k^0 + V_{kk}' + \sum_{k'}{}' \frac{V_{k'k}' V_{kk'}'}{E_k^0 - E_{k'}^0} \tag{4.6}$$

Here $V_{k'k}'$ is the matrix element of V', defined as

$$V_{k'k}' = \int_0^L \varphi_{k'}^{0*} V'(x)\varphi_k^0 \, dx \tag{4.7}$$

We now substitute for $\varphi_{k'}^0$ and φ_k^0 and write, because of the periodicity of $V'(x)$, the Fourier expansion

$$V'(x) = \sum_{n=-\infty}^{\infty} a_n \exp\left(\frac{2\pi j n x}{a}\right) \tag{4.8}$$

where $a_0 = 0$. We see that

$$V_{k'k}' = \frac{1}{L} \int_0^L \exp\left(-jk'x\right)\left[\sum_{n=-\infty}^{\infty} a_n \exp\left(\frac{2\pi j n x}{a}\right)\right] \exp\left(jkx\right) dx \tag{4.8a}$$

Therefore, $V_{kk}' = 0$, since $a_0 = 0$. $V_{k'k}'$ is zero unless

$$-k' + \frac{2\pi n}{a} + k = 0, \qquad n = \pm 1, \pm 2 \tag{4.8b}$$

and for $k' = k + 2\pi n/a$ it is

$$V_{k'k}' = a_n, \qquad k' \neq k \tag{4.8c}$$

Consequently,

$$E_k = V_0 + \frac{\hbar^2}{2m}k^2 - \sum_{n=1}^{\infty} \frac{|a_n|^2}{\frac{\hbar^2}{2m}\left[\left(k + \frac{2\pi n}{a}\right)^2 - k^2\right]} \tag{4.9}$$

Usually this series converges rapidly. Difficulty arises only when

$$k + \frac{2\pi n}{a} = -k \quad \text{or} \quad k = -\frac{n\pi}{a}, \qquad n = \pm 1, \pm 2, \ldots \tag{4.9a}$$

for in that case the nth term in Eq. (4.9) becomes infinite. For all other values of k the series (4.9) gives a finite sum. Near $k = n\pi/a$ the perturbation method thus gives inaccurate results.

What does the condition (4.9a) mean? In the unperturbed problem the momentum p of the electron is $\hbar k$ and the wavelength λ of the electron is $h/p = 2\pi/k$. Substituting (4.9a) yields

$$\lambda = \frac{2a}{n} \quad \text{or} \quad 2a = n\lambda \tag{4.9b}$$

This is the condition for Bragg reflection for electron motion in a direction perpendicular to atomic planes at distance a. Electron waves can thus propagate through the crystal unless Bragg reflection occurs; this is a very sensible result.

If condition (4.9a) is satisfied, $E^0[k + (2\pi n/a)] = E^0(-k)$. In other words, the system is degenerate, and the application of second-order perturbation theory is not allowed. We must then repeat the first-order perturbation method for the degenerate case. According to the Appendix, Eq. (A.23), this gives, if $k' = k + (2\pi n/a) = -k$,

$$\begin{vmatrix} E_k^{(1)} + V'_{kk} & V'_{kk'} \\ V'_{k'k} & E_k^{(1)} + V'_{k'k'} \end{vmatrix} = 0 \quad \text{or} \quad E_k^{(1)} = \pm|V'_{kk'}| = \pm|a_n| \quad (4.10)$$

since $V'_{kk} = V'_{k'k'} = 0$ in our calculation. The perturbing potential therefore removes the degeneracy at $k = \pi n/a$ and gives, in first-order approximation, a splitting $2|a_n|$ between the energy levels. The second-order approximation cannot be applied near $k = \pi n/a$.

The more accurate calculation of Sec. 4.2b shows that dE/dk is always $\geqq 0$ and is zero at $k = \pi/a$. There is a discontinuity in the energy at $k = \pi/a$; dE/dk is positive for $\pi/a < k < 2\pi/a$ and is zero at $k = \pi/a$ and $2\pi/a$. And so one can go on. Figure 4.3 shows both the unperturbed and the perturbed energy as a function of the wave number k.

We thus see that the effect of the periodic potential is to change the continuous distribution in electron energy into allowed and forbidden energy regions.

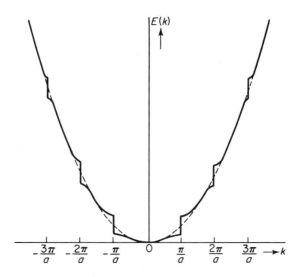

Fig. 4.3. Energy E plotted as a function of the wave number k. The dashed line gives the dependence of E upon k for the free electron case. Discontinuities in the energy E occur for $k = \pm n\pi/a$ ($n = 0, 1, 2, \ldots$).

4.2b. Floquet's Theorem; Bloch Functions

The differential equation

$$\frac{d^2\varphi}{dx^2} + \frac{2m}{\hbar^2}[E - V(x)]\varphi = 0 \tag{4.11}$$

where $V(x)$ is periodic in x with period a, has been studied extensively by mathematicians. According to *Floquet's theorem*, the solution can be written in the form

$$\varphi(x) = \exp(\mu x)u(\mu, x) \tag{4.12}$$

where $u(\mu, x)$ is periodic in x with period a.

For some values of the energy E, the constant μ is imaginary, and φ may be written

$$\varphi_k(x) = \exp(jkx)u(k, x) \tag{4.13}$$

These wave functions represent modulated propagating waves, running in the positive X direction (for $k > 0$) or in the negative X direction (for $k < 0$). The energy values for which such waves occur thus represent *allowed* energies for the electron. The wave functions (4.13) of these electrons are called *Bloch functions*. In this case, $u(k, x)$ is a periodic function of k with period $2\pi/a$.

For other values of the energy E, the constant μ is real. Such wave functions do not represent propagating waves, and hence the associated motion of the electron is not allowed. The energy values for which this is the case represent *forbidden* energies for the electron. We thus see that Floquet's theorem explains quite naturally the existence of allowed and forbidden energy bands.

We observe that $\varphi_k(x)$ does not change if k is replaced by $k + 2\pi n/a$ ($n = \pm 1, \pm 2, \ldots$). For that reason we can restrict the values of k to the range $-\pi/a \leq k \leq \pi/a$. If we do so, then for each k there are an infinite number of functions $u(k, x)$, each with a different energy $E(k)$. However, for a given energy band and a given k, $u(k, x)$ is unique.

Suppose that we restrict the value of k to the range $-\pi/a \leq k \leq \pi/a$. The quantity k thus introduced is called the *reduced wave vector*. We see that $E(k) = E(-k)$, since it does not make any difference to the energy whether the electron moves in the positive or in the negative X direction. Since $E(k)$ is a well-behaved function of k, this means that $dE/dk = 0$ at $k = 0$ for any energy band.

If we allow the value of k to assume all values from $-\infty$ to $+\infty$, then, for a given energy band, $E(k)$ must be a periodic function of k with period $2\pi/a$. This means that $dE/dk = 0$ for $k = \pm\pi/a$. For if this were not the case, then the function $E(k)$ would not be well behaved at $k = \pm\pi/a$.

We may thus choose either to represent the various energy bands all within the k range $-\pi/a \leq k \leq \pi/a$ (reduced wave vector representation),

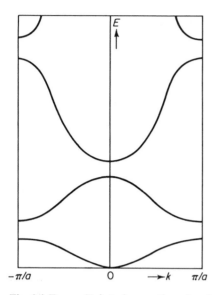

Fig. 4.4. Energy E plotted versus the reduced wave vector k. The energy gaps occur alternately at $k = \pm\pi/a$ and at $k = 0$.

or to restrict the lowest energy band to $-\pi/a \leq k \leq \pi/a$, the next higher one to the ranges $-2\pi/a \leq k \leq -\pi/a$ and $\pi/a \leq k \leq 2\pi/a$, and so on (wave vector representation). In the latter case we obtain the picture in Fig. 4.3; here the energy gaps occur at $k = \pm n\pi/a \, (n = 1, 2, \ldots)$. Figure 4.4 gives the reduced wave vector representation. Here the energy gaps occur where the function $E(k)$ for the one band and for the next higher band has an extreme value. The energy gaps then occur alternately at $k = \pm\pi/a$ and $k = 0$ (see Fig. 4.4).

The problem of energy bands in solids is often demonstrated in textbooks with the help of the Kronig–Penney model. We discuss this model in a few problems at the end of this chapter.

4.2c. Electrons in a Periodic Potential; Effective Mass; Holes

Since $\varphi_k(x)$ given by Eq. (4.13) describes the motion of an electron in a periodic field, the associated time-dependent wave equation is

$$\psi(x, t) = \exp\left(-j\frac{E}{\hbar}t\right)\varphi_k(x) = \exp\left(-j\frac{E}{\hbar}t + jkx\right)u(k, x) \quad (4.14)$$

Hence the velocity v of the electron should correspond to the group velocity of the wave; since $\omega = E/\hbar$,

$$v = \frac{d\omega}{dk} = \frac{d(E/\hbar)}{dk} = \frac{1}{\hbar}\frac{dE}{dk} \quad (4.15)$$

As is seen from Fig. 4.5b, v is negative for negative k, indicating that the electron moves in the negative X direction.

When an electric field F is applied, during a small time dt the electron gains an energy

$$dE = -eFv\,dt = -\frac{eF}{\hbar}\frac{dE}{dk}\,dt \quad (4.16)$$

But since $dE = (dE/dk)\,dk$, the rate of change of the wave vector k is

$$\frac{dk}{dt} = -\frac{eF}{\hbar} \quad (4.17)$$

The acceleration a is obtained by differentiating v with respect to t:

$$a = \frac{dv}{dt} = \frac{1}{\hbar} \frac{d^2E}{dk^2} \frac{dk}{dt} = -\frac{eF}{\hbar^2} \frac{d^2E}{dk^2} \tag{4.18}$$

as we find by substituting Eq. (4.17). Defining the effective mass of the electron by the relation $a = -eF/m^*$, we obtain

$$m^* = \frac{\hbar^2}{d^2E/dk^2} \tag{4.19}$$

This condition is pictured in Fig. 4.5. It shows the energy plotted versus k (Fig. 4.5a), the velocity plotted versus k (Fig. 4.5b), and the effective mass (Fig. 4.5c) plotted versus k. The effective mass is infinite when $E(k)$ has an inflection point; it is positive for the lower part of the band and negative for the upper part. We shall now see what this means.

If we have a number of electrons in the band, and the ith electron is moving with a velocity v_i, then $I = -e \cdot \sum_i v_i$. If a band is completely filled, the current is zero, since for every electron moving with a velocity v_i there is one moving in the opposite direction with the same speed; this is true because in a filled band for every electron with a wave vector k there is another electron with a wave vector $-k$. For a filled band, we write

$$I = -e \sum_i v_i = 0 \tag{4.20}$$

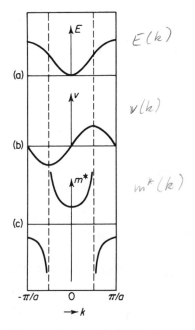

Fig. 4.5. Energy, velocity, and effective mass as a function of the reduced wave vector k. The dashed lines correspond to the inflection points in the $E(k)$ curve. [From A. J. Dekker, *Solid State Physics*, Prentice-Hall, Inc., Englewood Cliffs, N.J., 1957, reprinted by permission.]

Now if the jth electron were missing from the upper part of the band and a current I were flowing, then

$$I = -e \sum_{i \neq j} v_i = -e \sum_i v_i + e v_j = e v_j \tag{4.20a}$$

since $\sum_i v_i = 0$ for a filled band. The missing electron thus behaves like a positive charge. Such a positive charge is called a "hole."

Since an electron in the upper part of a band has a negative effective mass m^*, removing an electron from the upper part of the band corresponds to *adding* mass. A hole thus has a positive effective mass $m_p^* = -m^*$.

4.2d. Optical Transitions

Optical transitions are determined by the absolute square value of the matrix elements

$$m_{kk'}^0 = -\frac{e}{L} \int_0^L \varphi_k^*(x) x \varphi_{k'}(x)\, dx \qquad (4.21)$$

If we now substitute the Bloch expression for $\varphi_k^*(x)$ and $\varphi_{k'}(x)$ and use reduced wave vectors,

$$m_{kk'}^0 = -\frac{e}{L} \int_0^L \exp\left[j(k' - k)x \right] x u_k^*(x) u_{k'}(x)\, dx \qquad (4.21a)$$

The integrand oscillates extremely rapidly with x, unless $k = k'$. We may thus surmise that the integral averages to zero unless $k = k'$. Hence band-to-band optical transitions are restricted by the selection rule

$$k' = k \qquad (4.22)$$

Transitions occur from the bottom of a filled band to the top of the conduction band and from the top of a filled band to the bottom of the conduction band. In a one-dimensional insulator, the gap width can thus be determined by optical absorption measurements. This is not necessarily the case in a three-dimensional crystal, where the minimum quantum energy for optical absorption may not coincide with the energy difference between the highest occupied level in the filled band and the lowest unoccupied level in the conduction band.

4.3. THREE-DIMENSIONAL LATTICE

4.3a. Bloch Functions in the
Three-Dimensional Lattice

Having discussed the theory of the one-dimensional lattice, we now consider briefly the three-dimensional theory. In the wave equation for the electron

$$\nabla^2 \varphi + \frac{2m}{\hbar^2}[E - V(\mathbf{r})]\varphi = 0 \qquad (4.23)$$

the potential energy $V(\mathbf{r})$ is now periodic in $\mathbf{r}$ with a periodicity characteristic of the lattice. In analogy with Floquet's theorem, the solution may be written in the form

$$\varphi_\mu(\mathbf{r}) = \exp(\boldsymbol{\mu} \cdot \mathbf{r}) u(\boldsymbol{\mu}, \mathbf{r}) \qquad (4.24)$$

for electron energies that give rise to nonpropagating waves, and

$$\varphi_k(\mathbf{r}) = \exp(\mathbf{jk} \cdot \mathbf{r}) u(\mathbf{k}, \mathbf{r}) \qquad (4.25)$$

for electron energies that give rise to propagating waves. Here $u(\boldsymbol{\mu}, \mathbf{r})$ and $u(\mathbf{k}, \mathbf{r})$

are periodic in **r** with the lattice periodicity and **k** is the wave vector. Functions of the type (4.25) are called three-dimensional Bloch functions. The first functions thus correspond to *forbidden* energies of the electrons and the second functions to allowed energies. Again the existence of allowed and forbidden energy bands follows naturally from the wave equation applied to a periodic structure.

It is more difficult to picture the forbidden bands here than in the one-dimensional case, since the electron energy E in **k** space is now a function $E(k_1, k_2, k_3)$ of the components of the wave vector **k**. We shall see that this leads to *Brillouin zones*.

4.3b. Brillouin Zones

The problem is now to find where the discontinuities in the function $E(k_1, k_2, k_3)$ occur. Consider electrons that travel perpendicularly to atomic planes designated by their Miller indices (h, k, l). Let adjacent atomic planes have a distance d. Let (l_1, l_2, l_3) (Fig. 4.6) be the direction cosines of this direction of propagation. Bragg reflection occurs if d is equal to an integral number times a half-wavelength; when this happens, discontinuities in the allowed electron energies occur.

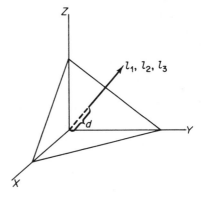

Just as in the one-dimensional case, where Bragg reflection occurred when $kx = n\pi$ for $x = d$, so in this case Bragg reflection occurs if

$$\mathbf{k} \cdot \mathbf{r} = k_1 x + k_2 y + k_3 z = n\pi$$

at

$$x = l_1 d, \qquad y = l_2 d, \qquad z = l_3 d$$

$$(4.26)$$

Fig. 4.6. Atomic plane with Miller indices (h, k, l) and direction cosines of the normal to the plane l_1, l_2, l_3. The plane has a distance d to the origin.

The components of the vector **k** thus satisfy the equation

$$k_1 l_1 + k_2 l_2 + k_3 l_3 = \frac{n\pi}{d} \qquad (4.27)$$

This represents a plane in **k** space having its normal in the direction (l_1, l_2, l_3) and having distance $n\pi/d$ from the origin.

There are, of course, two parallel planes in **k** space for each set of atomic planes in the crystal; they have a distance $2n\pi/d$. Since, for a given set of atomic planes, the wave can travel in opposite directions, the plane with its normal in the direction $(-l_1, -l_2, -l_3)$ satisfies the equation

$$k_1 l_1' + k_2 l_2' + k_3 l_3' = \frac{n\pi}{d} \qquad (4.27\text{a})$$

Here $l_1' = -l_1$, $l_2' = -l_2$, and $l_3' = -l_3$.

In more complicated lattices it may happen that certain Bragg reflections do not occur. For example, in the body-centered cubic lattice, the $(n, 0, 0)$ and (n, n, n) reflections are missing for odd n. In the face-centered cubic lattice the $(n, 0, 0)$ and $(n, n, 0)$ reflections are missing for odd n. We should retain only those planes in **k** space that are associated with *true* Bragg reflections (nh, nk, nl). Only at those planes in **k** space can discontinuities in the allowed energies occur.

We obtain all possible sets of equivalent planes in **k** space by interchanging and inverting the Miller indices. For example, there are three equivalent possibilities $(1, 0, 0)$, $(0, 1, 0)$, and $(0, 0, 1)$, and hence six associated planes in **k** space. There are six equivalent possibilities $(1, 1, 0)$, $(1, \bar{1}, 0)$, $(1, 0, 1)$ $(1, 0, \bar{1})$, $(0, 1, 1)$, and $(0, 1, \bar{1})$, and hence twelve associated planes in **k** space. There are four equivalent possibilities $(1, 1, 1)$, $(\bar{1}, 1, 1)$, $(1, \bar{1}, 1)$, and $(1, 1, \bar{1})$, and hence eight associated planes in **k** space. For each of these possibilities, the set of planes thus obtained intersect and form a solid body in **k** space. Energy discontinuities occur at the faces of these solid bodies. The smallest solid body of this kind is called the *first Brillouin zone*.

To construct the first Brillouin zone, we start with the lowest Bragg reflection (nh, nk, nl) and construct the solid body in **k** space associated with it. We then find the planes in **k** space associated with the next Bragg reflection $(n'h', n'k', n'l')$. If this set of planes does not truncate the first solid body mentioned above, that body corresponds to the first Brillouin zone. If these planes *do* truncate the solid body, then the first Brillouin zone corresponds to the solid body minus the truncated parts.

Figures 4.7, 4.8, and 4.9 show the first Brillouin zones for the simple cubic, the body-centered cubic, and the face-centered cubic lattice. The first, a cube, and the second, a rhombododecahedron, are not truncated, whereas the third is a truncated octahedron. The first Brillouin zone of the diamond structure also corresponds to Fig. 4.9.

It will now be shown with the help of an example that energy bands

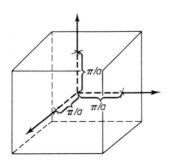

Fig. 4.7. First Brillouin zone of the simple cubic lattice. Also shown are the normals to the $(1, 0, 0)$, $(0, 1, 0)$, and $(0, 0, 1)$ planes in **k** space.

$\mathbf{b}_1 = r_0\,(1,1,0)$
$\mathbf{b}_2 = r_0\,(1,0,1)$
$\mathbf{b}_3 = r_0\,(0,1,1)$

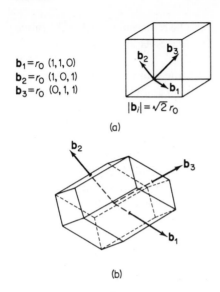

$|\mathbf{b}_i| = \sqrt{2}\,r_0$

(a)

(b)

Fig. 4.8. The first Brillouin zone of the body-centered cubic lattice; also shown are the normals to the $(1, 1, 0)$, $(1, 0, 1)$, and $(1, 1, 0)$ planes in **k** space. [From M. Sachs, *Solid State Theory*, McGraw-Hill Book Company, New York, copyright 1963. By permission of McGraw-Hill Book Company.]

$\mathbf{c}_1 = (2\pi/r_0)\,(\tfrac{1}{2}, \tfrac{1}{2}, -\tfrac{1}{2})$
$\mathbf{c}_2 = (2\pi/r_0)\,(\tfrac{1}{2}, -\tfrac{1}{2}, \tfrac{1}{2})$
$\mathbf{c}_3 = (2\pi/r_0)\,(-\tfrac{1}{2}, \tfrac{1}{2}, \tfrac{1}{2})$

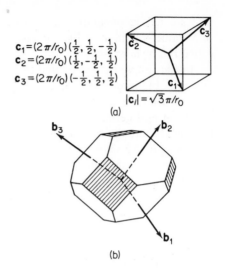

$|\mathbf{c}_i| = \sqrt{3}\,\pi/r_0$

(a)

(b)

Fig. 4.9. The first Brillouin zone for the face-centered cubic lattice and the diamond lattice. Also shown are the normals to the $(\bar{1}, 1, 1)$, $(1, \bar{1}, 1)$, and $(1, 1, \bar{1})$ planes in **k** space. [From M. Sachs, *Solid State Theory*, McGraw-Hill Book Company, New York, copyright 1963. By permission of McGraw-Hill Book Company.]

in two-dimensional solids can overlap. To that end we make the first Brillouin zones of a two-dimensional square structure (Fig. 4.10) and we draw the constant-energy curves in these zones. At E_1 the curve is a circle, at E_2 it is a distorted circle, at E_3 it is a set of quarter-circles with the corners at their centers. Especially if the gap width is not too large, it may happen that E_3 is an allowed energy in the second Brillouin zone. The energy bands corresponding to these two Brillouin zones then overlap. Since the overlap occurs already in a two-dimensional crystal, it should also occur in a three-dimensional one.

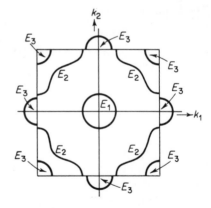

Fig. 4.10. Constant energy curves in the two-dimensional simple square lattice. At E_3 there are holes in the first Brillouin zone and electrons in the next zone.

    Just as in the one-dimensional case all allowed energy bands can be represented within the range of **k** values $-\pi/a \leq k \leq \pi/a$, so in the three-dimensional case all allowed energy bands can be represented within the range of **k** values of the first Brillouin zone. The wave vector thus introduced is called the *reduced* wave vector.

In the simple case presented in Fig. 4.10, the minimum energy occurred at **k** = 0. In more complicated cases, as in silicon and germanium, the energy minima can lie in a different part of the Brillouin zone. For the valence band of silicon and germanium the maximum lies at **k** = 0, but for the conduction band there are several equivalent energy minima, and they lie far away from **k** = 0. We shall see that this has an effect upon the optical absorption.

There is another complication in the case of silicon and germanium: the energy bands are degenerate and consist of three overlapping bands. We shall see that this results in holes with different effective masses: "light" holes and "heavy" holes.

Finally, something should be said about the number of energy states in the first Brillouin zone. We shall do so for a simple cubic lattice. If the electrons are s electrons, the number of states is $2N$, where N is the number of atoms in the volume.

Since the *number* of allowed energy states is independent of the presence of the periodic potential (only the *position*, not the *number*, is affected by it!), we replace the periodic potential energy by a fixed potential. The electrons are then quasi-free (s electrons). If we assume periodic boundary conditions in the X, Y, and Z directions with period L, then the wave function is

$$\varphi_\mathbf{k}(\mathbf{r}) = \exp(jk_1 x)\exp(jk_2 y)\exp(jk_3 z) \tag{4.28}$$

The boundary conditions applied at $x = L$, $y = L$, and $z = L$ yield

$$k_1 L = 2\pi n_1, \qquad k_2 L = 2\pi n_2, \qquad k_3 L = 2\pi n_3 \tag{4.28a}$$

where n_1, n_2, and n_3 are integral. But for the first Brillouin zone, if a is the lattice spacing,

$$-\frac{\pi}{a} \leqq k_1 \leqq \frac{\pi}{a}, \qquad -\frac{\pi}{a} \leqq k_2 \leqq \frac{\pi}{a}, \qquad -\frac{\pi}{a} \leqq k_3 \leqq \frac{\pi}{a} \tag{4.28b}$$

so that there are L/a different values for each n_1, n_2, and n_3. In total this corresponds to L^3/a^3 possibilities. Since the volume a^3 contains one atom, L^3/a^3 corresponds to the N atoms in the volume. But each set of (n_1, n_2, n_3) has two possible orientations of the electron spin. The Brillouin zone thus has N possible **k** values, and a filled band has $2N$ electrons, if the electrons are s electrons. For p and d electrons the wave function is not of the form (4.28), and the situation is more complicated.

4.3c. Tight Binding Approximation

In this model we start with a blown-up version of the lattice and then squeeze it together to its normal size. We assume that the electrons of each atom interact with only the nearest neighbors.

We start looking for a solution of the wave equation

$$\nabla^2 \varphi + \frac{2m}{h^2}[E - V(\mathbf{r})]\varphi = 0 \qquad (4.29)$$

that represents a progressive wave. If the atoms are far apart, and the electrons are s electrons, the wave equation becomes

$$\nabla^2 \varphi_0 + \frac{2m}{h^2}[E_0 - V_0(\mathbf{r} - \mathbf{r}_i)]\varphi_0 = 0 \qquad (4.29a)$$

corresponding to the electron associated with the ith atom. The total wave function is then a linear combination of these atomic wave functions

$$\varphi(\mathbf{r}) = \sum_{i=1}^{N} c_i \varphi_0(\mathbf{r} - \mathbf{r}_i) \qquad (4.30)$$

where the summation is taken over all the atoms (N) in the crystal. To make this a progressive wave of the Bloch type, we put $c_i = \exp(j\mathbf{k}\cdot\mathbf{r}_i)$. This gives as a sensible zero-order wave function

$$\varphi_\mathbf{k}(\mathbf{r}) = \sum_{i=1}^{N} \exp\,(j\mathbf{k}\cdot\mathbf{r}_i)\varphi_0(\mathbf{r} - \mathbf{r}_i) \qquad (4.30a)$$

According to first-order perturbation theory, the energy of the electron is then

$$E = E_0 + V'_{\mathbf{kk}} \qquad (4.31)$$

where $V'_{\mathbf{kk}}$ is the matrix element of the perturbation potential $[V(\mathbf{r}) - V_0(\mathbf{r} - \mathbf{r}_i)]$. This gives

$$V'_{\mathbf{kk}} = \frac{\displaystyle\int \varphi_\mathbf{k}^*(\mathbf{r})[V(\mathbf{r}) - V_0(\mathbf{r} - r_i)]\varphi_\mathbf{k}(\mathbf{r})\,dV}{\displaystyle\int \varphi_\mathbf{k}^*(\mathbf{r})\varphi_\mathbf{k}(\mathbf{r})\,dV} \qquad (4.31a)$$

[The denominator is introduced because the wave function $\varphi(\mathbf{r})$ is not normalized.] If we assume that the individual atomic wave functions are normalized, the denominator has the value N, at least if the atomic wave functions do not overlap very much. We then have

$$V'_{\mathbf{kk}} = \frac{1}{N} \sum_i \sum_l \exp\,[j\mathbf{k}\cdot(\mathbf{r}_l - \mathbf{r}_i)] \int \varphi_0^*(\mathbf{r} - \mathbf{r}_l)[V(\mathbf{r}) - V_0(\mathbf{r} - \mathbf{r}_i)]\varphi_0(\mathbf{r} - \mathbf{r}_i)\,dV$$
$$(4.32)$$

We first sum over all l for a given i. This gives only a significant effect for $l = i$ and for the nearest neighbors of the ith atom. Next we sum over all i's. Since each atom gives the same contribution, the total effect is N times the contribution of a single atom. Introducing $\mathbf{r}' = \mathbf{r} - \mathbf{r}_i$ as a new variable, we get

$$V'_{kk} = \int \varphi_0^*(\mathbf{r}')[V(\mathbf{r}') - V_0(\mathbf{r}')]\varphi_0(\mathbf{r}') \, dV$$

$$+ \sum_l \exp{(j\mathbf{k} \cdot \boldsymbol{\rho}_l)} \int \varphi_0^*(\mathbf{r}' + \boldsymbol{\rho}_l)[V(\mathbf{r}') - V_0(\mathbf{r}')]\varphi_0(\mathbf{r}') \, dV \qquad (4.32a)$$

where the summation is extended over the nearest neighbors and $\boldsymbol{\rho}_l = \mathbf{r}_l - \mathbf{r}_i$. Putting

$$\int \varphi_0^*(\mathbf{r}')[V(\mathbf{r}') - V_0(\mathbf{r}')]\varphi_0(\mathbf{r}') \, dV = -\alpha$$

$$\int \varphi_0^*(\mathbf{r}' + \boldsymbol{\rho}_l)[V(\mathbf{r}') - V_0(\mathbf{r}')]\varphi_0(\mathbf{r}') \, dV = -\gamma$$

$$(4.32b)$$

where γ is independent of l (the contributions of all the neighbors are equal!), yields

$$V'_{kk} = -\alpha - \gamma \sum_l \exp{(j\mathbf{k} \cdot \boldsymbol{\rho}_l)} \qquad (4.33)$$

For the simple cubic lattice of lattice spacing a, each ion has six neighbors with $\boldsymbol{\rho}_l = (\pm a, 0, 0), (0, \pm a, 0), (0, 0, \pm a)$; hence

$$V'_{kk} = -\alpha - 2\gamma(\cos{k_1 a} + \cos{k_2 a} + \cos{k_3 a}) \qquad (4.33a)$$

If the crystal is now squeezed together, α and γ increase and the range of values of V'_{kk} broadens. The atomic energy level broadens into a band of energy values, the width of which increases with decreasing lattice spacing. This broadening comes about because of the interaction of each electron with neighboring atoms, and because of the introduction of the factors $\exp{(j\mathbf{k} \cdot \mathbf{r}_l)}$, which transforms the atomic wave function into a propagating wave. Because of this propagating-wave aspect, the interaction is really an interaction with *all* the atoms.

We have discussed here the case of bands formed by s electrons. A similar treatment can be given for bands formed by p and d electrons; Sachs's book gives details and further references. It is interesting to note that the s, p, or d character of the outer electrons is retained in this model. This explains the existence of selection rules in the emission and absorption processes.

In the case of s bands the minimum in the energy lies at $\mathbf{k} = 0$. For the p bands or d bands this is not the case; here the energy minima lie at different points in the Brillouin zone. This occurs, for example, in the conduction bands of silicon and germanium. In the valence band of silicon and germanium, however, the top of the valence band (that is, the point where the energy has a maximum) lies at $\mathbf{k} = 0$.

The p bands and d bands are always degenerate. If the interaction between the orbital motion of the electrons and the electron spin is taken into account (spin–orbit coupling), then for $l \neq 0$ there are states with $j = l - \frac{1}{2}$ and with $j = l + \frac{1}{2}$ with slightly different energies. As a consequence, the

p and d bands will consist of several bands of slightly different energies partially overlapping each other. This is the case, for example, for the valence bands of silicon and germanium.

4.3d. Effective Mass Tensor; Cyclotron Resonance

In a three-dimensional crystal in the most complicated case the effective mass must be replaced by a mass tensor. We shall see that

$$\left[\frac{1}{m^*}\right]_{ij} = \frac{1}{\hbar^2}\frac{\partial^2 E}{\partial k_i \partial k_j} \tag{4.34}$$

The proof is relatively simple: In analogy with (4.15),

$$v_i = \frac{1}{\hbar}\frac{\partial E}{\partial k_i} \tag{4.35}$$

$$\alpha_i = \frac{dv_i}{dt} = \sum_j \frac{1}{\hbar}\frac{\partial^2 E}{\partial k_i \partial k_j}\frac{dk_j}{dt}$$
$$= \sum_j -\left[\frac{1}{m^*}\right]_{ij}eF_j \tag{4.36}$$

But in analogy with (4.17), one would expect

$$\frac{d\mathbf{k}}{dt} = -\frac{e\mathbf{F}}{\hbar} \tag{4.37}$$

Substituting into (4.36) and equating the corresponding F_j terms yields (4.34).

For electrons at the bottom of the conduction band or for holes at the top of the valence band, one may assume that $E(\mathbf{k})$ has a minimum or maximum, respectively, at the edge of the band. If the edge of the band occurs at $\mathbf{k} = \mathbf{k}_0$, then we may write the Taylor expansion for $\mathbf{k}$ values near $\mathbf{k} = \mathbf{k}_0$:

$$E(\mathbf{k}) = E(\mathbf{k}_0) + \frac{1}{2}\sum_i \sum_j \left(\frac{\partial^2 E}{\partial k_i \partial k_j}\right)_{\mathbf{k}_0} \Delta k_i \Delta k_j \tag{4.38}$$

where $\Delta \mathbf{k} = \mathbf{k} - \mathbf{k}_0$. But since this is a positive-definite quantity (for electrons in the conduction band) or a negative-definite quantity (for electrons in the valence band), one can choose the axes for the $\mathbf{k}$'s such that only terms with Δk_1^2, Δk_2^2, and Δk_3^2 remain. In such cases there are only three effective masses, and the constant-energy surfaces in $\mathbf{k}$ space are ellipsoids around $\mathbf{k} = \mathbf{k}_0$.

It may be that the ellipsoid has rotational symmetry around one of its axes. The constant-energy surfaces are then spheroids, and there are only two different terms in the mass tensor,

$$\begin{pmatrix} m_t & 0 & 0 \\ 0 & m_t & 0 \\ 0 & 0 & m_l \end{pmatrix}$$

a transverse mass m_t and a longitudinal mass m_l. This occurs, for example, in the conduction bands of silicon and germanium.

These constant-energy surfaces and the associated effective masses can be explored by means of cyclotron resonance experiments. If the material under study is kept at liquid-helium temperature, so that collisions between the carrier and the lattice are relatively rare, and a static magnetic field is applied, then the carrier will rotate in circles with an angular frequency[†]

$$\omega_0 = \frac{eB}{m^*} \tag{4.39}$$

If in addition a microwave electric field is applied at right angles to the magnetic field, energy will be absorbed, and this absorbed energy is a maximum at the frequency ω_0. Therefore, by measuring microwave absorption as a function of the applied magnetic field B and noting at what field the energy absorption is a maximum, one can determine the effective mass m^*.

In the case of spheroidal energy surfaces the effective mass thus obtained depends on the orientation of the magnetic field B. This is easy to see as follows. Let the Z direction point in the direction of the axis of the spheroid, and let the field $\mathbf{B}$ make an angle θ with the Z axis in the XZ plane. Then the equation of motion of the electron is[‡]

$$\frac{d\mathbf{p}}{dt} = -e\mathbf{v} \times \mathbf{B} \tag{4.40}$$

where $\mathbf{v}$ has components p_x/m_t, p_y/m_t, and p_z/m_l, and the field has components $B \sin\theta$, 0, and $B \cos\theta$. Substituting into (4.40) and putting $\omega_t = eB/m_t$ and $\omega_l = eB/m_l$ yields

$$-j\omega p_x - \omega_t \cos\theta p_y = 0$$
$$\omega_t \cos\theta p_x - j\omega p_y - \omega_l \sin\theta p_z = 0 \tag{4.41}$$
$$\omega_t \sin\theta p_y - j\omega p_z = 0$$

which has a solution only if the following determinant is zero:

$$\begin{vmatrix} -j\omega & -\omega_t \cos\theta & 0 \\ \omega_t \cos\theta & -j\omega & -\omega_l \sin\theta \\ 0 & \omega_t \sin\theta & -j\omega \end{vmatrix} = 0 \tag{4.42}$$

This yields

$$\omega^2 = \omega_0^2 = \omega_t^2 \cos^2\theta + \omega_t\omega_l \sin^2\theta \tag{4.42a}$$

[†] This follows from $m^*\omega_0^2 r = evB = e\omega_0 rB$.

[‡] If $\mathbf{i}$, $\mathbf{j}$, and $\mathbf{k}$ are unit vectors in the X, Y, and Z directions, respectively, then

$$\mathbf{v} \times \mathbf{B} = \begin{vmatrix} \mathbf{i} & \mathbf{j} & \mathbf{k} \\ v_x & v_y & v_z \\ B_x & B_y & B_z \end{vmatrix}$$

The effective mass is then given by

$$\left(\frac{1}{m^*}\right)^2 = \frac{\cos^2 \theta}{m_t^2} + \frac{\sin^2 \theta}{m_t m_l} \qquad (4.43)$$

By measuring m^* as a function of θ, one can determine m_t and m_l. In this way we find that $m_l = 0.98m$ and $m_t = 0.18m$ for silicon, whereas for germanium $m_l = 1.57m$ and $m_t = 0.082m$; here m is the mass of a free electron. Figure 4.11 shows the location of the energy minima in silicon and germanium. In silicon they lie in the (001), (010), and (001) axes; in germanium they lie in the (111), ($\bar{1}$11), ($1\bar{1}$1), and ($11\bar{1}$) directions.

The valence band in silicon and germanium is degenerate. This degeneracy is partly removed by the interaction of the electrons with the periodic lattice and partly by spin–orbit coupling. As a consequence, there are *three* valence bands. Because of the first effect, two coincide for $\mathbf{k} = 0$, but differ for $\mathbf{k} \neq 0$. The third has a slightly different energy at $\mathbf{k} = 0$ because of spin–orbit coupling; the $\mathbf{k} = 0$ level lies 0.035 V below the other two in silicon and 0.28 V below the other two in germanium.

There are thus three types of holes in the germanium and silicon valence bands. In silicon they have effective masses of $0.49m$, $0.16m$, and $0.24m$, respectively, whereas for germanium the effective masses are $0.28m$, $0.044m$, and $0.077m$. For silicon the energy difference is so small that the third valence band can be well populated with holes at room temperature. For germanium the energy difference is much larger, and hence the third valence band has a negligible hole population at room temperature.

4.3e. Optical Transitions

For optical transitions the same selection rule holds as for the one-dimensional case: *In unassisted optical transitions the reduced wave vector $\mathbf{k}$ does not change.* The proof is the same as for the one-dimensional case.

Besides unassisted transitions there are also phonon-assisted transitions. Phonons are quantized lattice vibrations. In absorption the electron absorbs a quantum and one or more phonons to go to a higher energy state with different reduced wave vector $\mathbf{k}$. In emission the electron emits a quantum and one or more phonons to go to a lower unoccupied energy state with different reduced wave vector $\mathbf{k}$. Since phonon-assisted transitions require the availability of phonons, these transitions are less probable than the unassisted ones (see Sec. 4.4 for a discussion of phonons).

Suppose that the valence band and the conduction band of a semiconductor have their respective band maximum and band minimum at $\mathbf{k} = 0$. Then unassisted transitions back and forth are possible, and the absorbed quantum can also be reemitted. The band gap and the "optical gap" then coincide (Fig. 4.12a). It may be shown that such semiconductor materials can be used for solid-state lasers. If, however, the conduction band has its

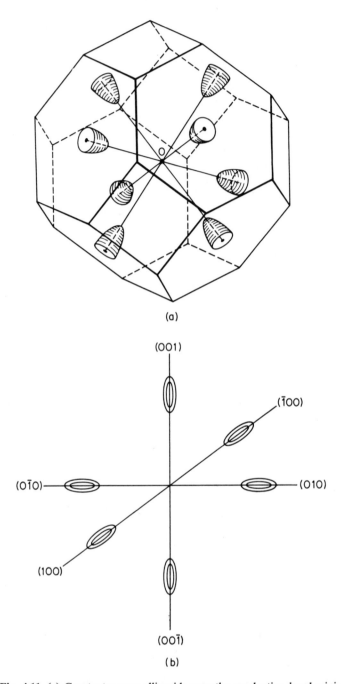

(a)

(001)

(ī00)

(0ī0)

(010)

(100)

(00ī)

(b)

Fig. 4.11. (a) Constant energy ellipsoids near the conduction band minimum in germanium (from A. Nussbaum, *Semiconductor Device Physics*, Prentice-Hall, Inc., Englewood Cliffs, N.J., 1962; by permission of Prentice-Hall, Inc.). (b) Constant energy ellipsoids near the conduction band minimum in silicon. [From J. H. Moll, *Physics of Semiconductors*, McGraw-Hill Book Company, New York, copyright 1964. By permission of McGraw-Hill Book Company.]

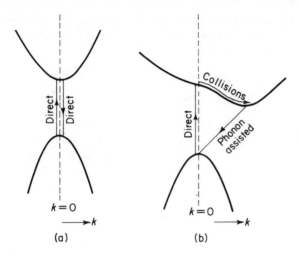

Fig. 4.12. (a) Direct transitions in emission and absorption. (b) Direct transition in absorption and indirect transition in emission.

minimum at $\mathbf{k} \neq 0$, then the most likely absorption process is $\mathbf{k} = 0 \rightarrow \mathbf{k} = 0$. But then the electron in the valence band loses energy rapidly until it reaches the bottom of the conduction band, when it emits light by a phonon-assisted process. The absorbed quantum therefore cannot be reemitted (Fig. 4.12b). It may be shown that such semiconductor materials cannot be used for solid-state lasers. Moreover, the optical gap for unassisted transitions and the band gap do not coincide in this case. GaAs and similar compounds belong to the first group, whereas germanium and silicon belong to the second group.

4.4. LATTICE VIBRATIONS; PHONONS

4.4a. Lattice Vibrations

We refer to the vibrations of the individual atoms making up a crystal around their equilibrium positions as lattice vibrations. Each atom can vibrate in three independent directions, which are the X, Y, and Z directions. At sufficiently high temperatures the average vibrational energy per atom is therefore $3 \times 2 \times \frac{1}{2}kT = 3kT$. The factor 2 is used because the vibrating atom has both kinetic energy and potential energy; the average for each independent direction of vibration is $\frac{1}{2}kT$ for each.

Actually, the vibrating atoms cannot be treated as independent because of the interaction of each atom with its neighbors, and, through it, with all the other atoms in the crystal. As a consequence, a whole band of frequencies is possible, rather then the few discrete frequencies of the isolated atom.

There are two ways of treating the problem of small oscillations, both classically and wave mechanically. The first method consists in writing down the total energy H in terms of the positional coordinates and velocities of the individual atoms. Let the crystal contain N atoms, and let the positional coordinates be denoted by

$$(x_1 \cdots x_N, y_1 \cdots y_N, z_1 \cdots z_N) \tag{4.44}$$

and the individual velocity components by

$$(\dot{x}_1 \cdots \dot{x}_N, \dot{y}_1 \cdots \dot{y}_N, \dot{z}_1 \cdots \dot{z}_N) \tag{4.45}$$

For relatively small oscillations the total energy H is then a quadratic function of these coordinates and velocities. It may now be shown that a coordinate transformation to a new set of coordinates Q_s ($s = 1 \ldots 3N$) can be found so that the H function can be written as

$$H = \sum_{s=1}^{3N} (a_s \dot{Q}_s^2 + b_s Q_s^2) \tag{4.46}$$

The set of coordinates satisfying this relationship is said to be the *normal coordinates* of the system; they are appropriately chosen linear combinations of the original positional coordinates. It is shown in textbooks of classical mechanics that such a transformation is always possible.

Since the total energy is now a sum of $3N$ terms of the form $(a_s \dot{Q}_s^2 + b_s Q_s^2)$, it follows that the system can be treated† as $3N$ uncoupled oscillators of frequency $\omega_s = (b_s/a_s)^{1/2}$. Classically, the equation of motion of these oscillators is thus

$$\ddot{Q}_s + \omega_s^2 Q_s = 0 \tag{4.47}$$

which has as solutions $\exp(\pm j\omega_s t)$. The most general vibration is thus a linear superposition of these separate normal modes.

In wave mechanical terms, one can use the method of separation of variables and separate the wave equation into $3N$ wave equations for the individual normal modes. Each mode can thus have energies

$$E_{vs} = \hbar\omega_s(v_s + \tfrac{1}{2}) = hv_s(v_s + \tfrac{1}{2}) \tag{4.48}$$

$v_s = 0, 1, 2 \ldots$, so that the total vibrational energy of the crystal may be expressed as

$$E_v = h \sum_{s=1}^{3N} v_s(v_s + \tfrac{1}{2}) \tag{4.48a}$$

The other approach to the problem of small oscillations in crystals consists in writing the wave motion as a superposition of plane waves. One must now distinguish between *longitudinal waves,* in which the motion of the atoms is *in* the direction of propagation, and *transverse shear waves,* in

---

† This is a simple extension of the one-dimensional case. According to Sec. 2.1b, the energy in that case is

$$H = \tfrac{1}{2}m\dot{x}^2 + \tfrac{1}{2}fx^2$$

and the frequency of oscillation is $\omega_0 = (f/m)^{1/2}$.

which the motion of atoms is *perpendicular to* the direction of propagation. Since there are two independent directions of vibration of the atoms in the plane perpendicular to the direction of propagation, there are two independent shear wave modes. In total there are thus three independent modes of wave propagation, one longitudinal and two transverse.

For each of these modes we may now write the following wave:

$$\mathbf{R} = \mathbf{A} \exp\left[j(\mathbf{K}\cdot\mathbf{r}) - j\omega t \right] \qquad (4.49)$$

where $\mathbf{R}$ is the displacement of an atom whose equilibrium position is at $\mathbf{r}$, $\mathbf{K}$ is a wave vector, and ω is the frequency of oscillation. By applying the proper boundary conditions, one finds that only certain frequencies ω are allowed; these are, of course, the frequencies ω_s introduced before. A value K_s of the wave vector is associated with each frequency ω_s; the wavelength λ_s associated with ω_s is $2\pi/K_s$. One can now evaluate ω as a function of K for a particular direction of $\mathbf{K}$. The form of the relationship between ω and K is very similar to the E–k relationship for electrons in a crystal.

Let us investigate the problem first for a simple cubic crystal with N elementary cells and, hence, $3N$ normal coordinates of the system and $3N$ frequencies ω_s. In this case one has one longitudinal mode and two independent shear modes. The ω, K_x relationship is pictured in Fig. 4.13a for waves traveling in the X direction; $\omega = 0$ for $K_x = 0$. There is now one curve for the longitudinal mode and one curve for the two shear modes, which in this case have the same ω for a given K_x. The maximum value of K_x is K_{xm}. Now ω/K is the *phase velocity* of the displacement waves and $d\omega/dK$ the *group velocity*. For $K_x = K_{xm}$ the group velocity is zero; thus one has standing waves. For small values of K the value of $d\omega/dK \simeq \omega/K$, and this corresponds to the two velocities of acoustic waves, the longitudinal and the transverse shear. For that reason these modes are called the *acoustic modes*.

If there are two atoms per unit cell, the bands split into two band systems. The lowest (ω, K_x) curves again correspond to the acoustic modes; the curves for $K_{xm} < K_x < 2K_{xm}$ have higher frequencies and correspond to the *optical modes*. The meaning of this term will soon become clear.

As in the case of electrons in a crystal, one can introduce the reduced wave vector K, which extends from zero to K_{xm}; the longitudinal and transverse optical modes then lie above the acoustic modes, as shown in Fig. 4.13b.

Suppose that the two atoms per elementary cell are positive and negative ions (polar crystals). Then for the frequency ω_0, due to the optical mode corresponding to $K_x = 0$, the oscillation consists of a wave in which the positive ions move in phase and the negative ions move in unison in opposite phase. An oscillating dipole moment of frequency ω_0 is therefore formed, which can couple strongly with electromagnetic waves of the same frequency. Thus a strong absorption in polar crystals at the frequency $f_0 = \omega_0/2\pi$ occurs; this is called the *reststrahl* frequency.

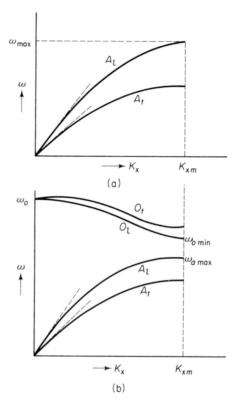

Fig. 4.13. (a) $\omega - K$ relations in a crystal with one atom per elementary cell. (b) $\omega - K$ relations in a crystal with two atoms per elementary cell.

If there are N_a atoms per elementary cell, there are $3N_a$ modes, one longitudinal and two transverse acoustical modes, and $N_a - 1$ bands of optical modes, each consisting of one longitudinal and two transverse optical modes. The total number of modes is thus always finite, in contrast with the case of electron waves in a crystal, where there are an infinite number of bands.

4.4b. Lattice Scattering; Phonons

We shall now show that, in the interaction between electrons and lattice waves, energy and momentum are conserved; that is, if $\mathbf{k}$ is the electron wave vector before the interaction and $\mathbf{k}'$ the electron wave vector after the interaction, then

$$\hbar(\mathbf{k} - \mathbf{k}') = \pm\hbar\mathbf{K}_s \qquad (4.50)$$

where $\mathbf{K}_s$ is the wave vector of a lattice wave of frequency ω_s, and

$$E(\mathbf{k}) - E(\mathbf{k}') = \pm\hbar\omega_s \qquad (4.51)$$

In other words, the lattice can take up quantized amounts of momentum $\hbar\mathbf{K}_s$ and quantized amounts of energy $\hbar\omega_s$, or it can give up the momentum $\hbar\mathbf{K}_s$ and the energy $\hbar\omega_s$. These amounts may be written as h/λ_s for the momentum and hf_s for the energy. A *photon* has an energy hf and a momentum $hf/c = h/\lambda$. There is thus a direct analogy between light waves and lattice waves. The quantum of lattice vibrations is called a *phonon*, and it is said that in the interaction of the electron with the lattice a phonon is either *emitted* or *absorbed*.

We shall prove Eqs. (4.50) and (4.51). Equation (4.51) seems immediately obvious, since a harmonic oscillator of frequency ω_s can only change its energy by amounts $\hbar\omega_s$. The other equation, however, needs proof. To do so we observe that in the lattice wave of wave vector $\mathbf{K}'_s$ the rarefactions and compressions repeat themselves in a set of planes perpendicular to $\mathbf{K}_s$ that are a distance $\lambda_s = 2\pi/K_s$ apart. We now look at wavelets scattered from one such plane. The reflected waves are all in phase if (see Fig. 4.14a)

$$\frac{a \sin \varphi}{\lambda} = \frac{a \sin \theta}{\lambda'} \tag{4.52}$$

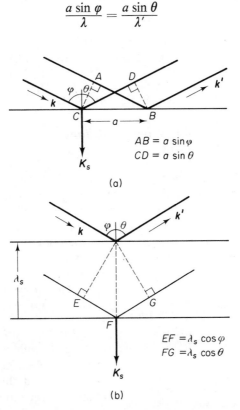

(a)

(b)

Fig. 4.14. (a) Phase difference of a wave scattered from a constant phase plane perpendicular to $\mathbf{K}_s$. (b) Phase difference in a wave scattered from two planes perpendicular to $\mathbf{K}_s$ at a distance λ_s apart.

where φ and θ are the angles that the incoming and scattered waves make with $\mathbf{K}_s$, $\lambda = 2\pi/k$, and $\lambda' = 2\pi/k'$. That is, the components of the wave vector along the plane perpendicular to $\mathbf{K}_s$ are equal, so that $\mathbf{k} - \mathbf{k}'$ has only a component in the direction of $\mathbf{K}_s$.

We now look at two planes that are a distance $\lambda_s = 2\pi/K_s$ apart. Here there should also be no destructive interference, and this requires that (see Fig. 4.14b)

$$\frac{\lambda_s \cos \varphi}{\lambda} + \frac{\lambda_s \cos \theta}{\lambda'} = 1 \qquad (4.53)$$

or $k \cos \varphi + k' \cos \theta = |\mathbf{k} - \mathbf{k}'| = K_s$.
Therefore,

$$\mathbf{k} - \mathbf{k}' = \mathbf{K}_s \qquad (4.50a)$$

from which (4.50) follows.

If the scattering process occurs for a lattice wave traveling in the opposite direction, we find

$$\mathbf{k} - \mathbf{k}' = -\mathbf{K}_s \qquad (4.50b)$$

so that a phonon of momentum $\hbar K_s$ can either be *given to* the lattice wave, as in Eq. (4.50a), or *taken from* the lattice wave, as in Eq. (4.50b).

4.5. CONTACT PROBLEMS IN METALS

4.5a. Vacuum Level; Work Function; Escape Condition

Figure 4.15 shows the energy-level diagram of a metal. It gives as reference levels the bottom of the conduction band and the vacuum level. The bottom of the conduction band represents the energy of an electron at rest *in* the metal; the vacuum level represents the energy of an electron at rest *outside* the metal. The energy difference between these levels is denoted by W.

We have already introduced the Fermi level E_f. In metals at $T = 0$ it represents the highest filled energy level. The energy difference

$$\phi = W - E_f \qquad (4.54)$$

is called the *work function*. It represents the minimum energy needed to take an electron out of the metal at $T = 0$, and thus determines the minimum quantum energy hf_{min} needed to produce photoemission.

The energy W determines the

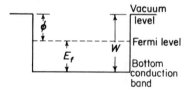

Fig. 4.15. Energy level diagram of electrons in a conductor. The vacuum level and the Fermi level are also shown, energies being measured with respect to the bottom of the conduction band.

escape condition of an electron from the metal. We observe that only the electron velocity component v_x perpendicular to the surface is effective in the escape; the velocity components v_y and v_z are parallel to the surface and cannot contribute. The escape condition is therefore

$$\tfrac{1}{2}mv_x^2 \geqq W \tag{4.55}$$

This condition plays an important part in problems of thermionic emission, photoemission, and secondary electron emission.

4.5b. Contact Potential

If metals 1 and 2 are brought into contact, electrons will flow from the one metal to the other until the Fermi levels are at equal height (Fig. 4.16a).

Let the metals have energy differences W_1 and W_2 between the vacuum level and the bottom of the conduction band, respectively. Let the respective work functions be denoted by ϕ_1 and ϕ_2, and let $\phi_1 > \phi_2$. Let the position

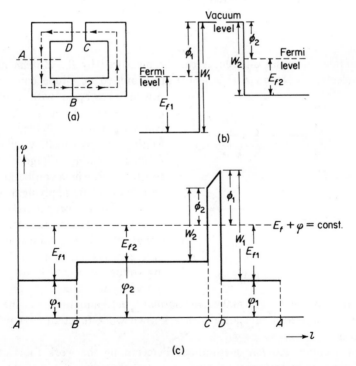

(a)

(b)

(c)

Fig. 4.16. (a) Two metals 1 and 2 in contact at B. (b) Energy level diagram of the metals before contact. (c) Potential energy of an electron at rest along the dotted path shown in part (a). There is a Galvani potential difference at B and a contact potential difference between D and C. The path length l is taken from the point A.

of the Fermi level with respect to the bottoms of the two conduction bands be denoted by E_{f1} and E_{f2} (Fig. 4.16b). At contact, electrons will flow from metal 2 to metal 1 until equilibrium is established. Metal 2 becomes positively charged and metal 1 becomes negatively charged (Fig. 4.16c). Then metal 2 has a potential V_2 with respect to metal 1, so that

$$\phi_1 = \phi_2 + eV_2 \quad \text{or} \quad V_2 = \frac{\phi_1 - \phi_2}{e} \qquad (4.56)$$

since the Fermi levels are at equal height. This potential difference is known as the *contact* potential difference or *Volta* potential difference. Note that $V_2 > 0$ if $\phi_1 > \phi_2$. This potential difference is actually measured (see below).

Another potential difference, the *Galvani* potential difference, is introduced as follows. We see from Fig. 4.16c that an amount of work $E_{f1} - E_{f2}$ must be done to bring an electron that is at rest in metal 1 into the condition where it is at rest in metal 2. This difference in potential energy $E_{f1} - E_{f2}$ can be described by the *Galvani potential difference*

$$V_{c1} - V_{c2} = \frac{E_{f1} - E_{f2}}{e} \qquad (4.57)$$

Contact-potential differences can be measured by the *Zisman* method. Two electrodes A and B (Fig. 4.17) are mounted in a tube envelope and

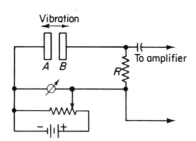

forced to vibrate with respect to one another. As a result of the contact-potential difference, a fluctuating current will flow, giving rise to an alternating voltage across the resistance R, which can be amplified and detected. By applying a bucking voltage V of the proper polarity and magnitude, we can reduce this ac signal to zero. The contact-potential difference is then equal to the bucking voltage V necessary to give zero output; an accuracy of 10^{-4} V

Fig. 4.17. Measurement of the contact potential difference with the help of the Zisman method described in the text.

may be obtained. To make an *absolute* determination of the work function, we need a substance with a known work function as a reference electrode.

An accurate *absolute* method for determining the work function of a metal consists in measuring the *photoelectric threshold*. If $f_{\min}$ is the lowest frequency at which photoemission takes place, the work function follows from the condition

$$hf_{\min} = \phi \qquad (4.58)$$

REFERENCES

JONES, H., *The Theory of Brillouin Zones and Electronic States in Crystals.* North-Holland Publishing Company, Amsterdam, 1960.

LONG, D., *Energy Bands in Semiconductors.* John Wiley & Sons, Inc. (Interscience Division), New York, 1968.

SACHS, M., *Solid State Theory.* McGraw-Hill Book Company, New York, 1963.

Review Papers

HERMAN, F., *Rev. Mod. Phys.*, **30**, 102 (1958).

LAX, B., *Rev. Mod. Phys.*, **30**, 122 (1958).

NUSSBAUM, A., "Crystal Symmetry, Group Theory, and Band Structure Calculations," F. Seitz and D. Turnbull, eds., *Solid State Physics*, vol. 18, p. 165. Academic Press, Inc. New York, 1966.

PROBLEMS

1. In a one-dimensional crystal with a distance of 2.5 Å between atomic planes, find the free-electron energy at which the first Bragg reflection occurs.

Answer: $E = h^2/(8ma^2) = 6.0$ eV.

2. If the potential energy of the electrons in the crystal of problem 1 is $V(x) = -10\,e$ for $-\frac{1}{4}a < x < \frac{1}{4}a$, $V(x) = -6e$ for $\frac{1}{4}a < x < \frac{3}{4}a$, and so on, where $-e$ is the electron charge, find by first-order perturbation theory:
(a) The position of the first band minimum in electron volts.
(b) The position of the first band maximum in electron volts.
(c) The gap width between the first and the second band.
Hint: Make a Fourier analysis of $V(x)$. Observe that $V_{kk} \neq 0$ in this case.

Answer: (a) -8.0 eV; (b) -4.5 eV; (c) 5.1 eV.

3. Solve the wave equation for an electron moving in the following potential field (Kronig–Penney model),

$$V(x) = 0 \quad \text{for} \quad 0 < x < a, \qquad V(x) = V_0 \quad \text{for} \quad a < x < b$$

and periodically repeated outside that interval. Show that for $E < V_0$ it leads to the following equation:

$$\frac{\beta^2 - \alpha^2}{2\alpha\beta} \sinh \beta b \sin \alpha a + \cosh \beta b \cos \alpha a = \cos k(a + b)$$

where $\alpha^2 = 2mE/\hbar^2$, $\beta^2 = 2m(V_0 - E)/\hbar^2$. *Hint:* Substitute $\varphi = \exp(jkx)u_k(x)$

and match the wave function and its first derivative at $x = 0$, $x = -b$, and also at $x = a$.

4. Show that the equation of problem 3 reduces to

$$\frac{mAa}{\hbar^2} \frac{\sin \alpha a}{\alpha a} + \cos \alpha a = \cos ka$$

if b goes to zero and V_0 to infinity, such that $V_0 b = A$ remains constant.

5. Find for a simple cubic crystal the equation of a plane in **k** space associated with the nth-order Bragg reflection against the (h, k, l) planes in the crystal.

Answer:

$$\frac{h}{(h^2 + k^2 + l^2)^{1/2}} k_1 + \frac{k}{(h^2 + k^2 + l^2)^{1/2}} k_2 + \frac{l}{(h^2 + k^2 + l^2)^{1/2}} k_3$$

$$= \frac{n\pi}{a}(h^2 + k^2 + l^2)^{1/2}$$

6. Evaluate the expression

$$\sum_l \exp{(j\mathbf{k} \cdot \mathbf{\rho}_l)}$$

in the tight binding approximation for the simple cubic crystal by first evaluating the coordinates of the nearest neighbors of the atom at the origin $(0, 0, 0)$, and so find the components of $\mathbf{\rho}_l$.

Answer: $2(\cos k_1 a + \cos k_2 a + \cos k_3 a)$.

7. Repeat problem 6 for a body-centered cubic lattice.

Answer: $8(\cos \frac{1}{2} k_1 a \cos \frac{1}{2} k_2 a \cos \frac{1}{2} k_3 a)$.

8. Repeat problem 6 for the face-centered cubic lattice.

Answer: $4(\cos \frac{1}{2} k_2 a \cos \frac{1}{2} k_3 a + \cos \frac{1}{2} k_3 a \cos \frac{1}{2} k_1 a + \cos \frac{1}{2} k_1 a \cos \frac{1}{2} k_2 a)$.

9. In a certain semiconducting material having an electron concentration n, the time constant τ for collisions is assumed to be independent of the velocity of the electrons. In that case the motion of an electron under the influence of an external force $\mathbf{F}_{ex}$ is described by the equation

$$\dot{\mathbf{p}} + \frac{\mathbf{p}}{\tau} = \mathbf{F}_{ex}$$

Starting from here, show that if an ac electric field is applied in the X direction $[\mathbf{F}_{ex} = -ieF_{x0} \exp{(j\omega t)}]$, then

$$v_x = \frac{-e\tau/m^*}{1 + j\omega\tau} F_{x0} \exp{(j\omega t)}$$

10. If an ac electric field $F_{x0} \exp(j\omega t)$ is applied in the X direction and a static magnetic field is applied in the Z direction $(\mathbf{B} = \mathbf{k}B)$, then the equation of motion of problem 9 gives

$$v_x = -\frac{(e\tau/m^*)(1 + j\omega\tau)}{1 + 2j\omega\tau + (\omega_0^2 - \omega^2)\tau^2} F_{x0} \exp(j\omega t)$$

$$v_y = \frac{\omega_0\tau}{1 + j\omega\tau} v_x$$

where $\omega_0 = eB/m^*$ is the cyclotron frequency.

5

Semiconductors

5.1. VARIOUS TYPES OF SEMICONDUCTORS

5.1a. Intrinsic Semiconductors

Suppose the energy gap E_g between the filled band and the conduction band of an insulator is relatively small. If so, the substance will be a good insulator at low temperatures, but at sufficiently high temperatures the crystal will start to conduct. The reason is that, because of the interaction between the electrons in the filled band and the lattice vibrations, some may gain enough energy to be transferred from the valence band into the conduction band.

Conductors that have this property are called *intrinsic semiconductors;* the name "intrinsic" implies that the semiconducting property is a characteristic of the pure material. In the next section we shall discuss several cases in which the semiconducting property of the material is due to impurities or deviations from stoichiometry. Such semiconductors are called *extrinsic semiconductors.*

The following list of elements of the fourth group of the periodic table illustrates this point.

Element	E_g	Property
C (diamond)	5.4 eV	Insulator
Si	1.1 eV	Semiconductor
Ge	0.7 eV	Semiconductor
Sn (α Sn)	0.08 eV	Semiconductor
Pb	—	Conductor

We see a gradual transition from insulating properties through semiconductor properties to metallic properties in this group.

Another characteristic of intrinsic semiconductors is that the current is carried by *two types of carriers*. One type is, of course, the electrons in the conduction band. But for every electron in the conduction band, an electron must be missing from the valence band. Such a vacant spot in the valence band is called a *hole*. We have seen that a hole acts in many respects like a positive charge, is mobile, and thus takes part in the conduction process (see Chapter 4).

5.1.b N-Type and P-Type Semiconductors

In intrinsic semiconductors the current is carried by two types of carriers: electrons and holes. In many respects it would be much more convenient to have semiconductors with only one type of carrier, either electrons or holes. Semiconductors in which the current is carried predominantly by holes are called *p-type semiconductors* (*p* for positive); if the current is carried predominantly by electrons, the semiconductor is said to be *n-type* (*n* for negative). We investigate first the electron structure of these types of semiconductors.

We can make ionic crystals into *n*-type or *p*-type semiconductors by introducing slight deviations from stoichiometry. Such substances, called *excess* or *defect semiconductors*, are discussed in Sec. 5.1c. Valence-bond crystals can be made into *n*-type or *p*-type semiconductors by the addition of impurities; these *impurity* semiconductors are discussed in Sec. 5.1d.

Consider a semiconductor at $T = 0$ having the valence band completely filled and the conduction band completely empty, and having in addition *occupied energy levels* slightly below the bottom of the conduction band. One then has only to increase the temperature very slightly in order to raise the electrons bound to those occupied levels into the conduction band; for that reason these levels are called *donor levels*. Electrons occupying these donor levels are bound to fixed positions in the crystal, and are unlike the energy levels of the conduction band, which extend through the whole crystal; these donor levels are thus *localized*.

Let N_d be the number of donor levels per cubic meter. At very low temperature the number of electrons in the conduction band is much smaller than N_d, but its value increases rapidly with temperature; in that temperature range the conductivity will thus increase with temperature. At a relatively low temperature n becomes equal to N_d; for higher temperatures $n = N_d$ in a wide temperature range. At a high temperature electrons can be raised by thermal energy from the valence band into the conduction band; the semiconductor becomes intrinsic and n increases very rapidly with increasing temperature. Here n is the free electron concentration.

Consider next a semiconductor at $T = 0$, having the valence band completely filled and the conduction band completely empty and also having *unoccupied energy levels* slightly above the top of the valence band. One then has only to raise electrons from the valence band to these unoccupied energy levels to leave free holes behind in the valence band. These unoccupied energy levels are called *acceptor levels*, and they are localized. The behavior of this semiconductor resembles the previous case, except in the type of carriers.

5.1c. Excess and Defect Semiconductors

We consider here the structure of excess and defect semiconductors. Let the ionic crystal be of the type M^+X^- and let extra M atoms be added. There are two possibilities:

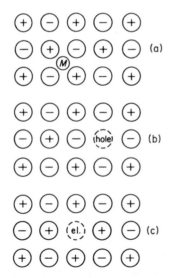

Fig. 5.1. Various types of lattice defects in an $M^+ X^-$ crystal. (a) Interstitial atom. (b) Positive ion vacancy occupied by a trapped hole. (c) Negative ion vacancy occupied by a trapped electron.

1. The extra M atoms occupy *interstitial positions* (we have seen that there are unoccupied interstitial holes even in close-packed structures). The valence electron of such an M atom is often very loosely bound to it; the extra M atoms thus form *donor levels* close to the bottom of the conduction band (Fig. 5.1a). This condition seems to occur in ZnO with excess Zn.

2. The excess M atoms occupy sites of M^+ ions; this is possible only if one *negative ion vacancy* is formed for each M atom added. The valence electrons of the excess M atoms will then be bound to the X^- ion vacancies. Such a binding means that the electron is shared by the M^+ ions surrounding the vacancy (Fig. 5.1c). This is often not a very strong binding, so that this case also gives rise to donor levels

close to the bottom of the conduction band. This condition seems to prevail in BaO with oxygen vacancies.

An X^- ion vacancy occupied by an electron is also called a *color center* or *F center* for the following reason. In good insulators the energy levels of these electrons may lie relatively deep below the bottom of the conduction band; optical absorption may then occur in the visible region of the spectrum owing to the transition of an electron from such an F center to the conduction band. The optical absorption in the visible region shows up as a coloration of the crystal; hence the name *color center* for this type of lattice disturbance.

If extra X atoms are added, usually only the second type of singularity occurs, because the X atoms are too big to fit into interstitial holes in the structure. That is, in general, the X atoms will occupy the normal position of the X^- ions. For each added X atom there must thus be one *positive-ion vacancy*. Since there are more X atoms than M atoms, there must be some electrons missing here and there. A missing electron is shared by the negative ions surrounding a positive-ion vacancy, so that this configuration can be considered a "trapped hole" (Fig. 5.1b). Often only a slight amount of energy is needed to bring an electron from a normal X^- ion to the X atoms surrounding such a positive-ion vacancy. That is, only a slight amount of energy is needed to create a *free* hole; or, in other words, this configuration gives rise to *unoccupied* energy levels close to the top of the filled band.

In good insulators, again, these unoccupied energy levels may lie considerably *above* the top of the valence band. Optical absorption may thus occur in the visible part of the spectrum, causing a coloration of the crystal (color centers). Such singularities are usually called *V centers*, to distinguish them from the F centers discussed earlier.

Cu_2O, ZnO, and NiO, when heated under oxygen pressure, take up an excess of oxygen; semiconductors formed that way are called *oxidation* semiconductors. ZnO and TiO_2 when heated in a vacuum or a reducing atmosphere lose some of their oxygen; semiconductors formed that way are called *reduction* semiconductors. For example, TiO_2 has a specific resistance of 10^8 Ω-m and $TiO_{1.9995}$ has a specific resistance of 0.10 Ω-m; this example illustrates that small deviations from stoichiometry may give large changes in electrical properties.

5.1d. Impurity Semiconductors

In these semiconductors the donor or acceptor levels are provided by impurities. Selenium, for instance, may be "activated" by dissolving bromine in the melt. The best known examples of this type of semiconductor are doped elements from the fourth group of the periodic table, in particular silicon and germanium. These crystals have a diamond structure; each atom has four outer electrons that are shared by its four close neighbors to form a covalent type of bond.

If impurity atoms from group V of the periodic system of elements (P, As, Sb) that have five outer electrons are added, they will take the place of some of the regular atoms of the lattice. Four of their outer electrons are used to form the four covalent bonds with their neighbors, leaving one loosely bound electron behind. These impurity atoms thus give rise to occupied donor levels close to the bottom of the conduction band; the semiconductor therefore becomes n type. The impurity atom and the extra electron form together a hydrogen-like structure; the electron moves in the field $e/4\pi\epsilon_0\epsilon r^2$ of the impurity atom, where ϵ is the dielectric constant of the medium, so that there will be a series of bound energy levels at energies $E_n = -13.60(m_n^*/m)/\epsilon^2 n^2$ V [$n = 1, 2, \ldots$; compare Chapter 2, Eq. (2.49)]; the level with $n = 1$ (ground state) is the most important one. In Si and Ge, ϵ is quite large, so that those donor levels are close to the bottom of the conduction band; the binding energies of the fifth electron to the group V impurity atoms phosphorus, arsenic, and antimony are 0.0120, 0.0127, and 0.0096 eV in germanium and 0.044, 0.049, and 0.039 eV in silicon, respectively.

If impurity atoms from group III of the periodic system of elements (B, Al, Ga, In) are added, they take the place of some of the regular atoms of the lattice. Their three outer electrons are used to make three covalent bonds, but the electron for the fourth covalent bond is missing; a loosely bound hole is thus left behind in the vicinity of the impurity atom. A "free hole" would be created by bringing an electron from one of the regular Si atoms to the impurity atom. Very little energy is needed to bring this about. This means that the added impurity atoms form unoccupied acceptor levels close to the top of the filled band; the semiconductor therefore becomes p type. The hole that is loosely bound to the impurity atom can be considered as a hydrogen-like structure, a positive charge moving in the field $e/4\pi\epsilon_0\epsilon r^2$ of the negatively charged impurity atom. As in the previous case, one would thus expect these acceptor levels to lie $13.60(m_p^*/m)/\epsilon^2 n^2$ eV above the top of the valence band. The binding energies for the hole in this state for group III impurities such as boron, aluminum, gallium, and indium are 0.0104, 0.0102, 0.0108, and 0.0112 eV in germanium and 0.045, 0.057, 0.065, and 0.16 eV in silicon, respectively.

The interesting point about these impurity semiconductors is that their properties can be so well controlled. The *type* of conduction (p or n type) depends upon the *kind* of impurity added; the *magnitude* of the conductivity depends upon the *concentration* of the impurity atoms. One needs only a very small amount of impurity in order to produce a semiconductor of 10^{-2} Ω-m resistivity (compare Sec. 6.1). This has necessitated novel methods of purification of these elements.

We next consider what happens if the impurity concentration is increased.[†] We do so first for an n-type semiconductor; for a p-type semiconductor the behavior is complementary.

If the impurity concentration becomes relatively large, the impurity levels become an impurity *band*, and the (small) conduction in the band is due to the "hopping" of electrons from one impurity site to the next impurity site. At about the same time the bottom of the conduction band develops a "tail" extending toward the impurity band. For higher concentrations, the impurity band and the tail expand, and for $N_d > 10^{19}$ donors/cm³ the two bands merge. In addition the energy gap width E_g decreases with increasing donor concentration for $N_d > 10^{18}$/cm³. We shall see in Chapter 16 that this has an important effect on the base current flow in transistors with a very high impurity concentration in the emitter.

At high acceptor concentration the acceptor levels in p-type material form a band, and the top of the valence band develops a tail extending toward the impurity band. For $N_a > 10^{19}$/cm³ the two bands merge, and the gap width decreases with increasing acceptor concentration for $N_a > 10^{18}$/cm³.

Silicon carbide (SiC) has the same crystal structure, but a much larger gap width (2.3 eV) than pure silicon. The compound can have very high purity and can be made p type or n type by doping. P-type silicon carbide can be made by adding Al as an impurity; the donors causing n-type conductivity are not well known.

5.1e. Intermetallic Compounds

The semiconductors silicon and germanium crystallize in the diamond structure. In this structure each atom is surrounded by four like neighbors, and the four outer electrons of each atom form electron-pair (covalent) bonds with each neighbor. A similar structure exists in compounds of the type AB, where A is an element of group III of the periodic table and B is an element of group V. Each atom of one type is surrounded by four neighbors of the other type; this structure, which shows otherwise close similarity with the diamond structure, is known as the zincblende (ZnS) structure. These compounds resemble the crystals of the group IV elements also in that they are semiconductors. This is at first sight the more remarkable, since the constituent elements can be metallic; for that reason those compounds are called *intermetallic compounds*.

The formation of four electron-pair bonds around each A atom and around each B atom requires the transfer of one electron from a B atom to an A atom. The AB bond is therefore not truly covalent, but is partly ionic. This results in a larger forbidden gap width than would be expected other-

---

† D. D. Kleppinger and F. A. Lindholm, *Solid State Electronics*, **14**, 199, 407 (1971).

wise. Compare, for instance, Si with the compound AlP, formed by its neighbors in the periodic table. AlP has a gap width of about 2.5 eV, whereas Si has a gap width of only 1.1 eV, thus illustrating the influence of the partly ionic character of the bond.

The following table lists the properties of the various semiconducting elements, of βSiC, and of various III–V compounds.

Table of Semiconductor Properties[a]

Compound	Energy Gap at Room Temperature (eV)	μ_n (in $cm^2/V\,S$)	μ_p (in $cm^2/V\,S$)
C (diamond)	5.4	1,800	1,400
Si	1.107	1,900	500
Ge	0.67	3,800	1,820
αSn	0.08	2,500	2,400
βSiC	2.3	4,000	—
AlP	2.5	—	—
AlAs	2.16	1,200	420
AlSb	1.60	200–400	550
GaP	2.24	300	100
GaAs	1.35	8,800	400
GaSb	0.67	4,000	1,400
InP	1.27	4,600	150
InAs	0.36	33,000	460
InSb	0.165	78,000	750

[a] For further details, see *CRC Handbook of Tables for Applied Engineering Science*, The Chemical Rubber Co., Cleveland, Ohio, 1970; or H. Wolf, *Semiconductors*, John Wiley & Sons, Inc. (Interscience Division), New York, 1971.

The crystals can be made p type by replacing some of the group III atoms by group II atoms: they can be made n type by replacing some of the group V atoms by group VI atoms. GaAs is the most promising compound.

In the group III–group V compounds each atom has on the average four electrons per atom and forms electron-pair types of bonds with a cubic structure. Such compounds are called isoelectronic compounds. Other examples are the group II–group VI compounds such as ZnSe and CdTe. Here the ionic character of the bond is stronger than in the group III–group V compounds, resulting in a larger gap width. A similar but even stronger trend exists in group I–group VII compounds such as CuBr and AgI.

Another group of similar compounds consists of one atom of group I, one atom of group III, and two atoms of group VI; they give in total 16 outer electrons per molecule, or on the average four electrons per atom, and form electron-pair types of bond with a cubic structure. $AgInTe_2$ is an example of the group.

Compounds of the type C_2D, where C represents an element of group II and D an element of group IV, are also semiconducting. Examples are Mg_2Si, Mg_2Ge, and Mg_2Sn. Some of these elements thus belong to the intermetallic compound group.

5.2. PREPARATION OF SEMICONDUCTOR MATERIALS

5.2a. Purification and Preparation of Semiconducting Material

The semiconducting material used in the manufacture of semiconductor devices should have high purity initially and should subsequently be doped as required by controlled amounts of prescribed impurities. To get an idea of the purity required, consider n-type germanium with a resistivity of 1 Ω cm = 10^{-2} Ωm. The impurity concentration needed (see Chapter 6) is $N_d = 1.7 \times 10^{21}/m^3$, corresponding to an impurity concentration of 1 part in 10^7. This is the concentration of *wanted* impurity atoms; the concentration of *unwanted* impurities should be appreciably lower to ensure the production of semiconducting material of prescribed conductivity. To obtain such a degree of purity, a powerful new refining method has been employed, known as *multiple zone melting.*Its operation is based upon the following principles.

Consider a sample of molten semiconducting material having an impurity concentration C_m. If the melt is cooled, solid material will freeze from the melt. It is found experimentally that the frozen material has an impurity concentration C that is much smaller than C_m; the ratio

$$k = \frac{C}{C_m} \tag{5.1}$$

is found to be independent of C_m for a wide range of values as long as C_m is relatively small. The constant k, known as the *distribution constant* or *segregation constant*, is a characteristic of the impurity atoms dissolved in the melt. Since the material freezing from the melt has an impurity concentration much smaller than the impurity concentration C_m of the melt, the value of C_m will gradually increase if the freezing proceeds, and as a consequence the impurity concentration in the solid material just frozen will also increase in the process (Fig. 5.2a).

To discuss this more closely, consider a cylindrical sample of molten material of unit cross-sectional area and length L having initially a uniform concentration C_{m0} of a certain impurity. Now let the sample freeze from the bottom up, and at a certain instant let a certain length x_0 be solidified. If it is assumed that no diffusion occurs in the solid phase and that the impurity concentration C_m of the melt, although it changes with time, is uniform at all times, then the concentration C_m of the liquid phase and the concentration C_x of the solid phase are given by the equations (Fig. 5.2b)

$$C_m = C_{m0}\left(1 - \frac{x_0}{L}\right)^{k-1}, \qquad C_x = kC_{m0}\left(1 - \frac{x}{L}\right)^{k-1}, \qquad 0 < x < x_0$$

$$(5.2)$$

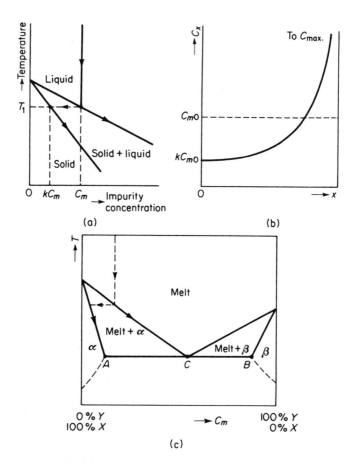

(a)   (b)

(c)

Fig. 5.2. (a) Phase diagram of material containing an impurity. Cooling down from the liquid phase of impurity concentration C_m, the solid phase starts to segregate at a temperature T_1 with an impurity concentration kC_m, where $k \ll 1$. The impurity concentration in the liquid phase gradually increases if the freezing continues, and so will the impurity concentration in the solid phase. (b) Impurity distribution in a solid that has frozen from the liquid phase from the bottom up. C_{max} is the maximum impurity concentration. (c) Phase diagram of melt XY. The α phase is a solid solution of X into Y, and the β phase is a solid solution of Y into X; C is the eutectic point.

If $C_x = C_{x0}$ at $x = x_0$, then $C_{x0} = kC_m$ for all x_0, as expected. That these formulas are correct is most easily seen from the condition that the total amount of impurity must be constant at all times. Mathematically this is expressed by the condition

$$\int_0^{x_0} C_x \, dx + C_m(L - x_0) = C_{m0}L \qquad (5.3)$$

for all values of x_0. Expression (5.2) satisfies this condition, as is seen by substitution of Eq. (5.2) into Eq. (5.3).

The phase diagram and the impurity distribution of the solid material after freezing are shown in Figs. 5.2a and b. Note that C_x becomes infinite at $x = L$ according to (5.2). This is obviously impossible, for C_x is a *relative concentration* defined as

$$C_x = \frac{\text{number of impurity atoms per unit volume}}{\text{total number of atoms per unit volume}} \qquad (5.4)$$

Hence C_x can at best become equal to unity. This actually comes about because k depends upon C_m for large concentrations. This problem is illustrated in Fig. 5.2c, the phase diagram of a melt containing a mixture of X and Y atoms. Let the melt contain a small Y concentration initially. If the melt is cooled, a solid solution of Y in X will segregate out; since $k < 1$, the impurity concentration of the melt will increase until the *eutectic point* C of the phase diagram is reached. Upon further cooling small crystals of a solid solution of Y in X and of a solid solution of X in Y will segregate out simultaneously. In some cases the eutectic point C may lie extremely close to $C_m = 0$ per cent or $C_m = 100$ per cent; one of the constituents then has a very small solubility into the other constituent. It may also happen that point A of the phase diagram lies at the 100 per cent line; in that case the impurity is said to have complete solubility in the solid.

We are now able to understand the process of *zone melting*. In the process a bar of semiconducting material moves slowly through the heater coil of an induction heater producing enough heat to melt the material locally. During the process the liquid zone moves along the bar, retaining most of the impurities, leaving the refrozen material behind in a purer state (Fig. 5.3). The impurity atoms will thus be concentrated toward the right-hand side of the bar. If we repeat the process many times (this is most easily done by using several heater coils, one behind the other), almost all impurities can be concentrated in a small section at the right side of the bar. Removing that section, we are left with a bar of very pure material. This process is known as *multiple zone refining*.

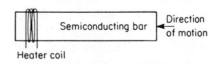

Fig. 5.3. Zone melting. An induction heater coil melts a small segment of a semiconducting bar that moves slowly through the coil. The liquid segment carries most of the impurities with it.

Having obtained pure material, we must add a prescribed amount of impurity and distribute it uniformly through the solid. This is achieved by the method of *zone leveling*. First, the proper amount of wanted impurity is added to the molten zone at the end of a very pure bar of semiconducting material and then the bar is put through a single cycle of zone melting. Let the impurity be added at $x = 0$ and let l be the length of the molten zone. Let C_i be the impurity concentration in the molten zone if this zone is at $x = 0$. The concentration in the molten zone then gradually decreases as the molten zone moves through the bar; but if the length l of the molten zone is not too small and the value of k is relatively small (for example, 0.01 or less), the decrease in concentration in the molten zone will be hardly noticeable even for a relatively long bar. A calculation shows that the concentration C in the melted part of the bar varies as

$$C = C_i \exp\left(-\frac{kx}{l}\right) \tag{5.5}$$

(except for the last zone, of course). The proof is simple. If the molten zone moves over a distance dx, the change in concentration is

$$dC = -\frac{k\, dx\, AC}{Al}$$

where A is the cross section of the zone. Integration yields Eq. (5.5). Therefore, the concentration in the refrozen part varies as

$$C = kC_i \exp\left(-\frac{kx}{l}\right) \tag{5.5a}$$

Even this variation can be avoided if we let the length l of the molten zone decrease with increasing x such that the concentration in the molten zone is kept constant. If l_0 is the zone length at $x = 0$, then this is the case if $C_i l + kC_i x = C_i l_0$, or

$$l = l_0 - kx \tag{5.5b}$$

Other methods are discussed in the next section.

5.2b. Preparation of Single Crystals

To make semiconductor devices in mass production in a reproducible manner, we must control the physical characteristics to an extent that cannot be achieved with polycrystalline material. The main reason is the existence of lattice imperfections, chiefly of the dislocation type, which have a strong influence on the physical properties of the material.

A *dislocation* is a stacking fault in the crystal. Obviously many of these stacking faults occur at the grain boundaries; at these boundaries rows of dislocations, known as *dislocation lines*, are found, which change the physical characteristics markedly. For example, if a dislocation line occurs in *n*-type material, the material in its immediate neighborhood may become p-type. These difficulties are avoided by producing single-crystal material.

One way of making single crystals is the *pulling* method. We shall discuss this method using germanium. A germanium single-crystal seed of controlled orientation is dipped into a germanium melt, part of the seed is melted to get rid of its strained portions, and a germanium single crystal is pulled progressively from the melt as it grows on the seed. The rate of lifting of the crystal is usually kept equal to the average rate of linear growth, thus resulting in an approximately planar solid–liquid interface at approximately a constant level above the surface of the melt. To obtain the proper impurity concentration, the required amount of donor or acceptor impurity is added to the melt. To obtain uniformity in planes perpendicular to the growth axis, the melt is stirred by rapid rotation of the crystal around its growth axis; sometimes simultaneous vibration is applied along this axis at moderately high speed (Fig. 5.4).

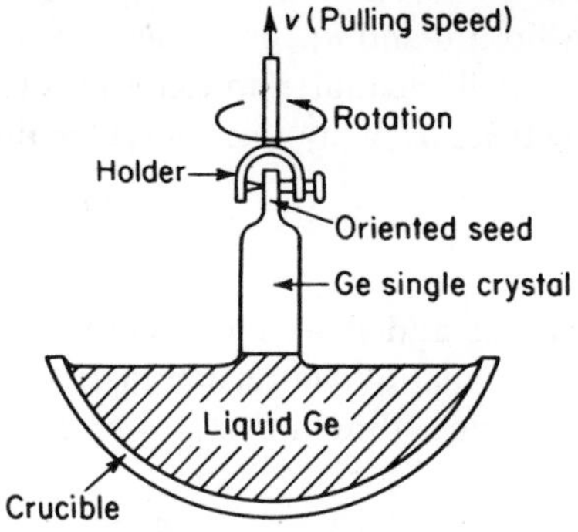

Fig. 5.4. Crystal pulling from the liquid phase. Starting with a single-crystal seed, a single crystal is pulled from the melt with a uniform speed v. To obtain uniformity, the melt is stirred by rapid rotation of the crystal around its growth axis.

Single crystals may also be made with the help of the zone-melting technique. At the moment when the molten zone is at the one end of a zone-melted semiconducting bar, a single-crystal seed of controlled orientation is dipped into the molten zone and the bar is put through a zone-melting cycle. The material freezing from the liquid zone is then a single crystal having the same orientation as the seed.

5.3. CARRIER CONCENTRATIONS IN SEMICONDUCTORS

5.3a. Equilibrium Considerations

We have seen how in intrinsic material conduction comes about by thermal generation of hole–electron pairs. The same processes are at work in extrinsic material, and as a consequence some holes will be present in n-type material and some electrons in p-type material. One can thus distinguish between *majority carriers* (holes in p-type material and electrons in n-type material) and *minority carriers* (holes in n-type material and electrons in p-type material). Let us now calculate the equilibrium concentrations of these minority carriers.

Let p be the hole concentration, n the electron concentration, and $n_i = p_i$ the carrier concentration in intrinsic material. The rate g of direct hole–electron pair generation is independent of p and n, for there are an almost infinite number of bound electrons in the valence band that can be shaken loose by the lattice vibrations. The inverse process, recombination, takes one hole–electron pair out of the sample, because a hole and an electron unite either under emission of quanta or in a radiationless transition. It is obvious that for direct recombination the rate of recombination will be proportional to both p and n; for if there are twice as many holes, the number of holes that will encounter an electron is twice as large, and the same is true if there are twice as many electrons. One thus has the rate equation (Sec. 6.3)

$$\frac{dp}{dt} = \frac{dn}{dt} = g - \rho pn \qquad (5.6)$$

where g and ρ are constants. This equation is not valid for generation and recombination via recombination centers (Sec. 6.3d).

At equilibrium $dp/dt = dn/dt = 0$, and hence

$$pn = \frac{g}{\rho} = \text{constant} = n_i^2 \qquad (5.7)$$

Since g and ρ are independent of p and n, the product pn is independent of p and n; it is a constant that may be equated to n_i^2. The parameter n_i represents the intrinsic carrier concentration. It depends only on the temperature and on the material, but not on the impurities causing the extrinsic conductivity. As will be shown in Sec. 5.3b,

$$np = n_i^2 = 2\left(\frac{2\pi m_n^* kT}{h^2}\right)^{3/2} 2\left(\frac{2\pi m_p^* kT}{h^2}\right)^{3/2} \exp\left(-\frac{E_g}{kT}\right) \qquad (5.8)$$

where E_g is the gap width of the semiconductor. Since this is a thermodynamical result, it is independent of the validity of Eq. (5.6).

We now calculate p and n for n-type material. Let N_d be the donor concentration. If there were no hole–electron pair generation, then n would be equal to N_d. Since hole–electron pairs are generated, $n = N_d + p$. Equation (5.7) thus yields

$$pN_d + p^2 = n_i^2 \qquad (5.7a)$$

which has the solution $p = p_0$ and $n_0 = N_d + p_0$. Solving Eq. (5.7a) for p yields

$$p = p_0 = -\tfrac{1}{2}N_d + (\tfrac{1}{4}N_d^2 + n_i^2)^{1/2} \qquad (5.9)$$

Two cases are of special importance: (1) the nearly intrinsic case, $N_d \ll n_i$; (2) the strongly extrinsic case, $N_d \gg n_i$. In the first case $\tfrac{1}{4}N_d^2 \ll n_i^2$ and hence

$$p_0 \simeq -\tfrac{1}{2}N_d + n_i, \qquad n_0 = N_d + p_0 \simeq \tfrac{1}{2}N_d + n_i \qquad (5.9a)$$

In the second case, since $(1 + x)^{1/2} \simeq 1 + \frac{1}{2}x$ for small x,

$$p_0 = -\frac{1}{2}N_d + \frac{1}{2}N_d\left(1 + \frac{4n_i^2}{N_d^2}\right)^{1/2}$$

$$\simeq -\frac{1}{2}N_d + \frac{1}{2}N_d\left(1 + \frac{2n_i^2}{N_d^2}\right) = \frac{n_i^2}{N_d} \qquad (5.9b)$$

$$n_0 = N_d + p_0 \simeq N_d$$

The same reasoning can be repeated for p-type material with an acceptor concentration N_a.

It may also happen that both donor and acceptor concentrations are present (counterdoping). If N_d is again the donor concentration and N_a the acceptor concentration, then the material is n-type if $N_d > N_a$ and p-type if $N_d < N_a$. If $N_d > N_a$, N_a electrons are used to ionize the acceptors and $N_d - N_a$ are left free. Since p hole–electron pairs are generated thermally, Eq. (5.7) becomes

$$p(N_d - N_a + p) = n_i^2 \qquad (5.10)$$

At very low temperatures there are practically no minority carriers and even the number of majority carriers becomes temperature-dependent, because no longer are all donors and acceptors ionized. Let us investigate this for n-type material. In equilibrium the processes

neutral donor $\rightleftharpoons$ ionized donor $+$ electron

balance. Let N_d be the donor concentration and n the electron concentration. The rate of ionization will then be proportional to the neutral donor density $N_d - n$. The rate of recombination will be proportional to *both* the number n of free electrons and the number n of ionized donors—that is, it is proportional to n^2. Therefore,

$$\frac{dn}{dt} = \alpha(N_d - n) - \beta n^2 \qquad (5.11)$$

where α and β are constants that depend only on the temperature and the location of the donor levels. In equilibrium $dn/dt = 0$, or

$$\frac{n^2}{N_d - n} = \frac{\alpha}{\beta} \qquad (5.12)$$

A calculation shows that this equation may be written as

$$\frac{n^2}{N_d - n} = 2\left(\frac{2\pi m_n^* kT}{h^2}\right)^{3/2} \exp\left(-\frac{E_0}{kT}\right)$$

$$= 2.51 \times 10^{19}\left(\frac{Tm_n^*}{300m}\right)^{3/2} \exp\left(-\frac{E_0}{kT}\right) \qquad (5.12a)$$

where E_0 is the energy difference between the bottom of the conduction band and the donor level. The reader should note the similarity to Eq. (5.8). A similar equation applies to p-type material.

5.3b. Carrier Densities in P-Type and N-Type Material; Fermi Level

What was said about the probability function $P(E)$ in Chapter 3 also holds for semiconductors. The main difference here is that E_f is negative. Hence in n-type material, where the electrons have a binding energy E_0 to the donor ions, the donor levels lie at a distance E_0 below the bottom of the conduction band. If $-E_0 - E_f > 4kT$, almost all donor levels will be empty and the electrons will all be in the conduction band. If $-E_0 - E_f < -4kT$, almost all donor levels will be filled and the conduction band will be nearly empty (Fig. 5.5a).

Similar considerations hold for p-type material. Let the forbidden energy gap have a width E_g, and let the holes have a binding energy of E_0' to the acceptor ions. The acceptor levels then lie at a distance E_0' above the top of the valence band. If $E_g - E_0' + E_f > 4kT$, almost all acceptor levels will be filled, and the hole density in the valence band will equal the acceptor density. If $E_g - E_0' + E_f < -4kT$, almost all acceptor levels will be empty, and the hole density in the valence band will be nearly zero (Fig. 5.5b).

We can now show that the density of electrons in the conduction band is

$$
n = 2\left(\frac{2\pi m_n^* kT}{h^2}\right)^{3/2} \exp\left(\frac{E_f}{kT}\right) = N_c \exp\left(\frac{E_f}{kT}\right)
$$

$$
= 2.51 \times 10^{19} \left(\frac{m_n^* T}{m \times 300}\right)^{3/2} \exp\left(\frac{E_f}{kT}\right) \qquad \text{per cm}^3
$$

(5.13)

if $-E_f > 4kT$, whereas the density of holes in the valence band is

$$
p = 2\left(\frac{2\pi m_p^* kT}{h^2}\right)^{3/2} \exp\left(-\frac{E_g + E_f}{kT}\right)
$$

$$
= N_v \exp\left(-\frac{E_g + E_f}{kT}\right)
$$

(5.14)

$$
= 2.51 \times 10^{19} \left(\frac{m_p^* T}{m \times 300}\right)^{3/2} \exp\left(-\frac{E_g + E_f}{kT}\right) \qquad \text{per cm}^3
$$

if $E_g + E_f > 4kT$. Here E_f is the Fermi level ($E_f < 0$) and E_g is the gap width; both are in joules. The quantity $4kT$ at $T = 300°K$ has the value 1.66×10^{-20} J $= 0.104$ eV. The parameters N_c and N_v are defined as

$$
N_c = 2\left(\frac{2\pi m_n^* kT}{h^2}\right)^{3/2}
$$

(5.13a)

$$
N_v = 2\left(\frac{2\pi m_p^* kT}{h^2}\right)^{3/2}
$$

(5.14a)

We first prove Eq. (5.13). Since

$$
P(E) = \frac{1}{1 + \exp\left[(E - E_f)/kT\right]}
$$

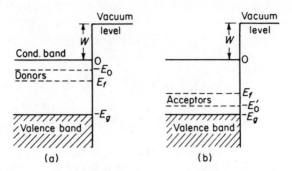

Fig. 5.5. (a) Energy level diagram of an n-type semiconductor. (b) Energy level diagram of a p-type semiconductor.

$P(E)$ may be written

$$P(E) = \exp\left(-\frac{E - E_f}{kT}\right) \tag{5.15}$$

if $E - E_f > 4kT$ for all energy levels in the conduction band. This certainly is the case if $-E_f > 4kT$. The density Δn of electrons of energy between E and $E + \Delta E$ in the conduction band may be written

$$\Delta n = 8\pi\sqrt{2}\,\frac{m_n^{*3/2}}{h^3}E^{1/2}\exp\left(-\frac{E - E_f}{kT}\right)\Delta E \tag{5.16}$$

The density n of electrons in the conduction band is obtained by integration over all available energy levels in the conduction band. This means that the limits of integration are 0 and ∞. Consequently,

$$\begin{aligned}
n &= 8\pi\sqrt{2}\,\frac{m_n^{*3/2}}{h^3}\int_0^\infty E^{1/2}\exp\left(-\frac{E - E_f}{kT}\right)dE \\
&= 8\pi\sqrt{2}\,\frac{m_n^{*3/2}}{h^3}(kT)^{3/2}\exp\left(\frac{E_f}{kT}\right)\frac{1}{2}\sqrt{\pi}
\end{aligned} \tag{5.16a}$$

since

$$\int_0^\infty x^{1/2}\exp\left(-x\right)dx = \frac{1}{2}\sqrt{\pi}$$

and $x = E/kT$. Equation (5.16a) is generally valid if $-E_f > 4kT$. Rearranging Eq. (5.16a) gives Eq. (5.13).

Next we prove Eq. (5.14). Since

$$\Delta n = P(E)Z(E)\,\Delta E$$

is the density of occupied energy levels in the valence band with energies between E and $(E + \Delta E)$, the number Δp of holes per cubic centimeter with energies between E and $(E + \Delta E)$ is

$$\Delta p = [1 - P(E)]Z(E)\,\Delta E \tag{5.17}$$

since each hole corresponds to an empty electron level in the valence band. But $Z(E)$ is now

$$Z(E) = 8\pi\sqrt{2}\,\frac{m_p^{*3/2}}{h^3}(-E_g - E)^{1/2} \tag{5.17a}$$

and

$$1 - P(E) = \frac{\exp\left[(E - E_f)/kT\right]}{1 + \exp\left[(E - E_f)/kT\right]}$$

$$\simeq \exp\left(\frac{E - E_f}{kT}\right) \tag{5.17b}$$

if $-E + E_f > 4kT$ for all holes in the valence band. This certainly is the case if $E_g + E_f > 4kT$. The density Δp of holes of energy between E and $(E + \Delta E)$ in the valence band may thus be written

$$\Delta p = 8\pi\sqrt{2}\,\frac{m_p^{*3/2}}{h^3}(-E_g - E)^{1/2}\exp\left(\frac{E - E_f}{kT}\right)\Delta E \tag{5.18}$$

The density p of holes in the valence band is obtained by integration over all available energy levels in the valence band. This means that the limits of integration are $-\infty$ and $-E_g$. Consequently,

$$p = 8\pi\sqrt{2}\,\frac{m_p^{*3/2}}{h^3}\int_{-\infty}^{-E_g}(-E_g - E)^{1/2}\exp\left(\frac{E - E_f}{kT}\right)dE$$

$$= 8\pi\sqrt{2}\,\frac{m_p^{*3/2}}{h^3}(kT)^{3/2}\exp\left(-\frac{E_g + E_f}{kT}\right)\frac{1}{2}\sqrt{\pi} \tag{5.18a}$$

as is found by substituting $x = (-E - E_g)/kT$ as a new variable; this leads to the same integral as in the previous case. Equation (5.18a) is generally valid if $E_g + E_f > 4kT$. Rearranging Eq. (5.18a) yields Eq. (5.14).

When Eqs. (5.13) and (5.14) are both valid, we obtain an equation for the intrinsic carrier density n_i by putting $np = n_i^2$. This yields Eq. (5.8).

$$n_i^2 = np = 4\left(\frac{2\pi mkT}{h^2}\right)^3\left(\frac{m_n^* m_p^*}{m^2}\right)^{3/2}\exp\left(-\frac{E_g}{kT}\right) \tag{5.19}$$

Substituting the various numerical values and expressing E_g in electron volts yields

$$n_i = 2.51 \times 10^{19}\left(\frac{m_n^* m_p^*}{m^2}\right)^{3/4}\left(\frac{T}{300}\right)^{3/2}\exp\left(-\frac{eE_g}{2kT}\right) \quad \text{per cm}^3 \tag{5.19a}$$

The derivation of (5.13) holds for strongly ionized donors. In the case of weakly ionized donors the position of the Fermi level is

$$E_f = -\frac{1}{2}E_0 + \frac{1}{2}kT\ln\frac{N_d}{N_c}, \qquad N_c = 2\left(\frac{2\pi m_n^* kT}{h^2}\right)^{3/2} \tag{5.20}$$

This follows immediately from (5.12a) and (5.13a). If the donors are weakly ionized, $n \ll N_d$, and hence from (5.12a)

$$n = (N_c N_d)^{1/2} \exp\left(-\frac{\frac{1}{2}E_0}{kT}\right) = N_c \exp\left(\frac{E_f}{kT}\right)$$

or
$$E_f = -\frac{1}{2}E_0 + \frac{1}{2}kT \ln\frac{N_d}{N_c}$$

In the same notation Eq. (5.13) may be written

$$E_f = kT \ln\frac{N_d}{N_c} \tag{5.21}$$

Similar considerations apply to p-type material.

In most semiconductors there are thus two characteristic temperatures, $T_{\min}$ and $T_{\max}$. For $T > T_{\max}$, $n \simeq p \simeq n_i$ and the material is practically intrinsic. For $T_{\min} < T < T_{\max}$, n or p is practically equal to the impurity concentration. For $T < T_{\min}$, n or p decreases as $\exp(-E_0/2kT)$ or $\exp(-E_0'/2kT)$, because fewer and fewer donors or acceptors are ionized at lower and lower temperatures. This is important for an understanding of the temperature dependence of the conductivity of semiconductors (Fig. 5.6).

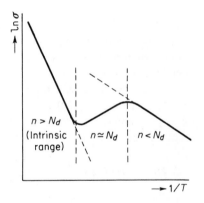

The only exception to this rule occurs when the impurity density is very large. In n-type material with $N_d > 10^{18}$–$10^{19}/\text{cm}^3$ the donor levels and also the Fermi level will lie in the conduction band. Some energy levels in the conduction band are then filled at all times and the material has nearly metallic character. There is now no $T_{\min}$, and, because of the large carrier density in the conduction band, $T_{\max}$ is extremely large.

Fig. 5.6. The value $\ln \sigma$ plotted against $1/T$ for an n-type semiconductor, showing the three ranges of conductivity described by $n > N_d$ (intrinsic), $n \simeq N_d$, and $n < N_d$. A similar curve holds for p-type semiconductors.

In p-type material with $N_a > 10^{18}$–$10^{19}/\text{cm}^3$, the acceptor levels lie in the valence band and so does the Fermi level. Some energy levels in the valence band are permanently empty (that is, they are permanently filled with holes) and the material has nearly metallic character. There is again no $T_{\min}$, and $T_{\max}$ is extremely large.

Semiconductor materials in which the Fermi level lies in the conduction band for n-type material, or in the valence band for p-type material, are called *degenerate semiconductors*. It goes without saying that Eqs. (5.13) and

(5.14) are then no longer valid, but it is beyond the scope of this book to derive the position of the Fermi level for these cases.

5.4. CONTACT PROBLEMS

5.4a. Energy-Level Diagram; Work Function

Figure 5.5 shows the energy-level diagram for an *n*-type and a *p*-type material. The energy difference W between the vacuum level and the bottom of the conduction band is usually called the *electron affinity* and denoted by χ. We take the bottom of the conduction band as reference level here.

If the zero point is taken with respect to the bottom of the conduction band as a reference level, then the work function is defined as

$$\phi = \chi - E_f \tag{5.22}$$

as in a metal, but the difference is that the Fermi level is now negative. It should be noted that $-E_f$ is obviously larger for a *p*-type than for an *n*-type semiconductor, so that a *p*-type sample of a given material has a larger work function than an *n*-type sample of the same material. This is important for thermionic emission.

It should be noted that the minimum quantum energy needed to produce a photoelectron at $T = 0$ is $\chi + E_0$ for *n*-type material and $\chi + E_g - E_0'$ for *p*-type material. In contrast with metals, the minimum photon energy for photoelectric effect is hardly ever equal to ϕ.

5.4b. Metal–Semiconductor Contacts

When two substances are brought into contact, a redistribution of charge occurs; finally a new equilibrium condition is reached in which the Fermi levels of the two substances are at equal heights (Chapter 3). This rule holds not only for contacts between two metals but also for the contact between a metal and an *n*-type or a *p*-type semiconductor.

Owing to the redistribution of charge, a dipole layer will be formed at the contact. In a metal–metal contact this dipole layer is always caused by *surface charges* on both sides of the contact; such a contact is an *ohmic contact*, since the electrons can move freely from the one metal into the other. In a metal–semiconductor contact, however, the contact may either be *ohmic* or *rectifying*. The latter is a contact in which the current flows much more easily in one direction than in the opposite. We shall see that this is intimately connected with the electronic energy-level diagram of the two substances.

Consider a contact between a metal and an *n*-type semiconductor (Fig.

5.7). Let the donor concentration in the semiconductor be relatively large, and let almost all donors be ionized at room temperature. Let ϕ_m be the work function of the metal, ϕ_s the work function of the semiconductor, and χ_s the electron affinity of the semiconductor.

Let us first consider the case $\phi_m > \phi_s$. The situation before contact is shown in Fig. 5.7a; the Fermi level of the semiconductor is then above the Fermi level of the metal by an amount $\phi_m - \phi_s$. After contact, an exchange of charge occurs; electrons from the surface layer of the semiconductor enter the metal, leaving ionized donors behind in that surface layer. After the exchange of charge is completed, the Fermi levels in both materials are at the same height; this means that the energy levels in the bulk semiconductor are lowered by the amount $\phi_m - \phi_s$. As a consequence, a potential barrier is formed at the surface. On the semiconductor side the height of this barrier is thus $\phi_m - \phi_s$; on the metal side, since the Fermi levels are at the same height, the height of the barrier is $(\phi_m - \phi_s) + (\phi_s - \chi_s) = \phi_m - \chi_s$ (Fig. 5.7b). Expressed in volts, the height of the barrier follows from

$$eV_{\text{dif}} = \phi_m - \phi_s \qquad (5.23)$$

The quantity V_{dif} is known as the *diffusion potential*. The potential at the interior of the semiconductor, taken with respect to the metal surface, is V_{dif}.

This potential difference is maintained by the electric dipole layer at the contact. Since the positive charge at the semiconductor side of the contact is caused by ionized donors having a density much smaller than the density of the ionized atoms in a metal, and since these donors are bound to fixed positions, the positive charge does not occur as a *surface charge* but as a *distributed charge* instead. For that reason the surface layer of this metal–

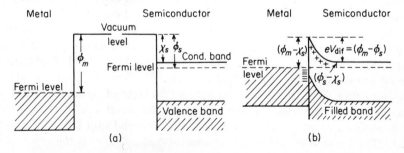

Fig. 5.7. Energy-level diagram of a metal n-type semiconductor contact with $\phi_m > \phi_s$, where ϕ_m and ϕ_s are the work functions of metal and semiconductor, respectively. (a) Energy-level diagram before contact. (b) Energy-level diagram after contact. The Fermi level is now at equal height; V_{dif} is the diffusion potential.

semiconductor contact is called the *space-charge layer*. Because of the potential barrier at the contact, the surface layer is also known as the *barrier layer*. The thickness d of this layer depends upon the concentration of ionized donors and upon the value of the diffusion potential V_{dif}.

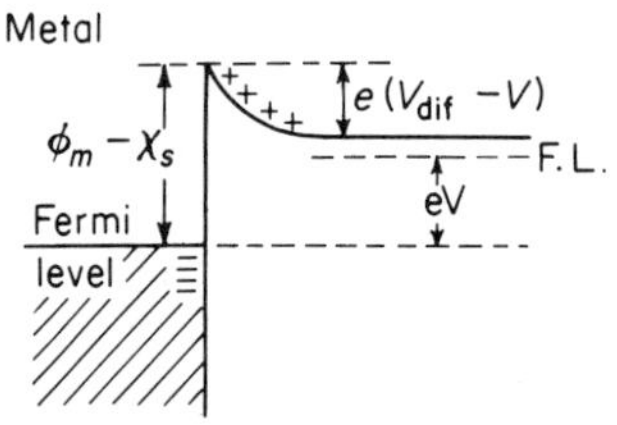

Fig. 5.8. To show that the metal-semiconductor contact of Fig. 5.7 is rectifying. V_{dif} is the diffusion potential. A voltage $-V$ is applied to the semiconductor. The Fermi level in the bulk semiconductor has shifted by an amount eV.

Owing to thermal agitation, some electrons of the metal will have enough energy to cross the potential barrier into the semiconductor, and some electrons of the semiconductor will have enough energy to cross the potential barrier into the metal. In equilibrium this gives rise to equal and opposite currents I_0 crossing the barrier.

If a voltage $-V$ is applied to the semiconductor (Fig. 5.8), the barrier for electrons going from left to right has not changed, and hence the corresponding current from right to left has not changed either. But since the energy levels in the conduction band have been *raised* by an amount eV, the barrier for electrons going from right to left has been *lowered* by an amount eV; as a consequence the corresponding current from left to right has changed by a factor exp (eV/kT). Consequently, the characteristic is

$$I = I_0 \left[\exp \left(\frac{eV}{kT} \right) - 1 \right] \tag{5.24}$$

This is a rectifying contact; for $V \gg kT/e$ the current is large and positive, and for $V \ll -kT/e$ the current is small and almost equal to $-I_0$. The first bias condition is called *forward* bias and the second *back* bias.

Up to now we have considered the case $\phi_m > \phi_s$. If $\phi_m < \phi_s$, no rectifying barrier is formed. The situation before contact is shown in Fig. 5.9a; the Fermi level of the semiconductor is *below* the Fermi level of the metal by an amount $\phi_s - \phi_m$. An exchange of charge takes place after contact; electrons flow from the metal into the surface layer of the semiconductor, leaving a positive surface charge behind on the metal side of the contact and causing a negative surface charge at the semiconductor side of the contact.† The Fermi level in the semiconductor bulk material is thereby raised

---

† A positive charge in the surface layer of the semiconductor can only be a *distributed charge*, because only a limited concentration of ionized donors is available. A negative charge in the surface layer of the semiconductor can be considered as a *surface charge*, since so many *unoccupied* energy levels are available for electrons in the conduction band.

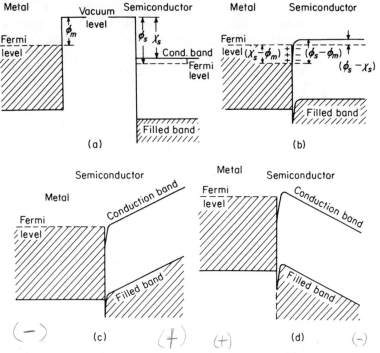

Fig. 5.9. Energy-level diagram of a metal n-type semiconductor contact with $\phi_m < \phi_s$ where ϕ_m and ϕ_s are the work functions of metal and semiconductor, respectively. (a) Energy-level diagram before contact. (b) Energy-level diagram after contact. This contact is ohmic. (c) A negative voltage is applied to the contact. (d) A positive voltage is applied to the contact.

by an amount $\phi_s - \phi_m$ (Fig. 5.9b). If a voltage V is applied, this potential difference is not taken up in the contact area as in the previous case, but is distributed across the bulk semiconductor (Figs. 5.9c and d).

We see that the electrons can move across the barrier without much difficulty, especially if $\phi_s - \chi_s$ is relatively small. The contact may thus be considered an *ohmic contact.*

We have hereby deduced a simple rule for determining whether a metal-semiconductor contact is an ohmic or rectifying contact. *The contact is ohmic if $\phi_m < \phi_s$, and rectifying if $\phi_m > \phi_s$.* For a rectifying contact one would expect the magnitude of the saturated current I_0 to decrease with increasing value of $\phi_m - \phi_s$; if two metals are compared with different values of ϕ_m, then the one with the largest value of ϕ_m should give the smallest value of I_0.

These predictions do not always work out in practice. It often happens—for example, in the case of germanium point-contact diodes—that the diode characteristic is almost independent of the work function of the metal. The reason for this behavior is that the semiconductor may have a natural surface barrier. One way in which this can happen is that a large number of electron energy levels, the *surface states*, are located at the surface of the semiconductor. Many of these surface states are occupied, leaving a distributed positive charge due to the ionized donors behind in the surface layer. Making contact with metals of different work functions then means that a different portion of the occupied surface states are emptied into the metal; we see that this does not change the space-charge barrier at the surface (Fig. 5.10).

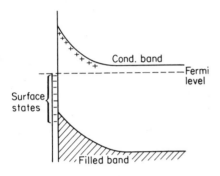

An ohmic contact also occurs if the metal and the semiconductor can alloy; it is then possible to make a true weld between the two materials with a gradual transition from the one material to the other. This occurs, for instance, if a gold wire is welded to *p*-type germanium. The same wire welded to *n*-type material produces a rectifying contact, however (gold-bonded diode).

Fig. 5.10. Energy-level diagram of an *n*-type semiconductor with surface states. The filled surface states may make a metal-semiconductor contact rectifying even if the condition $\phi_m > \phi_s$ is not satisfied.

It will be shown in Chapter 14 that the barrier becomes transparent if the donor concentration in the surface is above a certain critical value. One may thus also make an ohmic contact by making the surface layer of the material much more strongly *n* type than the interior.

After our discussion of the metal–*n*-type semiconductor contact, the discussion of the metal–*p*-type semiconductor contact can be brief. Let ϕ_m again be the work function of the metal, ϕ_s the work function of the semiconductor, and E_s the depth of the top of the filled band of the semiconductor below the vacuum level. The contact is now *rectifying* if $\phi_m < \phi_s$ and *ohmic* if $\phi_m > \phi_s$.

Let us consider the latter case first. The situation before contact is shown in Fig. 5.11a; the Fermi level in the semiconductor is *above* the Fermi level in the metal by an amount $\phi_m - \phi_s$. After contact, an exchange of charge takes place; electrons flow out of the semiconductor, leaving a positive surface charge (due to holes) behind on the semiconductor side and a negative *surface charge* on the metal side, thereby lowering the Fermi level in the semiconductor by an amount $\phi_m - \phi_s$ (Fig. 5.11b). Holes from the semiconductor can readily move into the metal and be neutralized almost instantly (because of the high electron concentration), and for opposite

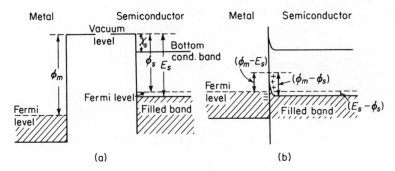

Fig. 5.11. Energy-level diagram of a metal p-type semiconductor contact. (a) Energy-level diagram before contact. (b) Energy-level diagram after contact if $\phi_m > \phi_s$. The contact is ohmic.

polarity of the applied voltage the holes formed thermally in the conduction band of the metal can readily move into the semiconductor. The contact is thus ohmic.

We now consider the case $\phi_m < \phi_s$. Let E_s again be the depth of the top of the filled band below the vacuum level and let all acceptors be ionized. Before contact, the Fermi level of the semiconductor is *below* the Fermi level of the metal by an amount $\phi_s - \phi_m$. After contact, electrons flow from the metal into the semiconductor until the Fermi levels of the semiconductor and the metal are at equal height. The surface layer of the semiconductor will thereby become negatively charged, and since this negative charge is caused by ionized acceptors, the charge is distributed through a *space charge layer* of thickness d. Since the energy levels in the bulk semiconductor have been raised by an amount $\phi_s - \phi_m$, the surface barrier for holes on the semiconductor side is (Fig. 5.12a)

$$\phi_s - \phi_m = eV_{\text{dif}} \tag{5.25}$$

where V_{dif} is again the diffusion potential; the potential at the interior of the semiconductor, taken with respect to the metal surface, is $-V_{\text{dif}}$. The potential barrier for holes on the metal side of the contact is therefore

$$(\phi_s - \phi_m) + (E_s - \phi_s) = (E_s - \phi_m)†$$

Owing to thermal agitation, some holes of the semiconductor will gain enough energy to cross the potential barrier into the metal, and some of the holes generated thermally in the metal will have enough energy to cross the potential barrier into the semiconductor, thus giving rise to two equal and opposite currents I_0 crossing the barrier.

---

† The figure plots the energy of an electron; the energy plot for a hole has opposite sign. A potential barrier for the holes corresponds to a minimum in the electron energy.

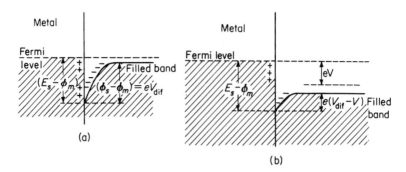

Fig. 5.12. The energy-level diagram of a metal *p*-type semiconductor contact if $\phi_m < \phi_s$; V_{dif} is the diffusion potential (a) No voltage applied. (b) A voltage $V > 0$ is applied to the semiconductor. The Fermi level in the bulk semiconductor is shifted by an amount eV.

If a voltage V is applied to the semiconductor (Fig. 5.12b), the hole current flowing from left to right has not changed; but the hole current flowing from right to left has changed by a factor exp (eV/kT), since all the energy levels in the semiconductor have been lowered by an amount eV. Hence the barrier height for holes going from right to left has been lowered by an amount eV. Consequently, if the direction from right to left is taken as the direction for positive current flow, the characteristic will be

$$I = I_0\left[\exp\left(\frac{eV}{kT}\right) - 1\right]$$

(5.26)

As before, this is a rectifying contact.

5.4c. Semiconductor–Semiconductor Contacts

If an *n*-type and a *p*-type sample of the same semiconductor material are brought into contact, electrons will flow from the *n*-type material to the *p*-type material until the Fermi levels are at equal height. We shall see that such a contact has rectifying characteristics.

Such rectifying contacts are usually made not by bringing two samples into contact but by diffusing *n*-type impurities into *p*-type material or *p*-type impurities into *n*-type material. As an end result, a structure is obtained that is partly *p* type and partly *n* type; such a structure is called a *pn* junction or *pn* diode. Figure 5.13 shows the energy-level diagram of such a diode, in which there is a sudden change from *p*-type to *n*-type material at $x = 0$. We see that there is a region of positive space charge for $-x_1 < x < 0$ due to ionized donors and a region of negative space charge for $0 < x < x_2$ due to ionized acceptors.

Owing to the space-charge region, the *n* region has a positive potential

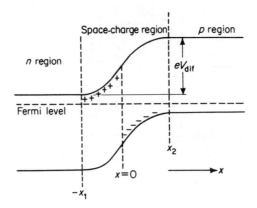

Fig. 5.13. Potential distribution in a *p-n* junction at equilibrium. The space-charge region extends between $x = -x_1$ and $x = +x_2$.

with respect to the *p* region at zero external bias. This potential difference is called the *diffusion potential* and is denoted by V_{dif}. An expression for V_{dif} is calculated in Chapter 15.

That this is a rectifying contact can be seen as follows. Let the direction of positive current flow be from right to left. In equilibrium there are two equal and opposite electron currents I_{n0} flowing to the left and to the right. The current flowing from right to left is caused by electrons moving from left to right; these electrons have to climb a potential-energy barrier of height eV_{dif}. The current flowing from left to right is caused by electrons generated in the *p* region and going downhill to the *n* region. The latter current does not change if a potential V is applied to the *p* region, but the first current depends on V as exp (eV/kT) since the potential-energy barrier is lowered by an amount eV; hence the probability for crossing the barrier varies as exp (eV/kT). Consequently, the characteristic is

$$I_n = I_{n0}\left[\exp\left(\frac{eV}{kT}\right) - 1\right] \tag{5.27}$$

and a similar expression holds for the hole currents. As before, this is a rectifying characteristic.

We now turn to contacts between samples of *different* semiconducting material. Actually such contacts are made, not by pressing two samples together, but by growing an *n*-type or *p*-type compound on another *p*-type or *n*-type compound of approximately the same lattice spacing. Such junctions are called *heterojunctions*. There are four possibilities: *nn* junctions, *np* junctions, *pn* junctions, and *pp* junctions. We shall give a few examples of such junctions.

Figure 5.14 shows two *n*-type semiconductors with work functions ϕ_1

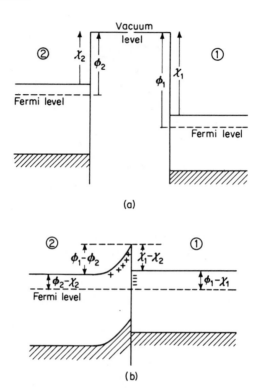

Fig. 5.14. (a) Two *n*-type semiconductors before contact. (b) The same semiconductors form a heterojunction after contact, with a space-charge region on one side of the contact.

and ϕ_2; it is here assumed that $\phi_2 < \phi_1$ and $\chi_2 < \chi_1$ (Fig. 5.14a). If the two materials are brought into contact, electrons will flow from semiconductor 2 to semiconductor 1 until the Fermi levels are at equal heights (Fig. 5.14b). There is now a potential energy barrier of height $\chi_1 - \chi_2$ on the side of semiconductor 1 and a barrier of height $\phi_1 - \phi_2$ on the side of semiconductor 2. In other words, this is a rectifying contact, even though it is of the *nn* type.

Figure 5.15 shows an *n*-type and a *p*-type semiconductor with work functions ϕ_1 and ϕ_2; it is here assumed that $\phi_2 > \phi_1$. If the two materials are brought into contact, electrons will flow from the *n*-type to the *p*-type semi-conductor, leaving ionized donors behind on the *n* side of the contact and giving rise to ionized acceptors on the *p* side. Since the gap widths are dif-ferent, there is no complete match at the contact—neither between the bottoms of the conduction band nor between the tops of the valence band. None-theless, this is also a rectifying contact.

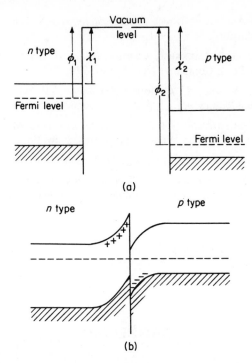

Fig. 5.15. (a) A *p*-type and an *n*-type semiconductor before contact. (b) The two materials form a heterojunction with space-charge regions on each side of the contact.

The investigation of some of the other band configurations in heterojunctions is left as a problem for the reader.

REFERENCES

BLAKEMORE, J. S., *Semiconductor Statistics*. Pergamon Press, Inc., Elmsford, N.Y., 1962.

CONWELL, E. M., "Properties of Silicon and Germanium—I and II," *Proc. IRE*, **40**, 1327 (1952); **46**, 1281 (1958).

GATOS, H. C., ed., *Properties of Elemental and Compound Semiconductors*, Metallurgical Society Conferences, vol. 5. John Wiley & Sons, Inc., New York, 1960.

GREIG, D., *Electrons in Metals and Semiconductors*. McGraw-Hill Book Company, New York, 1969.

GRUBEL, R. O., ed., *Metallurgy of Elemental and Compound Semiconductors*, Metallurgical Society Conferences, vol. 12. John Wiley & Sons, Inc., New York, 1961.

HILSUM, C., and A. C. ROSE-INNES, *Semiconducting* III–V *Compounds*. Pergamon Press, Inc., Elmsford, N.Y., 1961.

KANE, P. F., and G. B. LARRABEE, *Characterization of Semiconductor Materials*. McGraw Hill Book Company, New York, 1970.

LAX, B., and J. G. MAVROIDS, "Cyclotron Resonance," F. Seitz and D. Turnbull, eds., *Solid State Physics*, vol. 11, p. 261. Academic Press, Inc., New York, 1960.

SCHROEDER, J. B., ed., *Metallurgy of Semiconductor Materials*, Metallurgical Society Conferences, vol. 15. John Wiley & Sons, Inc., New York, 1962.

SMITH, R. A., *Semiconductors*. Cambridge University Press, New York, 1959.

WILLARDSON, R. K., and A. C. BEER, *Semiconductors and Semimetals*. Academic Press, Inc., New York. The yearly volumes contain many review papers.

WOLF, H. F., *Semiconductors*. John Wiley & Sons, Inc. (Interscience Division), New York, 1971.

WOLF, H. F., *Silicon Semiconductor Data*. Pergamon Press, Inc., Elmsford, N.Y., 1969.

PROBLEMS

1. In an n-type semiconductor $N_d = 10^{15}$ cm^3, $m_n^* = m$, and the donor levels lie 0.05 eV below the bottom of the conduction band. Find the temperature at which $n = \frac{1}{2}N_d$. *Hint:* To solve the transcendent equation resulting from Eq. (5.12a), try $T = 60°$K for the $(300/T)^{3/2}$ term. Since the exponent varies much faster than this term, only a rough value of this term is sufficient.

Answer: $T = 63°$K.

2. If $m_n^* = m$ and all donors are ionized, find the donor concentration N_d at which the Fermi level lies 0.10 eV below the bottom of the conduction band ($T = 300°$K).

Answer: $N_d = 5.3 \times 10^{17}$/cm^3.

3. If $m_n^* = m_p^* = m$, and $E_g = 1.10$ eV, find the intrinsic carrier concentration at $T = 300°$K.

Answer: $n_i = 1.4 \times 10^{10}$/cm^3.

4. Find the long-wavelength absorption limit for the semiconductor of problem 3.

Answer: 1.12 μm.

5. Show that the energy levels of a donor can be found from the Bohr theory of the hydrogen atom (problem 2.5), if we replace m by m_n^* and ϵ_0 by $\epsilon\epsilon_0$, where ϵ is the dielectric constant of the medium. Show by applying the results of problem 2.5 that

$$E_n = -\frac{13.60 m_n^*/m}{\epsilon^2 n^2} \text{ eV} \quad r_n = \frac{0.53\epsilon n^2}{m_n^*/m} \text{ Å}$$

Here E_n is the energy and r_n the radius of the nth Bohr orbit. Calculate the value of E_1 and of r_1 for silicon ($m_n^* = 1.1\ m$, $\epsilon = 12$).

Answer: $E_1 = -0.104$ eV, $r_1 = 5.8$ Å.

6

Semiconductor Electronics

6.1. CURRENT FLOW IN SEMICONDUCTORS

Current flow in semiconductors can occur in two ways: (1) when an electric field is applied, and (2) when a gradient in the carrier concentration is maintained. The first causes a current flow due to *carrier drift;* the second gives current flow due to *carrier diffusion.*

6.1a. Current Flow Due to Drift

We begin with current flow due to carrier drift in an applied electric field F. The carriers are accelerated by the electric field, gain energy, and, on the average, lose this energy during collisions with the lattice, in which they are scattered in random directions.

Let us first consider hole conduction. Let the carrier have a charge e and an effective mass m_p^*, and let it have an initial velocity component v_0 in the direction of the field F. The acceleration of the carrier is thus $(e/m_p^*) F$ in the direction of the field. If, for a particular carrier, τ happens to be the time after the last collision with the lattice, then the velocity of the carrier is

$$v_\tau = v_0 + \frac{e}{m_p^*} F \tau \tag{6.1}$$

We now average over all initial velocities v_0. If the initial velocity distribution is isotropic, the average value of v_0 is zero. Hence

$$\bar{v}_\tau = \frac{e}{m_p^*} F\tau \tag{6.1a}$$

Next we average over τ. It follows from this that the average drift velocity $\mathbf{u}_d$ is proportional to $\mathbf{F}$ and may be written as (see Sec. 6.2b)

$$\mathbf{u}_d = \frac{e}{m_p^*} \mathbf{F}\tau_0 = \mu_p \mathbf{F} \tag{6.2}$$

where τ_0 is the time constant associated with the collision process and $\mathbf{u}_d$ has the same direction as the field $\mathbf{F}$. The quantity μ_p is known as the hole mobility:

$$\mu_p = \frac{e}{m_p^*} \tau_0 \tag{6.2a}$$

It is expressed in square centimeters per volt second. The hole current density $\mathbf{J}_p$, caused by the average drift velocity $\mathbf{u}_d$, is thus

$$\mathbf{J}_p = e\mathbf{u}_d p = e\mu_p p\mathbf{F} \equiv \sigma_p \mathbf{F} \tag{6.3}$$

where p is the hole density. The quantity

$$\sigma_p = e\mu_p p \tag{6.3a}$$

is called the *hole conductivity*. Since J_p is proportional to F and $\mathbf{J}_p$ is in the direction of $\mathbf{F}$, Ohm's law holds.

Next consider electron conduction. In this case the charge of the carriers is $-e$; the effective mass is m_n^*. Going through the same calculation as before, for the average drift velocity $\mathbf{u}_d$ one obtains

$$\mathbf{u}_d = -\left(\frac{e}{m_n^*}\right)\mathbf{F}\tau_0 = -\mu_n \mathbf{F} \tag{6.4}$$

The quantity μ_n is known as the *electron mobility:*

$$\mu_n = \frac{e}{m_n^*} \tau_0 \tag{6.4a}$$

It is a positive quantity expressed in square centimeters per volt second. The electron current density $\mathbf{J}_n$ is thus

$$\mathbf{J}_n = -e\mathbf{u}_d n = e\mu_n n\mathbf{F} \equiv \sigma_n \mathbf{F} \tag{6.5}$$

where n is the electron density. The quantity

$$\sigma_n = e\mu_n n \tag{6.5a}$$

is known as the *electron conductivity*. Again, Ohm's law is valid.

If both electrons and holes are present, the current density $\mathbf{J}$ consists of a hole term and an electron term:

$$\mathbf{J} = \mathbf{J}_p + \mathbf{J}_n = (e\mu_p p + e\mu_n n)\mathbf{F} = \sigma\mathbf{F} \tag{6.6}$$

so that the conductivity σ becomes

$$\sigma = e\mu_p p + e\mu_n n \qquad (6.6a)$$

The electron and hole mobilities depend on the carrier concentration. They are constant at relatively low impurity concentrations, but decrease with increasing impurity concentrations because of scattering of the carriers by the ionized impurity centers. The electron and hole mobilities in germanium at room temperature and at relatively low concentrations are, respectively, 3900 and 1900 $cm^2/(Vs)$. For silicon the corresponding values are 1350 and 480 $cm^2/(Vs)$. It can be proved theoretically that the temperature dependence of the mobility is of the form constant $T^{-3/2}$, where T is the absolute temperature (Sec. 6.2b).

We can now understand the behavior of the conductivity as a function of temperature. If the logarithm of the conductivity σ (in siemens per centimeter) of a sample is plotted versus $1/T$, then an exponential dependence of σ upon $1/T$ is observed at very high temperatures and at very low temperatures. There is an intermediate region where σ increases with increasing value of $1/T$ (Fig. 5.6).

The intermediate region is the temperature region in which the carrier concentration has a fixed value because all donors or acceptors are ionized. The carrier concentration is equal to the donor concentration N_d in n-type material and equal to the acceptor concentration N_a in p-type material. The temperature dependence of σ is here caused by the temperature dependence of the mobility. The high-temperature region is the region where the material becomes intrinsic; here σ depends exponentially upon $1/T$ because the intrinsic carrier concentration n_i depends exponentially upon $1/T$. The low-temperature region is the region where not all donors or acceptors are ionized. In that region the number of free electrons or holes depends exponentially upon $1/T$.

The two slopes of the $\ln \sigma$ versus $1/T$ curves are quite different. We have seen that they are related to the difference in the binding energy of electrons in valence bands and of electrons (or holes) in impurity atoms.

We now solve a few problems.

Problem 1. The intrinsic resistivity of germanium at 300°K is 47 Ω cm. What is the intrinsic carrier concentration?

Answer

$$\sigma_i = e n_i (\mu_n + \mu_p)$$

$$n_i = \frac{\sigma_i}{e(\mu_n + \mu_p)} = \frac{1}{47 \times 1.6 \times 10^{-19} \times 5800} = 2.3 \times 10^{13}/cm^3$$

Problem 2. What is the donor concentration in n-type germanium of 1 Ω cm resistivity at 300°K?

Answer

$$N_d = \frac{\sigma}{e\mu_n} = \frac{1}{1.6 \times 10^{-19} \times 3900} = 1.60 \times 10^{15}/\text{cm}^3$$

6.1b. Current Flow Due to Diffusion

Next we turn to current flow by diffusion. If there is no field applied, the carriers move in random directions, collide with the lattice, and are scattered in random directions. How then can there be a net current flow? This can happen if there is a gradient in the carrier concentration.

First consider a hole conductor. Let p be the hole concentration and let there be a gradient ∇p in this concentration with components $\partial p/\partial x$, $\partial p/\partial y$, and $\partial p/\partial z$. The particle current density is then $-D_p \nabla p$, as will be shown below. The proportionality factor D_p is called the *hole diffusion constant* and is expressed in square centimeters per second. Thus we find the electrical current density $\mathbf{J}_p$ due to diffusion of holes by multiplying the particle current density by the hole charge e:

$$\mathbf{J}_p = -eD_p \nabla p \tag{6.7}$$

To prove this equation, we look in the direction of the X axis and make cross sections at x and $x + \Delta x$ of area ΔA. On the left side of the section Δx the concentration is p and on the right side $p + (\partial p/\partial x) \Delta x$ (Fig. 6.1) for small Δx. Each particle has an equal chance of going right or left, but since there are more particles at the right than at the left if $\partial p/\partial x > 0$, there will be a net particle flow from right to left proportional to ΔA and to $\partial p/\partial x$. The particle current density thus has an X component that is proportional to $-\partial p/\partial x$, for it has a direction from right to left; the proportionality factor is denoted by D_p.

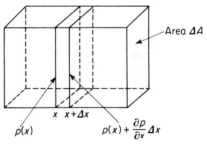

Fig. 6.1. Diagram for the derivation of the diffusion current density.

The same is true for the Y and Z components of the particle current density. Consequently, Eq. (6.7) has been proved.

Next consider an electron conductor. Applying the same reasoning as before and bearing in mind that the electrons have a charge $-e$, we get an electron current density $\mathbf{J}_n$:

$$\mathbf{J}_n = eD_n \nabla n \tag{6.8}$$

where ∇n is the electron density gradient and D_n is the electron diffusion constant.

If there are both carrier drift and carrier diffusion, then Eqs. (6.3) and

(6.5) must be combined with Eqs. (6.7) and (6.8), respectively. For the hole and electron current densities, this yields

$$\mathbf{J}_p = e\mu_p p\mathbf{F} - eD_p \nabla p \tag{6.9}$$

$$\mathbf{J}_n = e\mu_n n\mathbf{F} + eD_n \nabla n \tag{6.10}$$

Considerable diffusion current can flow if the gradient is large. For example, consider p-type germanium, in which the hole density drops linearly by $10^{15}/cm^3$ over a distance of 1 μm (10^{-4} cm). Then, since $D_p = 49$ cm^2/s,

$$J_p = eD_p\frac{\partial p}{\partial x} = 1.6 \times 10^{-19} \times 49 \times \frac{10^{15}}{10^{-4}} = 80 \text{ A/cm}^2$$

Diffusion-current densities of this magnitude can occur easily in semiconductor devices.

In an inhomogeneous semiconductor, n and p are functions of position and hence there is a concentration gradient. But in equilibrium no current flows, and hence an electric field must be set up so that the drift current balances the diffusion current everywhere. If $\psi(x, y, z)$ is the resulting potential distribution, one would expect $p(x, y, z)$ and $n(x, y, z)$ to be determined by a Boltzmann factor, so that

$$p(x, y, z) = p(0, 0, 0) \exp\left(-\frac{e\psi}{kT}\right)$$
$$n(x, y, z) = n(0, 0, 0) \exp\left(\frac{e\psi}{kT}\right) \tag{6.11}$$

where $p(0, 0, 0)$ and $n(0, 0, 0)$ are the electron and hole concentrations at the origin.

Applying this to a pn junction, which makes it a one-dimensional problem, shows that

$$p(x) = p(0) \exp\left(-\frac{e\psi}{kT}\right) \tag{6.11a}$$

where $p(0)$ is the hole concentration at $x = 0$ and $\psi(x)$ is the potential at x with respect to $x = 0$. We thus see that the equilibrium hole concentration p_n in the n region is related to the equilibrium hole concentration N_a in the p region by

$$p_n = N_a \exp\left(-\frac{eV_{\text{dif}}}{kT}\right) \tag{6.11b}$$

Here N_a is the acceptor concentration in the p region and V_{dif} is the diffusion potential, that is, the potential between the bulk p and n regions.

In equilibrium the net hole current density J_{px} must be zero. Hence, according to Eq. (6.9), for the one-dimensional case,

$$e\mu_p Fp - eD_p\frac{dp}{dx} = 0$$

But since $F = -d\psi/dx$, this equation may be written

$$-\mu_p p \, d\psi = D_p \, dp \quad \text{or} \quad -\frac{\mu_p}{D_p} d\psi = \frac{dp}{p}$$

with $p = p(0)$ at $x = 0$. The equation has the solution

$$p(x) = p(0) \exp\left[-\left(\frac{\mu_p}{D_p}\right)\psi(x)\right]$$

Equating this to (6.11a) yields the Einstein relation

$$\frac{\mu_p}{D_p} = \frac{e}{kT} \tag{6.12}$$

between μ_p and D_p. A similar relation can be proved for electrons.

6.1c. Current Equations as a Modified Version of Ohm's Law

By introducing the *electrochemical potential* or quasi-Fermi level for electrons and holes for nonequilibrium cases, one may write Eqs. (6.9) and (6.10) as an extended version of Ohm's law. To that end we define the chemical potentials for electrons and holes. To avoid confusion with the mobilities μ_n and μ_p, we add a subscript c and denote the chemical potentials for electrons and holes by μ_{cn} and μ_{cp} and the electrochemical potentials or quasi-Fermi levels by $\bar{\mu}_n = \mu_{cn} - e\psi$ and $\bar{\mu}_p = \mu_{cp} - e\psi$, where ψ is the potential. Note that this corresponds to the definition (3.66).

The definitions for μ_{cp} and μ_{cn} are

$$\mu_{cp} = kT \ln \frac{n_i}{p}, \qquad p = n_i \exp\left(-\frac{\mu_{cp}}{kT}\right) \tag{6.13}$$

$$\mu_{cn} = kT \ln \frac{n}{n_i}, \qquad n = n_i \exp\left(\frac{\mu_{cn}}{kT}\right) \tag{6.14}$$

Furthermore, introducing $D = (kT/e)\mu$, we may write Eqs. (6.9) and (6.10) as

$$J_{px} = -e\mu_p p \frac{d\psi}{dx} + \mu_p p \frac{d\mu_{cp}}{dx} = \mu_p p \frac{d\bar{\mu}_p}{dx} = \sigma_p \frac{d(\bar{\mu}_p/e)}{dx} \tag{6.15}$$

$$J_{nx} = -e\mu_n n \frac{d\psi}{dx} + \mu_n n \frac{d\mu_{cn}}{dx} = \mu_n n \frac{d\bar{\mu}_n}{dx} = \sigma_n \frac{d(\bar{\mu}_n/e)}{dx} \tag{6.16}$$

where $\sigma_p = e\mu_p p$ and $\sigma_n = e\mu_n n$ are the hole and electron conductivities, respectively. In nonequilibrium $\bar{\mu}_p$ and $\bar{\mu}_n$ are generally unequal. Only in equilibrium do they coincide and then they are independent of x, since the electrochemical potential is independent of x in that case. The quantities $\bar{\mu}_n/e$ and $\bar{\mu}_p/e$ are called the Fermi potentials or quasi-Fermi levels for electrons and holes, respectively.

We must show finally that Eqs. (6.13) and (6.14) make sense in equilibrium. To that end we consider a pn junction at zero current and set the zero point of the potential $\psi(x)$ at the intrinsic point where $p = n = n_i$.

We then have, according to (6.11),

$$p = n_i \exp\left(-\frac{e\psi}{kT}\right) \tag{6.17}$$

and in the same way

$$n = n_i \exp\left(\frac{e\psi}{kT}\right) \tag{6.18}$$

Comparing Eq. (6.17) with Eq. (6.13) and Eq. (6.18) with Eq. (6.14), we see that

$$\mu_{cp} = \mu_{cn} = e\psi(x) \quad \text{or} \quad \bar{\mu}_p = \bar{\mu}_n = 0 \tag{6.19}$$

The definitions (6.13) *and* (6.14) *thus fix the zero point of* $\bar{\mu}_p$ *and* $\bar{\mu}_n$. Since the current flow is determined by the *derivatives* of $\bar{\mu}_p$ and $\bar{\mu}_n$, this does not matter.

For most practical purposes, it is a matter of taste whether one wants to use the set of equations (6.9) and (6.10) or (6.15) and (6.16). Both are equivalent as long as the Einstein relation $D = (kT/e)\mu$ holds. If this relation does not hold, Eqs. (6.9) and (6.10) are the more fundamental ones.

6.2. BOLTZMANN TRANSPORT EQUATION

For an accurate evaluation of the mobility and diffusion constant of current carriers and for a calculation of Hall effect and magnetoresistance (Chapter 20), it is convenient to use the Boltzmann transport equation. We shall first derive this equation and then indicate how it is used.

6.2a. Derivation of the Boltzmann Transport Equation

Let an electron in a solid at the instant t have the position x, y, z and the momentum components p_x, p_y, p_z, and let

$$f(p_x, p_y, p_z, x, y, z, t) \, dp_x \, dp_y \, dp_z \, dx \, dy \, dz$$

be the number of electrons in the volume element $dx \, dy \, dz$ that at the instant t have momentum in the range p_x, p_y, p_z to $p_x + dp_x$, $p_y + dp_y$, $p_z + dp_z$.

In the steady-state condition

$$\frac{df}{dt} = 0 \tag{6.20}$$

but in nonequilibrium situations there are forces at work on the electrons. These are forces of electric and magnetic fields, forces of diffusion, and forces due to collisions with the lattice. Thus we may write

$$\frac{df}{dt} = \frac{\partial f}{\partial t}\bigg|_{\text{forces}} + \frac{\partial f}{\partial t}\bigg|_{\text{diffusion}} + \frac{\partial f}{\partial t}\bigg|_{\text{collisions}} = 0 \tag{6.21}$$

If we can now evaluate these terms, we have an equation for f.

The electric and magnetic forces change the momentum at the rate $\dot{\mathbf{p}}$; this vector has components $\dot{p}_x$, $\dot{p}_y$, $\dot{p}_z$. Consequently, during the time Δt the change in f caused by the forces is

$$\frac{\partial f}{\partial t}\bigg|_{\text{forces}} \Delta t = -\frac{\partial f}{\partial p_x}\Delta p_x - \frac{\partial f}{\partial p_y}\Delta p_y - \frac{\partial f}{\partial z}\Delta p_z$$

or
$$\frac{\partial f}{\partial t}\bigg|_{\text{forces}} = -\nabla_p f \cdot \dot{\mathbf{p}} \tag{6.22}$$

where $\nabla_p f$ is a vector with components $\partial f/\partial p_x$, $\partial f/\partial p_y$, and $\partial f/\partial p_z$.

Diffusion caused by the motion of carriers changes the position during the time Δt by $\Delta x = p_x \Delta t/m^*$, $\Delta y = p_y \Delta t/m^*$, and $\Delta z = p_z \Delta t/m^*$. The corresponding change in f is

$$\frac{\partial f}{\partial t}\bigg|_{\text{diffusion}} \Delta t = -\frac{\partial f}{\partial x}\frac{p_x \Delta t}{m^*} - \frac{\partial f}{\partial y}\frac{p_y \Delta t}{m^*} - \frac{\partial f}{\partial z}\frac{p_z \Delta t}{m^*}$$

or
$$\frac{\partial f}{\partial t}\bigg|_{\text{diffusion}} = -\frac{\mathbf{p} \cdot \nabla_r f}{m^*} \tag{6.23}$$

where $\nabla_r f$ is a vector with components $\partial f/\partial x$, $\partial f/\partial y$, and $\partial f/\partial z$.

Collisions tend to bring the system back to the thermal equilibrium state present in the absence of applied fields. It is customary to put

$$\frac{\partial f}{\partial t}\bigg|_{\text{collisions}} = -\frac{f(\mathbf{p}, \mathbf{r}) - f_0}{\tau(\mathbf{p}, \mathbf{r})} \tag{6.24}$$

where f_0 is the distribution function for thermal equilibrium, and τ is a time constant, indicating the rate at which f relaxes toward f_0. Equation (6.24) makes sense, for if $\tau(\mathbf{p}, \mathbf{r})$ is independent of $\mathbf{p}$ and $\mathbf{r}$, a sudden removal of the forces should give a decay to equilibrium of the form

$$(f - f_0)_t = (f - f_0)_{t=0} \exp\left(-\frac{t}{\tau}\right) \tag{6.24a}$$

and that is exactly what Eq. (6.24) describes.

Equation (6.21) may be written

$$-\dot{\mathbf{p}} \cdot \nabla_p f - \left(\frac{\mathbf{p}}{m^*}\right) \cdot \nabla_r f - \frac{(f - f_0)}{\tau} = 0 \tag{6.25}$$

This is the form of Boltzmann's transport equation needed in our discussion. If electric and magnetic fields are applied, then for electrons

$$\dot{\mathbf{p}} = -e(\mathbf{F} + \mathbf{v} \times \mathbf{B}) \tag{6.26}$$

where $\mathbf{F}$ is the electric field, $\mathbf{B}$ the magnetic induction, and $\mathbf{v}$ the electron velocity.

In the particular case of small deviations from equilibrium, very little error ensues if we replace $\nabla_p f$ and $\nabla_r f$ by $\nabla_p f_0$ and $\nabla_r f_0$. The solution of

(6.25) is then seen to be

$$f - f_0 = -\tau \dot{\mathbf{p}} \cdot \nabla_p f_0 - \tau\left(\frac{\mathbf{p}}{m^*}\right) \cdot \nabla_r f_0 \qquad (6.25a)$$

This solution must be applied with considerable caution, however.

6.2b. Derivation of the Current Equation in Semiconductors

If an electric field $\mathbf{F}$ is applied in the Z direction and if the gradient in the carrier concentration has only a Z component, then (6.25a) may be written

$$f - f_0 = +eF_z\tau\frac{\partial f_0}{\partial p_z} - \frac{p_z\tau}{m^*}\frac{\partial f_0}{\partial z} \qquad (6.27)$$

when $|\mathbf{F}| = F_z$. In an n-type semiconductor the velocity distribution is nearly Maxwellian, and hence we write, if the constant-energy surfaces are spherical,

$$f_0 = \frac{n}{(2\pi m^* kT)^{3/2}} \exp\left(-\frac{p^2}{2m^* kT}\right) \qquad (6.28)$$

where $p^2 = p_x^2 + p_y^2 + p_z^2$. We thus have

$$\int_{-\infty}^{\infty}\int_{-\infty}^{\infty}\int_{-\infty}^{\infty} f_0 \cdot dp_x\, dp_y\, dp_z = \int_0^{\infty} f_0 \cdot 4\pi p^2\, dp = n \qquad (6.29)$$

and the average of an arbitrary function $g(p)$ of p is

$$\langle g(p)\rangle = \frac{1}{n}\int_{-\infty}^{\infty}\int_{-\infty}^{\infty}\int_{-\infty}^{\infty} g(p)f_0\, dp_x\, dp_y\, dp_z = \frac{1}{n}\int_0^{\infty} g(p)f_0 \cdot 4\pi p^2\, dp \qquad (6.30)$$

Since the field is applied in the Z direction, the current flows in the Z direction. Putting $J = J_z$, we have

$$\begin{aligned}
J_z &= \int_{-\infty}^{\infty}\int_{-\infty}^{\infty}\int_{-\infty}^{\infty} -ev_z f\, dp_x\, dp_y\, dp_z \\
&= \int_{-\infty}^{\infty}\int_{-\infty}^{\infty}\int_{-\infty}^{\infty} -\frac{e}{m^*}p_z(f - f_0)\, dp_x\, dp_y\, dp_z
\end{aligned} \qquad (6.31)$$

since

$$n\langle p_z\rangle = \int_{-\infty}^{\infty}\int_{-\infty}^{\infty}\int_{-\infty}^{\infty} p_z f_0\, dp_x\, dp_y\, dp_z = 0 \qquad (6.32)$$

Substituting (6.27) and (6.28) into (6.31), bearing in mind that

$$\frac{\partial f_0}{\partial p_z} = -\frac{p_z}{m^* kT}f_0, \qquad \frac{\partial f_0}{\partial z} = \frac{1}{n}\frac{\partial n}{\partial z}f_0 \qquad (6.33)$$

and introducing polar coordinates p, ϑ, and ϕ instead of p_x, p_y, and p_z, we obtain

$$J_z = \left(\frac{1}{3}\frac{e^2 F_z}{m^{*2} kT} + \frac{1}{3}\frac{e}{m^* n}\frac{\partial n}{\partial z}\right)\int_0^{\infty} p^2\tau f_0 \cdot 4\pi p^2\, dp \qquad (6.34)$$

The proof is as follows. Substituting (6.27) into (6.31) yields

$$J_z = -\frac{e^2 F_z}{m^*} \int_{-\infty}^{\infty} \int_{-\infty}^{\infty} \int_{-\infty}^{\infty} p_z \tau \frac{\partial f_0}{\partial p_z} dp_x \, dp_y \, dp_z$$

$$+ \frac{e}{m^*} \int_{-\infty}^{\infty} \int_{-\infty}^{\infty} \int_{-\infty}^{\infty} p_z^2 \tau \frac{\partial f_0}{\partial z} dp_x \, dp_y \, dp_z$$

$$\int_{-\infty}^{\infty} \int_{-\infty}^{\infty} \int_{-\infty}^{\infty} -p_z \tau \frac{\partial f_0}{\partial p_z} dp_x \, dp_y \, dp_z = \frac{1}{m^* kT} \int_{-\infty}^{\infty} \int_{-\infty}^{\infty} \int_{-\infty}^{\infty} p_z^2 \tau f_0 \, dp_x \, dp_y \, dp_z$$

$$\int_{-\infty}^{\infty} \int_{-\infty}^{\infty} \int_{-\infty}^{\infty} p_z^2 \tau \frac{\partial f_0}{\partial z} dp_x \, dp_y \, dp_z = \frac{1}{n} \frac{\partial n}{\partial z} \int_{-\infty}^{\infty} \int_{-\infty}^{\infty} \int_{-\infty}^{\infty} p_z^2 \tau f_0 \, dp_x \, dp_y \, dp_z$$

$$\int_{-\infty}^{\infty} \int_{-\infty}^{\infty} \int_{-\infty}^{\infty} p_z^2 \tau f_0 \, dp_x \, dp_y \, dp_z = \int_0^{\infty} \int_{-\pi/2}^{\pi/2} \int_0^{2\pi} \tau p^2 \cos^2 \vartheta f_0 p^2 \, dp \sin \vartheta \, d\vartheta \, d\varphi$$

$$= \tfrac{1}{3} \int_0^{\infty} p^2 \tau f_0 \cdot 4\pi p^2 \, dp$$

since

$$\int_{-\pi/2}^{\pi/2} \cos^2 \vartheta \sin \vartheta \, d\vartheta = \tfrac{2}{3} \quad \text{and} \quad \int_0^{2\pi} d\varphi = 2\pi$$

Since the integral is equal to $n\langle p^2 \tau \rangle = nm^{*2}\langle v^2 \tau \rangle$, Eq. (6.34) may be written[†]

$$J_z = e\frac{\langle v^2 \tau \rangle}{3kT} nF_z + e\frac{\langle v^2 \tau \rangle}{3}\frac{\partial n}{\partial z} = e\mu_n nF_z + eD_n \frac{\partial n}{\partial z} \qquad (6.34a)$$

Hence

$$\mu_n = \frac{e\langle v^2 \tau \rangle}{3kT}, \qquad D_n = \frac{\langle v^2 \tau \rangle}{3} \qquad (6.35)$$

so that

$$\frac{\mu_n}{D_n} = \frac{e}{kT}$$

This is again the *Einstein relation* between μ_n and D_n.

We have here proved the current equation for electrons. The current equation for holes is proved in the same manner.

In the particular case that τ is independent of v and equal to τ_0, we have

$$\mu_n = \frac{e}{m_n^*}\tau_0, \qquad D = \frac{kT}{m_n^*}\tau_0 \qquad (6.36)$$

which corresponds to Eq. (6.4a). For isotropic scattering and constant free path length l, we have $\tau = l/v$; hence

$$\mu_n = \frac{el\langle v \rangle}{3kT} = el\left(\frac{8}{9\pi m_n^* kT}\right)^{1/2} \qquad (6.37)$$

[†] The reader should notice that the second half of Eq. (6.34a) remains valid if f_0 is not Maxwellian. The current equations (6.9) and (6.10) are generally valid even for non-Maxwellian particles.

since, according to Eq. (6.28),

$$\langle v \rangle = \frac{\langle p \rangle}{m_n^*} = \frac{1}{nm_n^*} \int_0^\infty p f_0 \cdot 4\pi p^2 \, dp = \left(\frac{8kT}{\pi m_n^*}\right)^{1/2} \tag{6.37a}$$

The case of constant free path length l holds, for example, for lattice scattering in semiconductors.

We now calculate the collision cross section for lattice vibrations. If we look at the vibration of an atom in a plane perpendicular to the Z direction of motion, then according to the equipartition law

$$\tfrac{1}{2} f_x \overline{x^2} = \tfrac{1}{2} f_y \overline{y^2} = \tfrac{1}{2} kT \tag{6.38}$$

where f_x and f_y are the binding constants for vibrations in the X and Y directions in that plane. Since the vibrations in the X and Y directions are independent, the collision cross section for lattice vibrations would be (compare Sec. 6.3 for the meaning of the term "collision cross section")

$$\sigma_c = \text{const} \,(\overline{x^2} \cdot \overline{y^2})^{1/2} = \text{const} \cdot kT \tag{6.38a}$$

and hence, since $l \propto 1/\sigma_c$,

$$l = \frac{\text{const}}{\sigma_c} = \frac{\text{const}}{kT} \tag{6.38b}$$

This relationship is also obtained from a detailed wave-mechanical analysis, which also derives the value of the constant. It breaks down at low temperatures, because the equipartition law does not hold in that case.

Substituting into (6.37) yields

$$\mu_n = \mu_{nl} = \frac{\text{const}}{T^{3/2}} \tag{6.38c}$$

This temperature dependence is found in some semiconductors, but it is not always valid since other scattering mechanisms can play a part. We observe that for materials in which the mobility is proportional to $T^{-3/2}$ the intrinsic conductivity σ_i is exactly proportional to $\exp(-eE_g/2kT)$. The reason is that $\sigma_i = e(\mu_n + \mu_p)n_i$, and n_i is proportional to $T^{3/2} \exp(-eE_g/2kT)$, as is seen from Eq. (5.19a).

One reason why Eq. (6.38c) may not hold is that scattering due to the ionized impurity centers can predominate. It is easily seen that the free path length should be inversely proportional to the density N_i of the ionized impurity centers. A detailed calculation shows that

$$\mu_n = \mu_{ni} = \frac{\text{const} \cdot T^{3/2}}{N_i} \tag{6.39}$$

which is quite a different temperature dependence than lattice scattering. Because the mobility is inversely proportional to N_i, the effect of the ionized impurity centers predominates at high impurity densities and low temperatures. This explains, for example, why the mobility in semiconductors decreases with increasing impurity concentration.

If both lattice scattering and impurity scattering are important, the mobility μ_n is approximately given by the equation

$$\frac{1}{\mu_n} = \frac{1}{\mu_{nl}} + \frac{1}{\mu_{ni}} \tag{6.39a}$$

The reason is that if two processes with time constants τ_1 and τ_2 act simultaneously, then the net time constant follows from [compare Eq. (6.43a)]

$$\frac{1}{\tau} = \frac{1}{\tau_1} + \frac{1}{\tau_2} \tag{6.39b}$$

The temperature dependence of the impurity scattering mobility can be understood as follows. A classical treatment of the scattering problem, which corresponds to the well-known Coulomb scattering problem of nuclear physics, shows that the time constant τ_i of impurity scattering varies as v^3, so that low-energy electrons are scattered much more strongly. Consequently,[†]

$$\langle v^2 \tau_i \rangle = \text{const} \langle p^5 \rangle = \text{const } T^{5/2} \tag{6.40}$$

and therefore

$$\mu_i = \frac{e \langle v^2 \tau_i \rangle}{3kT} = \text{const } T^{3/2} \tag{6.40a}$$

The lattice-scattering time constant τ_l varies as $1/v$, and the impurity-scattering time constant τ_i varies as v^3. For a given v, it is still true that

$$\frac{1}{\tau} = \frac{1}{\tau_l} + \frac{1}{\tau_i} \tag{6.40b}$$

but since the τ's in (6.39b) are obtained by averaging over v, one cannot expect Eqs. (6.39b) and (6.39a) to be accurate under all circumstances.

6.3. RECOMBINATION OF HOLE–ELECTRON PAIRS

6.3a. Direct Recombination

Equation (5.6) not only predicts equilibrium conditions, but it also describes how the carrier concentration varies with time if there is originally an excess carrier concentration. Suppose that by some mechanism excess hole–electron pairs are added, changing n from n_0 to $n_0 + \Delta n_0$ and p from

---

[†]
$$\langle p^5 \rangle = \frac{1}{(2\pi m^* kT)^{3/2}} \int_0^\infty p^5 \exp\left(-\frac{p^2}{2m^* kT}\right) \cdot 4\pi p^2 \, dp$$

$$= \frac{2^{1/2}}{\pi^{1/2}} (2m^* kT)^{5/2} \int_0^\infty u^7 \exp(-u^2) \, du$$

which varies as $T^{5/2}$, as stated. Here $u = p/(2m^* kT)^{1/2}$.

p_0 to $p_0 + \Delta n_0$. Since $g = \rho p_0 n_0$, and p_0 and n_0 are constants, Eq. (5.6) yields

$$\frac{d \Delta n}{dt} = \rho[p_0 n_0 - (p_0 + \Delta n)(n_0 + \Delta n)] \simeq -\rho(p_0 + n_0) \Delta n \quad (6.41)$$

if Δn is sufficiently small so that the term $-\rho(\Delta n)^2$ can be neglected. If we substitute

$$\rho(p_0 + n_0) = \frac{1}{\tau} \quad (6.41a)$$

Eq. (6.41) may be written

$$\frac{d \Delta n}{dt} = -\frac{\Delta n}{\tau} \quad (6.41b)$$

With the initial condition $\Delta n = \Delta n_0$ at $t = 0$, the solution is

$$\Delta n = \Delta n_0 \exp\left(-\frac{t}{\tau}\right) \quad (6.42)$$

as can be seen by substitution. The quantity τ is called the *lifetime of the excess carriers*. If Δn_0 is comparable to n_0 and p_0, Eq. (6.41) becomes non-linear and the behavior of the carrier concentration is more complex.

Because of the nature of Eq. (6.41b), we can easily see how it will be modified if there are two competing processes with time constants τ_1 and τ_2. Then with both processes at work

$$\frac{d \Delta n}{dt} = -\frac{\Delta n}{\tau_1} - \frac{\Delta n}{\tau_2} = -\frac{\Delta n}{\tau} \quad (6.43)$$

so that the lifetime τ is related to τ_1 and τ_2 as

$$\frac{1}{\tau} = \frac{1}{\tau_1} + \frac{1}{\tau_2} \quad (6.43a)$$

We apply this as follows. In many semiconductors the rate of recombination at the surface is much larger than in the interior. If τ_v is the lifetime for volume recombination and τ_s is the lifetime for surface recombination,† for the actual lifetime τ due to the combination of both processes, we have

$$\frac{1}{\tau} = \frac{1}{\tau_v} + \frac{1}{\tau_s} \quad (6.43b)$$

The lifetime τ_s depends upon the sample geometry and upon the physical condition of the surface. Its magnitude can be changed by surface operations such as etching and sandblasting, and where it predominates (is faster) Eq. (6.43b) becomes

$$\frac{1}{\tau} = \frac{1}{\tau_s} \quad (6.43c)$$

---

† It is supposed here that surface recombination can be described in good approximation by an exponential decay $\exp(-t/\tau_s)$.

6.3b. Electron and Hole Traps; Recombination Centers

Suppose that an n-type semiconductor has *localized electron energy levels* far below the bottom of the conduction band and far above the top of the filled band that are normally *empty*. Such empty localized energy levels can act as *electron traps;* free electrons wandering through the crystal may enter the neighborhood of such a localized lattice disturbance and be captured, with release of the excess energy (as by radiation). Any unoccupied energy level of that character may act as an electron trap, but acceptor levels close to the top of the filled band are usually filled so that they do not do so.

In a similar way one can have *hole traps* in a p-type semiconductor. They represent localized electron energy levels far below the bottom of the conduction band and far above the top of the filled band that are normally *filled*. A free hole entering the neighborhood of such a localized lattice disturbance may disappear because the electron in that occupied energy level may make a transition to that hole in the filled band under release of the energy difference. Any occupied energy level of that character may act as a hole trap, but donor levels close to the bottom of the conduction band are usually empty and thus do not act as hole traps.

Many semiconductor devices are based upon the principle of injection of holes into n-type material or injection of electrons into p-type material. Such injected carriers are called *minority carriers* to distinguish them from the *majority carriers*. These minority carriers will sooner or later recombine with the majority carriers and thus be permanently removed. In principle such a recombination might take place in a single step according to the process

free electron + free hole $\rightleftarrows$ electron bound in valence band

but in many cases such a recombination occurs in a two-step process involving *recombination centers*.

A *recombination center* is a center that can act as an electron trap when empty and as a hole trap when filled. Let holes be injected into n-type material and let a particular center be occupied by an electron. It may then act as a hole trap; in the trapping process the electron combines with the free hole, with release of the excess energy, thus leaving the center empty. The center can now act as an electron trap; in the trapping process an electron from the conduction band makes a transition to the empty center, with release of the excess energy. One hole–electron pair is thus taken out of circulation after completion of such a two-step process (Fig.6.2).

In principle all electron and hole traps can act as recombination centers; the only distinction is in the recombination probabilities involved. If a center has a large recombination probability with free electrons when empty and a small recombination probability with free holes when filled, it acts

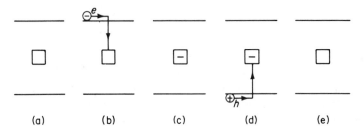

Fig. 6.2. Operation of a recombination center. (a) Empty (neutral) level.
(b) Electron captured by center. (c) Center now negatively charged. (d)
Hole captured by center. (e) Original (neutral) center restored.

mainly as an *electron trap*. If the opposite is true, the center acts mainly
as a *hole trap*. If both recombination probabilities are comparable, the center
acts as a *recombination center*.

In the case of back-biased space-charge regions, the centers may act as
generation centers, alternately generating a free electron and a free hole.

6.3c. Kinetics of Electron Traps

To illustrate some useful concepts concerning the influence of traps in
semiconductors, we use a simple example consisting of N_t electron traps per
cubic meter at a depth E_n below the bottom of the conduction band, n free
electrons per cubic meter, and n_t trapped electrons per cubic meter, and
assume that the traps interact only with electrons in the conduction band
according to the process

$$\text{free electron} + \text{empty trap} \rightleftharpoons \text{filled trap}$$

We assume further that the traps do not interact with electrons in the valence
band, nor with free holes in that band.†

If the distribution of free and trapped electrons is disturbed in one way
or the other, the return to equilibrium is described by the equations (Fig.
6.3a)

$$\frac{dn}{dt} = -vs(N_t - n_t)n + v_0 \exp\left(-\frac{eE_n}{kT}\right)n_t$$

$$\frac{dn_t}{dt} = -\frac{dn}{dt} = +vs(N_t - n_t)n - v_0 \exp\left(-\frac{eE_n}{kT}\right)n_t$$

(6.44)

where v is the thermal velocity of the free carriers and s is the "capture cross

---

† This example occurs if the semiconductors contain both donors and acceptors, with
the donors predominating, in addition to electron traps. If the donor levels are close to
the bottom of the conduction band and the acceptor levels are close to the top of the
valence band, all the donor and acceptor ions will be ionized and the remaining electrons
will be distributed over the conduction band and the traps.

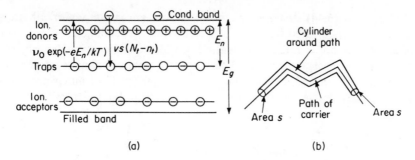

Fig. 6.3. (a) Energy level diagram showing ionized donors and acceptors, the remaining electrons being distributed over the traps and the conduction band. The probabilities for exciting a trapped electron and for capturing an electron are also shown. (b) A cylinder of cross-sectional area s is drawn around the path of the carrier, to show that the capture probability of the electron is $vs(N_t - n_t)$.

section" of the traps, usually expressed in square meters. The equations for dn/dt and for dn_t/dt are not independent. Adding the two equations yields

$$\frac{d}{dt}(n + n_t) = 0 \quad \text{or} \quad n + n_t = n_0 \tag{6.44a}$$

where n_0 is the total number of electrons in the "free" and the "trapped" states. We thus have to prove only that free electrons disappear by trapping at the rate $vs(N_t - n_t)n$ and that trapped electrons are released at the rate $v_0 \exp(-eE_n/kT)n_t$.

The first term in dn/dt can be understood as follows. Let a free carrier and a trap recombine if their average distance is less than r_0. We draw a cylinder of radius r_0 around the path; then recombination will occur if an empty trap is contained inside the cylinder (Fig. 6.3b). The cross-sectional area $s = \pi r_0^2$ of the cylinder is called the *capture cross section* of the trap; the volume of the cylinder traced per second is vs. The density of empty traps is $N_t - n_t$, and hence the number of recombinations per free carrier per second is $vs(N_t - n_t)$. Multiplying by n, we obtain the net trapping rate per second, which proves the first term of dn/dt.

To understand the second term of dn/dt, we observe that a trapped electron is released by interaction with the lattice vibrations. The release frequency v of a trapped electron is

$$v = v_0 \exp\left(-\frac{eE_n}{kT}\right) \tag{6.44b}$$

where v_0 is a vibration frequency, usually of the order of $10^{12}/s$; $\exp(-eE_n/kT)$ is a Boltzmann factor, representing the probability that the trapped electron gains an energy E_n per vibration period. The net rate of release of trapped carriers is thus vn_t, corresponding to the second part of dn/dt.

Equation (6.44) describes phenomena that are common for all traps; there is always a rate of capture of free carriers, depending upon a capture cross section, and a rate of thermal release of trapped carriers described by an activation energy (here E_n). There are two time constants τ_1 and τ_2 involved in the process, τ_1 being the average "free time" of an electron, and τ_2 being the average "captured time" of an electron. We see from (6.44) that

$$\frac{1}{\tau_1} = vs(N_t - n_t), \qquad \frac{1}{\tau_2} = v_0 \exp\left(-\frac{eE_n}{kT}\right) \qquad (6.45)$$

The quantity τ_2 is independent of the occupancy of the traps, whereas τ_1 depends upon the trap occupancy.

We may also calculate the equilibrium concentrations from (6.44) by putting $dn/dt = dn_t/dt = 0$. After some rearrangement of terms, this yields

$$n = n_0 \frac{\tau_1}{\tau_1 + \tau_2}, \qquad n_t = n_0 \frac{\tau_2}{\tau_1 + \tau_2} \qquad (6.45a)$$

These equations are only useful as long as $n_t \ll N_t$, so that τ_1 is independent of n_t. Otherwise, it is better not to use the quantity τ_1 at all.

Introducing τ_1 and τ_2 into Eq. (6.44), adding a carrier density Δn_0, and bearing in mind that $\Delta n + \Delta n_t = \Delta n_0$, we have

$$\frac{d\,\Delta n}{dt} = -\frac{\Delta n}{\tau_1} + \frac{\Delta n_t}{\tau_2} = -\Delta n\left(\frac{1}{\tau_1} + \frac{1}{\tau_2}\right) + \frac{\Delta n_0}{\tau_2} = -\frac{\Delta n}{\tau} + \frac{\Delta n_0}{\tau_2} \qquad (6.46)$$

where τ is the time constant of the added carriers

$$\frac{1}{\tau} = \frac{1}{\tau_1} + \frac{1}{\tau_2} \qquad (6.46a)$$

Finally, we calculate some numerical values for τ_1 and τ_2. First, we shall discuss the capture cross section s. A commonly reported capture cross section is 10^{-19} m². This is comparable to the cross-sectional area of an atom; one would expect such a value for a neutral trap. Very small capture cross sections, as low as 10^{-26} m², have been reported in certain cases; these small cross sections are ascribed to charged traps *repelling* the carrier. Very large capture cross sections, as high as 10^{-17} m², have been ascribed to charged traps *attracting* the carrier; this capture cross section is temperature-dependent.†

Next we shall discuss the value of N_t. Values as low as 10^{17}/m³ have been reported for highly purified single crystals of Ge; less highly purified crystals have $N_t = 10^{20}$ to 10^{22}/m³; and an upper limit may be taken as

---

† The average kinetic energy of a carrier due to its motion perpendicular to a certain direction is kT; the potential energy of a carrier of charge $-e$ at a distance r from an attractive trap of charge $+e$ in a medium of dielectric constant ϵ is $-e^2/(4\pi\epsilon_0\epsilon r)$, where $\epsilon_0 = 8.85 \times 10^{-12}$ F/m. The carrier will be absorbed if the sum of the potential and kinetic energy is negative. The capture cross section $s = \pi r_0^2$, where r_0 follows from $e^2/(4\pi\epsilon\epsilon_0 r_0) = kT$.

$10^{25}/m^3$. We now calculate τ_1 under the assumption that nearly all traps are empty; much larger values may be obtained if nearly all traps are filled.

Substituting median values ($N_t = 10^{21}/m^3$, $s = 10^{-19}$ m^2) and putting $v = 10^5$ m/s yields $\tau_1 \simeq 10^{-7}$ s; the values of τ_1 may range from 10^4 s for very pure samples with traps of small cross section ($N_t = 10^{17}/m^3$, $s = 10^{-26}$ m^2) to values of 10^{-13} s for very impure samples with traps of very large cross section ($N_t = 10^{25}/m^3$, $s = 10^{-17}$ m^2). The time constant τ_2 is small if E is small (shallow traps) and very large if E is large (deep traps).

The assumption that the empty electron trap does not interact with electrons of the valence band and that the filled electron trap does not interact with free holes is not justified for recombination centers. Hence we shall not use this assumption in the next section.

6.3d. Kinetics of Recombination Centers; The Shockley–Read Theory

In this section we shall follow the treatment given by Shockley and Read, since we need their expression for the recombination rate of electrons and holes via recombination centers (Fig. 6.4).

We first calculate the rate of trapping electrons from the conduction band. If n is the electron concentration and f_t the fraction of traps occupied by electrons, the capture rate R_c may be written in the form

$$R_c = C_n n(1 - f_t) \tag{6.47}$$

where C_n gives the capture rate per electron if all traps are empty. The rate of emission R_e from the traps will be proportional to f_t; hence

$$R_e = C_n' f_t \tag{6.47a}$$

In equilibrium

$$R_c = R_e \quad \text{or} \quad C_n' = \frac{C_n n_0(1 - f_{t0})}{f_{t0}} \tag{6.48}$$

where n_0 and f_{t0} are the equilibrium values of n and f_t. Since f_{t0} follows from the Fermi function, we have

$$\frac{1}{f_{t0}} = \exp\left(\frac{E_t - E_f}{kT}\right) + 1, \qquad \frac{1 - f_{t0}}{f_{t0}} = \exp\left(\frac{E_t - E_f}{kT}\right) \tag{6.48a}$$

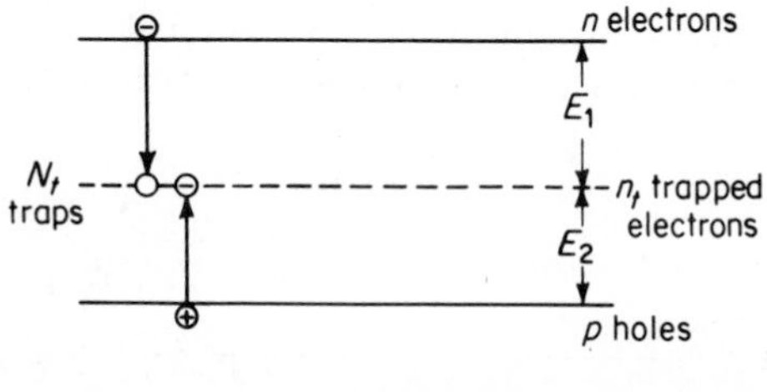

Fig. 6.4. Schematic energy level diagram of a recombination center. $f_t = n_t/N_t$.

where E_t is the energy of the trapping level. Also, if the system is nonde-generate [see Eq. (5.13)],

$$n_0 = N_c \exp\left(\frac{E_f}{kT}\right), \qquad N_c = 2\left(\frac{2\pi m_n^* kT}{h^2}\right)^{3/2} \tag{6.49}$$

Consequently,

$$C_n' = C_n N_c \exp\left(\frac{E_t}{kT}\right) = n_1 C_n \tag{6.49a}$$

where $n_1 = N_c \exp(E_t/kT)$ is the density of electrons in the conduction band when the Fermi level coincides with the trapping level. The net rate of trapping may thus be written as

$$R_n = R_c - R_e = C_n[(1 - f_t)n - n_1 f_t] \tag{6.50}$$

where C_n is the capture rate per electron if all traps are empty. In exactly the same way the net trapping rate for holes may be written in the form

$$R_p = C_p[f_t p - p_1(1 - f_t)] \tag{6.51}$$

where $p_1 = N_v \exp[-(E_g + E_t)/kT]$ is the density of holes in the valence band when the Fermi level coincides with the trapping level, and C_p the capture rate per hole if all traps are filled.

If we now have a disturbing influence that creates hole–electron pairs at the constant rate R, and steady-state conditions have been established, then

$$R = R_n = R_p \tag{6.52}$$

Substituting (6.50) and (6.51), we find that

$$f_t = \frac{nC_n + p_1 C_p}{C_n(n + n_1) + C_p(p + p_1)} \tag{6.52a}$$

because $C_n[n - f_t(n + n_1)] = C_p[f_t(p + p_1) - p_1]$.

Substituting (6.52a) back into equations (6.50) or (6.51) and bearing in mind that $n_1 p_1 = n_i^2$, the recombination rate R may be written as†

$$R = \frac{pn - n_i^2}{(n + n_1)\tau_{p0} + (p + p_1)\tau_{n0}} \tag{6.53}$$

where $\tau_{n0} = 1/C_n$ and $\tau_{p0} = 1/C_p$. This is the *Shockley–Read formula*.

To find the meaning of τ_{p0}, we consider n-type material having a relatively small deviation from equilibrium. That is, we put $p = p_0 + \Delta p, n = n_0 + \Delta p$ (space-charge neutrality, Sec. 6.5), and require that $n_0 \gg p_0, n_0 \gg n_1$, and

---

† Since

$$R = C_n[n - f_1(n + n_1)] = C_n\left[n - \frac{(nC_n + p_1 C_p)(n + n_1)}{C_n(n + n_1) + C_p(p + p_1)}\right]$$

$$= \frac{C_n C_p[n(p + p_1) - (n + n_1)p_1]}{C_n(n + n_1) + C_p(p + p_1)} = \frac{(np - n_1 p_1)C_n C_p}{C_n(n + n_1) + C_p(p + p_1)}$$

dividing by $C_n C_p$ and making the suggested substitutions yields (6.53).

$n_0 \gg \Delta p$. Since $n_0 p_0 = n_i^2$, Eq. (6.53) becomes

$$R \simeq \frac{(n_0 + p_0) \Delta p}{n_0 \tau_{p0}} \simeq \frac{\Delta p}{\tau_{p0}} \qquad (6.53\text{a})$$

Hence τ_{p0} is the *lifetime of added holes* in relatively strongly n-type material. In the same way τ_{n0} is the *lifetime of added electrons* in relatively strong p-type material. These are called Shockley–Read (S.R.) lifetimes.

When large numbers of minority carriers are injected into the material, then $pn \gg n_i^2$ and $p \simeq n$. Then, since $n \gg n_1$ and $p \gg p_1$,

$$R \simeq \frac{p^2}{p \tau_{p0} + p \tau_{n0}} \simeq \frac{p}{\tau_{p0} + \tau_{n0}} \qquad (6.53\text{b})$$

The time constant $(\tau_{p0} + \tau_{n0})$ is called the *lifetime at high injection*.
The rate equations of the system of Fig. 6.4 are

$$\frac{dn}{dt} = -a(N_t - n_t)n + bn_t \qquad (6.54)$$

$$\frac{dp}{dt} = -cn_t p + d(N_t - n_t) \qquad (6.55)$$

with the condition that $n + n_t - p$ is a constant for the system. Here a, b, c, and d are constants, N_t is the trap density, and n_t the trapped electron density. This set of equations has two time constants. If $N_t \ll n$, or $N_t \ll p$, or both, there is a short time constant τ_1 and a long, predominant time constant τ_2, corresponding to the S.R. lifetime. For strongly n-type material the predominant lifetime is $\tau_{p0} = (cn_{t0})^{-1}$; for strongly p-type material the predominant lifetime is $\tau_{n0} = [a(N_t - n_{t0})]^{-1}$, where n_{t0} is the equilibrium trapped electron density. The Shockley–Read lifetime evaluated earlier is thus the predominant time constant of the system.

6.4. CONTINUITY EQUATIONS

6.4a. Derivation of the Continuity Equations

If in a semiconductor sample there is an excess concentration of minority carriers, then the excess carrier concentration can disappear in two ways: by recombination with majority carriers and by minority carrier flow. We now investigate this problem in more detail.

Consider first the case of excess holes in n-type material. Let p_n be the equilibrium hole concentration in the material, let p be the actual hole concentration, and let τ_p be the hole lifetime in the material. If there is no net flow of holes, and if $p \ll n$, then the rate of change of the hole concentration will be [see, for example, Eq. (6.41b)]

$$\frac{\partial p}{\partial t} = -\frac{p - p_n}{\tau_p}$$

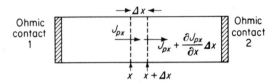

Fig. 6.5. Illustration of the derivation of the continuity equation.

If there is a net flow of holes, then a term must be added. Consider current flow in the X direction. If $J_{px}(x)$ is the X component of the hole current density at x, then the X component of that current density at $x + \Delta x$ is (Fig. 6.5)

$$J_{px}(x + \Delta x) = J_{px}(x) + \frac{\partial J_{px}}{\partial x} \Delta x$$

for small Δx. The net number of holes arriving per second in a volume element of unit cross section and width Δx is $-(\partial J_{px}/\partial x)\,\Delta x$ (the minus sign is used because more holes leave the section than arrive if $\partial J_{px}/\partial x > 0$). Dividing by Δx, we write the net rate of increase in the hole density due to current flow in the X direction as

$$-\frac{1}{e} \frac{\partial J_{px}}{\partial x} \tag{6.56}$$

If J_{py} and J_{pz} are the Y and Z components of the hole current density, then they give contributions

$$-\frac{1}{e} \frac{\partial J_{py}}{\partial y} \quad \text{and} \quad -\frac{1}{e} \frac{\partial J_{pz}}{\partial z} \tag{6.56a}$$

to the net rate of increase in the hole density. The total rate of change of the hole concentration is thus

$$\frac{\partial p}{\partial t} = -\frac{p - p_n}{\tau_p} - \frac{1}{e} \nabla \cdot \mathbf{J}_p \tag{6.57}$$

where
$$\nabla \cdot \mathbf{J}_p = \frac{\partial J_{px}}{\partial x} + \frac{\partial J_{py}}{\partial y} + \frac{\partial J_{pz}}{\partial z} \tag{6.57a}$$

This equation is known as the *continuity equation for holes in n-type material.* It has the following simple interpretation. The minority carrier density in a small volume element ΔV can decrease in two ways: either by recombination with majority carriers (first term), or because more carriers are swept out of the volume than arrive at it (second term).

Next consider the case of excess electrons in *p*-type material. If n_p is the equilibrium electron concentration in that material, n the actual electron concentration, and τ_n the electron lifetime, then one has instead of Eq. (6.57)

$$\frac{\partial n}{\partial t} = -\frac{n - n_p}{\tau_n} + \frac{1}{e} \nabla \cdot \mathbf{J}_n \tag{6.58}$$

where J_n is the electron current density. The second term has a different sign from the one in Eq. (6.57), because the charge of the carriers is now $-e$ instead of $+e$. The equation is known as the *continuity equation for electrons in p-type material.* Equations (6.57) and (6.58) together with Eqs. (6.9) and (6.10) give the starting point for the theory of current flow in semiconductor devices.

6.4b. Diffusion Equations

It sometimes happens that the drift terms in the current densities are negligible in comparison with the diffusion terms. Equations (6.9) and (6.10) may then be written

$$\mathbf{J}_p = -eD_p\,\nabla p \tag{6.9a}$$

$$\mathbf{J}_n = +eD_n\,\nabla n \tag{6.10a}$$

Substituting into Eqs. (6.57) and (6.58) yields

$$\frac{\partial p}{\partial t} = -\frac{p - p_n}{\tau_p} + D_p\,\nabla^2 p \tag{6.59}$$

$$\frac{\partial n}{\partial t} = -\frac{n - n_p}{\tau_n} + D_n\,\nabla^2 n \tag{6.60}$$

where ∇^2 stands for $\nabla \cdot \nabla = \partial^2/\partial x^2 + \partial^2/\partial y^2 + \partial^2/\partial z^2$. In the one-dimensional case these equations become

$$\frac{\partial p}{\partial t} = -\frac{p - p_n}{\tau_p} + D_p\frac{\partial^2 p}{\partial x^2} \tag{6.59a}$$

$$\frac{\partial n}{\partial t} = -\frac{n - n_p}{\tau_n} + D_n\frac{\partial^2 n}{\partial x^2} \tag{6.60a}$$

These equations are known as *diffusion equations* (including recombination).

In the steady-state case, $\partial p/\partial t$ and $\partial n/\partial t$ are zero; hence the equations for the one-dimensional case become

$$\frac{\partial^2 p}{\partial x^2} = \frac{p - p_n}{D_p\tau_p} = \frac{p - p_n}{L_p^2} \tag{6.61}$$

$$\frac{\partial^2 n}{\partial x^2} = \frac{n - n_p}{D_n\tau_n} = \frac{n - n_p}{L_n^2} \tag{6.62}$$

where $L_p = (D_p\tau_p)^{1/2}$ and $L_n = (D_n\tau_n)^{1/2}$ are called the *diffusion lengths* for electrons and holes, respectively. We shall see that these lengths represent the average penetration depth of diffusing minority carriers [Eq. (15.30)].

The diffusion equations can be used for solving transient problems. Suppose, for example, that one has an infinite semiconductor bar of n-type material, extending from $-\infty$ to $+\infty$. At $t = 0$, P_0 excess hole–electron pairs are generated at $x = 0$. If the excess holes have a lifetime τ_p, then the differential equation for the excess hole density p' is

$$\frac{\partial p'}{\partial t} = -\frac{p'}{\tau_p} + D_p \frac{\partial^2 p'}{\partial x^2}$$ (6.63)

This equation must be solved under the following initial conditions:

1. $p'(x, t) = 0$ at $t = \infty$ for all x.
2. $p'(x, t) = P_0 \, \delta(x)$ at $t = 0$, where $\delta(x)$ is the delta function with the properties that $\delta(x) = 0$ for $x \neq 0$, $\delta(x) = \infty$ for $x = 0$, and

$$\int_{-\infty}^{\infty} \delta(x)\, dx = 1$$

We see by substitution that the equation

$$p'(x, t) = \frac{P_0 \exp\left(-t/\tau_p\right)}{2(\pi D_p t)^{1/2}} \exp\left(-\frac{x^2}{4D_p t}\right)$$ (6.64)

satisfies both the differential equation and the initial conditions, so that it represents the transient solution of the problem.

We now make the simplifying assumption that τ_p is very large; that is, the carriers do not decay by recombination. Then Eq. (6.64) tells us that the probability $P(x, t)$ that the carrier will travel a distance x during the time t is

$$P(x, t) = \frac{\exp\left(-x^2/4D_p t\right)}{2(\pi D_p t)^{1/2}}$$ (6.64a)

Hence the mean square distance traveled by the holes during the time t is

$$\overline{x^2} = \int_{-\infty}^{\infty} x^2 P(x, t)\, dx = 2D_p t$$ (6.65)

This equation was first derived by Einstein and holds for an arbitrary one-dimensional diffusion process without recombination.

6.5. SPACE CHARGE IN SEMICONDUCTORS

6.5a. Gauss's Law and Poisson's Equation

According to Gauss's law, a space-charge distribution $\rho(x, y, z)$ is accompanied by a field strength $\mathbf{F}$ such that

$$\nabla \cdot \mathbf{F} = \frac{\rho}{\epsilon \epsilon_0}$$ (6.66)

where ϵ is the relative dielectric constant of the medium and $\epsilon_0 = 8.85 \times 10^{-12}$ F/m. Since the field strength $\mathbf{F}$ is related to the potential $\psi(x, y, z)$ by the definition

$$\mathbf{F} = -\nabla \psi$$ (6.67)

Eq. (6.66) may be written

$$\nabla^2 \psi = \frac{\partial^2 \psi}{\partial x^2} + \frac{\partial^2 \psi}{\partial y^2} + \frac{\partial^2 \psi}{\partial z^2} = -\frac{\rho}{\epsilon \epsilon_0} \qquad (6.68)$$

This is called *Poisson's equation.*

In a semiconductor in equilibrium a net space charge may be present. But according to Gauss's law this space charge is accompanied by an electric field **F**. Because of this field there will be a drift term in the electric current. But since there can be no net current flow in equilibrium, the drift term in the current must be exactly balanced by a diffusion current. Only in non-equilibrium cases can a net current flow.

Space charge is generally present in *n*-type or *p*-type semiconductors with a nonuniform donor or acceptor distribution. In *n*-type material with a donor distribution $N_d(x, y, z)$, the net space-charge density $\rho(x, y, z) = e(N_d + p - n)$, where p is the hole concentration and n the electron concentration. In *p*-type material with an acceptor distribution $N_a(x, y, z)$, the net space-charge density $\rho(x, y, z) = -e(N_a + n - p)$. Hence, for the two cases, Gauss's law is

$$\mathbf{V} \cdot \mathbf{F} = \frac{e}{\epsilon \epsilon_0}(N_d + p - n) \quad \text{or} \quad \mathbf{V} \cdot \mathbf{F} = -\frac{e}{\epsilon \epsilon_0}(N_a + n - p) \qquad (6.68a)$$

This can be applied, for example, to a *pn* junction to calculate the potential distribution in the space-charge region (see Chapter 15).

6.5b. Relaxation Effects

We shall now investigate what would happen if a small number of majority carriers were injected into a material of dielectric constant ϵ and conductivity σ. It will be shown that the charge involved will be distributed over the surface in a time comparable to $\tau_d = \epsilon \epsilon_0 / \sigma$. This time constant is called the *dielectric relaxation time*, and the redistribution of charge is known as *dielectric relaxation.*

The proof is carried out for *n*-type material. It will be assumed that the material is uniform with an equilibrium concentration n_0, and that the injected carrier density Δn is small in comparison with n_0. The only space charge of concern, $\rho_n = -e\,\Delta n$, is caused by injected carriers, and, according to Gauss's law, it gives rise to an electric field **F**. Since the carriers are majority carriers, recombination can be ignored.

According to Eq. (6.58), the continuity equation may be written

$$-e\frac{\partial n}{\partial t} = \frac{\partial \rho_n}{\partial t} = -\mathbf{V} \cdot \mathbf{J}_n \qquad (6.69)$$

But according to Eq. (6.10),

$$\mathbf{J}_n = e\mu_n n\mathbf{F} + eD_n \nabla n \qquad (6.70)$$

where it must be taken into account that $n = n_0 + \Delta n$ and that $\mathbf{F}$ is caused by Δn. Neglecting second-order terms involving $\Delta n \mathbf{F}$, we thus have

$$\mathbf{J}_n = e\mu_n n_0 \mathbf{F} + eD_n \nabla n = \sigma_n \mathbf{F} + eD_n \nabla n \qquad (6.70a)$$

Carrying out the differentiation yields

$$\nabla \cdot \mathbf{J}_n = \frac{\partial J_{nx}}{\partial x} + \frac{\partial J_{ny}}{\partial y} + \frac{\partial J_{nz}}{\partial z} = \sigma_n \nabla \cdot \mathbf{F} + eD_n \nabla^2 n \qquad (6.71)$$

where $\sigma_n = e\mu_n n_0$ is the equilibrium conductivity of the sample. Since $\nabla \cdot \mathbf{F} = \rho_n / \epsilon\epsilon_0$, according to Gauss's law, Eq. (6.71) may be written

$$\frac{\partial \rho_n}{\partial t} = -\frac{\sigma_n}{\epsilon\epsilon_0}\rho_n - eD_n \nabla^2 n \equiv -\frac{\rho_n}{\tau_d} + D_n \nabla^2 \rho_n \qquad (6.72)$$

where $\tau_d = \epsilon\epsilon_0 / \sigma_n$ is the dielectric relaxation time of the material. We thus obtain a differential equation in ρ_n, which can be solved under the appropriate boundary conditions.

In the particular case that the excess carrier density is spread uniformly over the sample, $\nabla^2 \rho_n = 0$, so that

$$\frac{\partial \rho_n}{\partial t} = -\frac{\rho_n}{\tau_d} \qquad (6.72a)$$

with $\rho_n = \rho_{n0}$ at $t = 0$. The solution of this equation is

$$\rho_n = \rho_{n0} \exp\left(-\frac{t}{\tau_d}\right) \qquad (6.73)$$

If $\rho_{n0}(x, y, z)$ depends on x, y, and z, the last term in Eq. (6.72) can no longer be neglected and Eq. (6.73) is no longer correct. Nevertheless, it remains true that τ_d is the time constant involved in the relaxation process, and that it is still a good measure of the time needed to reestablish equilibrium.

As an example, consider n-type germanium of $1\ \Omega$ cm resistivity. Since $\epsilon = 16$, and $D_n = 100\ \text{cm}^2/\text{s}$,

$$\tau_d = \frac{8.85 \times 10^{-12} \times 16}{100} = 1.4 \times 10^{-12}\text{s}$$

so that the dielectric relaxation time is very short indeed. Only in high-resistivity material do these relaxation effects become easily observable.

Consider next the case of n-type material into which holes are injected. Since the electron concentration is so much larger than the equilibrium hole concentration, the injected hole charge will not distribute itself rapidly over the n-type material, but instead the electron concentration will rearrange itself in a time of the order of the dielectric relaxation time τ_d to maintain space-charge neutrality internally. Afterwards the excess hole concentration spreads out gradually by diffusion (there is a hole concentration gradient) and disappears by recombination.

We thus see that in an n-type or p-type semiconductor in equilibrium *there is always a tendency to maintain space-charge neutrality internally, although there may be charge on the surface.* This is important for an understanding of the operation of semiconductor devices.

6.5c. Space-Charge Neutrality

It will now be shown with the help of an example that in inhomogeneous n- or p-type semiconductors the space charge is relatively weak, so that in most calculations involving equilibrium it may be neglected. Only in the calculation of the field itself must we take the space charge into account. In other words, *it may usually be assumed that there is approximate space-charge neutrality in an inhomogeneous n- or p-type semiconductor.*

This is not true, of course, if a sample is partly p- and partly n-type. In that case there are important space-charge effects at the boundary between the p and n regions. Outside those boundary regions, however, the statement is true.

To illustrate the effect of inhomogeneities in the donor and acceptor concentrations, consider a piece of n-type material that has a linearly decreasing electron concentration

$$n(x) = n_0\left(1 - \frac{ax}{w}\right) \tag{6.74}$$

for $0 \leq x \leq w$, with $a < 1$. This is illustrated in Fig. 6.6. The question is now, "What is the donor distribution $N_d(x)$ corresponding to Eq. (6.74)?" We could also have assumed that the *donor* concentration was a linear function of x; this is a much more difficult problem to solve, however, and the solution does not shed more light on the problem.

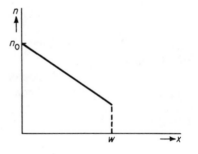

Fig. 6.6. Illustration of the non-uniform electron concentration of Eq. (6.74).

We assume that $n(x) \gg n_i$ everywhere, even at $x = w$. The hole concentration is then negligible everywhere. In equilibrium $J_n = 0$; hence

$$e\mu_n nF + eD_n\frac{\partial n}{\partial x} = 0$$

or

$$F = -\frac{D_n}{\mu_n}\frac{1}{n}\frac{\partial n}{\partial x} = \frac{kT}{e}\frac{a/w}{1 - ax/w} \tag{6.75}$$

According to Gauss's equation (6.68a),

$$N_d - n = \frac{\epsilon\epsilon_0}{e}\frac{\partial F}{\partial x} = \frac{\epsilon\epsilon_0 kT}{e^2}\frac{a^2/w^2}{(1 - ax/w)^2} \tag{6.76}$$

To demonstrate that $N_d - n \ll n$, we take the extreme case that $n(x)$ varies from $10^{16}/cm^3$ at $x = 0$ to $10^{15}/cm^3$ at $x = w = 10^{-3}$ cm. Then $a = 0.9$. We assume that the material is germanium ($\epsilon = 16$) and that $T = 300°K$. Since $N_d - n$ is largest at $x = w$, we calculate this quantity at that point.

Substituting the values at $x = w$, we get

$$N_d - n = \frac{16 \times 8.85 \times 10^{-12} \times 1.38 \times 10^{-23} \times 300}{(1.6 \times 10^{-19})^2} \cdot \frac{0.81 \times 10^{10}}{10^{-2}}$$

$$= 1.85 \times 10^{19}/m^3$$

which corresponds to $N_d - n = 1.85 \times 10^{13}/cm^3$. This is quite small in comparison with n, so that $N_d(x) \simeq n(x)$ in very good approximation. Hence even in this case we find that space-charge neutrality is approximately maintained.

6.5d. Ambipolar Effects

We shall now investigate drift and diffusion in samples in which both electrons and holes are present. Unless there are true space-charge regions, as in a *pn* junction, space-charge neutrality is approximately preserved. One must be careful with this rule, however, for although it is always essentially possible to put

$$p - p_0 = n - n_0, \qquad \frac{\partial p}{\partial t} = \frac{\partial n}{\partial t}, \qquad \nabla p = \nabla n, \qquad \nabla^2 p = \nabla^2 n$$

it is not proper to substitute $p = n$ in Poisson's equation, for that would give a constant field $\mathbf{F}$.

We now derive the diffusion equations for the case when an electric field is present. We start from the continuity equations

$$\frac{\partial p}{\partial t} = -\frac{p - p_0}{\tau_p} - \frac{1}{e} \nabla \cdot \mathbf{J}_p$$

$$\frac{\partial n}{\partial t} = -\frac{n - n_0}{\tau_p} + \frac{1}{e} \nabla \cdot \mathbf{J}_n$$

$$(6.77)$$

and the current equations

$$\mathbf{J}_p = e\mu_p p\mathbf{F} - eD_p \nabla p, \qquad \mathbf{J}_n = e\mu_n n\mathbf{F} + eD_n \nabla n \qquad (6.78)$$

Substituting Eq. (6.78) into Eq. (6.77) yields

$$\frac{\partial n}{\partial t} = -\frac{n - n_0}{\tau_p} + \mu_n \mathbf{F} \cdot \nabla n + \mu_n n \nabla \cdot \mathbf{F} + D_n \nabla^2 n \qquad (6.79)$$

$$\frac{\partial p}{\partial t} = -\frac{p - p_0}{\tau_p} - \mu_p \mathbf{F} \cdot \nabla p - \mu_p p \nabla \cdot \mathbf{F} + D_p \nabla^2 p \qquad (6.80)$$

To eliminate the $\nabla \cdot \mathbf{F}$ term we multiply the first equation by $\mu_p p$ and the second by $\mu_n n$, add, and substitute the space-charge neutrality conditions in the remaining terms. This gives, since $p - p_0 = n - n_0$

$$(\mu_p p + \mu_n n)\frac{\partial p}{\partial t} = -(\mu_p p + \mu_n n)\frac{p - p_0}{\tau_p} - \mu_n \mu_p (n - p)\mathbf{F} \cdot \nabla p$$
$$+ (\mu_p D_n p + \mu_n D_p n) \nabla^2 p$$

Dividing by $\mu_p p + \mu_n n$ yields

$$\frac{\partial p}{\partial t} = -\frac{p - p_0}{\tau_p} - \mu_a \mathbf{F} \cdot \nabla p + D_a \nabla^2 p \qquad (6.81)$$

where

$$\mu_a = \frac{\mu_n \mu_p (n - p)}{\mu_n n + \mu_p p}, \qquad D_a = \frac{\mu_p D_n p + \mu_n D_p n}{\mu_n n + \mu_p p} = \frac{D_n D_p (p + n)}{D_n n + D_p p} \qquad (6.81a)$$

The quantities μ_a and D_a are called the *ambipolar mobility* and the *ambipolar diffusion* constant, respectively. Note that $\mu_a = \mu_p$ and $D_a = D_p$ for $p \ll n$. For intrinsic material, $\mu_a = 0$ and the $\mathbf{F} \cdot \nabla p$ term disappears.

6.6. HOT ELECTRON EFFECTS

At high field strengths the mobility μ_n and the diffusion constant D_n of electrons in semiconductors become field-dependent. Shockley[†] calculated the effect under the assumption of a displaced Maxwellian velocity distribution of the carriers with a higher effective temperature T_e, whereas Yamashita and Watanabe[‡] showed that the velocity distribution approaches the Druyvesteyn distribution, known from the theory of gas discharges, at very high fields. We give here an outline of their approach.

The electron energy may be written

$$E = \frac{1}{2}m^* v^2 = \frac{\hbar^2 k^2}{2m^*} \qquad (6.82)$$

where v is the (isotropic) velocity of the carriers, $\mathbf{k}$ is the electron wave vector with components k_x, k_y, and k_z, and m^* is the effective electron mass. If an electric field F is applied in the X direction, the Boltzmann transport equation is

$$-\frac{eF}{\hbar}\frac{\partial}{\partial k_x}f(\mathbf{k}, \mathbf{r}) + \left(\frac{\partial f}{\partial t}\right)_{\text{coll}} = 0 \qquad (6.83)$$

where $f(\mathbf{k}, \mathbf{r})$ is the distribution in $\mathbf{k}$. Expanding $f(\mathbf{k}, \mathbf{r})$ with respect to k_x and terminating with the first-order term yields

$$f(\mathbf{k}, \mathbf{r}) = f_0(E, x) + k_x g(E, x) \qquad (6.84)$$

---

† W. Shockley, *Bell System Tech. J.*, **30**, 4 (1951), as revised in S. Wang, *Solid State Electronics*, McGraw-Hill Book Company, New York, 1966, pp. 225–231.

‡ J. Yamashita and M. Watanabe, *Progr. Theoret. Phys. (Japan)*, **12**, 443 (1954).

Equation (6.83) thus becomes

$$\frac{he}{m^*}Fk_x\frac{\partial f_0}{\partial E} + \frac{e}{h}Fg + \frac{he}{m^*}Fk_x^2\frac{\partial g}{\partial E} + \left(\frac{\partial f}{\partial t}\right)_{\text{coll}} = 0 \qquad (6.85)$$

$(\partial f/\partial t)_{\text{coll}}$ can be described in terms of the collision integral that describes the interaction of the electrons with acoustical phonons. The integrand of that integral is then expanded with respect to the phonon energy $h\omega$ up to second-order terms, and the integration is carried out. After some manipulations, this yields

$$\left(\frac{\partial f}{\partial t}\right)_{\text{coll}} = \frac{2(2m^*)^{1/2}u^2E^{1/2}}{l}\left[E\frac{\partial^2 f_0}{\partial E^2} + \left(2 + \frac{E_0}{k_0 T}\right)\frac{\partial f_0}{\partial E} + \frac{2}{k_0 T}E\right]$$

$$- \frac{1}{l}\frac{2^{1/2}}{m^{*1/2}}E^{1/2}g \qquad (6.86)$$

Here k_0 is Boltzmann's constant, T the lattice temperature, u the (isotropic) longitudinal sound velocity, and l the electron free path length, which is independent of the energy and of the field F.

Substituting (6.86) into (6.85) and separating the symmetric and antisymmetric terms in k_x yields the equations

$$g(E, x) = \frac{hl}{(2m^*E)^{1/2}}eF\frac{\partial f_0}{\partial E} \qquad (6.87)$$

$$E\frac{\partial^2 f_0}{\partial E^2} + \left(2 + \frac{E}{k_0 T}\right)\frac{\partial f_0}{\partial E} + \frac{2}{k_0 T}f_0 = -\frac{l}{hu^2(2m^*)^{1/2}}\left[\frac{eF}{2E^{1/2}}g + \frac{1}{3}eFE^{1/2}\frac{\partial g}{\partial E}\right]$$

$$(6.88)$$

Substituting (6.87) into (6.88) and rearranging terms yields

$$(E + pk_0 T)\frac{\partial^2 f_0}{\partial E^2} + \left(2 + \frac{E}{k_0 T} + \frac{pk_0 T}{E}\right)\frac{\partial f_0}{\partial E} + \frac{2}{k_0 T}f_0 = 0 \qquad (6.89)$$

where

$$p = \frac{eF^2l^2}{6m^*u^2k_0 T} \qquad (6.89a)$$

From this equation, $f_0(E)$ can be determined. For $p = 0$ it yields, of course, the Maxwellian velocity distribution; but for $pk_0 T \gg E$ and $p \gg 1$, we may write the equation as

$$pk_0 T\frac{\partial^2 f_0}{\partial E^2} + \left(\frac{E}{k_0 T} + \frac{pk_0 T}{E}\right)\frac{\partial f_0}{\partial E} + \frac{2}{k_0 T}f_0 = 0 \qquad (6.89b)$$

As is seen by substitution, this equation has the solution

$$f_0 = N\exp\left(-\frac{E^2}{2pk_0^2 T^2}\right) \qquad (6.90)$$

where $E = \frac{1}{2}m^*v^2$ and N is a normalization factor. This is the *Druyvesteyn distribution.*

Since $\tau = l/v$ is the average time between collisions, the definitions for μ_n and D_n are†

$$\mu_n = \frac{2}{3}\frac{e}{m^*}l\left\langle\frac{1}{v}\right\rangle, \qquad D_n = \frac{1}{3}l\langle v\rangle \qquad (6.91)$$

where $\langle\ \rangle$ means averaging over the velocity distribution. Carrying out the averaging with the help of (6.90) yields

$$\mu_n = \mu_0\left(\frac{F_c}{F}\right)^{1/2}, \qquad D_n = 0.927D_0\left(\frac{F}{F_c}\right)^{1/2}, \qquad F_c = 1.514\frac{u}{\mu_0} \qquad (6.92)$$

Evaluating the electron temperature T_e from the definition $\mu_n kT_e = eD_n$ and bearing in mind that $\mu_0 kT = eD_0$ yields

$$T_e = 0.927\left(\frac{F}{F_c}\right)T \qquad (6.92a)$$

Shockley's expressions for μ_n, D_n, T_e, and F_c are similar but with slightly different numerical factors; they are in closed form, however.

REFERENCES

CONWELL, E. M., *High-Field Transport in Semiconductors*. Academic Press, Inc., New York, 1967.

FISTUL, V. I., *Heavily Doped Semiconductors*. Plenum Publishing Corporation, New York, 1969.

MILNES, A. G., *Deep Impurities in Semiconductors*. John Wiley & Sons, Inc. (Interscience Division), New York, 1973.

MOLL, J. H., *Physics of Semiconductors*. McGraw-Hill Book Company, New York, 1964.

SHOCKLEY, W., *Electrons and Holes in Semiconductors*. Van Nostrand Reinhold Company, New York, 1950.

SMITH, R. A., *Semiconductors*. Cambridge University Press, New York, 1959.

WANG, S., *Solid State Electronics*. McGraw-Hill Book Company, New York, 1966.

PROBLEMS

1. A semiconductor material has electron and hole mobilities μ_n and μ_p, respectively, with $\mu_p < \mu_n$. When the conductivity is considered as a function of the hole

---

† The definition of D_n follows from (6.35), whereas the definition of μ_n follows from the electron current density $J = e\mu_n Fn$.

concentration p, show that the minimum value σ_{min} of the conductivity σ can be written as

$$\sigma_{min} = \sigma_i 2 \frac{(\mu_n \mu_p)^{1/2}}{\mu_n + \mu_p}, \quad \text{when } p = n_i \left(\frac{\mu_n}{\mu_p}\right)^{1/2}$$

where n_i and σ_i refer to intrinsic material.

2. In an infinite semiconducting bar stretching from $x = -\infty$ to $x = +\infty$ a constant electric field F_0 is maintained, pointing in the positive X direction. At $t = 0$, P_0 hole–electron pairs are generated at $x = 0$. The differential equation of these excess holes is

$$\frac{\partial p'}{\partial t} = -\frac{p'}{\tau_p} - \mu_a F_0 \frac{\partial p'}{\partial x} + D_a \frac{\partial^2 p'}{\partial x^2}$$

which must be solved under the initial conditions $p'(x, t) = P_0\, \delta(x)$ at $t = 0$, where $\delta(x)$ is the delta function, and $p'(x, t) = 0$ at $t = \infty$ for all x. Show by substitution and by comparison of the boundary conditions that the solution of this equation is

$$p'(x, t) = \frac{P_0 \exp\,(-t/\tau_p)}{2(\pi D_a t)^{1/2}} \exp \left[-\frac{(x - \mu_a F_0 t)^2}{4 D_a t} \right]$$

3. Discuss the solution of problem 2 for the case when the material is intrinsic.

4. Show that D_a and μ_a are equal to the minority carrier values if the material is relatively far from being intrinsic.

5. Figure 6.7 shows an n-type semiconducting bar in which a field $F_0 = V/L$ is maintained. A light flash creates hole–electron pairs at $x = 0$ and $t = 0$. Because of the applied field, these hole–electron pairs drift downstream with the ambipolar drift mobility. At $x = d$, a biased point is located that measures p' at $x = d$ by collecting holes (collector contact). From the solution of problem 2 find at what time the current at the point contact will be a maximum. Determine the hole mobility from the measurement of the current $i(t)$ flowing to the point contact (Shockley–Haynes experiment). For a good measurement the lifetime τ_p must be so large that $\exp\,(-t/\tau_p)$ is relatively close to unity.

Answer: $t = d/(\mu_a F_0)$.

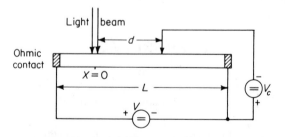

Fig. 6.7. Shockley–Haynes experiment.

6. Show from the solution of problem 2 that the probability $P(x, t)$ that a carrier will travel a distance x during the time t by diffusion is

$$P(x, t) = \frac{1}{2(\pi D_a t)^{1/2}} \exp\left(-\frac{x^2}{4 D_a t}\right)$$

To simplify matters, put $\tau_p = \infty$ and $F_0 = 0$.

7. Show that the mean square distance traveled by the minority carriers owing to one-dimensional diffusion is

$$\overline{x^2} = 2 D_a t$$

8. Show that in three-dimensional diffusion this equation is replaced by

$$\overline{r^2} = 6 D_a t$$

9. Starting from problem 9, Chapter 4, by equating v_x to the ac drift velocity, show that the high-frequency conductivity is

$$\sigma = \frac{e^2 n \tau / m^*}{1 + j\omega\tau}$$

10. Starting from problem 10, Chapter 4, by equating v_x to the ac drift velocity, show that the ac conductivity is

$$\sigma = \frac{(e^2 n \tau / m^*)(1 + j\omega\tau)}{1 + 2j\omega\tau + (\omega_0^2 - \omega^2)\tau^2}$$

For fixed ω and $\omega\tau \gg 1$ (low temperature), show that σ versus B has a resonance at $\omega_0 = \omega$ (cyclotron resonance).

7

Thermionic Emission
and Field Emission

7.1. THERMIONIC EMISSION

7.1a. Richardson's Equation

Thermionic emission usually refers to the escape of electrons from a hot surface into a vacuum. If the surface is used as a *cathode* and all the emitted electrons are collected, the cathode is said to give *saturated emission;* the emitted current density in that case is called the *saturated current density* J_s. The equation expressing J_s in terms of the cathode temperature T and the cathode work function ϕ is called *Richardson's equation*.

Figure 7.1a shows the energy-level diagram of a metal–vacuum system. Here W is the energy difference between an electron at rest outside the metal and an electron at rest in the metal, whereas E_f is the energy difference between the Fermi level and the bottom of the conduction band. The work function ϕ is defined as $\phi = W - E_f$.

Figure 7.1b shows the energy-level diagram of a metal–semiconductor system. Here W' denotes the energy difference between the highest point of the barrier and the bottom of the conduction band in the metal, whereas $E_b = W' - E_f$ is called the barrier height.

We can now extend the thermionic emission concept so that it includes the "emission over the barrier" from metal to semiconductor in the metal–semiconductor contact, and we can evaluate Richardson's equation for that case, too. It is convenient to evaluate this equation for the latter system and then obtain the case of emission into a vacuum by replacing W' by W and E_b by ϕ.

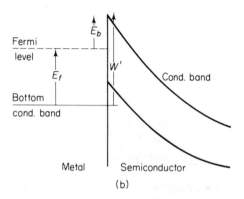

Fig. 7.1. (a) Energy level diagram of the metal-vacuum system. (b) Energy level diagram of the metal-semiconductor system.

We introduce a coordinate system with the X axis perpendicular to the interface. We assume that in the metal the electrons have an effective mass m_1^* and velocity components v_{x1}, v_{y1}, and v_{z1}, whereas in the semiconductor the electrons have an effective mass m_2^* and velocity components v_{x2}, v_{y2}, and v_{z2}. The escape condition for an electron in the metal is then $\frac{1}{2}m_1^*v_{x1}^2 \geqq W'$. The electrons lose the energy W' when passing through the interface; hence one would expect the energy due to motion in the X direction to be conserved. For the motion in the Y and Z directions one would expect the Y and Z components of the momentum to be conserved when the electron passes through the interface. Hence

$$\tfrac{1}{2}m_1^*v_{x1}^2 = \tfrac{1}{2}m_2^*v_{x2}^2 + W', \qquad m_1^*v_{y1} = m_2^*v_{y2}, \qquad m_1^*v_{z1} = m_2^*v_{z2} \qquad (7.1)$$

For the "fast" electrons that can escape, one must certainly require $E \geqq W'$. Since $W' - E_f \gg kT$, we may substitute in Eq. (3.54)

$$\left[\exp\left(\frac{E - E_f}{kT}\right) + 1\right]^{-1} \simeq \exp\left(\frac{-E + E_f}{kT}\right)$$

Hence the density ΔN of the fast electrons with velocity components between v_{x1} and $(v_{x1} + \Delta v_{x1})$, v_{y1} and $(v_{y1} + \Delta v_{y1})$, and v_{z1} and $(v_{z1} + \Delta v_{z1})$ is, according to Eq. (3.54),

$$\Delta N = \frac{2}{h^3} \exp\left[\frac{E_f - \frac{1}{2}m_1^* v_{x1}^2 - \frac{1}{2}m_1^* v_{y1}^2 - \frac{1}{2}m_1^* v_{z1}^2}{kT}\right] m_1^* \Delta v_{x1} m_1^* \Delta v_{y1} m_1^* \Delta v_{z1}$$

(7.2)

The number of electrons arriving per second at 1 m² of surface with velocity components between v_{x1} and $(v_{x1} + \Delta v_{x1})$, v_{y1} and $(v_{y1} + \Delta v_{y1})$, and v_{z1} and $(v_{z1} + \Delta v_{z1})$ is therefore

$$v_{x1}\,\Delta N = \frac{2}{h^3} \exp\left(\frac{E_f}{kT}\right) \exp\left(-\frac{m_1^* v_{x1}^2}{2kT}\right) m_1^* v_{x1}\,\Delta v_{x1}$$

$$\times \exp\left(-\frac{m_1^* v_{y1}^2}{2kT}\right) m_1^*\,\Delta v_{y1} \exp\left(-\frac{m_1^* v_{z1}^2}{2kT}\right) m_1^*\,\Delta v_{z1}$$

(7.3)

Changing over to the velocity components v_{x2}, v_{y2}, and v_{z2} of the escaping electrons we thus have for the number Δn of electrons emitted per square meter per second, with velocity components between v_{x2} and $(v_{x2} + \Delta v_{x2})$, v_{y2} and $(v_{y2} + \Delta v_{y2})$, and v_{z2} and $(v_{z2} + \Delta v_{z2})$,

$$\Delta n = \frac{2}{h^3} \exp\left(\frac{E_f - W'}{kT}\right) \exp\left(-\frac{m_2^* v_{x2}^2}{2kT}\right) m_2^* v_{x2}\,\Delta v_{x2}$$

$$\times \exp\left(-\frac{m_2^* v_{y2}^2}{2kT}\right) m_2^*\,\Delta v_{y2} \exp\left(-\frac{m_2^* v_{z2}^2}{2kT}\right) m_2^*\,\Delta v_{z2}$$

(7.4)

where v_{x2} can vary between 0 and ∞, v_{y2} and v_{z2} between $-\infty$ and $+\infty$, and $W' - E_f = E_b$.

The contribution ΔJ_s of these electrons to the electron current density J_s is therefore

$$\Delta J_s = e\,\Delta n \tag{7.4a}$$

and J_s is obtained by summing over all possible velocity intervals. Replacing, as usual, the summation by an integration, we have†

$$J_s = 2e\frac{m_2^{*3}}{h^3} \exp\left(-\frac{E_b}{kT}\right) \int_0^\infty \exp\left(-\frac{m_2^* v_{x2}^2}{2kT}\right) v_{x2}\,dv_{x2}$$

$$\int_{-\infty}^\infty \exp\left(-\frac{m_2^* v_{y2}^2}{2kT}\right) dv_{y2} \int_{-\infty}^\infty \exp\left(-\frac{m_2^* v_{z2}^2}{2kT}\right) dv_{z2}$$

(7.5)

$$= \frac{4\pi e m_2^* k^2 T^2}{h^3} \exp\left(-\frac{E_b}{kT}\right)$$

---

† The calculation involves the following integrals:

$$\int_{-\infty}^\infty \exp\left(-\frac{m_2^* v_{y2}^2}{2kT}\right) dv_{y2} = \int_{-\infty}^\infty \exp\left(-\frac{m_2^* v_{z2}^2}{2kT}\right) dv_{z2} = \left(\frac{2\pi kT}{m_2^*}\right)^{1/2}$$

as is found by putting $u = v_{y2}\,(m_2^*/2kT)^{1/2}$ as a new variable, and

$$\int_0^\infty v_{x2} \exp\left(-\frac{m_2^* v_{x2}^2}{2kT}\right) dv_{x2} = \frac{kT}{m_2^*}$$

as is found by putting $u' = m_2^* v_{x2}^2/2kT$ as a new variable.

This is Richardson's equation for emission from a metal into a semiconductor. The corresponding equation for emission into a vacuum is obtained by putting $m_2^* = m$ and replacing E_b by ϕ, so that

$$J_s = \frac{4\pi emk^2}{h^3} T^2 \exp\left(-\frac{\phi}{kT}\right) = AT^2 \exp\left(-\frac{\phi}{kT}\right) \tag{7.6}$$

where the constant A has the value

$$A = \frac{4\pi emk^2}{h^3} = 1.20 \times 10^6 \text{A/ m}^2/\text{degree}^2 \tag{7.6a}$$

The formula was derived under the assumption that all electrons arriving with a velocity $v'_{x1} > (2W'/m_1^*)^{1/2}$ will escape. Classically, this is indeed the case. But wave mechanically one must bear in mind that the wave length $\lambda_1 = h/m_1^* v_1$ inside the metal differs from the wavelength $\lambda_2 = h/m_2^* v_2$ outside the metal; as a consequence, a partial reflection of the electron wave arriving at the surface would be expected. Denoting the average transmission coefficient of the arriving electrons by t, Eq. (7.6) may be written

$$J_s = tAT^2 \exp\left(-\frac{\phi}{kT}\right) \tag{7.7}$$

Usually t is close to unity, so that the error made by ignoring the effect is not very significant.

Equation (7.5) is needed in the discussion of the thermionic emission model of the (I, V) characteristic of the metal–semiconductor contact.

7.1b. Velocity Distribution of the Emitted Electrons

For some applications we need to know the velocity distribution of the electrons emitted by a thermionic cathode. To simplify the notation, we shall replace m_2^* by m and v_{x2}, v_{y2}, and v_{z2} by v_x, v_y, and v_z. The integrals involved in the following calculations are of the same type as those shown in the last footnote.

We obtain ΔN_x, the number of electrons emitted per second with an X component of the velocity between v_x and $(v_x + \Delta v_x)$, by integrating (7.4) with respect to v_y and v_z. This gives

$$\begin{aligned} \Delta N_x &= \frac{4m^2\pi kT}{h^3} \exp\left(-\frac{\phi}{kT}\right) \exp\left(-\frac{\frac{1}{2}mv_x^2}{kT}\right) v_x \,\Delta v_x \\ &= \frac{J_s}{e} \exp\left(-\frac{\frac{1}{2}mv_{x'}^2}{kT}\right) \Delta\left(\frac{\frac{1}{2}mv_x^2}{kT}\right) \end{aligned} \tag{7.8}$$

The number of electrons, ΔN_y, emitted per second with a component of the velocity between v_y and $(v_y + \Delta v_y)$ can be found by integration of (7.4) with respect to v_x and v_z; this yields

$$\Delta N_y = \frac{J_s}{e} \sqrt{\frac{m}{2\pi kT}} \exp\left(-\frac{\frac{1}{2}mv_y^2}{kT}\right) \Delta v_y \tag{7.9}$$

A corresponding equation holds for the velocity component v_z.

The velocity distribution in v_y and v_z is thus of the form of Eq. (3.24). This is a *Maxwellian velocity distribution;* noting that the conditions for the equipartition law hold in this case, we have

$$\tfrac{1}{2}m\overline{v_y^2} = \tfrac{1}{2}m\overline{v_z^2} = \tfrac{1}{2}kT \tag{7.10}$$

The velocity distribution in v_x contains an extra term v_x; we shall call it a *modified Maxwellian distribution.* Since $J_s/e = N$ is the number of electrons emitted per second, the probability that an electron is emitted with a velocity between v_x and $(v_x + \Delta v_x)$ is

$$\Delta P(v_x) = \frac{\Delta N_x}{N} = \exp\left(-\frac{\tfrac{1}{2}mv_x^2}{kT}\right)\Delta\left(\frac{\tfrac{1}{2}mv_x^2}{kT}\right) \tag{7.11}$$

Consequently, the average energy due to v_x is

$$\frac{1}{2}m\overline{v_x^2} = \sum \frac{1}{2}mv_x^2\,\Delta P(v_x)$$

$$= \int_0^\infty \frac{1}{2}mv_x^2 \exp\left(-\frac{\tfrac{1}{2}mv_x^2}{kT}\right) d\left(\frac{\tfrac{1}{2}mv_x^2}{kT}\right) = kT \tag{7.12}$$

according to the definition of Chapter 3; the summation is again replaced by an integration. The value (7.12) is *twice as large* as the value expected from the equipartition law. This is because the distribution is not Maxwellian; the number of fast electrons is enhanced in comparison to the slow ones. The average energy due to v_y and v_z is of course $\tfrac{1}{2}kT$ for each [Eq. (7.10)].

For some discussions we also need $\overline{v_x}$:

$$\overline{v_x} = \sum v_x\,\Delta P(v_x) = \int_0^\infty v_x \exp\left(-\frac{\tfrac{1}{2}mv_x^2}{kT}\right) d\left(\frac{\tfrac{1}{2}mv_x^2}{kT}\right) = \left(\frac{\pi}{4}\cdot\frac{2kT}{m}\right)^{1/2} \tag{7.13}$$

so that

$$\frac{1}{2}m\overline{v_x^2} = kT \quad \text{and} \quad \frac{1}{2}m(\overline{v_x})^2 = \frac{\pi}{4}kT = \frac{\pi}{4}\left(\frac{1}{2}m\overline{v_x^2}\right) \tag{7.13a}$$

7.1c. Cooling Effect

Energy is needed to bring an electron from the inside of the cathode to the outside. The electron emission will thus tend to cool the cathode.

Let I_s be the emission current. We introduce $\phi' = \phi/e$, the work function in electron volts, and observe that an energy $e\phi'$ is needed to bring an electron from the Fermi level to the vacuum level. To bring I_s/e electrons outside the metal per second with zero energy thus requires a power $I_s\phi'$. But the electrons do not leave the metal with zero energy; their average energy is

$$E = \tfrac{1}{2}m\overline{v_x^2} + \tfrac{1}{2}m\overline{v_y^2} + \tfrac{1}{2}m\overline{v_z^2} = 2kT \tag{7.14}$$

according to Eqs. (7.12) and (7.10).

Consequently, the total power needed to bring (I_s/e) electrons out per second with an average energy of $2kT$ is

$$P = I_s\left(\phi' + \frac{2kT}{e}\right) \tag{7.15}$$

If the cathode temperature is kept constant, this extra heater power has to be supplied if the cathode starts to emit. The effect is easily measurable in diodes at large emission currents, and according to (7.15) it may be used for a determination of ϕ', the work function in electron volts.

7.1d. Further Discussion of Richardson's Equation

Replacing ϕ by $\epsilon\phi'$, where ϕ' is the work function in electron volts, and neglecting the correction factor of (7.7), we write Richardson's law as

$$J_s = AT^2 \exp\left(-\frac{e\phi'}{kT}\right) \tag{7.16}$$

To verify this equation experimentally, we plot $\ln(J_s/T^2)$ as a function of $1/T$; such a plot is called a *Richardson plot* (Fig. 7.2). Since

$$\ln\left(\frac{J_s}{T^2}\right) = \ln A - \frac{e\phi'}{kT} \tag{7.16a}$$

we obtain A by extrapolating the curve to $1/T = 0$ and find ϕ' by measuring the slope of the line.

Actually, one finds values of A that are sometimes quite different from the theoretical value of 1.20×10^6 A/m²/degree². One of the most important reasons is the temperature coefficient of the work function. In a limited range close to the operating point T_c, ϕ' can be represented by the equation (see Fig. 7.3)

$$\phi' = \phi'_0 + T\left(\frac{d\phi'}{dT}\right)_{T=T_c} = \phi'_0 + \alpha T \tag{7.17}$$

where α is the temperature coefficient of ϕ'. Substituting into (7.16), we obtain

$$J_s = A \exp\left(-\frac{e\alpha}{k}\right) T^2 \exp\left(-\frac{e\phi'_0}{kT}\right) = A'T^2 \exp\left(-\frac{e\phi'_0}{kT}\right) \tag{7.18}$$

For $\alpha > 0$, A' may be much smaller than A and ϕ'_0 will be somewhat smaller than the work function at the operating point. Since $e/k = 11,600$, a value of $\alpha = 10^{-4}$ (which is reasonable for metals; for oxide-coated cathodes α is much larger) gives $A'/A \simeq \frac{1}{3}$; larger values of α give a correspondingly larger effect. *For that reason it is usually more reliable to substitute $A = 1.20 \times 10^6$ A/m²/degree² and use Eq. (7.16) as a definition for ϕ'.* The results of this method can be compared with direct methods for measuring work function.

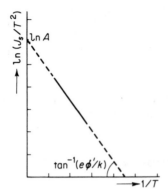

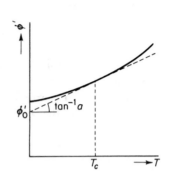

Fig. 7.2. Richardson plot in which $\ln (J_s/T^2)$ is plotted against $1/T$. The full-drawn curve represents the temperature range in which measurements can be performed. Extrapolation to $1/T = 0$ gives $\ln A$; the angle of the Richardson plot with the $1/T$ axis is $\tan^{-1}(e\phi'/k)$.

Fig. 7.3. Temperature dependence of ϕ'. At the operating point T_c, ϕ' can be approximated by a linear relationship of the form (7.17). It is shown in the text that this leads to measured values of A that differ from the theoretical value derived earlier.

7.2. SCHOTTKY EFFECT

According to the preceding discussion, the emission current should be independent of the field strength at the cathode. Experimentally, this is found to be untrue; the emission current is found to increase with increasing field strength at the cathode. This effect is called the *Schottky effect;* it indicates that the work function of the cathode depends upon the surface field strength. We shall see that this is caused by the attractive force between the cathode and an escaping electron.

In Sec. 7.1a we assumed that the conduction electrons moved in a potential well of depth W, and that the potential energy $\varphi_0(x)$ jumped abruptly from the value $\varphi_0(x) = -W$ for $x < 0$ to the value $\varphi_0(x) = 0$ at the surface $(x = 0)$. Actually, the change in $\varphi_0(x)$ at the surface cannot be so abrupt. For an electron of charge $-e$ at a distance x from the surface induces an image charge $+e$ at a distance $-x$ inside the surface, at least if the cathode is a conductor. The force F with which the electron is attracted by the cathode is thus given by Coulomb's law as

$$F = \frac{e^2}{4\pi\epsilon_0(2x)^2} \tag{7.19}$$

where $\epsilon_0 = 10^7/4\pi c^2 = 8.85 \times 10^{-12}$ F/m. The work that has to be done to bring an electron from the position $x = x_0$ to infinity is

$$\int_{x_0}^{\infty} F \, dx = \int_{x_0}^{\infty} \frac{e^2}{16\pi\epsilon_0 x^2} \, dx = \frac{e^2}{16\pi\epsilon_0 x_0} \tag{7.19a}$$

so that the potential energy $\varphi_0(x)$ of an electron at a distance x from the surface is

$$\varphi_0(x) = -\frac{e^2}{16\pi\epsilon_0 x} \tag{7.20}$$

This equation ceases to be valid for small x, for Eq. (7.20) approaches $-\infty$ if x goes to zero, whereas in fact $\varphi_0(x)$ approaches $-W$ if x goes to zero. Fortunately, we have to know only the value of $\varphi_0(x)$ for relatively large values of x (Fig. 7.4a).

Because of the dependence of $\varphi_0(x)$ upon distance, the work function of the cathode surface will depend upon the applied electric field. Let F_0 be the applied field strength at the surface, and let the field be homogeneous, at least in the immediate neighborhood of the surface. Then the potential energy $\varphi(x)$ of the electron becomes

$$\varphi(x) = -\frac{e^2}{16\pi\epsilon_0 x} - eF_0 x \tag{7.21}$$

which has a maximum value (Fig. 7.4b)

$$\varphi_{\max} = -\frac{e}{2}\sqrt{\frac{eF_0}{\pi\epsilon_0}}, \quad \text{at} \quad x = x_m = \sqrt{\frac{e}{16\pi\epsilon_0 F_0}} \tag{7.22}$$

According to these equations, $x_m = 2 \times 10^{-7}$ m at $F_0 = 10^4$ V/m, and $x_m = 2 \times 10^{-8}$ m at $F_0 = 10^6$ V/m. These distances are sufficiently large so that (7.21) should be valid.

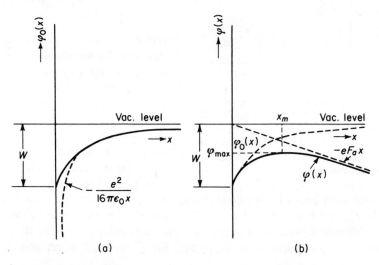

Fig. 7.4. (a) Potential energy of an electron at a distance x from the surface of the metal. $\varphi_0(x) = -W$ for $x < 0$; $\varphi_0(x) = -e^2/(16\pi\epsilon_0 x)$ for $x > 0$ and not too close to the surface. A deviation occurs for small x, since $\varphi_0(x)$ goes to $-W$ (and not to ∞) if x becomes zero. (b) Application of an electric field F_a shifts the maximum in the potential energy $\varphi(x)$ from $x = \infty$ to $x = x_m$ and lowers the work function by an amount $-\varphi_{\max}/e$.

If $e\phi_0'$ is the work function in joules without applied field, the work function $e\phi''$ in joules *with* applied field is thus equal to $e\phi_0' + \varphi_{max}$, or

$$\phi'' = \phi_0' - \frac{1}{2}\sqrt{\frac{eF_0}{\pi\epsilon_0}} = \phi_0' - 3.79 \times 10^{-5}\sqrt{F_0} \qquad (7.23)$$

where F_0 is in volts per meter and ϕ'' is in electron volts.

Substituting this result into (7.16), and denoting the emission current density for zero field strength by J_{s0}, we have, for the emission current density *with field*,

$$J_s = AT^2 \exp\left(-\frac{e\phi''}{kT}\right)$$
$$= J_{s0} \exp\left[+\frac{e(3.79 \times 10^{-5}\sqrt{F_0})}{kT}\right] \qquad (7.24)$$

Hence, plotting $\ln J_s$ against $\sqrt{F_0}$, we should get a straight line. This graph is called the *Schottky line* or the *Schottky plot*. Extrapolating this line to $F_0 = 0$ gives J_{s0} (Fig. 7.5).

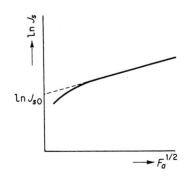

This discussion shows that Richardson's law actually holds only for the *emission current density J_{s0} without field:*

$$J_{s0} = AT^2 \exp\left(-\frac{e\phi_0'}{kT}\right) \qquad (7.25)$$

Therefore, *in order to make a correct Richardson plot, one should first make a number of Schottky plots for different temperatures, determine J_{s0} for each plot, and then plot (J_{s0}/T^2) against $1/T$.*

Up to now we have discussed only emission into a vacuum. For emission into a semiconductor similar equations apply, but some obvious modi-

Fig. 7.5. Schottky plot, in which $\ln J_s$ is plotted against $F_0^{1/2}$. Extrapolation to $F_0 = 0$ yields $\ln J_{s0}$. Deviations from the linear relationship at low field strength may be due to various causes (space charge, surface inhomogeneity, etc.).

fications must be made. We must replace the work function ϕ by the barrier height E_b, and hence $\phi' = \phi/e$ by $E_b' = E_b/e$, the barrier height in electron volts. Moreover, one must now take into account the relative dielectric constant ϵ in the semiconductor, so that Eq. (7.19) must be replaced by

$$F = \frac{e^2}{4\pi\epsilon_0\epsilon(2x)^2} \qquad (7.19a)$$

Thus we must replace ϵ_0 by $\epsilon_0\epsilon$ everywhere. If we look carefully at Eq. (7.23), we see that F_0 must be replaced by F_0/ϵ in the second half of Eq. (7.23) and hence in Eq. (7.24). Consequently,

$$J_s = J_{s0} \exp\left[\frac{e(3.79 \times 10^{-5}\sqrt{F_0/\epsilon})}{kT}\right] \tag{7.24a}$$

and according to Eq. (7.5), if E_b' is the barrier height in electron volts at zero field,

$$J_{s0} = \frac{m_1^*}{m} AT^2 \exp\left(-\frac{eE_b'}{kT}\right) \tag{7.25a}$$

The problem of emission into a semiconductor has hereby been solved.

7.3. FIELD EMISSION

For field strengths of the order of 10^9 V/m at the surface of a cathode, a new effect occurs; the cathode starts to emit electrons even at room temperature. The effect is called *field emission*; it is a consequence of the wave character of the electron. The effect has found an interesting application in the "field emission microscope."

7.3a. Field Emission   *tunnel effect*

At field strengths of the order of 10^9 V/m, the potential barrier at the surface has become very thin. The electrons then have a finite probability of passing through the surface barrier and being emitted. The effect is called *tunnel effect*.

We calculate the field emission for $T = 0$. Figure 7.6 shows the potential barrier for the case of a metal with image forces neglected. Then, if the X axis is perpendicular to the surface, the kinetic energy due to electron

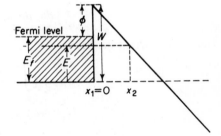

Fig. 7.6. Potential barrier for field emission from a metal point. $W = \phi + E_f$. The limits of integration, x_1 and x_2, are indicated for an energy E.

motion perpendicular to the surface inside the metal is $E = \frac{1}{2}m_1^* v_x^2$. The potential energy without image force is

$$V(x) = 0, \quad \text{for} \quad x < 0, \qquad V(x) = W - eF_0 x, \quad \text{for} \quad x > 0 \tag{7.26}$$

where F_0 is the field strength. Applying the theory of Appendix Sec. A.1c, we have

$$x_1 = 0, \qquad x_2 = \frac{W - E}{eF_0} \tag{7.26a}$$

Since

$$\int_{x_1}^{x_2} (W - E - eF_0 x)^{1/2}\, dx = \frac{2}{3}\frac{(W - E)^{3/2}}{eF_0} \tag{7.26b}$$

the transmission coefficient T of the barrier [Eq. (A.30)] is

$$T = \exp\left[-\frac{4}{3}\left(\frac{2m}{\hbar^2}\right)^{1/2}\frac{(W-E)^{3/2}}{eF_0}\right] \tag{7.27}$$

where m is the mass of the free electron in the vacuum; it is obvious that this mass must be used, for the tunneling described by (A.30) occurs in the vacuum and not in the metal. The theory of Sec. A.1c assumes that the tunneling occurs at constant energy, so for the energy E we can take the value $\frac{1}{2}m_1^* v_x^2 = p_x^2/2m_1^*$, where p_x is the momentum in the X direction and m_1^* the effective mass of the electron in the metal.

The same treatment can also be applied to tunnel emission into a semiconductor. The only differences are that the work function ϕ must be replaced by the barrier height E_b, and m must be replaced by m_2^*, the effective mass of the electron in the semiconductor. It is thus sufficient to treat the vacuum case.

We now bear in mind that the carrier density with momentum between p_x and $(p_x + dp_x)$, p_y and $(p_y + dp_y)$, and p_z and $(p_z + dp_z)$ is $(2/h^3)\cdot dp_x\, dp_y\, dp_z$, so that the number arriving at the surface per square meter per second is

$$v_x\left(\frac{2}{h^3}\right) dp_x\, dp_y\, dp_z, \qquad v_x > 0 \tag{7.28}$$

Hence the current density is

$$J = \frac{2e}{m_1^* h^3}\int\int\int p_x T(p_x)\, dp_x\, dp_y\, dp_z \tag{7.28}$$

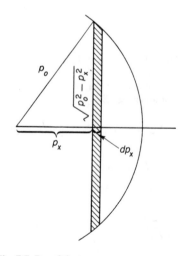

Fig. 7.7. Proof that
$$\int\int dp_y\, dp_z = \pi(p_o^2 - p_x^2)$$

where $T(p_x)$ is the transmission coefficient and the integration must be carried out for all electrons in the conduction band. That is, if E_f is the Fermi energy and $p_0 = (2m_1^* E_f)^{1/2}$, we must require that

$$p_x^2 + p_y^2 + p_z^2 \leqq 2m_1^* E_f = p_0^2 \tag{7.28a}$$

To integrate with respect to p_y and p_z, we turn to Fig. 7.7, which shows that the result is

$$\pi(p_0^2 - p_x^2)$$

Therefore, Eq. (7.28) may be written

$$J = \frac{2\pi e}{m_1^* h^3}\int_0^{p_0} T(p_x)p_x(p_0^2 - p_x^2)\, dp_x \tag{7.28b}$$

We now introduce $\theta = p_0 - p_x$ as a new variable. Since $T(p_0 - \theta)$ decreases rapidly with increasing θ, only contributions for small θ are significant. We may thus write, in good approximation,

$$p_x \simeq p_0, \qquad p_0^2 - p_x^2 = (p_0 + p_x)(p_0 - p_x) \simeq 2p_0\theta$$

$$(W - E)^{3/2} = \left(\phi + \frac{p_0^2 - p_x^2}{2m_1^*}\right)^{3/2} \simeq \phi^{3/2} + \frac{3}{2}\frac{\phi^{1/2}}{m_1^*}p_0\theta \qquad (7.29)$$

as is found from a Taylor expansion in θ. The lower limit for θ is of course 0; since $T(p_0 - \theta)$ decreases so rapidly with increasing θ, the upper limit of integration can be extended to ∞. Substituting into Eq. (7.28b) yields†

$$J = \frac{4\pi e p_0^2}{m_1^* h^3} \exp\left[-\frac{4}{3}\left(\frac{2m}{\hbar^2}\right)^{1/2}\frac{\phi^{3/2}}{eF_0}\right] \cdot \int_0^\infty \exp\left[-2\left(\frac{2m}{\hbar^2}\right)^{1/2}\frac{\phi^{1/2}}{m_1^* eF_0}p_0\theta\right]\theta \, d\theta$$

$$= \frac{m_1^*}{m}\frac{e^3 F_0^2}{8\pi h\phi}\exp\left[-\frac{4}{3}\left(\frac{2m}{\hbar^2}\right)^{1/2}\frac{\phi^{3/2}}{eF_0}\right] \qquad (7.30)$$

No large error is made if the factor m_1^*/m is neglected.

The result for the tunneling into a semiconductor is instead

$$J = \frac{m_1^*}{m_2^*}\frac{e^3 F_0^2}{8\pi h E_b}\exp\left[-\frac{4}{3}\left(\frac{2m_2^*}{\hbar^2}\right)^{1/2}\frac{E_b^{3/2}}{eF_0}\right] \qquad (7.30a)$$

as follows immediately from the discussion at the beginning of the section.

Expressing these current densities in MKS units (F_0 in volts per meter, $\phi' = \phi/e$ in volts, and $E_b' = E_b/e$ in volts), we obtain

$$J = \frac{m_1^*}{m} \cdot \frac{1.54 \times 10^{-6}F_0^2}{\phi'}\exp\left(\frac{-6.83 \times 10^9\phi'^{3/2}}{F_0}\right) \qquad (7.31)$$

and a similar expression for Eq. (7.30a).

If the image force is taken into account, Eq. (7.26) is replaced by

$$V(x) = 0, \quad \text{for } x < 0, \qquad V(x) = W - eF_0 x - \frac{e^2}{16\pi\epsilon_0 x}, \qquad (7.32)$$

for $x > 0$

In that case the barrier height changes from ϕ_0, the value without field, to the value

$$\phi = \phi_0 - \frac{1}{2}e\left(\frac{eF_0}{\pi\epsilon_0}\right)^{1/2} = \phi_0\left(1 - \frac{3.79 \times 10^{-5}F_0^{1/2}}{\phi_0'}\right) = \phi_0(1 - y) \qquad (7.32a)$$

according to Sec. 7.2, where

$$y = \frac{3.79 \times 10^{-5}F_0^{1/2}}{\phi_0'} \qquad (7.32b)$$

---

† $\displaystyle\int_0^\infty \exp(-a\theta)\theta \, d\theta = \frac{1}{a^2}$, where $a = 2\left(\frac{2m}{\hbar^2}\right)^{1/2}\frac{\phi^{1/2}}{m_1^* eF_0}p_0$

If we now repeat the calculation, we obtain numerically

$$J = \frac{m_1^*}{m} \frac{1.54 \times 10^{-6} F_0^2}{\phi_0'} \exp\left[-\frac{6.83 \times 10^9 \phi_0'^{3/2} v(y)}{F_0} \right] \tag{7.33}$$

where $v(y)$ is Nordheim's elliptic function, which varies slowly from the value $v(y) = 1$ for $y = 0$ to the value $v(y) = 0$ for $y = 1$. It should be noted that $y = 1$ corresponds to $\phi = 0$; in other words the barrier height disappears for $y = 1$. It is usually permitted to put $v(y) = 1$ unless great accuracy is required.

The case of field emission into a semiconductor is quite similar; the only difference is that ϵ_0 must be replaced by $\epsilon_0\epsilon$ everywhere, so that in the parameter y the field strength F_0 must be replaced by F_0/ϵ, according to Eq. (7.32a). Moreover, the expression in the exponential must be multiplied by $(m_2^*/m)^{1/2}$, and m_1^*/m becomes m_1^*/m_2^*.

In experimental investigations one usually studies the field emission from fine points (radius of less than 1 μm) of single-crystal material, such as tungsten; the fine point makes it easier to obtain a large field strength. Current densities up to 10^{12} A/m² have been observed. At those densities Eq. (7.33) is still satisfied, but there is no longer a linear relationship between F_0 and the applied voltage V. Measuring J as a function of V, one obtains Child's law ($\frac{3}{2}$ power law) for J at those current densities (compare Chapter 8).

7.3b. Field Emission Microscope

In the field emission microscope a large voltage V (10–20 kV) is applied between a fine single-crystal metal point of radius r and a fluorescent screen of radius R (Fig. 7.8), the point being in the center of the spherical screen. Since the potential distribution is spherical, electrons emitted by the point will move in radial direction. Atoms adsorbed on the point will give rise to a local decrease or increase of the work function ϕ', and hence to a local increase or decrease in field emission. Because of the occurrence of ϕ' in the exponential function in (7.33), a small change in ϕ' will cause a considerable change in the field emission density, which will show up on the fluorescent screen. Small spots of lower work function thus give bright spots on the screen.

The spherical geometry is the cause of a considerable magnification, because the electrons move in radial direction; it is easily seen that the magnification factor is R/r. If $R = 0.10$ m and $r = 10^{-8}$ m (100 Å), then the magnification factor is 10^7. A calculation shows that the resolving power of the microscope is very high and that it is possible to see individual molecules in this manner. This has been verified experimentally.

The field emission microscope is an important tool in the study of surface adsorption and allows the accurate measurement of the work function of

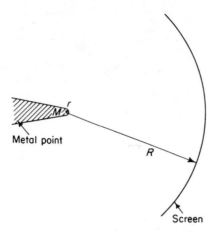

Fig. 7.8. The field emission microscope. A large voltage (10 to 20 kv) is applied between a fine, single-crystal, metal point and a fluorescent screen, which forms part of two concentric spheres of radii r(e.g., 1000 Å) and R(e.g., 10 cm), respectively. The amplification is R/r.

individual crystal faces. The reader is referred to the original literature on the subject.

7.3c. Field Ion Microscope

If in the field emission microscope the field is reversed and raised to about 10^{10} V/m, and if a small amount of helium is present, the helium atoms will be ionized at or near the surface. Owing to the field, they will then impinge upon the screen, where they will give an enlarged image of the surface.

The advantage of this configuration is that the helium atoms are much heavier than the electrons, so that the effects of tangential velocities and of diffraction (the wavelength $\lambda = h/mv$ is much smaller) are much less pronounced. As a consequence, much clearer and much more detailed pictures can be taken of the surface. Protruding atoms become clearly visible on such pictures.

Both the field emission microscope and the field ion microscope have become important tools in the study of surface conditions.

7.3d. Metal–Oxide–Metal Diode

By evaporating an aluminum strip on a glass plate, oxidizing the aluminum surface, and then evaporating another strip over it at about right angles (Fig. 7.9a), one obtains a metal–oxide–metal diode consisting of a thin oxide layer (20–200 Å thick) sandwiched between two metal electrodes (Fig. 7.9b). If a potential difference is maintained between the electrodes, current will flow by field emission (tunneling through the barrier).

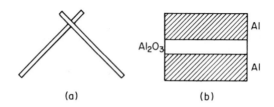

(a)   (b)

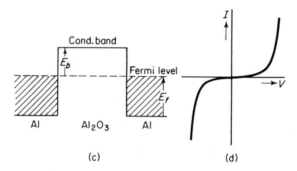

(c)   (d)

Fig. 7.9. (a) View of metal-oxide-metal device. (b) Cross section. (c) Energy level diagram at equilibrium. (d) Characteristic.

The theory of Sec. 7.3a is applicable, provided that the modifications mentioned in the text are made. Fields of the order of a few million volts per centimeter are needed to produce an appreciable current. If the two metal electrodes are identical, the characteristic is antisymmetrical (Fig. 7.9d). The temperature dependence of J is caused by the temperature dependence of E_b in this case.

In some units, especially at higher temperatures, the current may flow by "emission over the barrier" rather than by "tunneling through the barrier." The characteristic is then similar to the characteristic for the Schottky effect. Owing to the image force, the effective barrier height E_b'' will be less than the barrier height E_b' for zero applied voltage. The emission current is then

$$J = AT^2 \exp\left(-\frac{eE_b''}{kT}\right) \qquad (7.34)$$

where

$$E_b'' = E_b' - \frac{1}{2}\left(\frac{eF_0}{\pi\epsilon\epsilon_0}\right)^{1/2} = E_b' - 3.79 \times 10^{-5}\left(\frac{F_0}{\epsilon}\right)^{1/2} \qquad (7.34a)$$

This has a voltage dependence differing from the tunnel emission case and an appreciable temperature dependence.

Metal–oxide–metal diodes have the advantage that they can be made by evaporation techniques in a manner compatible with integrated circuitry.

Their chief disadvantage seems to be their sensitivity to overload; they easily burn out. For that reason they have not lived up to earlier expectations.

REFERENCES

JENKINS, J., *Basic Principles of Electronics*, vol. 1, Thermionics. Pergamon Press, Inc., Elmsford, N.Y., 1966.

MULLER, E. W., "Field Ionization and Field Ion Microscopy," in L. Marton, ed., *Advances in Electronics and Electron Physics*, vol. 13, pp. 83–181. Academic Press, Inc., New York, 1960.

PROBLEMS

1. An L-cathode (see Chapter 8) has a cathode area of 0.10 cm² and is operated at 1140°C. A continuous current of 1.0 A is drawn. Assuming the work function of the cathode to be 2.00 eV, calculate the extra heater power needed to offset the heat loss due to the electron emission.

Answer: 2.24 W.

2. Thoriated tungsten cathodes have an emission current density of 1 A/cm² at a cathode temperature of 1600°C.
(a) Calculate the true work function ϕ', assuming $A = 120$ A/cm²/degree².
(b) Measurements give $A' = 15$ A/cm² and $\phi_0' = 2.85$ eV. Calculate the temperature coefficient of the work function both from the difference between A' and A and from the difference between ϕ' and ϕ_0'.

Answer: 3.20 eV; 1.8×10^{-4} and 1.9×10^{-4} eV/degree.

3. According to the original publications an L-cathode has an emission current density of 1 A/cm² at 950°C and 100 A/cm² at 1320°C. Assuming the validity of Richardson's equation and $A = 120$ A/cm²/degree², calculate the following:
(a) The true work function at both temperatures.
(b) The temperature coefficient of the work function.
(c) The values of ϕ_0' and A'.

Answer: $\phi_{1220}' = 2.004$ eV; $\phi_{1590}' = 2.050$ eV; $\alpha = 1.24 \times 10^{-4}$ eV/degree; $\phi_0' = 1.85$ eV; $A' = 29$ A/cm²/degree².

4. Huxford measured the work function ϕ_0' and the emission constant A' for BaO cathodes and found $A' = 1.2 \times 10^{-2}$ A/cm²/degree² and $\phi_0' = 1.24$ eV. Calculate, under the usual assumption concerning A, the following:
(a) The temperature coefficient of the work function.

(b) Assuming that the cathode was operated at 1100°K, calculate the true work function at that temperature.

Answer: $\alpha = 7.9 \times 10^{-4}$ eV/degree; $\phi' = 2.11$ eV.

5. In the method for determining the cooling power of thermionic emission, as described in Sec. 7.1c, the filament direct voltage before emission current is drawn is V_f. After emission current is drawn a filament direct voltage $V_f + \Delta V_f$ is needed to keep the filament resistance R the same in the two cases.
(a) Calculate the cooling power, assuming $\Delta V_f \ll V_f$.
(b) How would you determine ΔV_f accurately?

Answer: $2V_f \Delta V_f / R$.

6. A saturated planar diode with oxide-coated cathode operates at a temperature of 800°K. It has a zero-field emission current of 1 mA and a cathode–anode distance of 0.1 mm. Assuming Schottky's law to be valid, calculate the emission current at $V_a = 100$ V.

Answer: 1.73 mA.

7. (a) Calculate the internal conductance of a saturated diode due to Schottky effect (electrode distance d, cathode temperature T).
(b) g can be accurately measured and so can the applied voltage $V_a = F_0 d$, but d is not measured so accurately. Eliminating d, show that the formula for Schottky effect may be written

$$I_s = I_{s0} \exp\left(\frac{2g V_a}{I_s}\right)$$

(c) Show how you would determine the zero-field emission current I_{s0} from measurements at a single operating voltage V_a. Sum up the advantages over the usual extrapolation method.

Answer: $g = 0.219 I_s / (dT \sqrt{F_0})$.

8. Apply the results of problem 7(a) to problem 6, and calculate the internal resistance of that saturated diode.

Answer: $R = 0.21$ MΩ.

9. (a) Calculate the conductance g of a field-emission diode. You may neglect the dependence of $v(y)$ upon F_0 and hence upon the applied voltage V_a. You should take into account that $dF_0/dV_a = F_0/V_a$ is a constant of the diode.
(b) Apply the result to the case $\phi' = 4.50$ eV, $I = 100$ μA, $V_a = 10^4$ V, $F_0 = 10^7$ v/cm, and $v(y) = 1$.

Answer: $g = \left[2 + \dfrac{6.83 \times 10^9 \phi'^{3/2} v(y)}{F_0}\right] \dfrac{I}{V_a}$; $g = 0.67$ μS.

8

Applications of
Thermionic Emission

8.1. CHARACTERISTIC OF A PLANAR THERMIONIC
DIODE

8.1a. Langmuir's Theory

To calculate the I_a, V_a characteristic of a planar diode we must solve
Poisson's equation

$$\frac{d^2V}{dx^2} = -\frac{\rho(x)}{\epsilon_0}$$
(8.1)

where $\rho(x)$ is the space-charge density and $\epsilon_0 = 8.85 \times 10^{-12}$ F/m, under
the initial conditions that the potential V is zero at the cathode ($x = 0$) and
equal to V_a at the anode ($x = d$), and that the current density is J_a. We must
now distinguish three parts of the characteristic.

1. The region where $dV/dx > 0$ everywhere. In that case all the emitted
electrons are collected by the anode, and the current density J_a is equal to
the emission current density J_s.

2. The region where $dV/dx < 0$ everywhere. In that case only those electrons that have an initial energy $V_0 > -V_a$ can arrive at the anode. According to Eq. (7.8) the number of electrons emitted per second with an energy
between V_0 and $V_0 + dV_0$ is

$$N(V_0)\, dV_0 = \frac{J_s}{e} \exp\left(-\frac{eV_0}{kT_c}\right) d\frac{eV_0}{kT_c}$$
(8.2)

where T_c is the cathode temperature. The anode current density is

$$J_a = \int_{-V_a}^{\infty} eN(V_0)\, dV_0 = J_s \int_{-eV_a/kT_c}^{\infty} \exp{(-u)}\, du = J_s \exp{\left(\frac{eV_a}{kT_c}\right)} \quad (8.3)$$

This gives therefore the exponential part of the characteristic.

3. The intermediate region where $dV/dx = 0$ somewhere between cathode and anode. Let this occur at $x = d_m$; then the potential V has a minimum $-V_m$ at that point. The current is now limited by the space charge. The calculation of this part of the characteristic is rather tedious.

Let $n(V_0)\, dV_0$ again be the energy distribution of the emitted electrons; then the space density $\rho(x)$ at a distance x from the cathode is

$$\rho(x) = -e \int \frac{n(V_0)\, dV_0}{v} \quad (8.4)$$

where v is the velocity of the electrons at x. The integration must be carried out over all electrons passing x. For $d_m \leqq x \leqq d$

$$\rho(x) = -e \int_{V_m}^{\infty} \frac{n(V_0)\, dV_0}{v} \quad (8.4a)$$

since only those electrons that pass the potential minimum $(V_0 > V_m)$ contribute to the space charge. For $0 \leqq x \leqq d_m$,

$$\rho(x) = -e \int_{V_m}^{\infty} \frac{n(V_0)\, dV_0}{v} - 2e \int_{-V}^{V_m} \frac{n(V_0)\, dV_0}{v} \quad (8.4b)$$

The first term comes from the electrons that pass the potential minimum, whereas the second term is due to electrons turning between x and d_m; these contribute twice to the space charge.

Introducing this into Poisson's equation (8.1), shifting the origin of the coordinate system from the cathode to the potential minimum, multiplying by $2\, dV$ and integrating, we obtain

$$\int_{-V_m}^{V} 2\frac{d^2V}{dx^2}\, dV = \int_{-V_m}^{V} d\left(\frac{dV}{dx}\right)^2 = \left(\frac{dV}{dx}\right)^2 = \frac{2}{\epsilon_0} \int_{-V_m}^{V} \rho(x)\, dV \quad (8.5)$$

Bearing in mind that

$$v = \left(\frac{2e}{m}\right)^{1/2} (V_0 + V)^{1/2} \quad (8.6)$$

and carrying out the integrations,† we get, for $0 \leqq x \leqq d_m$,

$$\left(\frac{dV}{dx}\right)^2 = \left(\frac{kT_c}{eL}\right)^2 \left[\exp{(\eta)} - 1 + \exp{(\eta)}\,\text{erf}\,(\eta^{1/2}) - 2\left(\frac{\eta}{\pi}\right)^{1/2}\right]$$
$$= \left(\frac{kT_c}{eL}\right)^2 h^-(\eta) \quad (8.5a)$$

---

† We omit here the details of the calculations, which are tedious and uninteresting, in order to concentrate on the essentials of the results. We refer those interested to Langmuir's original papers.

and, for $d_m \leqq x \leqq d$,

$$\left(\frac{dV}{dx}\right)^2 = \left(\frac{kT_c}{eL}\right)^2 \left[\exp(\eta) - 1 - \exp(\eta)\operatorname{erf}(\eta^{1/2}) + 2\left(\frac{\eta}{\pi}\right)^{1/2}\right]$$

$$= \left(\frac{kT_c}{eL}\right)^2 (h^+\eta)$$

(8.5b)

Here $\eta = e(V + V_m)/kT$, erf stands for the error function

$$\operatorname{erf}(x) = \frac{2}{\sqrt{\pi}} \int_0^x \exp(-u^2)\,du$$

and

$$\frac{1}{L} = \frac{2^{1/2}\pi^{1/4}J_a^{1/2}}{\epsilon_0^{1/2}(2e/m)^{1/4}(kT_c/e)^{3/4}} = 0.92 \times 10^6 \frac{J_a^{1/2}}{T_c^{3/4}}$$

(8.7)

Introducing the new dependent variable η and furthermore the new independent variable ξ by the definition

$$-\xi = \xi^- = \frac{d_m - x}{L}, \qquad \text{for } 0 \leqq x \leqq d_m$$

$$\xi = \xi^+ = \frac{x - d_m}{L}, \qquad \text{for } d_m \leqq x \leqq d$$

(8.8)

we may rewrite Eqs. (8.5a) and (8.5b) as

$$\left(\frac{d\eta}{d\xi^-}\right)^2 = h^-(\eta), \qquad \left(\frac{d\eta}{d\xi^+}\right)^2 = h^+(\eta)$$

(8.9)

Langmuir[†] solved these equations numerically and tabulated the results. For more extensive tables see Kleynen.[‡]

Denoting the values of ξ and η at cathode and anode by ξ_c^- and η_c and ξ_a^+ and η_a, respectively, we have

$$\xi_c^- = \int_0^{\eta_c} \frac{du}{[h^-(u)]^{1/2}} = \frac{d_m}{L}$$

$$\xi_a^+ = \int_0^{\eta_a} \frac{du}{[h^+(u)]^{1/2}} = \frac{d - d_m}{L}$$

(8.9a)

where

$$\eta_c = \frac{eV_m}{kT_c}, \qquad \eta_a = \frac{e(V_a + V_m)}{kT_c}$$

(8.9b)

and

$$J_a = e \int_{V_m}^{\infty} n(V_0)\,dV_0 = J_s \exp(-\eta_c)$$

(8.10)

The calculation of the characteristic now proceeds as follows. Since J_s is known, an assumed value of J_a gives η_c, L, and ξ_c^- from the tables. This yields d_m, which, in turn, gives ξ_a^+, and ξ_a^+ finally gives η_a and V_a according to the tables.

† I. Langmuir, *Phys. Rev.*, **21**, 419–435 (1923).

‡ P. H. J. A. Kleynen, *Philips Research Reports*, **1**, 81–96 (1946).

For relatively large values of V_a and relatively small values of J_a/J_s, asymptotic expansions may be obtained for ξ_c^- and ξ_a^+, since $h^-(\eta)$ and $h^+(\eta)$ may then be approximated as

$$h^-(\eta) = 2 \exp{(\eta)}, \qquad h^+(\eta) = 2\left(\frac{\eta}{\pi}\right)^{1/2} - 1 \qquad (8.11)$$

Substituting into (8.9a) and carrying out the integrations yields

$$\xi_c^- = \frac{d_m}{L} = \int_0^\infty \frac{du}{[h^-(u)]^{1/2}} - \int_{\eta_c}^\infty \frac{du}{[h^-(u)]^{1/2}}$$

$$\simeq 2.554 - 2^{1/2} \exp{\left(-\frac{1}{2}\eta_c\right)} \qquad (8.11a)$$

$$= 2.554 - 2^{1/2} \left(\frac{J_a}{J_s}\right)^{1/2}$$

from which d_m may be evaluated. Moreover,

$$\xi_a^+ = \frac{d - d_m}{L} \simeq \int_{\pi/4}^{\eta_a} \frac{du}{[2(u/\pi)^{1/2} - 1]^{1/2}}$$

$$\simeq \frac{\pi^{1/4}2^{3/2}}{3}\eta_a^{3/4}\left(1 + \frac{3}{4}\pi^{1/2}\eta_a^{-1/2} + \cdots\right) \qquad (8.11b)$$

as follows from a Taylor expansion of the integrand. Solving for J_a, one obtains, after some manipulations and with the help of Eq. (8.7),

$$J_a = \frac{4}{9}\epsilon_0\left(\frac{2e}{m}\right)^{1/2}\frac{(V_a + V_m)^{3/2}}{(d - d_m)^2}\left\{1 + 3\left[\frac{(\pi/4)kT_c/e}{V_a + V_m}\right]^{1/2} + \cdots\right\} \qquad (8.12)$$

Despite the approximations involved, this expression is surprisingly accurate.

The expression $[(\pi/4)kT_c/e]^{1/2}$ corresponds to the average value $\bar{v}_x$ of the x component of the electron velocity at the cathode or at the potential minimum.† The first term in (8.12) thus corresponds to what would be expected if the electrons passed the potential minimum with zero velocity, and the second term gives the correction caused by the average velocity of the electrons at the potential minimum.

If $\bar{v}_x$, V_m, and d_m were zero, Eq. (8.12) would become

$$J_a = \frac{4}{9}\epsilon_0\left(\frac{2e}{m}\right)^{1/2}\frac{V_a^{3/2}}{d^2} \qquad (8.13)$$

This is called *Child's law*. This expression is usually not very accurate and

---

† According to Eq. (7.13a),

$$\frac{1}{2}m(\bar{v}_x)^2 = \frac{\pi}{4}kT_c = eV_0' \quad \text{or} \quad V_0' = \frac{\pi}{4}\frac{kT_c}{e}$$

it is better to use Eq. (8.12) instead. The corrections due to the position and the depth of the potential minimum and to $\bar{v}_x$ are often quite large, especially for low values of V_a, and are of approximately equal importance.

Thus far we have neglected a possible difference in work function between cathode and anode. Let ϕ'_c and ϕ'_a be the cathode and anode work functions in electron volts, respectively; then all one must do is to replace V_a by $V_a + \phi'_c - \phi'_a$. Thus at a given potential V_a a larger current I_a is obtained if the contact potential difference $\phi'_c - \phi'_a$ can be made larger. This is important for thermionic energy conversion.

Figure 8.1 shows the potential distribution in the diode for the saturated, space-charge-limited, and exponential conditions. Figure 8.2 shows schematically log I_a plotted versus V_a. Two very significant points of the characteristic are the *exponential point* and the *saturation point*. The exponential point gives the boundary between the exponential and the space-charge-limited region of the characteristic and corresponds to $\xi_a^+ = 0$. The saturation point gives the boundary between the space-charge-limited region and the saturated region of the characteristic and corresponds to $\xi_c^- = 0$.

It is often thought that the space-charge-limited condition is always associated with a potential minimum. This is not the case. At high current densities, field emission becomes space-charge-limited because the space

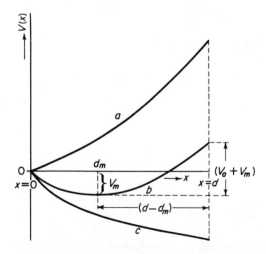

Fig. 8.1. Potential distribution in a diode for the saturated, space-charge-limited, and exponential parts of the characteristic (a, b, and c, respectively), if the velocity distribution of the electrons is taken into account. If the velocity distribution is disregarded, d_m and V_m become zero, so that $dV/dx = 0$ at $x = 0$ for the space-charge-limited diode (Sec. 8.1a).

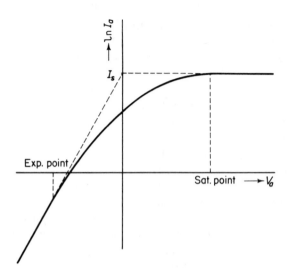

Fig. 8.2. Schematic diagram of the $I_a - V_a$ characteristic of a diode, showing the exponential region, the saturated region, the exponential point, and the saturation point. The anode and the cathode have the same work function in this example.

charge limits the *field strength* at the cathode and not because a potential minimum is developed. This is left as a problem for the reader (problem 8.6).

8.2. PRACTICAL CATHODES

8.2a. Metal and Atomic Film Cathodes

Metal cathodes have the disadvantage of a high work function (tungsten, 4.52 eV; tantalum, 4.13 eV) so that they give appreciable emission above 2000°K only. Since these metals have very high melting points (3640°K for W, 3300°K for Ta), they can be operated at those temperatures without difficulty. Tungsten cathodes are used in high-power tubes, since they stand up well against ion bombardment, a common phenomenon in those tubes, and since tungsten cleans up rapidly after contamination by various gases. Tantalum has the advantage of a lower work function, but it is easily contaminated by residual gases.

If atoms are adsorbed at the surface of a metal as positive ions, then the work function is lowered; if they are adsorbed as negative ions, the work function is raised. The reason is that the ion and its image charge form an electrical dipole of dipole moment M (coulomb-meter). According to electrostatics, the change in potential caused by a dipole layer having a dipole moment p per unit area (coulomb/meter) is p/ϵ_0. If there are N dipoles per unit area, the potential energy of an electron outside the cathode, taken with respect to the bottom of the conduction band, is thus

$$\varphi = W - \frac{eNM}{\epsilon_0} \tag{8.14}$$

Therefore, the work function in electron volts is

$$\phi' = \phi'_0 - \frac{NM}{\epsilon_0} \tag{8.14a}$$

where ϕ'_0 is the work function for the clean metal. Hence we see indeed that $\phi' < \phi'_0$ if $M > 0$ (positive ions adsorbed).

If the change in work function $\Delta\phi'$ is plotted as a function of the degree of coverage θ of the cathode (Fig. 8.3), it is found that $\Delta\phi'$ first varies linearly with θ, as expected from (8.14a), passes through a maximum at nearly full coverage, and then decreases again. The reason is that at higher degrees of coverage it becomes energetically more favorable for the atoms to be adsorbed as neutrals rather than as ions. Consequently, N may ultimately decrease with increasing degree of coverage.

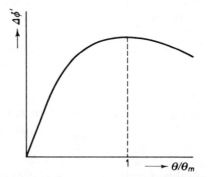

It would in principle be possible to deposit adsorbed atoms on top of the adsorbed ions, but since they are less strongly bound to the surface than the ions, they will tend to evaporate quickly at the operating temperature of the cathode. For that reason usually not much more than

Fig. 8.3. Change in work function $\Delta\phi'$ plotted against θ/θ_m (schematic), θ being the degree of coverage and θ_m the value of θ for which $\Delta\phi'$ is a maximum. In some cases $\Delta\phi'$ may be as large as a few electron volts.

a monatomic layer of atoms is adsorbed, hence the name *atomic film cathode*.

Usually tungsten is used as the base metal. One then has two possibilities. At low temperatures the tungsten surface is covered by a monatomic layer of oxygen, and if metal atoms are deposited afterward, they form a "metal ion on oxygen on tungsten" layer. At high temperatures the tungsten surface is clean, and if metal atoms are deposited afterward they form a "metal ion on tungsten" layer. The first layer has a lower work function and is more stable at higher temperatures than the latter. The most important atomic film cathodes are the thoriated tungsten cathodes, barium film cathodes, and cesium film cathodes.

Thoriated tungsten cathodes consist of tungsten with up to $1\frac{1}{2}$ per cent ThO_2 (thoria) added. The prepared cathode consists of a monatomic layer of Th on W; it gives an emission current density of 1 A/cm^2 at a temperature of 1600°C. The cathode is activated by "flashing"; that is, the filament is heated to a very high temperature (2800°K) for 1 or 2 minutes and then held at a temperature of 2100–2300°K for 15–30 minutes. At that temperature the thorium diffuses to the surface faster than it evaporates, so that a nearly monatomic layer builds up on the surface. After activation the filament is kept at the operating temperature of 1900°K, where it operates in a stable fashion for a long time. The thoriated tungsten cathode can be "poisoned" by residual gases, but emission properties can be restored by renewed activation. Heating the filament in a low-pressure hydrocarbon vapor (naphthalene or anthracene) at 1600°K changes the surface layer to W_2C, which has a much lower rate of thorium evaporation than pure tungsten; this "carbonization process" thus results in a much longer cathode life.

Barium film cathodes are of several types, including the L cathode and the impregnated cathode. In both of these the emitting layer consists of a nearly monatomic layer of barium on oxygen on tungsten, but they differ in the way the barium layer is formed. In the L cathode a molybdenum container, closed with a porous tungsten plug, contains a pill of $BaCO_3$ that is decomposed to BaO by heating and pumping. After the carbonate has been decomposed, free barium is formed in the reaction

$$6BaO + W \rightleftharpoons Ba_3WO_6 + 3Ba$$

This free barium then flows through the pores of the plug and covers its outside, forming what is nearly a monatomic layer of barium on oxygen on tungsten. In the impregnated cathodes porous tungsten is impregnated with a mixture of normal and basic barium aluminates, which is dispersed throughout the pores. In the activation process the tungsten reacts with the barium aluminates to form free barium that can migrate to and over the surface to form a monatomic layer of barium on oxygen on tungsten. The impregnated cathodes have the advantage of greater freedom of construction and simpler pumping and activation schedules.

The cesium film cathodes are not very important in themselves, but they illustrate the effect of monatomic surface layers upon emission. Figure 8.4 shows the emission current I_s of a clean tungsten filament, kept at a temperature T in a cesium atmosphere, as a function of time. This diagram shows that the time necessary for obtaining a maximum value of I_s is independent of temperature.

Figure 8.5 shows the logarithm of the emission current (ln I_s) plotted versus $10^3/T$ for a Cs–W and for a Cs–O–W layer on tungsten. At elevated cathode temperatures the emission corresponds to that of clean tungsten; at lower temperatures it corresponds to Cs–W or Cs–O–W, respectively. We see that W–O–Cs layers are more stable and have a higher emission than W–Cs layers. The experiments are simple to perform, since Cs has a vapor

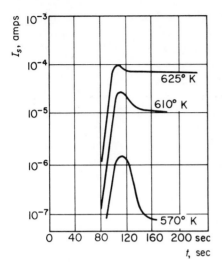

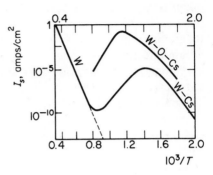

Fig. 8.4. Emission current I_s of a clean tungsten filament kept at a temperature T in a cesium atmosphere, as a function of time. The diagram shows that the time necessary for obtaining a maximum value of I_s is independent of the temperature. [J. A. Becker, *Phys. Rev.*, **28**, 341–361 (1926)].

Fig. 8.5. The values $\ln I_s$ plotted against $10^3/T$ for a Cs-W and a Cs-O-W cathode on tungsten. At elevated cathode temperatures, the emission corresponds to that of clean tungsten; at lower temperatures it corresponds to a Cs-W or Cs-O-W layer, respectively. Figs. 8.4, 8.5, and 8.6 are reproduced by permission from *The Oxide Coated Cathode*, Vol. 2, by G. Herrmann and S. Wagener (Chapman and Hall Ltd., London, 1951).

pressure of 5×10^{-6} mm Hg at room temperature, and higher pressures can be obtained by heating.

8.2b. Oxide-Coated Cathodes

Oxide-coated cathodes consist of a mixture of barium, strontium, and calcium oxides deposited on a nickel base. Since these oxides are not stable in air, the coatings are generally applied in the form of alkaline earth carbonates. Modern cathodes either are of the double-carbonate type (50 per cent $BaCO_3$, 50 per cent $SrCO_3$) or of the triple-carbonate type (50 per cent $BaCO_3$, 43 per cent $SrCO_3$, 7 per cent $CaCO_3$). The carbonates are decomposed by heating and pumping, leaving a $(Ba, Sr)O$ coating behind.

To be a good emitter, the cathode coating should have a low work function. Thus both the electron affinity χ (external work function) and the internal work function (difference between the bottom of the conduction band and the Fermi level) should be small. For the latter the material should be an *n*-type semiconductor with relatively shallow donors. The electron affinity can be lowered by adsorption of Ba atoms on the surface grains. Both of these desirable characteristics are obtained in the activation process.

In the activation process impurities in the nickel, called *reducing agents*, reduce the oxides to the free alkali metal. Commonly used reducing agents are Si, Al, Mg, Ti, and W; often only a few hundredths of 1 per cent of impurity is needed. The barium vapor will flow through the pores and partially cover the surface of the (Ba, Sr)O grains with Ba atoms. The adsorbed Ba atoms form donors at the surface of the grains; these donors gradually diffuse into the grains and cause a more or less homogeneous distribution of donors through the grains. Because of their size, the Ba ions do not move, but they form oxygen vacancies at the surface that diffuse into the grains. These vacancies are normally occupied by electrons and act as donors.

Some time is required before an equilibrium in the donor distribution is established. Consequently, vacuum tubes are usually not yet fully activated when taken from the pump. Further activation is obtained by heating at an elevated temperature and by drawing current. In the latter process the oxygen ions seem to move toward the surface, leaving an excess of alkali atoms (or oxygen ion vacancies) behind; this follows from the fact that the rate at which oxygen evaporates from the surface *increases* if current is drawn from the cathode.

If Richardson's law is assumed to be valid for the oxide coating and if one puts $A = 1.20 \times 10^6$ A/m²/degree, emission current density measurements may be directly translated into corresponding work functions. Figure 8.6 shows results obtained by Huber, indicating that the lowest work function of (Ba, Sr)O mixtures is obtained at about 50 per cent concentration of each of the constituents.

There is a difference between dc and pulsed-current emission. If the

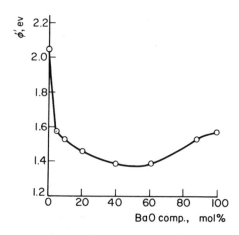

Fig. 8.6. Work function ϕ' in electron volts as a function of the BaO concentration of a (Ba, Sr)O coating with variable BaO content, calculated directly from the observed emission (H. Huber, Thesis, University of Berlin, 1941).

current is drawn in short pulses, the full emission potential of the cathode is used. But if dc current is drawn, the surface is depleted of donors (oxygen vacancies), and the energy-level diagram will curve upward near the surface, resulting in an increase in the effective work function.

8.3. THERMIONIC ENERGY CONVERSION

8.3a. Narrow-Spacing Vacuum Diode Converter

Since the I_a, V_a characteristic of a vacuum diode gives positive currents for negative anode voltages, a vacuum diode can *deliver* dc power for negative V_a. The amount of power is

$$P_a = -I_a V_a \qquad (8.15)$$

The design of practical energy converters now consists in optimizing (8.15).

First, let us neglect space-charge effects in the diode. Let ϕ_c' and ϕ_a' be the work functions of cathode and anode in electron volts, respectively, and let T_c and T_a be the absolute temperatures of cathode and anode. Then the emission current densities J_{sc} and J_{sa} of the cathode and anode are

$$J_{sc} = AT_c^2 \exp\left(-\frac{e\phi_c'}{kT_c}\right), \qquad J_{sa} = AT_a^2 \exp\left(-\frac{e\phi_a'}{kT_a}\right) \qquad (8.16)$$

The first requirement that must be made is that $J_{sa} \ll J_{sc}$. If T_c is given, this inequality settles the upper limit for T_a.

Suppose that T_a has been chosen properly. Then $J_a = J_{sc}$ for $V_a > \phi_a' - \phi_c'$ and $J_a = J_{sc} \exp\left[e(V_a + \phi_c' - \phi_a')/kT_c\right]$ for $V_a < \phi_a' - \phi_c'$. If $\phi_c' - \phi_a'$ is equal to at least a few times kT_c/e, the optimum power per unit area is approximately given by

$$P_{\text{opt}} = P_{\text{max}} = J_{sc}(\phi_c' - \phi_a') \qquad \text{W/m}^2 \qquad (8.17)$$

In practice the optimum power is less because of space-charge effects. According to Langmuir's theory, these effects become much more pronounced if the electrode distance d increases. For constant $\phi_c' - \phi_a'$ and given V_a, the value of J_a, and as a consequence the value of P_{opt}, decreases strongly with increasing d. This is shown quite clearly in Fig. 8.7. The smaller we can choose the value of d, the closer P_{opt} will approximate the value P_{max}.

The actual determination of P_{opt} is best done from Langmuir's theory of the diode. If T_c and ϕ_c' are known, J_{sc} is known; if d and $\phi_c' - \phi_a'$ are also known, the J_a, V_a characteristic can be determined. Plotting $-J_a V_a$ against $-V_a$, one can determine the optimum value P_{opt}.

Note that up to now we have neglected the resistance of the electrode lead wires. In practice this resistance gives some reduction in P_{opt}.

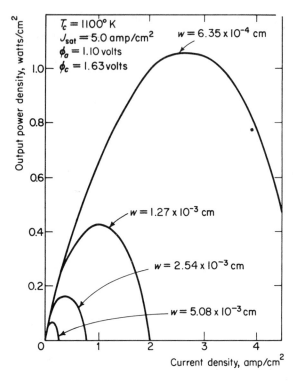

Fig. 8.7. Output power density versus current density for four different interelectrode spacings *w*. The conditions assumed for these calculations are shown at the top of the figure. (From S. W. Angrist, *Direct Energy Conversion*, Allyn and Bacon, Inc., Boston. Copyright 1965, reprinted by permission from the publisher.)

The heat balance of the thermionic energy converter can be set up as follows:

1. The power needed to emit the electrons. This gives a term

$$J_a\left(\phi_c' + \frac{2kT_c}{e}\right) \tag{8.18}$$

since the emitted electrons have an average kinetic energy $2kT_c/e$.

2. The power lost by radiation. This gives a term

$$\sigma(\epsilon_c T_c^4 - \epsilon_a T_a^4) \tag{8.19}$$

Here σ is the Stefan–Boltzmann constant, and ϵ_c and ϵ_a are the emissivities of cathode and anode, respectively.

3. The power lost by heat conduction. This gives a term

$$\kappa \frac{A}{L}(T_c - T_0) \tag{8.20}$$

where κ is the heat conductivity of the cathode lead, A is its cross-sectional area, L its length, and T_0 the temperature of the environment. This term is not negligible, for according to the Wiedemann–Franz law there is a relationship between heat conductivity and electrical conductivity of a metallic conductor; in order to obtain a relatively low electrical resistance of the leads one also gets a relatively large heat conductivity of the leads.

In well-constructed energy converters the first process predominates. Adding all the terms, one can determine the efficiency η of the thermionic energy converter as

$$\eta = \frac{P_{\text{opt}}}{P_{\text{loss}}} \tag{8.21}$$

where P_{loss} is the sum of the power terms discussed before.

8.3b. Cesium Diode Energy Converter

If cesium is admitted to the diode, the hot cathode produces cesium ions that can neutralize the space charge. Under optimum conditions practically all space charge can be neutralized; it is then not so necessary to keep the electrode spacings very small. One can distinguish between the low-pressure Cs diode and the high-pressure Cs diode.

In the low-pressure Cs diode the Cs atoms arriving at the hot cathode are ionized and then evaporated, giving rise to a cesium ion current J_i flowing from cathode to anode. If space charge is completely neutralized, and the field between cathode and anode is zero,

$$\frac{J_i}{v_i} = \frac{J_{sc}}{v_e} \tag{8.22}$$

where v_i is the ion velocity and v_e the electron velocity. Since the electrons and the ions have equal kinetic energy at all points, the ratio of their velocities is $(m_i/m_e)^{1/2}$, where m_e and m_i are the electron and ion masses, respectively, so that

$$J_i = J_{sc}\left(\frac{m_e}{m_i}\right)^{1/2} \tag{8.22a}$$

Since $m_e \ll m_i$, $J_i \ll J_{sc}$, so that J_i gives only a negligible reduction in the total current. Since J_{sc} is given, J_i is fully determined. But we may also write

$$J_i = \tfrac{1}{4}en\bar{u}_i \tag{8.22b}$$

where n is the cesium atom density and $\bar{u}_i$ the average ion velocity. It is assumed here that each cesium atom striking the cathode results in the emission of a Cs ion. Consequently, n, and hence the minimum Cs pressure, is fully prescribed.

In low-pressure Cs diodes collisions between electrons and Cs atoms are rare; in high-pressure Cs diodes collisions between the electrons and Cs atoms are quite frequent, so that the electron mean free path is much smaller than the electrode spacing. This has pronounced effects on the operation of the diode as an energy converter, but the details are beyond the scope of this book.

REFERENCES

Angrist, S. W., *Direct Energy Conversion*. Allyn and Bacon, Inc., Boston, 1965.

Chang, S. S. L., *Energy Conversion*. Prentice-Hall, Inc., Englewood Cliffs, N.J., 1963.

Herrmann, G., and S. Wagener, *The Oxide-Coated Cathode*, vols. 1 and 2. Chapman & Hall Ltd., London, 1951.

Houston, J. M., and H. F. Webster, "Thermionic Energy Conversion," in L. Marton, ed., *Advances in Electronics and Electron Physics*, vol. 17, pp. 125–206. Academic Press, Inc., New York, 1962.

Kaye, J., and J. A. Welsh, eds., *Direct Energy Conversion of Heat to Electricity*. John Wiley & Sons, Inc., New York, 1960.

Levine, S. N., ed., *Selected Papers on New Techniques for Energy Conversion*. Dover Publications, Inc., New York, 1961.

PROBLEMS

1. A planar diode is operating in the exponential part of its characteristic. The diode is found to have a conductance of 150 μS at a plate current of 10 μA.
(a) Calculate the cathode temperature.
(b) How would you find out that the tube was really operating in the exponential part of the characteristic? You cannot deduce it from a single measurement.

Answer: 773°K.

2. Assuming that the electrons in a planar space-charge-limited diode are emitted by the cathode with zero velocity, show that the transit time τ from cathode to anode is

$$\tau = \frac{3d}{(2e/m)^{1/2} V_a^{1/2}}$$

where d is the electrode spacing and V_a is the anode voltage. *Hint:* Make use of the following facts:

$$\tau = \int_0^d \frac{dx}{v}, \qquad v = \left(\frac{2e}{m}\right)^{1/2} V^{1/2}, \qquad V = V_a\left(\frac{x}{d}\right)^{4/3}$$

3. Huber measured the emission of two types of CaO cathodes at 750°K. The first CaO cathode was self-activated; that is, the bulk coating was pure CaO and Ca ions were adsorbed to the surface. He found $I_s = 1.0 \times 10^{-8}$ A/cm². The second CaO cathode was barium-activated; that is, the bulk coating was pure CaO and Ba ions were adsorbed to the surface. He found $I_s = 1.8 \times 10^{-4}$ A/cm².
(a) Is there any difference in internal work function?
(b) Calculate the difference in external work function.

Answer: (a) No; (b) 0.63 eV.

4. According to Huber, the work function of pure CaO cathodes is 2.37 eV and the work function of a BaO–CaO cathode is 1.53 eV. Calculate the difference in internal work function of the two cathodes, taking for the difference in external work function the value found in problem 3. Why is this allowed?

Answer: 0.21 eV.

5. In a planar field emission tube the current density J at the cathode is a function $J(F_0)$ of the field strength F_0 at the cathode. By solving Poisson's equation, under the assumption that $F = F_0$ at $x = 0$ and $V = V_a$ at $x = d$, show that the anode potential follows from the equation

$$d = \frac{4}{3b^2}[(a + bV_a^{1/2})^{3/2} - a^{3/2}] - \frac{4a}{b^2}[(a + bV_a^{1/2})^{1/2} - a^{1/2}]$$

where $a = F_0^2$ and $b = 4J/[\epsilon_0(2e/m)^{1/2}]$. You may assume that the electrons are emitted with zero velocity.

6. Show that the relation in problem 5 reduces to Child's law if $bV_a^{1/2} \gg a$ (space-charge-limited condition).

9

Photoemission

In photoemission in solids, slow electrons are emitted under the influence of light absorbed by the solid; the energy needed for escape of the electrons is supplied by the absorbed light quanta. Experimentally, it is found that the *photocurrent is proportional to the light intensity.* This is understandable, for if a photon is absorbed by the solid and its energy is imparted to an electron, then the electron will have the probability p that it can escape from the surface. Since the individual processes are independent, the photocurrent should be proportional to the number of absorbed quanta per second, as found by experiment.

The interaction between electrons and photons can be treated as a collision process in which the laws of conservation of momentum and of energy hold. The momentum law has to be applied with some caution, however; one has to take into account that both the crystal lattice and the crystal surface can take up momentum.

The process of photoemission can be split into two parts, which must be discussed separately:

1. The *production process*, in which the energy of the quantum is imparted to the lattice electrons.

2. The *escape process*, in which the lattice electrons that have excess energy travel toward the surface; if their energy is large enough, they can escape from the surface.

9.1. MOMENTUM TAKEN UP BY THE LATTICE AND BY THE CRYSTAL SURFACE

For the theory of the photoelectric effect we must know the amounts of momentum that can be taken up by the lattice and by the crystal surface.†

We first treat the problem of the momentum that can be taken up by the *lattice*. Consider a beam of electrons incident upon a crystal with an angle of incidence φ. For certain angles of incidence, *Bragg reflection* will occur when

$$2d \sin \varphi = n\lambda, \qquad n = 0, 1, 2, \ldots \qquad (9.1)$$

Here d is the lattice spacing, $\lambda = h/p$ is the wavelength of the electrons, and $p = mv$ is the momentum of the electrons.

We can also treat this problem as a collision process in which the energy of the electron and the component of its momentum parallel to the surface do not change, whereas the component perpendicular to the surface changes from $p \sin \varphi$ to $-p \sin \varphi$, or in total by the amount $2p \sin \varphi$ (Fig. 9.1). According to Eq. (9.1), this may be written

$$\Delta p = 2p \sin \varphi = n\frac{h}{d}, \qquad n = 0, 1, 2, \ldots \qquad (9.2)$$

as we find by substituting $\lambda = h/p$. This shows *that the lattice can take up discrete amounts of momentum in the direction perpendicular to the surface; these amounts are an integral number times* h/d. The same result would have been obtained for Bragg reflection of x rays.

Equation (9.2) holds when the problem is treated as one-dimensional.

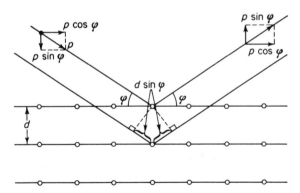

Fig. 9.1. Bragg reflection of electron waves or x rays treated as a collision process, to indicate that the lattice can take up amounts of momentum that are an integral number times h/d, where d is the lattice spacing.

---

† This is also important for the theory of secondary emission that is discussed in Chapter 10.

Actually the crystal is three-dimensional, so that appropriate components of momentum can be taken up in the directions perpendicular to three non-parallel crystal faces. The problem is simplest for a cubic crystal. Considering three perpendicular crystal faces and bearing in mind that the components of momentum that can be taken up in directions perpendicular to these crystal faces are integral numbers times h/d, we can extend Eq. (9.2) as follows (vector notation):

$$\Delta \mathbf{p} = \mathbf{n}\frac{h}{d} \tag{9.2a}$$

where $\mathbf{n}$ is a vector with integral components n_1, n_2, and n_3 in those three perpendicular directions. The reader will recognize that this corresponds to the following selection rule: *In optical transitions the reduced wave vector of the electron does not change.*

To find the amount of momentum that can be taken up by the *surface* of a crystal, we consider thermionic emission as a collision process between the lattice electrons and the surface. Suppose that the electrons can be treated as free. Let an electron move in a direction perpendicular to the surface with a velocity v_x; this direction will be considered the X direction. The energy E with respect to the bottom of the conduction band and its momentum p are (Fig. 9.2)

$$E = \tfrac{1}{2}mv_x^2, \qquad p = mv_x \tag{9.3}$$

If $E > W$, the electron escape with a velocity v'_x, such that its energy

$$E' = E - W = \tfrac{1}{2}mv_x'^2 \tag{9.4}$$

and the escaped electron has a momentum

$$p' = mv'_x = \sqrt{2m(E - W)} \tag{9.5}$$

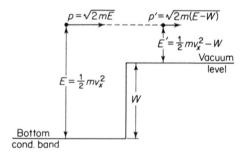

Fig. 9.2. Demonstration that a metal surface has to take up momentum if an electron is emitted. Before emission the electron is supposed to move perpendicularly to the surface with a kinetic energy E and a momentum $p = (2mE)^{1/2}$. After the electron is emitted, its kinetic energy is $(E - W)$ and its momentum is $p' = [2m(E - W)]^{1/2}$. The amount of momentum $(p - p')$ is taken up by the surface.

The change in momentum is thus

$$\Delta p = p - p' = \sqrt{2mE} - \sqrt{2m(E - W)} \tag{9.6}$$

If E changes continuously from $E = W$ to $E = \infty$, then Δp changes continuously from the maximum value $\sqrt{2mW}$ for $E = W$ to the minimum value of zero for $E = \infty$.

For $E < W$ the electron is reflected at the surface; in that case the surface takes up a momentum

$$\Delta p = 2mv_x = 2\sqrt{2mE} \tag{9.7}$$

which changes from the minimum value $\Delta p = 0$ for $E = 0$ to the value $\Delta p = 2\sqrt{2mW}$ for $E = W$. We thus conclude that *a crystal surface can take up continuous amounts of momentum between zero and* $2\sqrt{2mW}$.

9.2. VOLUME AND SURFACE PHOTOEFFECT

9.2a. Comparison of the Two Effects

In the surface photoeffect the electron is liberated directly by the incident light (Fig. 9.3a). Since the crystal surface can take up arbitrary amounts of momentum, the momentum law does not impose any restrictions on the quantum energy that can be imparted to the lattice electron. Therefore, in a metal at $T = 0$ the threshold energy for the surface photoelectric effect is

$$hf_0 = e\phi' \tag{9.8}$$

where ϕ' is the work function in electron volts. The reason is that in metals at $T = 0$ there are electrons with energies corresponding to the Fermi level E_f.

For $T > 0$, some energy levels for $E > E_f$ are also occupied, so that photoelectrons may be emitted for quantum energies $hf < hf_0$. If n quanta of energy hf arrive per unit area per second, the photoelectric current I is given by

$$I = enCT^2 S g\left[\frac{h(f - f_0)}{kT}\right] \tag{9.9}$$

where C is a materials constant, S the photocathode area and $g[h(f - f_0)/kT]$ is a universal function of $h(f - f_0)/kT$; here C and f_0

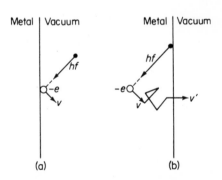

Fig. 9.3. (a) Surface photoemission. A photon strikes the surface and liberates an electron directly. (b) Volume photoemission. A photon is absorbed in the interior while its energy is imparted to a lattice electron. The electron describes a zigzag path inside the material, losing energy on its way, finally ending up at the surface with sufficient energy to escape.

are the only unknowns. By measuring I as a function of f and bringing the theoretical and experimental curves to coincidence, one obtains C, f_0, and hence ϕ'. This method is quite accurate.

In semiconductors or insulators there are no occupied energy levels in the neighborhood of the Fermi level; hence photoelectric emission cannot be used to evaluate ϕ'. One can determine the position of donor levels, acceptor levels, and of the top of the filled band in this way, however.

In volume photoeffect in metals the photoelectron is emitted internally, finds its way to the surface, and escapes if its energy is sufficiently large (Fig. 9.3b). In this case momentum must be taken up by the lattice. Considering for simplicity a one-dimensional lattice, the energy and momentum law give the equations (Fig. 9.4)

$$\frac{1}{2}mv^2 = \frac{1}{2}mv_0^2 + hf \tag{9.10}$$

$$mv = mv_0 \pm \frac{h}{d} \tag{9.11}$$

Here v_0 is the initial velocity of the lattice electrons and v the velocity after the interaction. The plus sign holds when mv_0 and h/d have the same direction, and the minus sign holds if they have opposite direction. The momentum hf/c of the quantum is very small and has been neglected. Solving for the quantum energy, we obtain

$$hf = \frac{1}{2m}\left[\left(\frac{h}{d}\right)^2 \pm mv_0\frac{h}{d}\right] \tag{9.12}$$

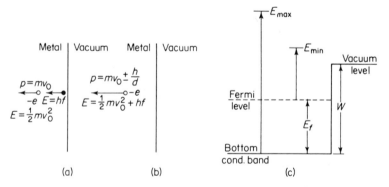

Fig. 9.4. Collision process between a photon and a lattice electron. (a) Before collision. (b) After collision. Energy and momentum are conserved in the collision process, a momentum h/d being taken up by the lattice; the momentum hf/c of the photon is so small that it can be disregarded. (c) The maximum energy E_{max} is absorbed by an electron at the bottom of the conduction band; the minimum energy $E_{min} - E_f$ is absorbed by an electron at the Fermi level.

and for the energy of the electron

$$E = \frac{1}{2}mv^2 = \frac{1}{2m}\left(\frac{h}{d} \pm mv_0\right)^2 \tag{9.13}$$

The smallest value of E is obtained for the minus sign. The expression thus found has its largest value $E_{max} = h^2/(2md^2)$ for $v_0 = 0$ (electrons at the bottom of the conduction band), and its minimum value for electrons at the Fermi level ($\frac{1}{2}mv_0^2 = E_f$), in which case

$$E_{min} = \frac{1}{2m}\left[\frac{h}{d} - (2mE_f)^{1/2}\right]^2 \tag{9.13a}$$

The electron can escape if

$$\tfrac{1}{2}mv_x^2 \geq W \tag{9.14}$$

where v_x is the velocity component perpendicular to the surface and W is the energy difference between the vacuum level and the bottom of the conduction band.

It would be surprising if $E_{min} + E_f$ corresponded to the energy W. The minimum quantum energy needed for an electron to escape the photocathode by volume photoeffect is thus generally larger than the work function ϕ of the metal.

The same considerations apply to a semiconductor or insulator. In the case when the band structure is simple, the minimum energy hf that can be absorbed corresponds to the gap width E_g between the bottom of the conduction band and the top of the valence band. The energy difference between the vacuum level and the bottom of the conduction band is now denoted by χ, so that Eq. (9.14) becomes

$$\tfrac{1}{2}mv_x^2 \geq \chi \tag{9.14a}$$

In addition, we must take into account that the electrons on their way to the surface are scattered by other electrons and by collisions with the lattice; both give a change in direction and a loss of energy. There is therefore no guarantee that condition (9.14a) is still satisfied when the photoelectron arrives at the surface.

We can also understand why surface photoemission depends on the *direction of polarization* of the incident light, whereas volume photoeffect usually does not. Classically and wave mechanically, the photoelectron should be ejected in the direction of the electric vector. Because of the previously mentioned scattering effects, this directivity is lost in the volume photoeffect. Surface photoeffect, however, will only occur if the electric vector has a component perpendicular to the surface. Experimentally, one finds that the full polarization effect shows up for smooth surfaces, whereas it is unimportant for rough surfaces.

There is one exception to this general rule. In clean single crystals of Ge or Si irradiated by ultraviolet light of about 5.8 eV of quantum energy, the light is so strongly absorbed that it penetrates only over about 100 Å. A considerable portion of the photoelectrons will then be able to escape without scattering, if ejected in the proper direction; hence these crystals show a *polarization* effect.†

9.2b. Theory of Volume Photoemission

We shall now formulate the theory of reflection photoemission in a manner quite similar to the theory of secondary electron emission to be discussed in Chapter 10. Let $n(x)\,dx$ be the number of photoelectrons generated between x and $(x + dx)$ per second, and let $g(x)$ be the probability that such electrons will escape. Then the photocurrent I_{ph} is

$$I_{ph} = e \int_0^\infty n(x)g(x)\,dx \qquad (9.15)$$

Now $n(x)\,dx$ is proportional to the amount of radiation $dP(x)$ absorbed in the section dx. The power $P(x)$ at x may be written

$$P(x) = P_0 \exp(-\alpha x) \qquad (9.16)$$

where P_0 is the incident power and α the power absorption coefficient, which is about 10^5 cm^{-1}. Assuming that all absorbed photons of quantum energy hf produce a photoelectron, we have

$$n(x)\,dx = \frac{|dP(x)|}{hf} = \frac{\alpha P_0}{hf} \exp(-\alpha x)\,dx \qquad (9.16a)$$

We now write

$$g(x) = g_0 h(x) \qquad (9.17)$$

where g_0 is the escape probability of electrons produced at the surface ($x \simeq 0$) and $h(x)$ is a kind of "absorption law" for the photoelectrons generated at x. The exact shape of the function $h(x)$ is not very important; therefore, since one can define an "escape depth," $1/\beta$, it is not unreasonable to write

$$h(x) = \exp(-\beta x) \qquad (9.17a)$$

The escape depth $1/\beta$ is of the order of 50–100 Å in metals, but may be much larger in semiconductors. Therefore, substituting into Eq. (9.15), we obtain

$$I_{ph} = eg_0\frac{\alpha P_0}{hf} \int_0^\infty \exp[-(\alpha + \beta)x]\,dx = \frac{eP_0}{hf}\,\frac{\alpha}{\alpha + \beta}g_0 \qquad (9.18)$$

We define the quantum efficiency of the photoemitter by

$$\eta = \frac{I_{ph}/e}{P_0/(hf)} = \frac{\alpha}{\alpha + \beta}g_0 \qquad (9.19)$$

---

† G. W. Gobeli, F. G. Allen, and E. O. Kane, *Phys. Rev. Letters 12*, 94 (1964); E. O. Kane, *Phys. Rev. Letters*, **12**, 97 (1964).

In metals α is of the order of 10^5 cm^{-1}, β is of the order of 10^6 cm^{-1} and g_0 is small, so that η is small. To improve the situation, one should make β smaller and g_0 larger. This is the case in well-chosen semiconductors.

The preceding theory holds for *reflection* photoemission. Many photoemissive cells, however, use *transmission* photoemission; in this case the theory must be somewhat modified, since an electron generated at x must now travel a distance $L - x$ in order to escape, where L is the thickness of the photoemissive layer. Instead of Eq. (9.15), we have

$$I_{ph} = e \int_0^L n(x)g(L - x)\, dx \tag{9.20}$$

where $n(x)\, dx$ is again given by (9.16a) and

$$g(L - x) = g_0 \exp\left[-\beta(L - x)\right] \tag{9.20a}$$

Consequently,

$$
\begin{aligned}
I_{ph} &= e\frac{\alpha P_0}{hf} g_0 \exp\left(-\beta L\right) \int_0^L \exp\left[-(\alpha - \beta)x\right] dx \\
&= e\frac{\alpha P_0}{hf} g_0 \frac{\exp\left(-\beta L\right) - \exp\left(-\alpha L\right)}{\alpha - \beta}
\end{aligned}
\tag{9.21}
$$

so that the quantum efficiency is

$$\eta = \alpha g_0 \frac{\exp\left(-\beta L\right) - \exp\left(-\alpha L\right)}{\alpha - \beta} \tag{9.21a}$$

There is now a thickness L for which η is a maximum. Clearly, this is the case if $d\eta/dL = 0$, or

$$-\beta \exp\left(-\beta L\right) + \alpha \exp\left(-\alpha L\right) = 0 \quad \text{or} \quad L = \frac{\ln \alpha - \ln \beta}{\alpha - \beta} \tag{9.21b}$$

for $\alpha \neq \beta$.†

9.2c. Escape Probability g_0 at the Surface

To understand why the escape probability g_0 is so low in metals and in many semiconductors and insulators, we investigate the following model. Let the electrons arrive at the surface with an energy

$$E = \frac{1}{2}mv^2 = \frac{1}{2}m(v_x^2 + v_y^2 + v_z^2) \tag{9.22}$$

and let all directions of arrival be equally likely; because of the scattering processes involved this is quite probable.

Now these electrons must satisfy condition (9.14). Putting $v_x = v \cos \theta$,

---

† The case $\alpha \simeq \beta$ requires special treatment. See problem 7 of this chapter.

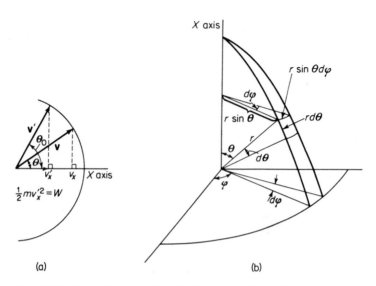

Fig. 9.5. (a) Two electrons with velocities **v** and **v′** have the same magnitude ($v′ = v$) but different directions. The electron arriving at the surface with the velocity **v′** (making an angle θ_0 with the x axis) can just barely escape. (b) Shows that the area of a surface element on a sphere of unit radius in polar coordinates is given by $d\Omega = \sin\theta\, d\theta\, d\varphi$.

where θ is the angle between the direction of arrival and the normal to the surface, we have (Fig. 9.5a)

$$\frac{1}{2}mv^2\cos^2\theta \geqq W \quad \text{or} \quad \cos\theta \geqq \left(\frac{W}{E}\right)^{1/2} \tag{9.23}$$

Putting

$$\cos\theta_0 = \left(\frac{W}{E}\right)^{1/2} \tag{9.23a}$$

condition (9.23) may be written as $\theta \leqq \theta_0$.

Since all directions of arrival are equally likely, we have to find the solid angle of a cone with top angle θ_0; that is, we have to calculate the surface area that is cut out of a sphere of unit radius by this cone. Introducing polar coordinates θ and φ and bearing in mind that the small solid angle $d\Omega$ is given by (Fig. 9.5b)

$$d\Omega = \sin\theta\, d\theta\, d\varphi$$

we have

$$\text{solid angle} = \int d\Omega = \int_0^{\theta_0}\sin\theta\, d\theta \int_0^{2\pi} d\varphi = 2\pi(1 - \cos\theta_0) \tag{9.24}$$

Dividing by 4π, the solid angle of full space, we find therefore ($E \geqq W$)

$$g_0 = \frac{1}{2}(1 - \cos\theta_0) = \frac{1}{2}\left[1 - \left(\frac{W}{E}\right)^{1/2}\right] \tag{9.25}$$

which is very small if E is close to W. Assuming that $E - W \ll W$, this may be written

$$g_0 \simeq \frac{E - W}{4W} \tag{9.25a}$$

Taking $E - W = 1$ eV and $W = 15$ eV yields $g_0 = \frac{1}{60}$, so that the photo-emission effect is indeed small in this case.

The same theory applies to semiconductors and insulators, provided that W is replaced by the electron affinity χ. This yields

$$g_0 = \frac{1}{2}\left[1 - \left(\frac{\chi}{E}\right)^{1/2}\right] \tag{9.26}$$

We thus see that g_0 can be large if χ is quite small. Actually, surfaces have been made for which χ is negative. In that case Eq. (9.26) does not hold, since it only applies for $\chi > 0$. The escape probability can then be near unity, for an electron that had the wrong direction initially can be scattered back toward the surface and escape.

This does not mean, however, that all generated photoelectrons can escape, so that $\eta = 1$. For practical reasons this kind of photoemissive material must be p-type, so that electrons on their way to the surface may be absorbed by recombination with holes.

9.2d. Conditions for High-Yield Photoemission

We now list the necessary conditions for a large photoelectric yield in insulators and semiconductors (Fig. 9.6):

1. The escape probability g_0 for electrons generated near the surface must be large. Thus the electron affinity χ must be small or preferably negative. The latter condition is met for cesium-covered GaAs and Si.

2. The escape depth $1/\beta$ of the photoelectrons must be large, or β relatively small. Thus the energy losses suffered by the photoelectrons on their way to the surface must be small. Let E_g be the gap width of the semiconductor; then for an electron energy $E > E_g$ the electrons will quickly lose their energy in band-to-band excitation. For $E < E_g$ the electrons can only lose energy in amounts of units $\hbar\omega_s$ owing to inter-action with the lattice vibrations. The electrons can then come from

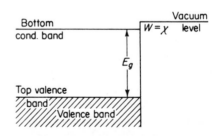

Fig. 9.6. Conditions for a high photoemission (and high secondary emission) yield. The bottom of the conduction band should be close to the vacuum level ($W = \chi$ small) and the gap width E_g should be large in comparison with χ.

great depth, of the order of 1 μm or more, suffer many collisions on their way to the surface, and still escape if χ is small or negative.

3. The semiconductor or insulator should have a large power absorption coefficient α for the light. In fact, one should require $\alpha > \beta$. For $h\nu \simeq E_g$ (absorption edge of the crystal), α is very small; but its value increases rapidly with increasing frequency and may be as large as 0.5–1.0×10^6 cm^{-1} for $h\nu$ equal to a few times E_g.

Insulators and semiconductors with $E_g < \chi$ have a low photoelectric yield because photoelectrons with $E > E_g$ quickly lose their energy by band-to-band transitions, so that they can no longer escape. This is the case for silicon and germanium without surface treatment.

The condition $\chi \leq 0$ can be achieved by covering the surface of the semiconductor or insulator with a material that lowers the work function considerably. Cesium is the best material for this purpose, but usually the cesium is absorbed to the surface via oxygen atoms, so that the surface is most probably covered with an -O–Cs layer.

It is highly desirable that the semiconductor used in the high-yield photocathodes be p type. For if the material were n type and $\chi \leq 0$, there would be a very large thermionic emission from the cathode. To prevent this, the Fermi level should lie as low as possible, which is the case if the semiconductor is heavily doped p-type material.

We now illustrate the semiconductor photocathode for three specific cases:

1. The first is the *flatband* case in which the energy bands of the material are not bent at the surface (Fig. 9.7a). This situation occurs when the potential difference between the surface and the bulk semiconductor is zero. In this case g_0 is a maximum, but β is not zero, since the electrons in the conduction band will recombine with holes left in the valence band; consequently, the escape depth $1/\beta$ will be equal to the diffusion length L_n of the electrons in the p-type material. According to Chapter 15, $L_n = (D_n \tau_n)^{1/2}$, where D_n is the diffusion constant for electrons and τ_n the electron lifetime in the p-type material. Values of α considerably larger than β, and escape depths of the order of a few μm, can be achieved in this manner.

2. As a second case we take the situation in which the band bends down toward the surface. This is the case if the surface potential $\psi(0)$ is positive with respect to the bulk semiconductor; it comes about because of a positive surface charge and a distributed negative charge due to ionized acceptors. We assume for this example that $\chi > 0$, but that the vacuum level lies lower than the bottom of the conduction band in the bulk semiconductor (Fig. 9.7b). The theory of the surface layer follows closely the discussion of similar layers in metal–semiconductor diodes (Chapter 14).

As long as the bottom of the conduction band in the bulk material lies above the vacuum level, the *escape depth* is not affected. But the electrons passing through the space-charge layer on their way to the surface will lose energy; hence, if $\chi > 0$, some of them may not be able to escape, that is, g_0 is affected by the energy-loss process. To overcome this difficulty, the space-charge region should be made thinner, which can be achieved by making the material more strongly p type.

The threshold quantum energy is now still equal to E_g, and the dark current due to thermionic emission is larger than in the previous case, since the Fermi level at the surface is closer to the vacuum level because of the band bending.

3. In n-type material band bending in the opposite direction can occur if $\psi(0)$ is negative. There is then a negative surface charge at the surface and a distributed positive charge due to ionized donors inside the material. If $\chi > 0$, the barrier height is $e|\psi(0)| + \chi$, and the threshold quantum energy is $E_g + e|\psi(0)| + \chi$, which is much larger than in the previous case. The escape probability of the photoelectrons and the photoelectric threshold are therefore seriously affected. This seems to occur in K_3Sb and Na_3Sb photocathodes.

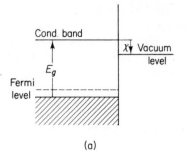

(a)

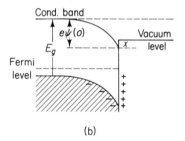

(b)

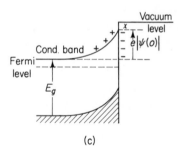

(c)

Fig. 9.7. a) Energy level diagram for the flatband case. b) Energy level diagram if the bands bend down (p-type material). c) Energy level diagram if the bands bend up (n-type material).

In practice quantum yields of up to 40 per cent can be achieved with the best photocathodes. Response into the near infrared is feasible, since the minimum quantum energy is given by the gap width E_g' in electron volts,

$$hf_{\min} = eE_g', \qquad \lambda_{\max} = \frac{hc}{eE_g'} \tag{9.27}$$

where $\lambda_{\max}$ is the long wavelength threshold. For silicon this gives a threshold wavelength of 11,000 Å. Longer threshold wavelengths can be achieved with properly chosen III–V compounds.

One should bear in mind, however, that for decreasing gap widths the thermionic emission becomes quite significant. This is certainly the case for silicon photocathodes covered with an –O–Cs monolayer, where the thermionic emission current is as high as 10^{-9} A/cm^2 at room temperature. One can eliminate the effect by cooling the photocathode, but this complicates the detection arrangement.

9.3. HIGH-YIELD PHOTOCATHODES

The present high-yield photocathodes are all of the volume photoeffect type; they have superseded the earlier selective photoeffect cathodes, which probably exhibited a type of surface photoeffect.[†]

Modern cathodes are Ag–O–Cs cathodes, cesium antimonide cathodes, multialkali cathodes, silicon, and III–V compounds with a –Cs or –O–Cs layer on the surface. For details we refer to Sommer's book.

9.3a. Ag–O–Cs Cathode

The [Ag]–Cs$_2$O–Cs photocathode (silver base, Cs$_2$O deposited upon it, and Cs absorbed on the surface) has a photoemissive response with a maximum at 6080 Å. The photoeffect is even more pronounced in layers of the type [Ag]–Cs$_2$O, Ag, Cs–Cs (silver base with Cs$_2$O layer, silver atoms dispersed through the layer, and Cs atoms absorbed on active places inside the layer and on the surface); there is now a long-wavelength response up to 12,000 Å. The maximum sensitivity occurs for layers about a few hundred angstroms thick, so that the effect cannot qualify as a surface effect but must be considered a volume effect.

The chief use of the Ag–O–Cs photoemitter is as a photocathode for the near-infrared region. Under this condition the photoelectric yield is between 0.1 and 0.5 per cent between 9000 and 4000 Å and drops at longer wavelengths, becoming quite small at 12,000 Å. The thermionic emission ranges from 10^{-11} to 10^{-14} A/cm^2, depending on the sample.

There is still no complete understanding of the Ag–O–Cs photoemission. The phenomenon cannot be understood as a simple semiconductor photoemission process, since the band gap of Cs$_2$O is considerably larger than the photoelectric threshold. Quite likely, the photoelectrons are produced in the dispersed silver particles; since the reflected light can be intercepted by other particles, a larger percentage of the incident light would be effective in producing photoelectrons. If the photoelectrons escape through the Ag–Cs$_2$O interface, the photoelectric threshold would be determined by the barrier

---

† R. Suhrmann and H. Theissing, *Z. Phys.* **55**, 701 (1929).

height of the Ag–Cs$_2$O contact.
The photoelectrons may then travel
through the Cs$_2$O layer and escape
from the Cs$_2$O–Cs surface.

9.3b. Cesium Antimonide Photocathodes

The active substance on an
antimony–cesium photosurface is
the intermetallic compound Cs$_3$Sb,
cesium antimonide. As indicated by
measurements on thin Cs$_3$Sb films
on quartz, the substance has the
properties of an intrinsic semiconductor. The photoelectric yield can
be quite high, up to 20 per cent.
The best layers are *p* type.

According to Sommer, the gap
width in Cs$_3$Sb is 1.60 eV and the
value of χ is 0.45 eV. There is some
photoemission below 2.05 eV, which
is attributed to photoexcitation from
defect levels above the top of the
valence band. Cooling reduces the
population of these defect levels
and hence the photoemission below
2.05eV quantum energy (Fig. 9.8).

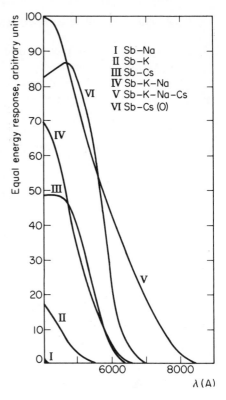

Fig. 9.8. Spectral response curves of alkali
metal-antimony photocathode (taken from
L. Marton, Methods of Experimental Physics, Vol. 6B, Academic Press, Inc., New
York, 1959).

9.3c. Multialkali Photocathodes

Multialkali photocathodes are of the form Na$_2$KSb–Cs; they have a lower
gap width than Cs$_3$Sb and hence cover the visible part of the spectrum better
than most cathodes. According to Sommer, the band gap of Na$_2$KSb–Cs is
about 1.0 eV and the photoelectric threshold is about 1.55 eV, indicating an
effective value of χ of about 0.55 eV. The layers are *p* type (Fig. 9.8).

There is also some photoemission below the true photoelectric threshold,
which is again attributed to photoexcitation from defect levels.

Na$_3$Sb has a band gap of 1.1 eV and a photoelectric threshold of 3.0–3.5
eV. Probably, the band picture of Fig. 9.7c is appropriate here. These layers
are *n* type. The connection between low electron affinity and *p*-type conduction is physically significant.

9.3d. Ga$_{1-x}$ In$_x$ As Photoemitters

Fisher et al.† have published curves for the photoelectric yield of Ga$_{1-x}$In$_x$As alloys covered with an –O–Cs layer about 1 monolayer thick (Fig. 9.9). It should be noted that GaAs has a photoelectric yield of about 40 per cent at somewhat higher photon energies, and that $g_0 > \frac{1}{2}$. The threshold of the photoeffect increases to longer wavelengths if the In concentration is increased (increasing x). The photoelectric threshold is approximately equal to the band gap, indicating a negative electron affinity situation. The surface escape probability g_0 (B in Fig. 9.9) is largest for GaAs and decreases with increasing x, probably caused by band bending.

Photocathodes have also been made of the type III–V substrate –Cs$_2$O–Cs. The substrate and the Cs$_2$O form a heterojunction, and complicated

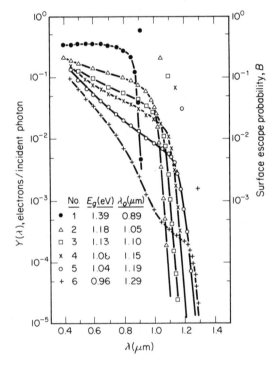

No.	E_g(eV)	$\lambda_0(\mu m)$
• 1	1.39	0.89
△ 2	1.18	1.05
□ 3	1.13	1.10
× 4	1.08	1.15
○ 5	1.04	1.19
+ 6	0.96	1.29

Fig. 9.9. Spectral yield curves of $Ga_{1-x}In_xAs(OCs)$ cathodes of different bandgaps showing band-gap limited emission to $\lambda \simeq 1.3\mu$ meter. Also shown are the corresponding values of B, the surface escape probability. (From D. G. Fisher, R. E. Enstrom and B. F. Williams, Appl. Phys. Letters, **18**, 371–373, 1971.)

---

† D. G. Fisher, R. E. Enstrom, and B. F. Williams, *Appl. Phys. Letters*, **18**, 371 (May 1, 1971).

energy diagrams can arise. The Cs layer is nearly a monatomic layer that lowers the work function.

9.3e. High-Yield Silicon

Martinelli[†] has published curves of the photoelectric yield of heavily doped silicon surfaces covered with a thin –O–Cs layer. He found that the photoelectric yield paralleled the absorption as long as the optical absorption coefficient was relatively small. This is as expected from Eq. (9.19), for if $\alpha \ll \beta$

$$\eta \simeq g_0 \frac{\alpha}{\beta} = g_0 \alpha L \qquad (9.28)$$

where $L = 1/\beta$ is the escape depth, then η is proportional to α. From the data Martinelli deduced that $L = 5.5$ μm and $g_0 = 0.18$; this relatively low escape probability is a consequence of band bending. The threshold wavelength is about 11,000 Å.

The drawback of these high-yield surfaces is that the thermionic emission can be as high as 10^{-9} A/cm^2 at room temperature.

9.4. NOISE IN PHOTOCATHODES

If the arrival of light quanta at the surface is a series of independent, random events, the same should be true for the emission of photoelectrons. One would thus expect full shot noise of the photocurrent I; that is,

$$\overline{i^2} = 2eI\,\Delta f \qquad (9.29)$$

Here Δf is the frequency band under consideration, e the electron charge and $\sqrt{\overline{i^2}}$ the current generator representing the noise.

REFERENCES

GÖRLICH, P., "Recent Advances in Photoemission," in L. Marton, ed., *Advances in Electronics and Electron Physics*, vol. 11, pp. 1–30. Academic Press, Inc., New York, 1959.

LARACH, S., ed., *Photoelectric Materials and Devices*. Van Nostrand Reinhold Company, New York, 1965.

MARTON, L., ed., *Methods of Experimental Physics*, vol. 6, part B, *Solid State Physics*. Academic Press, Inc., New York, 1959.

SOMMER, A. H., *Photoemissive Materials: Preparation, Properties and Uses*. John Wiley & Sons, Inc., New York, 1968.

---

† R. U. Martinelli, *Appl. Phys. Letters*, **16**, 261 (April 1, 1970).

VERNIER, P., *L'Emission photoelectrique et ses applications*. Dunod Editeur, Paris, 1963. (Distribution in the U.S.A. by Gordon and Breach, Science Publishers, Inc., New York.)

PROBLEMS

1. A metal photoemitter has a work function of 2.5 eV, and the bottom of the conduction band is 7.5 eV below the vacuum level. Calculate the wavelength at which the surface photoeffect starts.

Answer: 4950 Å.

2. The photoemitter of problem 1 is irradiated by ultraviolet light. It is found that volume photoeffect starts at $\lambda = 2500$ Å. Evidence shows that this absorption corresponds to electrons excited from the Fermi level to a higher unoccupied level. Calculate the following:
(a) The energy of the excited electron with respect to the bottom of the conduction band.
(b) The probability that an electron arriving at the surface with that energy will escape.

Answer: 9.95 eV; 0.132.

3. A semiconductor photoemitter has the bottom of its conduction band 0.5 eV below the vacuum level and has numerous donor levels at 1.0 eV below the bottom of the conduction band; the gap between the filled band and the conduction band is 4.0 eV.
(a) Calculate the wavelength limit due to photoemission from the donor levels. You may assume that there are no momentum restrictions for the absorption of quanta by electrons bound in donor levels.
(b) At what wavelength does the surface effect due to excitation from the filled band start?

Answer: 8270 Å; 2760 Å.

4. The photoemitter of problem 3 is illuminated by light of 6500-Å wavelength. Assuming again that there are no momentum restrictions on the absorption of quanta by electrons bound to donor levels, calculate the following:
(a) The energy of the excited electrons, taken with respect to the bottom of the conduction band.
(b) The maximum energy of the electrons escaping from the target.
(c) The escape probability of an electron arriving at the surface with the energy mentioned under part (a).

Answer: 0.91 eV, 0.41 eV, 0.259.

5. Let the primary photoelectric processes in problem 4 (ejection of an electron from a donor level into the conduction band) have a quantum efficiency of unity, and let the light be absorbed in a sufficiently thin layer so that all electrons have roughly the same probability of escaping from the target. Calculate the photoelectric yield of the target, and give reasons why its value is half the value found in problem 4(c).

Answer: 0.130.

6. (a) The photoemitter of problem 3 has a cathode area of 10 cm^2. Assuming that the Fermi level lies halfway between the donor levels and the bottom of the conduction band and that Richardson's A-constant is 120 A/cm^2/degree^2, calculate the dark current due to thermionic emission at 300°K.
(b) Using the photoelectric yield of problem 5, calculate how much light is needed to give a photocurrent equal to the dark current, if the light used has a wavelength of 6500 Å.

Answer: 1.6×10^{-9} A; 2.4×10^{-8} W.

7. Find the maximum thickness and the maximum quantum efficiency for a transmission-type photocathode with $\alpha = \beta$.

Answer: $L = 1/\alpha$, $\eta = g_0/(2.72)$.

10

Secondary
Electron Emission

When electrons with sufficient kinetic energy hit the surface of a solid, the solid emits electrons. These electrons are called *secondary* electrons; the bombarding electrons are called *primary* electrons. In this chapter we review the most important features of the process. Some of its applications are discussed in Chapter 13.

The problem has many features in common with the volume photoeffect, in that one must distinguish between the *production process* and the *escape process*. The production process is somewhat different, of course, since the interaction of the fast primary electrons with the lattice electrons supplies the necessary energy to the latter to escape from the crystal. The escape process is the same, however, in both cases; the only difference is that in the case of secondary electron emission the energy of the escaping electrons is generally higher.

In the photoelectric case one distinguishes between the volume effect and the surface effect. One might therefore think that interaction of the fast primary electrons with the electrons at the surface might produce a kind of surface secondary emission effect. The best calculations show, however, that this gives only a negligible contribution to the total secondary emission; the volume effect thus predominates.

10.1. PHYSICAL CHARACTERISTICS OF SECONDARY EMISSION

10.1a. Energy Distribution of the Secondaries

If primary electrons of energy E_{p0} strike the surface of a solid, the emitted electrons consist of three groups (Fig. 10.1):

1. *Slow secondaries* with energies less than about 20 V. These electrons are called *true* secondaries. Usually the energy distribution of these slow electrons has a maximum at an energy of a few electron volts; the exact value may depend somewhat on the target used as a secondary emitter. The distribution can be roughly approximated by a Maxwellian energy distribution,

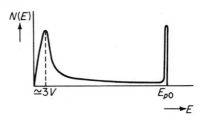

Fig. 10.1. Energy distribution of secondary electrons. The peak at low energies is due to true secondaries; the peak at $E \simeq E_{p0}$ is due to reflected electrons.

$$\Delta N = CE^{1/2} \exp\left(-\frac{E}{E_0}\right) \Delta E \qquad (10.1)$$

In a Maxwellian energy distribution of particles at a temperature T_e, one would have $E_0 = kT_e$; since E_0 is of the order of a few electron volts, the slow electrons can be considered as having a Maxwellian velocity distribution at an equivalent temperature of roughly 50,000°K. It is interesting to note that for metals *the distribution (10.1) is independent of the primary energy*, whereas there are indications that in some insulators the relative number of low-energy true secondaries increases with increasing primary energy.

Besides the energy distribution, it is also important to know the *angular* distribution of the slow secondaries. The best experiments give approximately a cosine distribution; this means that the number of emitted secondaries per second emitted in the solid angle formed by two cones of angles θ and $\theta + \Delta\theta$ with their axes perpendicular to the surface is (Fig. 10.2)

$$\Delta N = NP(\theta)\,\Delta\theta = N \cos\theta\,\Delta\theta \qquad (10.2)$$

where N is the number of secondary electrons emitted per second.

2. *Electrons of intermediate energies.* This group is partly made up of primary electrons that have lost large amounts of energy and partly of lattice electrons that have gained a large amount of energy in a collision with a primary electron. It is impossible to discriminate between these possibilities.

3. *Back-scattered primaries*, with energies close to the primary energy E_{p0} (for example, $E > \frac{1}{2}E_{p0}$); their energy distribution has a sharp maximum for $E \simeq E_{p0}$ (reflected electrons). It is common practice to introduce the *reflection coefficient or back-scattering coefficient* γ, which is defined as

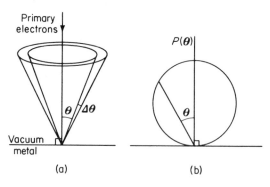

(a)   (b)

Fig. 10.2. (a) Solid angle between two cones of angles θ and $(\theta + \Delta\theta)$. The number of secondary electrons emitted within this solid angle has the cosine distribution of Eq. (10.2). (b) Cosine distribution for the emitted secondaries.

$$\gamma = \frac{\text{total number of back-scattered primaries}}{\text{total number of incident primaries}}$$

The quantity γ increases with increasing atomic number Z of the atoms in the target; its value is rather small for the light elements and increases to a value of about 0.50 for the heavy elements at high primary energies. The quantity γ also increases with increasing primary energy, reaching an asymptotic value for $E_{p0} > 100\,\text{kV}$. The value of $\gamma = 0.5$ for the heavy elements at high energies indicates that at some depth inside the material the angular distribution of the primary electrons has become isotropic; the decrease in γ for the lower primary energies indicates that the electrons have lost their energy before their distribution has become isotropic.

There is another process involved here that has recently become important for the identification of atoms in surface layers. This process is electron emission by *Auger effect*. We illustrate this effect with the help of the energy-level diagram of Fig. 10.3. Shown are the energy levels $-E_K$ of an electron

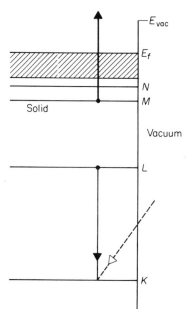

Fig. 10.3. Energy level diagram of an atom, showing the ejection of an Auger electron of energy $E_K - E_L - E_M$. (*Courtesy Physical Electronics Corporation.*)

in the K shell and $-E_L$ of an electron in the L shell; also shown is the energy level corresponding to zero energy. Now suppose that an electron in the K shell is removed by a collision with a primary electron. In the x-ray process an electron will go from energy level $-E_L$ to energy level $-E_K$ under emission of a photon of energy $E_K - E_L$. In the Auger process no x-ray quantum is emitted, but the difference in energy is imparted to another electron, say in the M shell, so that it is emitted with an energy $E_K - E_L - E_M$. It is obvious that emission from other occupied energy levels is also possible. The electrons are therefore emitted with characteristic energies that identify the atom that emits the electron. For that reason *Auger electron spectroscopy* has been introduced as a means of identifying atoms in or on surfaces (Chapter 13). The effect gives a small contribution to the secondary emission ratio.

10.1b. Dependence of the Secondary Emission upon the Primary Energy

The secondary emission is characterized by the secondary emission factor δ, which is defined as the average number of emitted secondaries per incident primary. The quantity $\delta - \gamma$ represents the average number of *slow secondaries* per primary.

If δ is plotted as a function of the primary energy E_{p0}, one finds that the dependence of δ upon E_{p0} is quite similar for all targets. For very low energies (< 20 to 40 eV for most targets) the curve shows a peculiar structure that is characteristic of the composition of the target. For larger values of E_{p0} the curve first varies linearly with E_{p0}, then δ increases more slowly with increasing E_{p0}, passes through a maximum value δ_{max} at $E_{p0} = E_{max}$, and decreases for $E_{p0} > E_{max}$. The value of E_{max} depends upon the composition of the target; it is usually of the order of several hundred volts (Fig. 10.4a).

If δ/δ_{max} is plotted against E_{p0}/E_{max}, one obtains a curve that is almost

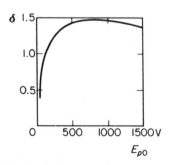

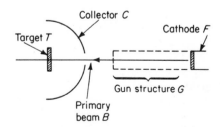

Fig. 10.4a. Secondary emission yield of silver. (Reproduced by permission from Bruining, *Physics and Applications of Secondary Electron Emission*, Pergamon Press, 1954.)

Fig. 10.4b. Schematic diagram of a tube suitable for secondary emission measurements.

independent of the target material; the targets may therefore be roughly characterized by the values of δ_{max} and E_{max}.

Most metals have values of δ_{max} that are less than 1.5; the alkali metals even have $\delta_{max} < 1$, unless the surface is oxidized. Carbon has $\delta_{max} \simeq 1$, but soot has $\delta_{max} \simeq \frac{1}{2}$ (see below). Some insulating and semiconducting compounds, such as alkali oxides, alkali halides, alkaline earth oxides, and cesium antimonide, have large values of δ_{max}; MgO, for instance, has $\delta_{max} \simeq 8$ to 20, depending upon the method of preparation. Some of these surfaces are thus suitable for applications requiring a large secondary emission factor.

The secondary emission factor δ can be measured with the help of the tube shown in Fig. 10.4b. A narrow electron beam B is formed with the help of the hot filament F and the electron gun G. The electron beam passes through the hole in the spherical collector and hits the target T; all the secondary electrons are collected by the collector C, if C has a higher potential than T; only the fast back-scattered electrons are collected if the potential of C is much lower than the potential of T. The tube thus permits measurement of δ and of γ.

By using very small primary currents, so that space-charge effects in the target-collector area can be neglected, one may determine the velocity distribution of the electrons from a retarding-field plot. A more accurate method makes use of magnetic deflection methods.

The method has to be modified if the secondary emission factor of insulating layers is measured. One must then pulse the beam.

Experiments indicate that the secondary emission coefficient δ for a very rough granular surface is smaller than for a microscopically plane surface. The reason is that part of the secondary electrons cannot escape from the pores between the grains. For the same reason, δ depends very little upon the angle of incidence of the primary electrons for those surfaces (Fig. 10.5).

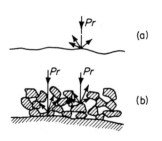

For plane surfaces the secondary emission factor is a minimum for normal incidence of the primary electrons, but δ increases with increasing angle of incidence and is a maximum for grazing incidence (see Fig. 10.7). An explanation will be given in Sec. 10.2.

The secondary emission factor of surfaces with a high initial value of δ is found to decrease under electron bombardment. The effect, which can

Fig. 10.5. (a) Secondary emission from a smooth surface; the secondary electrons can escape. (b) Secondary emission from a rough surface; many secondary electrons cannot escape. *Pr* indicates primary electron. (Reproduced by permission from Bruining, *Physics and Applications of Secondary Electron Emission*, Pergamon Press, 1954.)

probably be described by a change in the escape probability of the electrons, is accompanied by a decomposition of the surface.

10.1c. Manufacture of Targets with a High Secondary Emission Factor

Alkali and alkaline earth oxides have a high secondary emission factor δ and may therefore be used in secondary emission applications requiring a high δ; the most commonly used oxides are MgO, BeO, and Cs_2O.

High-yield photoemissive targets such as Cs_3Sb also have a very high secondary emission factor δ. The same is true for silicon and for III–V compounds covered with an –O–Cs monatomic layer. The reason is that the two processes have the same escape mechanism and that the escape probability of the slow secondaries is quite large. Secondary emission factors as high as 100–200 at 1 kV primary energy have been observed in the latter group of compounds.

10.2. THEORETICAL CONSIDERATIONS

10.2a. Explanation of the Observed Secondary Emission Effects

As in the case of photoemission, the observed behavior of secondary electron emitters can be explained in terms of the production mechanism and the escape mechanism. Let $n(x)\,\Delta x$ be the number of secondary electrons produced between x and $(x + \Delta x)$, and let $g(x)$ be the average probability of escape of a secondary electrons produced at the depth x. Let x_0 be the penetration depth of the primaries; x_0 increases with increasing primary energy. The function $g(x)$ decreases with increasing x and is independent of the primary energy. We therefore write

$$g(x) = g_0 h(x) \qquad (10.3)$$

where g_0 is the value of $g(x)$ at $x = 0$ and $h(0) = 1$, whereas $h(x)$ decreases monotonically with increasing x. Since the exact dependence of $h(x)$ does not have a great effect on the theory, one generally chooses $h(x) = \exp{(-\beta x)}$, where $1/\beta$ is the average escape depth of the secondaries. The secondary emission factor may now be expressed as

$$\delta = \int_0^{x_0} n(x) g(x)\, dx \qquad (10.4)$$

For very small primary energies the penetration depth is so small that $h(x) \simeq 1$ (Fig. 10.6a). In that case

$$\delta = g_0 \int_0^{x_0} n(x)\, dx = g_0 N \qquad (10.5)$$

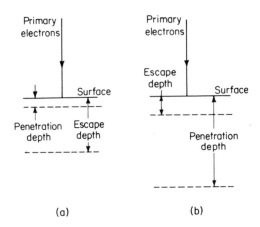

Fig. 10.6. (a) Secondary emission at low primary energies. The penetration depth is small in comparison with the escape depth; hence $g(x) \simeq g(0)$ in this case. (b) Secondary emission at high primary energies. The penetration depth is large in comparison with the escape depth; hence $n(x) \simeq n(0)$ in this case. Scale: one-third of (a).

where N is the total number of secondaries produced at the primary energy E_{p0}. If an energy E_1 is needed to produce a secondary, then $N = E_{p0}/E_1$, and

$$\delta = g_0 \frac{E_{p0}}{E_1} \tag{10.6}$$

so that δ is proportional to E_{p0}, as experiment requires.

If E_{p0} increases, the primary electrons penetrate more deeply into the target and $h(x) < 1$. One would thus expect δ to be smaller than the value given by Eq. (10.6), and the discrepancy should increase with increasing energy, in agreement with experiment.

At very high energies, the range of escape of the secondaries is much smaller than the penetration depth x_0 of the secondaries; that is $h(x) \simeq 0$ for values of x for which $n(x) \simeq n(0)$ (Fig. 10.6b). In that case Eq. (10.4) yields

$$\delta \simeq g_0 n(0) \int_0^\infty \exp(-\beta x)\, dx = g_0 \frac{n(0)}{\beta} \tag{10.7}$$

so that a large g_0 and a large escape depth $1/\beta$ have a very beneficial effect on δ; since $h(x_0) \simeq 0$, we could extend the upper limit of integration to ∞.

Since δ decreases with increasing E_{p0} for $E_{p0} > E_{\max}$, Eq. (10.7) gives that $n(0)$ must decrease with increasing E_{p0}. According to Sec. 10.2d, this is indeed the case.

We are now also able to understand the increase in δ with increasing angle of incidence (Fig. 10.7a). If the primary electrons have an angle of incidence θ and *travel in straight paths*, the path length through a strip of

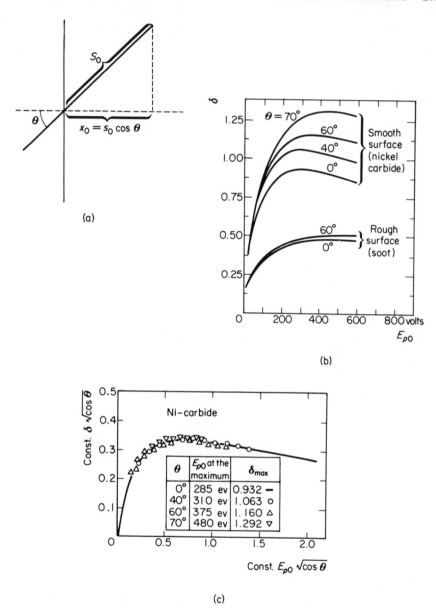

Fig. 10.7. (a) If θ is the angle of incidence of the primaries, then their penetration depth is $x_0 = s_0 \cos \theta$, where s_0 is the penetration depth for normal incidence. (b) Secondary emission factor δ of NiC and for soot as a function of the primary energy E_{p0} for different angles of incidence θ (Jonker). Note that δ for soot is independent of θ. (c) $\delta \sqrt{\cos \theta}$ plotted against $E_{p0}\sqrt{\cos \theta}$ for NiC, indicating that the various curves of part (b) now coincide (Jonker).

width Δx will be $\Delta s = \Delta x/\cos \theta$. If $n_0(x) \Delta x$ represents the number of electrons produced between x and $(x + \Delta x)$ for normal incidence, and s_0 is the range of the primaries for normal incidence, we have

$$n(x) = \frac{n_0(x)}{\cos \theta}, \qquad x_0 = s_0 \cos \theta \qquad (10.8)$$

Substituting into (10.4), (10.5), and (10.6) gives no increase in δ at low energies [compare (10.5)], but a considerable increase in δ at high energies [because $n_0(0)$ increases with increasing θ]. This agrees qualitatively with the experimental data (Fig. 10.7b). A calculation along the lines to be discussed in Sec. 10.2d shows that $\delta\sqrt{\cos \theta}$ is a universal function of $E_{p0}\sqrt{\cos \theta}$ for a given target. According to Fig. 10.7c, this prediction agrees well with experiment.† The predicted universal function itself does not agree well with experiment (see Fig. 10.9), because the assumption that the primary electrons travel in straight paths is not justified.

10.2b. Escape Mechanism

The two important parameters in the escape process are the surface escape probability g_0 and the escape depth $1/\beta$. The larger g_0 and the larger the escape depth are, the larger the secondary emission factor.

It has been found that δ increases with decreasing work function of the target. This can be explained as follows. In analogy with the discussion in Chapter 9, since the electrons arriving at the surface have an isotropic velocity distribution

$$g_0 = \frac{1}{2}\left[1 - \left(\frac{W}{E}\right)^{1/2}\right] \qquad (10.9)$$

where W is the distance between the bottom of the conduction band and the vacuum level, and $E > W$ the energy of the escaping secondaries. For metals the effect is small, since W is of the order of 10–15 V, and a change in W by 1 or 2 V will not change g_0 very strongly. In semiconductors or insulators, W must be replaced by χ, and

$$g_0 = \frac{1}{2}\left[1 - \left(\frac{\chi}{E}\right)^{1/2}\right] \qquad (10.10)$$

for $E > \chi$. Here χ is often quite small; hence g_0 is much larger and so is the effect of a decrease in χ.

This is the case for targets such as MgO. Targets of silicon or III–V compounds, when covered with an –O–Cs monolayer, have $\chi < 0$ so that the escape probability is quite large (Sec. 9.2c). In that case Eq. (10.10) is invalid, since all generated electrons can potentially escape when they do not recombine on their way to the surface.

We have seen that only semiconductors or insulators can give high

---

† J. L. H. Jonker, *Philips Research Reports*, **7**, 1–20 (1952).

secondary emission factors δ. The requirements that must be met are similar to those for photoemission.

1. The materials must have a small or negative electron affinity χ so that the factor g_0 can be large. Under favorable circumstances g_0 can be larger than $\frac{1}{2}$ (see Chapter 9). If $\chi < 0$, the energy bands should either be flat near the surface or should bend down toward the surface with a narrow space charge region at the surface; this requires that the material be the doped heavily p type. In the latter case, χ may be slightly positive, provided that the vacuum level lies below the bottom of the conduction band of the bulk semiconductor (Sec. 9.2d).

2. The secondary electrons must have a large escape depth $1/\beta$. Materials such as silicon with an –O–Cs monolayer on top have an escape depth of a few μm. If the energy E of the produced secondary electrons is larger than the gap width E_g, the electrons quickly lose their energy in band-to-band excitations, until $E < E_g$. After that, the electrons can only lose their energy to the lattice vibrations. But even if the electron energy is fully thermalized, they can still escape if $\chi < 0$. For that reason the escape depth is quite large, being determined by the diffusion length of electrons in p-type material (Sec. 9.2b).

3. The band gap E_g must be relatively small. For since the secondary electrons can produce new secondaries as long as their energy $E > E_g$, more secondary electrons are produced if E_g is small. This makes it understandable why silicon and III–V compounds covered with an –O–Cs monolayer have such a high secondary emission yield.

Figure 10.8 shows the secondary emission factor of silicon covered with an –O–Cs monolayer as a function of the primary energy. The secondary emission factor at low primary energies is large, since g_0 is large and E_1 is quite small in Eq. (10.6). The curve does not show a maximum at higher primary energies, because even at 20-kV primary energy, the penetration depth of the primaries is somewhat smaller than the escape depth of the secondaries. For the same reasons one would not expect a large dependence of δ on the angle of incidence of the primaries. Finally, one would expect many secondary electrons to have a low escape energy. The results for the transmission type secondary emission will be discussed in Sec. 10.3.

The preceding conditions are not met by all semiconductors and insulators. Clean germanium, for example, has a low secondary emission factor ($\delta_{max} \simeq 1.2$). The material has $E_g \simeq 0.7$ eV and $\chi > E_g$; the secondaries thus lose their energy quickly by exciting electrons from the valence band, so that they can no longer escape. It should be understood, of course, that a considerable improvement in secondary emission yield can be obtained by covering germanium with an –O–Cs monolayer, just as in the case of silicon.

High-yield silicon secondary emitters have the drawback that the thermionic emission of the targets is quite large at room temperature. The situa-

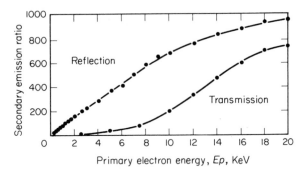

Fig. 10.8. Reflection and transmission secondary emission ratios obtained from the 4-5-μ-thick Si sample (From R. U. Martinelli, Applied Phys. Letters, *17*, 313–314, 1971).

tion can be improved considerably by taking targets with a larger gap width and less band bending. For these reasons GaAs and GaP targets covered with an –O–Cs monolayer give considerably less thermionic emission.

10.2c. Production Mechanism for the Secondary Electrons

In contrast with the electron–photon interaction in the volume photo-effect, an exchange of energy between fast primary electrons and slow lattice electrons is possible without the necessity of imparting momentum to the lattice; the energy law and the momentum law can always be satisfied in this case. We now calculate the probability that a fast primary electron produces a slow secondary.

Since there are six unknowns after the interaction (the velocity components of the primary and of the lattice electron) and only four equations (one for the energy, and one for each of the three components of the momentum), the conservation laws are not sufficient to solve the problem. We therefore use the following approximation method (Fig. 10.9). Suppose that the primary particle moving with a speed v is deflected so little that its path can be approximated by a straight line passing by a resting charge in the origin at a

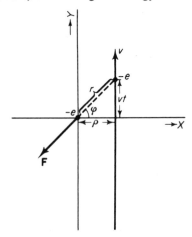

Fig. 10.9. Calculation of the interaction of fast primary electrons with lattice electrons. The lattice electron is initially at rest at the origin of the coordinate system; the primary is considered not to be deflected. p is the minimum distance of the primary to the origin.

minimum distance ρ. Measuring time from the point of closest distance ρ, we may express the distance r as

$$r = (\rho^2 + v^2t^2)^{1/2} \qquad (10.11)$$

and the force between the charges at that distance is, if $\epsilon_0 = 10^7/(4\pi c^2) = 8.85 \times 10^{-12}$ F/m, where c is the velocity of light,

$$F = \frac{e^2}{4\pi\epsilon_0 r^2} = \frac{e^2}{4\pi\epsilon_0(\rho^2 + v^2t^2)} \qquad (10.12)$$

which has an X component $F_x(t)$ that is always negative:

$$F_x = -\frac{e^2}{4\pi\epsilon_0 r^2}\cos\varphi = -\frac{e^2\rho}{4\pi\epsilon_0 r^3} \qquad (10.12a)$$

The Y component $F_y(t)$ is such that $F_y(t) = -F_y(-t)$. Thus after the interaction the net Y component of the momentum imparted to the lattice electron is zero, whereas the net X component Δp_x imparted to the electron is[†]

$$\Delta p_x = \int_{-\infty}^{+\infty} F_x(t)\, dt = \frac{-2e^2}{4\pi\epsilon_0 \rho v} \qquad (10.13)$$

The energy E_1 gained by the lattice electron is therefore

$$E_1 = \frac{1}{2m}(\Delta p_x)^2 = \frac{e^2}{(4\pi\epsilon_0)^2 E_p \rho^2} \qquad (10.14)$$

where $E_p = \frac{1}{2}mv^2$ is the energy of the incident primaries.

We now try to find the energy distribution of these electrons. Let N be the number of electrons per unit volume; then the number ΔN of electrons in a cylindrical shell with radii ρ and $(\rho + \Delta\rho)$ and unit length is

$$\begin{aligned}\Delta N &= N(2\pi\rho\,\Delta\rho) = \pi N\,\Delta\rho^2 = \frac{\pi Ne^4}{(4\pi\epsilon_0)^2 E_p}\,\Delta\frac{1}{E_1}\\[4pt] &= \frac{\pi Ne^4}{(4\pi\epsilon_0)^2 E_p}\,\frac{\Delta E_1}{E_1^2}\end{aligned} \qquad (10.15)$$

according to Eq. (10.14). Equation (10.15) represents the number of lattice electrons per unit length that gain an energy between E_1 and $(E_1 + \Delta E_1)$ in the interaction process. A more accurate calculation gives essentially the same result.

The energy loss of a primary electron per unit length is thus

$$\frac{dE_p}{dx} = -\sum E_1\,\Delta N = -\frac{\pi Ne^4}{(4\pi\epsilon_0)^2 E_p}\int\frac{dE_1}{E_1} \qquad (10.16)$$

where the summation has to be carried out over all energies E_1; as usual, this summation can be replaced by an integration.

[†] $\displaystyle\int_{-\infty}^{\infty}\frac{dt}{(\rho^2 + v^2t^2)^{3/2}} = \frac{1}{\rho^2 v}\int_{-\infty}^{\infty}\frac{du}{(1 + u^2)^{3/2}} = \frac{2}{\rho^2 v}$

where $u = vt/\rho$.

We must now determine the limits of integration. The upper limit is, of course, E_p, but what about the lower limit? For bound electrons with a binding energy E_i, the lower limit of integration should be E_i, but for the nearly free electrons in a metal one would at first sight expect a lower limit of zero, so that the integral would diverge. One can make the following observations, however.

1. According to Pauli's exclusion principle, all energy states for which $E < E_f$ are occupied. An electron having an initial energy E_i can absorb only an energy E_1 if $E_i + E_1 > E_f$. For $E_1 \simeq 0$ this condition is satisfied for very few electrons.

2. The force between two electrons was assumed to be given by (10.12) for all distances r. Actually, this equation is inaccurate for large r; since an electron will repel the other electrons in its immediate neighborhood, it will surround itself by a net positive charge, which screens its field at large distances. A calculation shows that the potential energy $V(r)$ of two electrons at a distance r can be written

$$V(r) = \frac{e^2}{4\pi\epsilon_0 r} \exp\left(-\lambda r\right) \tag{10.17}$$

and that $\lambda \simeq 10^{10}/\text{m}$. Small energy losses, which come from the interaction of electrons at large distances, are thus rare.

We can take both effects into account by choosing a suitable lower limit of integration E_i for the nearly free electrons, which can be considered as an "effective binding energy." Equation (10.16) may thus be written

$$\frac{dE_p}{dx} = -\frac{\pi e^4 N}{(4\pi\epsilon_0)^2 E_p} \sum_i \frac{N_i}{N} \ln \frac{E_p}{E_i} \tag{10.18}$$

where N_i/N is the fraction of electrons in each atom that has a binding energy E_i, and the summation is carried out over all electrons that have a binding energy $E_i < E_p$.

A wave-mechanical calculation gives essentially the same result. It is interesting to note that the simple classical approach gives so good an estimate of the energy loss. The preceding theory can be applied to both metals and insulators.

10.2d. Theory of the Secondary Emission Factor

To make Eq. (10.18) manageable, we approximate it as

$$\frac{dE_p}{dx} = -\frac{a}{E_p^\alpha} \tag{10.19}$$

where α is a very slow function of E_p, starting at $\alpha < 1$ and approaching unity for large E_p. We now treat α as a constant and integrate Eq. (10.19), with the condition $E_p = E_{p0}$ at $x = 0$, obtaining

$$E_p^{1+\alpha} = E_{p0}^{1+\alpha} - (1 + \alpha)ax \qquad (10.20)$$

This result is called the extended Whiddington law; Whiddington's original law is obtained by putting $\alpha = 1$. The range x_0 of the primary electrons is found by putting $E_p = 0$. This yields

$$E_{p0}^{1+\alpha} = (1 + \alpha)ax_0, \qquad x_0 = \frac{E_{p0}^{1+\alpha}}{(1 + \alpha)a} \qquad (10.20a)$$

so that Eq. (10.20) may be written

$$E_p^{1+\alpha} = E_{p0}^{1+\alpha}\left(1 - \frac{x}{x_0}\right) \qquad (10.21)$$

indicating that most of the energy loss occurs toward the end of the range. Measurements of the range of primaries in various materials for energies up to a few kilovolts give $\alpha \simeq 0.35$ to 0.50, whereas Whiddington's original law would give $\alpha = 1.00$.

To give a detailed theory of secondary emission, we assume that $n(x)$, the number of secondary electrons produced per unit length, is proportional to $-dE_p/dx$. That is, if b is a proportionality factor,

$$n(x) = \frac{ba}{E_p^\alpha} \qquad (10.22)$$

Substituting (10.22) into (10.4) and putting

$$g(x) = g_0 \exp(-\beta x) \qquad (10.23)$$

yields

$$
\begin{aligned}
\delta &= bg_0 \int_0^{x_0} \frac{a}{E_p^\alpha} \exp(-\beta x)\, dx \\
&= bg_0 \int_0^{E_{p0}} \exp\left[-\frac{\beta}{(1+\alpha)a}(E_{p0}^{1+\alpha} - E_p^{1+\alpha})\right] dE_p
\end{aligned}
\qquad (10.24)
$$

as we see by applying Eq. (10.20).

If this integral is evaluated numerically, we see that δ goes through a maximum at $E_{p0} = E_{max}$ and decreases for $E_{p0} > E_{max}$. If E_{p0} is larger than a few times E_{max}, we can find an approximate expression for δ by replacing a/E_p^α in the first half of (10.24) by a/E_{p0}^α. Equation (10.24) can then be integrated and yields

$$\delta = \frac{bg_0 a}{\beta E_{p0}^\alpha} \qquad (10.25)$$

so that $\delta = \mathrm{const}/E_{p0}^\alpha$. For very large values of E_{p0} ($E_{p0} > 10$ kV), the factor α approaches unity and δ varies as const/E_{p0}. The latter agrees with measurements above 10 kV, and for oxides it agrees even at lower energies, so that the whole curve $\delta(E_{p0})$ is reasonably well approximated by the result of

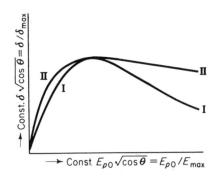

Fig. 10.10. Universal curve (δ/δ_{max}) plotted against (E_{po}/E_{max}). Curve I: theory; curve II: measurements for metals (J. L. H. Jonker, *Philips Research Reports*, **7**, 1–20, 1952).

(10.24). For metals it does not agree too well for $E_{max} < E_p < 10$ kV. This is evident from Fig. 10.10, where $\delta/\delta_{max} = f(E_{po}/E_{max})$ is compared with theory, using $\alpha = 1$. Choosing a smaller value of α would decrease the discrepancy but would not eliminate it.

This discrepancy most likely comes about because of back-scattered electrons. Some of the primary electrons entering the secondary emission target return to the surface and are reemitted. Upon their return to the surface region these electrons produce more secondaries. Since metals have a much lower secondary emission than the oxides mentioned, the effect of the back-scattered electrons is much more pronounced.

10.3. TRANSMISSION SECONDARY EMISSION

If primary electrons enter a metal or oxide target, and the thickness of the target is comparable with the penetration depth of the target, slow secondaries can escape on the opposite side. The phenomenon is known as *transmission secondary emission*.

At low primary energies the primaries do not penetrate far, and few secondaries can escape from the other side. At very high primary energies, all primaries pass through the target, and the secondary emission decreases with increasing primary energy. Somewhere in between there must be a primary energy at which the transmission secondary emission factor $\delta_t = I_{sec}/I_{prim}$ has a maximum value. This should occur at an energy where the thickness of the target is approximately equal to the penetration depth of the primary electrons (Fig. 10.8).

Very large transmission secondary emission factors δ_t have been observed in (porous) low-density films of KCl, obtained by evaporating low-density layers of KCl onto aluminum in an argon atmosphere. Figure 10.11 shows the cross section of such a secondary emission target; the Al_2O_3 and Al layers are needed for support. This supporting film can be mounted over a stainless steel support ring. Stable targets of 1-in. diameter can be obtained in this manner.

Figure 10.12 shows that with these targets maximum secondary emission factors $\delta_t > 50$ can be obtained for primary energies between 7 and 10 kV.

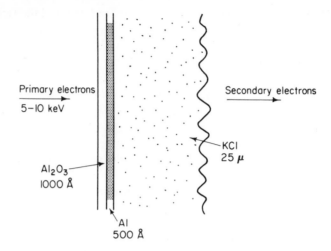

Fig. 10.11. Cross section of a low-density dynode (from G. W. Goetze, A. H. Boerio, and M. Green, *J. Appl. Phys.*, **35**, 482, 1964).

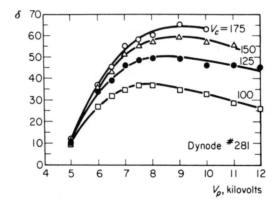

Fig. 10.12. The variation of gain with primary voltage (from G. W. Goetze, A. H. Boerio, and M. Green, *J. Appl. Phys.*, **35**, 482, 1964).

A sufficiently large collector voltage V_c, of the order of about 200 V, is needed to attain maximum secondary emission. Apparently, this voltage is required to extract the low-energy secondaries out of the porous KCl film.

Figure 10.8 showed the transmission secondary emission factor of a thin (4–5 μm thick) silicon target covered with an –O–Cs monolayer on the exit side of the secondary electrons[†]. A transmission secondary emission factor as high as 750 was obtained at a primary energy of 20 kV. This must again

---

[†] R. U. Martinelli, *Appl. Phys. Letters*, **17**, 313–314 (1971).

be attributed to the large escape depth of the secondaries and the large value of g_0.

10.4. SECONDARY EMISSION NOISE

If each primary electron produced exactly δ primaries, the electron multiplication process would be completely noiseless, and the signal-to-noise power ratio at the output would be equal to the similar ratio at the input. But if the number of secondary electrons fluctuates, additional noise will be generated, and the signal-to-noise power ratio will deteriorate. We now calculate this deterioration factor.

Let n_p primary electrons strike the target per second. If the primary electrons arrive at random instants,

$$\text{var } n_p = \overline{n_p^2} - (\bar{n}_p)^2 = \bar{n}_p$$

If β_n is the probability that n electrons are emitted by a primary, then

$$\delta = \sum_n n\beta_n = \bar{n} \tag{10.26}$$

It will be convenient to call

$$\overline{n^2} = \sum_n n^2\beta_n = \kappa\delta \tag{10.27}$$

We shall see that κ is a measure for the noise.

We now calculate the fluctuation in the number n_c of secondary electrons. To that end we put $n_p = (n_p - \bar{n}_p) + \bar{n}_p$. The primary fluctuation $n_p - \bar{n}_p$ gives a contribution

$$(\bar{n})^2 \text{ var } n_p = (\bar{n})^2\bar{n}_p = \delta^2\bar{n}_p \tag{10.28}$$

to var n_c (amplified shot noise), whereas the average number $\bar{n}_p$ gives a fluctuation

$$\bar{n}_p \text{ var } n = \bar{n}_p[\overline{n^2} - (\bar{n})^2] \tag{10.28a}$$

to var n_c. Consequently,

$$\text{var } n_c = \bar{n}_p\overline{n^2} = \bar{n}_p\kappa\delta = \kappa\bar{n}_c \tag{10.29}$$

Since var n_c would be equal to $\bar{n}_p\delta^2$ if the secondary emission process did not fluctuate, we see that *the signal-to-noise power ratio has deteriorated by the factor κ/δ in the secondary emission process.*

Experimentally, one finds $\kappa - \delta \simeq 1$ for E_{p0} of less than a few hundred electron volts, whereas it increases for higher E_{p0}. Hence

$$\frac{\kappa}{\delta} \simeq 1 + \frac{1}{\delta} \tag{10.30}$$

for low E_{p0}, decreases with increasing E_{p0} up to a few hundred electron volts, passes through a minimum, and increases with a further increase in E_{p0}. There is thus a minimum in the noise deterioration factor; it is very close to unity for high-yield targets.

REFERENCES

BRUINING, H., *Physics and Applications of Secondary Electron Emission*. Pergamon Press, Inc., Elmsford, N.Y., 1954.

DEKKER, A. J., *Solid State Physics*. Prentice-Hall, Inc., Englewood Cliffs, N.J., 1957.

HACHENBERG, O., and W. BRAUER, "Secondary Electron Emission from Solids," in L. Marton, ed., *Advances in Electronics and Electron Physics*, vol. 11, pp. 413–499. Academic Press, Inc., New York, 1959.

PROBLEMS

1. Calculate the average probability of escape for a secondary electron liberated inside a secondary emission target and arriving at the surface with an energy $E = 20$ V, if the direction of arrival is random. Carry out this calculation for a depth of the bottom of the conduction band below the vacuum level of 16 V and 15 V, and determine from the results how much the secondary emission factor of the target is changed if the work function is decreased by 1 V, taking the case $W = 16$ V as a reference.

Answer: 0.105; 0.134; factor 1.28.

2. A secondary emission target has its δ value measured at two temperatures at primary energies much larger than the value for maximum δ. Let T_1 and T_2 be those temperatures and let δ_1 and δ_2 be the secondary emission factors measured. Show that $\delta_1/\delta_2 = \beta_2/\beta_1$, where β_1 and β_2 are the absorption coefficients for the secondaries at the temperatures T_1 and T_2. You may assume an absorption function $h(x) = \exp{(-\beta x)}$ in both cases.

3. Show that if the scattering of the primary electrons is negligible, then at sufficiently large energies ($E_p \gg E_{max}$) the secondary emission factor will depend upon the angle of incidence θ in the following manner: $\delta = \delta_0/\cos \theta$, where δ_0 is the secondary emission factor for normal incidence. To what causes would you attribute deviations from this law?

4. Show with the help of the discussion of Sec. 10.2d that $\delta\sqrt{\cos \theta}$ is a universal function of $E_{p0}\sqrt{\cos \theta}$ for a given target. Assume that $\alpha = 1$.

5. Electrons with 1000-V primary energy have a penetration depth of 150 Å in platinum. Calculate Whiddington's *a* constant, assuming that $\alpha = 1$.

Answer: $\frac{1}{3} \times 10^{12}$ V²/cm.

6. What part of its initial energy has a particle lost after it has passed through the first half of its range? Assume that $\alpha = 1$.

Answer: The part 0.293.

11

Photoconductivity

11.1. PHOTOCONDUCTION

11.1a. Types of Photoconduction

If an insulator is irradiated with light of a sufficiently short wavelength, so that transitions from the valence band to the conduction band occur, the insulator becomes a conductor. This process is known as *photoconduction.* It is detected by making ohmic contacts to the insulator, applying a voltage, and measuring the current. The condition for this type of photoconduction is that the quantum energy

$$hf > eE_g \qquad (11.1)$$

The maximum wavelength for the photoconductive effect is therefore

$$\lambda_{\max} = \frac{hc}{eE_g} = \frac{1.24}{E_g} \, \mu m \qquad (11.1a)$$

Here E_g is the gap width in electron volts.

If a semiconductor is irradiated with light of sufficiently short wavelength so that transitions from the valence band to the conduction band occur, the conductivity of the material is increased (Fig. 11.1a). In these materials one must distinguish between the "dark conductivity," which is present without light, and the "photon-induced conductivity." The condition for this type of photoconductivity is again given by Eq. (11.1).

Actually, Eq. (11.1) is not correct in all cases. It is correct if the gap width

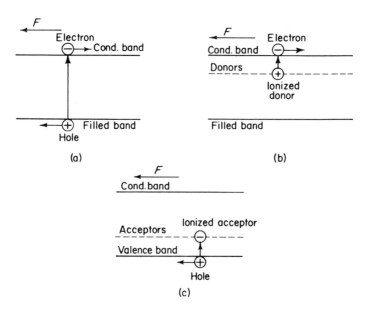

Fig. 11.1. (a) Excitation of an electron from the filled band, leaving a free hole behind. (b) Excitation of an electron from a deep-lying donor center, leaving an ionized donor behind. (c) Excitation of an electron from the valence band to a neutral acceptor, leaving a free hole behind.

corresponds to the energy difference for a direct transition. But if the minimum energy in the conduction band and the maximum energy in the valence band do not occur at the same value of the reduced wave vector, photoconduction becomes possible by phonon-assisted transitions.

There is another reason why Eq. (11.1) may be violated. In the absorption process a hole and an electron are formed. If both are completely free, both can take part in the conduction process. But there are also *exciton states*, in which an electron and a hole are bound together. At low temperatures this does not lead to photoconduction, but at sufficiently high temperatures the electron can be shaken loose from the hole and the exciton state is destroyed by thermal agitation. This results in a free electron and a free hole that can contribute to conduction. The binding energy of the exciton determines at what temperature this effect becomes important.

In the examples discussed thus far both free electrons and free holes are present, and both can contribute to the conduction process. We shall call this *intrinsic photoconductivity*. In *extrinsic photoconduction*, however, only *one* type of carrier is present. Consider, for example, an extrinsic semiconductor with deep-lying donor levels. Let E_0 be the energy difference between the donor level and the bottom of the conduction band in electron volts. At low temperatures almost all the electrons are bound to donor levels and the conductivity is small. If the material is irradiated by light of quantum energy

$$hf > eE_0 \qquad (11.2)$$

the conductivity of the material increases, because electrons make transitions from the neutral donor levels to the conduction band (Fig. 11.1b).

In an extrinsic semiconductor with deep-lying acceptor levels, low-temperature photoconductivity becomes possible (Fig. 11.1c) if

$$hf > eE_0' \qquad (11.3)$$

where E_0' is the energy difference between the acceptor level and the top of the valence band in electron volts. In this case free holes are formed when electrons make transitions from the valence band to the neutral acceptor levels.

One difference between intrinsic and extrinsic photoconductivity is that in the first case the light absorption is much stronger. In intrinsic photoconductivity the penetration depth of the light may be only of the order of 1000 Å. The conductivity is then confined to a relatively narrow region close to the surface, and it is sufficient to use relatively thin photoconducting layers.

In extrinsic photoconductivity, where the light absorption is proportional to the impurity density, the penetration depth of the light is much larger. The photoconducting samples must be made thicker than the penetration depth of the light, so that most of the light is utilized for the photoconduction process. Alternatively, for a given thickness of the photoconducting sample, one must choose the concentration of the required impurities in such a way that the penetration depth of the light is less than the thickness of the sample.

11.1b. Mechanism of Photoconduction

Consider a material having an equilibrium electron concentration $n = n_0$ and an equilibrium hole concentration $p = p_0$ when not irradiated by light. If the electron and hole mobilities are μ_n and μ_p, respectively, then the dark conductivity σ_d of the sample is

$$\sigma_d = e(n_0\mu_n + p_0\mu_p) \qquad (11.4)$$

For insulators, conduction in the dark is negligible; for intrinsic and extrinsic semiconductors, conduction in the dark may be significant unless the material is cooled.

We first investigate intrinsic photoconductivity. Let the sample be irradiated *uniformly* by E watts of light consisting of quanta of energy hf. Let R be the power reflection coefficient of the surface and η the quantum efficiency of the hole–electron pair production process. Then the number of hole–electron pairs produced per second is

$$Q = \eta\frac{E(1 - R)}{hf} \qquad (11.5)$$

If the electrons and holes thus created have lifetimes τ_n and τ_p, respectively, then the steady-state *added* numbers of carriers are

$$\Delta N = Q\tau_n, \qquad \Delta P = Q\tau_p \tag{11.6}$$

where ΔN is the number of added electrons and ΔP the number of added holes. In general, τ_n and τ_p will depend on Q, and hence on E; they should be considered as lifetimes of the added carriers. For insulator small signal photoconductors, ΔN and ΔP correspond to the total numbers of carriers; for semiconductor photoconductors these numbers of carriers are added to the "dark" numbers $N_0 = n_0 AL$ and $P_0 = p_0 AL$, where A is the cross section of the sample and L its length.

How does this change the conductance of the sample? The dark conductance

$$g_d = \sigma_d \frac{A}{L} = e\left(\frac{N_0}{AL}\mu_n + \frac{P_0}{AL}\mu_p\right)\frac{A}{L} = \frac{e(N_0\mu_n + P_0\mu_p)}{L^2} \tag{11.4a}$$

and hence the change in conductance due to the light is

$$\Delta g = \frac{e(\Delta N \mu_n + \Delta P \mu_p)}{L^2} = \frac{eQ(\mu_n\tau_n + \mu_p\tau_p)}{L^2} \tag{11.6a}$$

For insulator photoconductors this is the actual conductance of the sample.

In general, both the electrons and the holes are mobile so that both contribute to the current. It may well be, however, that $\mu_n\tau_n$ and $\mu_p\tau_p$ are of a different order of magnitude, because of differences either in mobility or in lifetime. In that case the main contribution to the increased conductivity comes from the type of carrier having the largest value of $\mu\tau$. This is found to occur actually in many photoconductors. In some of them the electrons give the main contribution to the current; in others the main contribution comes from the holes.

If V is the voltage applied to the sample, the change in current due to the incident light is

$$\Delta I = V \Delta g = \frac{eV(\mu_n\tau_n + \mu_p\tau_p)}{L^2}Q \tag{11.7}$$

For sensitive photoconductors the expression $(\mu_n\tau_n + \mu_p\tau_p)/L^2$ should be large. This means that either the mobilities should be large or the small signal lifetimes should be large, or both. Furthermore, the length L should be small; this is especially important for samples with low dark current.

The condition that τ_n and τ_p should be large for sensitive photoconductors holds for the change in conductance under steady illumination. If the sample is irradiated by alternating light, however, the situation is different. We discuss this problem for electrons; a similar discussion holds for holes.

Let the production rate (the generation rate) of hole–electron pairs have a time dependence $Q = Q_0 \exp(j\omega t)$; then

$$\frac{d\,\Delta N}{dt} = -\frac{\Delta N}{\tau_n} + Q_0 \exp(j\omega t) \tag{11.8}$$

Trying the solution $\Delta N = \Delta N_0 \exp(j\omega t)$, we get the equation

$$\left(j\omega + \frac{1}{\tau_n}\right)\Delta N_0 = Q_0, \qquad \Delta N_0 = \frac{Q_0 \tau_n}{1 + j\omega \tau_n} \tag{11.9}$$

and a similar expression holds for the holes. Consequently, Eq. (11.7) now becomes

$$\Delta I = \frac{eV}{L^2}\left(\frac{\mu_n \tau_n}{1 + j\omega \tau_n} + \frac{\mu_p \tau_p}{1 + j\omega \tau_p}\right)Q_0 \exp(j\omega t) \tag{11.10}$$

For $\omega \tau_n \ll 1$ and $\omega \tau_p \ll 1$, this reduces to Eq. (11.7); but for $\omega \tau_n \gg 1$ and $\omega \tau_p \gg 1$,

$$\Delta I = \frac{eV}{L^2}\frac{\mu_n + \mu_p}{j\omega}Q_0 \exp(j\omega t) \tag{11.10a}$$

The lifetimes have now disappeared from the expression, and the sensitivity of the material is determined by $\mu_n + \mu_p$.

The foregoing discussion holds for intrinsic photoconductors. It can easily be repeated for extrinsic photoconductors. To obtain the corresponding equations for n-type photoconductors, one puts $\mu_p = 0$ and $\tau_p = 0$ in Eqs. (11.4)–(11.10a). To obtain the corresponding equations for p-type semiconductors, one puts $\mu_n = 0$ and $\tau_n = 0$ in Eqs. (11.4)–(11.10a). This is left as a problem for the reader.

11.1c. Other Time Constants Involved in Photoconduction

Besides the small signal lifetime, the drift time τ_{dr} of the carriers is also of interest. We shall now see how this time constant enters into the considerations. If we consider that only one type of carrier gives a significant contribution to the conductance, then, if the sample has a length L,

$$\tau_{dr} = \frac{L}{u_d} = \frac{L}{\mu F} = \frac{L^2}{\mu V} \tag{11.11}$$

where u_d is the drift velocity, μ the mobility, F the field strength, and V the applied voltage. If we substitute this into the equivalent of Eq. (11.7), we obtain

$$\Delta I = e\frac{\mu V}{L^2}\tau Q = eQ\frac{\tau}{\tau_{dr}} \tag{11.12}$$

where τ is the lifetime of the carriers.

This has the following simple interpretation. If all generated carriers were collected, the current would be eQ. The factor

$$G = \frac{\tau}{\tau_{dr}} \tag{11.12a}$$

is called the *gain factor*; it can change from values much larger than unity for very sensitive photoconductors $(\tau \gg \tau_{dr})$ to values much smaller than unity for very insensitive photoconductors $(\tau \ll \tau_{dr})$.

The reason why τ can be much larger than τ_{dr} requires an explanation. Naïvely, one might think that a carrier is eliminated when it reaches the ohmic contact. But actually this not the case. If carriers drift toward the right, then new carriers are injected by the left ohmic contact when old carriers are taken out at the right ohmic contact, so that space-charge neutrality is maintained at all times. The conduction event, starting when the carrier is generated, hence does not stop when the carrier reaches an ohmic contact; the event terminates when a carrier is removed by recombination or by trapping. The ohmic contacts thus help in maintaining space-charge neutrality.

11.1d. Practical Photoconductors

Photoconductivity occurs in many insulators and semiconductors, but not all of them have reached technical importance. Technically most important are lead sulfide (PbS), lead selenide (PbSe), lead telluride (PbTe), strontium sulfide (SrS) activated by europium and samarium, cadmium sulfide (CdS), gold-doped and mercury-doped Si and Ge, various III–V compounds such as InSb, and lead-tin telluride and mercury-cadmium telluride.

It has sometimes been assumed for the commonly used thin-layer cells of PbS, PbSe, and PbTe that these layers consist of randomly distributed *pn* junctions and are not genuine photoconductors in the true sense of the word. It is now held, however, that they are true photoconductors in which the minority carriers (electrons) are trapped. The SrS and CdS cells are also true photoconductors, and so are mercury- and gold-doped Si and Ge and the III–V compounds.

In many recent applications of photoconductors, one wants response in the far infrared, for example, at wavelengths around 5 or around 10 μm, where the water vapor in the atmosphere transmits. For the first case, the intrinsic semiconductor InSb is very useful. For the second case, one can use an extrinsic semiconductor such as gold-doped germanium.

Lately, materials have become available that have a band gap dependent on composition. The most important ones are the lead–tin chalcogenide system, involving materials such as $Pb_{1-x}Sn_xTe$ and $Pb_{1-y}Sn_ySe$, and the mercury–cadmium telluride ($Hg_{1-x}Cd_xTe$) system.

We first discuss $Pb_{1-x}Sn_xTe$ (Fig. 11.2, 3). PbTe has a band gap of 0.18 eV; SnTe has a bandgap of 0.30 eV. The energy gap first decreases with increasing Sn concentration, goes through zero at an intermediate concentration, and then rises again. This situation is explained by a crossover of the energy states. If the energy states in the valence band of PbTe are denoted by L_6^+ and those in the conduction band by L_6^-, then the energy levels of the valence band of SnTe must be denoted by L_6^- and those in the conduction band by L_6^+ (Fig. 11.2). At an intermediate composition the energy states cross and the band gap is zero (Fig. 11.3).

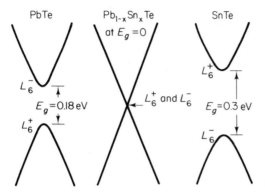

Fig. 11.2. Schematic presentation of the valence and conduction bands at 12°K for PbTe, for the composition at which the energy gap is zero, and for SnTe (J. O. Dimmick, I. Melngailis and A. J. Strauss, Physical Review Letters, **16**, 1193, 1966).

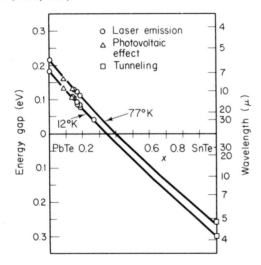

Fig. 11.3. Energy gap of $Pb_{1-x}Sn_xTe$ as a function of x. (From I. Melngailis and T. C. Harman, in Semiconductors and Semimetals, Vol. 5, Academic Press, Inc. 1970, by permission).

CdTe is a semiconductor with a bandgap of 1.60 eV, whereas HgTe is a semimetal. In this crystal structure, which is a zincblende structure, the Cd atoms can be replaced by Hg atoms. The gap width decreases with increasing mercury concentration and becomes practically zero for $x = 0.18$ (Fig. 11.4).

The details of these materials and their uses can be obtained from two recent review papers cited at the end of the chapter. Their advantage is that the gap width, and hence the upper wavelength response, can be designed

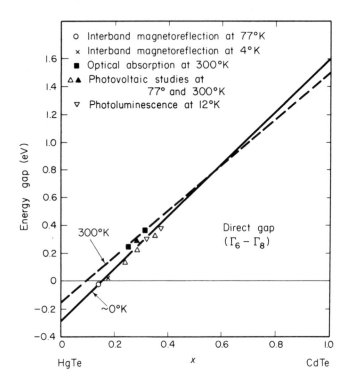

Fig. 11.4. Energy gap versus composition in Hg$_{1-x}$Cd$_x$Te. The solid line represents the dependence at $\simeq$ 0°K, the dashed line is for 300°K. The types of experiments giving the data points are indicated (From D. Long, Energy Bands in Semiconductors, Wiley and Sons, New York, 1968, by permission).

by choice of the composition. The materials can be used as intrinsic, or near intrinsic, photoconductors.

Since the materials can be doped by various impurities, one can also make the materials *p* type or *n* type according to doping. *Pn* junctions can therefore be made that can be used as far-infrared photovoltaic cells (Chapter 17). Because of the small gap width, the photoconductors or photovoltaic cells made of these materials must be cooled to a relatively low temperature to reduce the dark current.

For practical applications the following properties of the cell are important:

1. The "technical sensitivity" of the cell, expressed in microamperes per watt or microamperes per lumen. Obviously one wants this quantity to be large. Sensitivities far in excess of 1000 μa/lm are possible, but, as we saw before, the more sensitive the cell, the more sluggishly it will respond.

2. The time constant of the cell. This quantity is important in infrared

detectors in which the radiation is measured by chopping the light beam or in which scanning methods are used for detecting sources of infrared radiation; the time constant τ must then be smaller than the chopping time or scanning time. Time constants range from microseconds to many milliseconds.

3. The dark current of the cell. This quantity is determined by the gap width between the conduction band and the valence band in intrinsic photoconduction, by the gap between the donor levels and the conduction band in n-type extrinsic photoconductors, and by the gap between the acceptor levels and the valence band in p-type extrinsic photoconductors. If the dark current harms the operation of the cell, one can reduce it materially by cooling the cell.

4. The wavelength response of the photoconductivity. Infrared detectors should have a cutoff wavelength far in the infrared. This requires photoconductors with a small forbidden gap width. At 20°C the gap widths for PbS, PbTe, and PbSe are 0.40, 0.31, and 0.25 eV, respectively, leading to cutoff wavelengths of 3, 4, and 5 μm, respectively. Mercury-doped and gold-doped Ge and Si have even longer cutoff wavelengths. In these cases it is necessary to cool the cell to reduce the dark current. In PbS, PbSe, and PbTe this has the additional advantage of reducing the gap width, thus increasing the cutoff wavelength.

5. The noise properties of the cell. The absolute noise limit is set by the spontaneous fluctuations in the incident radiation. The photoconduction process adds to this noise, but, as will be seen in Sec. 11.3, this additional noise is not excessive for good cells. Often in poorer cells most of the noise is generated at the contacts; improvement of the contacts may then give a considerable decrease in noise level.

As expected from Ohm's law, the cell response increases linearly with the applied voltage V [compare, for example, Eq. (11.7)]. At high voltages the mobility may become field dependent, and then the current increases slower than Ohm's law predicts. Or it may happen that the cell starts to operate in a space-charge-limited mode of operation (Chapter 19); in this case the current may increase faster than Ohm's law predicts.

11.2 KINETICS OF PHOTOCONDUCTION

11.2a. Kinetics of Photoconduction in Extrinsic Photoconductors

As an example of an extrinsic photoconductor we treat here the case of an n-type extrinsic photoconductor with deep-lying donors. The case of p-type extrinsic photoconduction, which is also of practical significance, can be treated in the same manner.

The kinetic equation for the photoconduction process is

$$\frac{dn}{dt} = a(N_d - n) - bn^2 + q \qquad (11.13)$$

Here n is the density of carriers, N_d the density of donors, q the rate of generation of photoelectrons, and a and b constants. The first term in Eq. (11.13) gives the rate of spontaneous generation from the $N_d - n$ neutral donors, and the term $-bn^2$ gives the rate of recombination of n free electrons and n ionized donors. For the values of the constants a and b, see Sec. 6.3c.

The steady-state dark current follows from this equation by our putting $q = 0$ and $dn/dt = 0$. This yields

$$a(N_d - n) - bn^2 = 0 \quad \text{or} \quad n = n_0 = -\frac{a}{2b} + \left(\frac{a^2}{4b^2} + \frac{aN_d}{b}\right)^{1/2} \quad (11.14)$$

If a small amount of light is admitted, so that the increase Δn in electron density is small in comparison with the equilibrium density n_0, the photoresponse is linear. The steady-state equation is now

$$-a\,\Delta n - 2bn_0\,\Delta n + q = 0, \qquad \Delta n = \frac{q}{a + 2bn_0} = q\tau \qquad (11.15)$$

The response is thus linear in q, and the small-signal lifetime of the added carriers is $1/(a + 2bn_0)$.

If a large amount of light is admitted, so that $n \gg n_0$, recombination predominates over spontaneous generation and the steady-state equation may be approximated as

$$-bn^2 + q = 0 \quad \text{or} \quad n = \left(\frac{q}{b}\right)^{1/2} \qquad (11.16)$$

The current now varies as the square root of the light intensity, so that the response is sublinear.

Thus the conclusion is that in this case one obtains a linear response at low light levels and a sublinear response at high light levels.

11.2b. Kinetics of Photoconduction in Intrinsic Photoconductors

In this section we consider three simple cases: (1) direct recombination, (2) recombination via recombination centers, and (3) trapping of the minority carriers. A more complicated case is dealt with in Sec. 11.2c.

In the case of direct recombination the kinetic equation is

$$\frac{dn}{dt} = g_0 - pn^2 + q \qquad (11.17)$$

Here the first term gives the spontaneous generation rate of hole–electron pairs, the second term the recombination rate (bimolecular) of hole–electron pairs, and the third the external generation rate. We neglect here the possible presence of donor and acceptor impurities.

The discussion now proceeds along the same line as before. One can again evaluate the equilibrium carrier concentrations $n_0 = p_0$. The response is linear if $\Delta n = \Delta p \ll n_0$ and is sublinear if $n \gg n_0$. In the latter case, where spontaneous generation can be neglected, the steady-state equation becomes

$$-pn^2 + q = 0, \qquad n = \left(\frac{q}{p}\right)^{1/2} \tag{11.18}$$

so that the current again varies as the square root of the light intensity. This case may be of interest in InSb, $Pb_{1-x}Sn_xTe$, and $Hg_{1-x}Cd_xTe$.

If one substitutes $n = n_0 + \Delta n$ into (11.17), where n_0 is the equilibrium carrier concentration, for the small-signal lifetime of the bimolecular process one finds

$$\tau = 2pn_0 \tag{11.17a}$$

Often several processes in parallel determine the lifetime τ. In that case, if the ith process has a lifetime τ_i,

$$\frac{1}{\tau} = \sum_i \frac{1}{\tau_i} \tag{11.17b}$$

We refer the reader to Blakemore's book, cited at the end of the chapter, for details.

If the recombination is via recombination centers, the kinetic equations become

$$\frac{dn}{dt} = -vs_1(N_t - n_t)n + v_{01} \exp\left(-\frac{eE_1}{kT}\right)n_t + q \tag{11.19}$$

$$\frac{dp}{dt} = -vs_2n_tp + v_{02} \exp\left(-\frac{eE_2}{kT}\right)(N_t - n_t) + q \tag{11.20}$$

Here N_t is the density of the centers, n_t the density of the trapped electrons, v is the thermal speed of the carriers, s_1 and s_2 are the capture cross sections of the centers for electrons and holes, respectively, v_{01} and v_{02} are vibration frequencies of the order of 10^{12}/s, E_1 and E_2 are the energy differences between conduction band and center and between center and valence band, respectively, and q is the external generation rate. From these equations the unknown parameter n_t can be eliminated since $n + n_t - p = N_0$ is a constant at all times.

This problem is easily solved if the thermal generation terms in Eqs. (11.19) and (11.20) are negligible. The steady-state conditions are then

$$-vs_1(N_t - n_t)n + q = 0, \qquad -vs_2n_tp + q = 0 \tag{11.21}$$

$$n = \frac{q}{vs_1(N_t - n_t)}, \qquad p = \frac{q}{vs_2n_t} \tag{11.22}$$

As long as $n_t \ll N_t$, n varies linearly with q; as long as n_t is comparable with the dark trapped-electron density n_{t0}, p varies linearly with q. There is thus always a linear range of operation at low light intensities.

For larger values of q the situation is more complicated. We have as an additional equation

$$n + n_t - p = N_0 \tag{11.23}$$

but since Eqs. (11.21) involve nonlinear relationships, that set of equations is difficult to solve. We can, however, illustrate the behavior with a few examples.

If the holes have a much higher mobility than the electrons, the material is sublinear at higher light levels. For since n_t increases with increasing q, p increases slower than q. If the electrons have a much higher mobility than the holes, the material is linear as long as $n_t \ll N_t$. If n_t could become comparable to N_t, a superlinear range might occur when $N_t - n_t$ becomes sufficiently small and decreases with increasing q.

The question is, can $N_t - n_t$ become so small? We shall see that this is not the case. If the centers are true recombination centers, then s_1 and s_2 are comparable; we thus make little error by putting $s_1 = s_2 = s$. Equations (11.22) may then be written

$$n(N_t - n_t) = \frac{q}{vs}, \qquad pn_t = \frac{q}{vs} \tag{11.22a}$$

If $n + n_t \gg N_0$, Eq. (11.23) may be written

$$n + n_t = p \tag{11.23a}$$

Eliminating p and n from these equations yields

$$\frac{q}{vs} n_t + n_t^2(N_t - n_t) = \frac{q}{vs}(N_t - n_t) \tag{11.24}$$

from which it follows, since $n_t^2(N_t - n_t) > 0$, that

$$N_t - n_t > n_t \quad \text{or} \quad n_t < \tfrac{1}{2}N_t \tag{11.25}$$

Therefore, the semiconductor cannot become superlinear if s_1 and s_2 are comparable.

The case of trapped minority carriers is simple. The increase in majority carriers is proportional to the rate of generation q; hence the response is linear over a wide range of light levels. At very high light levels, however, the trapped minority carriers will begin to act as recombination centers for the majority carriers. When that happens, the majority-carrier density increases slower than linearly with q; hence the response becomes sublinear. This is, for example, the case in PbS photoconductors.

11.2c. The Case of CdS Photoconductors

In photoconductors such as CdS the situation is more complex. There are electron traps that can act as recombination centers at intermediate light levels and hole traps that can act as recombination centers at lower light levels. The energy-level diagram is shown in Fig. 11.5. It contains N_t electron

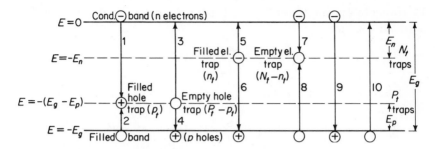

Fig. 11.5. Energy level diagram of a semiconductor with electron traps and hole traps. The various possible transitions are labeled by the numbers 1, . . . , 10. Electron, hole, and trap densities are also shown.

traps per cubic meter at an energy E_n below the bottom of the conduction band, and P_t hole traps per cubic meter at an energy E_p above the top of the filled band; the band gap is denoted by E_g. Let n be the free-electron density, n_t the trapped-electron density, p the free-hole density, and p_t the trapped-hole density. There are thus $N_t - n_t$ empty electron traps and $P_t - p_t$ empty hole traps per cubic meter.

The following processes can occur:

1. Recombination of a free electron and a trapped hole.
2. Thermal release of a trapped hole.
3. Thermal or optical creation of a free electron, leaving a trapped hole behind.
4. Trapping of a free hole.
5. Thermal release of a trapped electron.
6. Recombination of a trapped electron and a free hole.
7. Trapping of a free electron.
8. Thermal or optical creation of a free hole, leaving a trapped electron behind.
9. Direct recombination of a free hole and a free electron.
10. Thermal or optical generation of free hole–electron pairs.

Some of these processes are unimportant in particular cases. Process 9 is often unimportant because the probability of direct recombination is small. If the energies E_g, $(E_g - E_n)$, and $(E_g - E_p)$ are large, the thermal processes 3, 8, and 10 have a very small probability. The optical processes 3, 8, and 10 are important in photoconduction. Processes 2 and 5 may also occur if we irradiate the photoconductor with infrared light; this is sometimes important in experimental work (infrared quenching).

We now write down the equations of the photoconductive process under the assumption that q hole–electron pairs are created per second by process

10, owing to optical absorption, and that processes 3, 8, and 9 can be disregarded completely.

$$\frac{dn}{dt} = -vs_1 p_t n + v_{05} \exp\left(-\frac{eE_n}{kT}\right) n_t - vs_7 n(N_t - n_t) + q$$

$$\frac{dn_t}{dt} = -v_{05} \exp\left(-\frac{eE_n}{kT}\right) n_t + vs_7 n(N_t - n_t) - vs_6 n_t p$$

$$\frac{dp_t}{dt} = -vs_1 p_t n - v_{02} \exp\left(-\frac{eE_p}{kT}\right)(P_t - p_t) + vs_4 p(P_t - p_t)$$

$$\frac{dp}{dt} = v_{02} \exp\left(-\frac{eE_p}{kT}\right)(P_t - p_t) - vs_4 p(P_t - p_t) - vs_6 n_t p + q$$

(11.26)

Here s_1, s_4, s_6, and s_7 refer to capture cross section for processes 1, 4, 6, and 7, v_{05} and v_{02} are vibration frequencies ($\simeq 10^{12}$/s) involved in processes 5 and 2, and v is again the average thermal speed of the electrons. Equations (11.26) are not independent, of course; since holes and electrons appear and disappear in pairs, $n + n_t - p_t - p$ is constant, or the sum of the first two equations minus the sum of the latter two equations is zero.

The equilibrium condition is

$$\frac{dn}{dt} = \frac{dn_t}{dt} = \frac{dp_t}{dt} = \frac{dp}{dt} = 0 \qquad (11.26a)$$

which gives the equilibrium concentrations of n and p, and hence the magnitude of the photoconductivity as a function of the light intensity.

In CdS the deep-lying levels act mainly as hole traps (that is, $s_1 \ll s_4$), and the other centers act as a recombination center (that is, $s_6 \simeq s_7$). Nevertheless, at low light levels these centers are nearly inactive as recombination centers; since most of the holes are trapped in the deep-lying center, the recombination center acts mainly as an electron trap.

Because almost all current is carried by electrons and the thermal release of trapped electrons or trapped holes is very unlikely, the first of the equations (11.26) may be written

$$-vs_1 p_t n - vs_7 (N_t - n_t)n + q = 0 \quad \text{or} \quad n = \frac{q}{vs_7(N_t - n_t) + vs_1 p_t} \qquad (11.27)$$

At relatively low light levels $n_t \ll N_t$ and $v_1 s_1 p_t$ is small because p_t and s_1 are small. Consequently, n varies linearly with q (linear range of operation). At higher light levels it may happen that n_t becomes comparable to N_t, whereas $vs_1 p_t$ is still relatively small. Then, since $N_t - n_t$ decreases with increasing q, n will increase faster than q, and the semiconductor material becomes superlinear.

At still higher intensities p begins to increase sufficiently to make the electron traps act as recombination centers. As a consequence, $N_t - n_t$ may actually increase with increasing q. Assisting here is the fact that the term

$vs_1 p_t$ becomes comparable to $vs_7(N_t - n_t)$. As a consequence, n will increase slower than q, and the operation becomes sublinear.

The superlinear range is missing in poorly activated CdS crystals. Poor activation means that the number of deep-lying centers is small. There is then only *one* center that acts as a true recombination center ($s_6 \simeq s_7$). But we know already from the previous section that this will not give super-linearity.

At very high light levels, process 9 may begin to predominate. When that is the case, the theory of Sec. 11.2b applies and n and p vary as $q^{1/2}$.

The possible ranges of operation of CdS photoconductor are shown in Fig. 11.6.

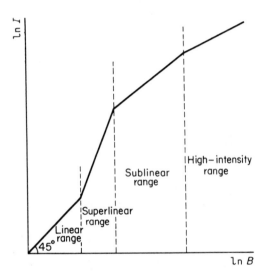

Fig. 11.6. Dependence of the photocurrent I upon the light intensity B, showing the linear, superlinear, sublinear, and high-intensity ranges.

11.3. NOISE IN PHOTOCONDUCTORS

As shown in the Appendix, the fluctuations in the rates of generation and recombination of carriers can be described by a shot-noise mechanism.

If the photoconductor has a negligible dark current, the fluctuation in the rate of generation is caused by the fluctuations in the arrival of incoming photons. This is described by a shot-noise process. Since in equilibrium the rate of recombination balances the rate of generation and both rates show shot noise, the recombination process gives as much noise as the generation process; hence the signal-to-noise power ratio of the device is decreased by a factor 2.

If the dark current is not negligible, then the fluctuation in the dark conductance is determined by the generation and recombination processes that establish this dark conductance. At low incident light levels, these fluctuations determine the signal-to-noise power ratio of the device. Again, the fluctuations can be described by shot-noise processes, but it is beyond the scope of this book to go into details. It should be clear, however, that a low dark conductivity is beneficial for the signal-to-noise power ratio.

11.4. BOMBARDMENT-INDUCED CONDUCTIVITY

Fast electrons or other charged particles, when incident upon a crystal, give rise to an increased conductivity of the material. The reason is that a large number of slow secondaries are produced per primary particle (for example, 10,000 or more for a primary particle with 10^6 eV energy). A few of those reach the surface with sufficient energy to escape from the target, thus giving rise to secondary emission. The great majority of the slow secondaries, however, contribute to the conductivity.

The theory of the effect is quite similar to that of photoconduction; it differs only in the mode of generating carriers. Usually, hole–electron pairs are formed, and both take part in the conduction process, as in the case of photoconductivity. The theory of Sec. 11.1 is thus fully applicable.

In some cases the current pulses due to individual primary particles may be so large that they can be further amplified by a wide-band amplifier and detected against the noise background. To discriminate better between pulses, one wants the pulses to be short, which is best achieved in materials with short lifetimes.

REFERENCES

GÖRLICH, P., "Problems of Photoconductivity," in L. Marton, ed., *Advances in Electronics and Electron Physics*, vol. 14, pp. 37–85. Academic Press, Inc., New York, 1961.

Proc. I.R.E., **43**, December 1955, contains many review papers on the subject with references to the earlier literature.

VAN VLIET, K. M., "Noise Limitations in Solid State Photodetectors," *Appl. Optics*, **6**, 1145–1169 (July 1967).

Books

BLAKEMORE, J. S., *Semiconductor Statistics*. Pergamon Press Inc., Elmsford, N.Y., 1962.

BUBE, R. H., *Photoconductivity of Solids*. John Wiley & Sons, Inc., New York, 1960.

KRUSE, P. W., A. D. MCGLAUCHLIN, and R. B. MCQUISTAN, *Elements of Infrared Technology*. John Wiley & Sons, Inc., New York, 1962.

LONG, D., *Energy Bands in Semiconductors*. John Wiley & Sons, Inc., New York, 1968.

MOSS, T. S., *Photoconductivity in the Elements*. Academic Press, Inc., New York, 1952.

PANKOVE, J. I., *Optical Processes in Semiconductors*. Prentice-Hall, Inc., Englewood Cliffs, N.J., 1971.

WILLARDSON, R. K., and A. C. BEER, *Semiconductors and Semimetals*, vol. 5, Infrared Detectors. Academic Press, Inc., New York, 1970.

Of special interest in this volume are the following papers:

KRUSE, P. W., "Indium Antimonide Photoconductive and Photoelectromagnetic Detectors," Chapter 2, 15–83.

LONG, D., and J. L. SCHMIT, "Mercury–Cadmium Telluride and Closely Related Alloys," Chapter 5, 175–255.

MELNGAILIS, I., and T. C. HARMAN, "Single-Crystal Lead–Tin Chalcogenides," Chapter 4, 111–174.

SOMMERS, H. S., Jr., "Microwave-Biased Photoconductive Detector," Chapter 11, 436–465.

PROBLEMS

1. A CdS photoconductor has a 5100-Å long-wavelength cutoff of the photoconductive process; the lifetime of the electrons is 10^{-3} s, the holes are trapped, and the electron mobility is 100 cm²/v s. The photoconductive cell is 1 mm long, 1 mm wide, and 0.1 mm thick, with electric contacts at the end, so that the receiving area for the radiation is 1 mm², whereas the area of the contacts is 0.1 mm². The cell is irradiated by violet light ($\lambda = 4096$ Å) of 1 mW/cm² intensity. Calculate the following for a quantum efficiency of unity:

(a) The number of hole–electron pairs generated per second.
(b) The increase in the number of electrons in the sample.
(c) The change Δg in conductance of the sample.
(d) The photocurrent produced if 50 V is applied to the sample.
(e) The gain factor of the photoconductive process.

Answer: 2.07×10^{13} pairs; 2.07×10^{10}; 3.3×10^{-5} S; 1.65 mA; 500.

2. The photoconductive cell mentioned in problem 1 has a dark conductance g_0 of 10^{-8} S. How much light intensity is needed to change the conductance of the cell by a factor of 2 (that is, $\Delta g = g_0$)?

Answer: 0.30×10^{-6} W/cm².

3.   The photoconductive cell discussed in problems 1 and 2 is used to detect extremely small amounts of radiation (P_0 W/cm²). To that end, the light beam is chopped at a frequency of 10 Hz and the ac signal is amplified and detected by a receiver having a bandwidth of 2 Hz. The input circuit consists of a 100-V battery connected to the photocell through a load resistor of 10^8 Ω (equal to the dark resistance of the cell).
(a) If the chopper does not work, calculate the change ΔV in output signal due' to the light beam, if $\Delta g \ll g_0$, where g_0 is the dark conductance of the cell.
(b) Calculate the rms ac signal, assuming square-wave modulation.

Answer: $8.3 \times 10^7 P_0$ V; $3.7 \times 10^7 P_0$ V.

4.   In a photoconductor, q hole–electron pairs are created per second by incident light. The holes are trapped immediately; the free electrons disappear by recombination with trapped holes.
(a) Show that

$$\frac{dn}{dt} = -an^2 + q$$

where n is the free-electron density and a is a constant.
(b) If the light is switched on at $t = t_0$, show that

$$n = \sqrt{\frac{q}{a}} \tanh [\sqrt{aq}(t - t_0)]$$

so that $n \longrightarrow \sqrt{q/a}$ for large t.
(c) After equilibrium has been attained, the light is switched off at $t = t_1$. That is, $q = 0$ for $t > t_1$ and $n = \sqrt{q/a}$ for $t = t_1$. Show that

$$n = \frac{\sqrt{q/a}}{1 + \sqrt{aq}(t - t_1)}$$

(d) Draw the two transients $n/\sqrt{q/a}$ as functions of $t - t_0$ and $t - t_1$, respectively, and compare the results.

5.   The term an^2 in the differential equation mentioned in problem 4 corresponds to the number of hole–electron pairs recombining per second. Each recombination is accompanied by the emission of a quantum of luminescent light. Using the results of problem 4, calculate the transient response of the luminescence, both when the primary light is switched on and when it is switched off.

12

Luminescence

Luminescence refers to the emission of light by a crystal under influence of energy imparted to it (by ultraviolet light, x rays, fast charged particles, and so on). If the time between the introduction of excitation energy and the emission of light is very short ($< 10^{-8}$ s), we usually speak of *fluorescence;* if it is longer, we usually speak of *phosphorescence* and refer to the crystals as *phosphors.* The light emission may persist for 10^{-7} s in very short persistence phosphors to minutes (or even hours) for very long persistence phosphors. With respect to the means by which luminescence is excited we use the following terms:

1. *Photoluminescence*, if the excitation is by light or x rays.
2. *Cathodoluminescence*, if the excitation is by fast electrons.
3. *Electroluminescence*, if the excitation is by strong direct or alternating electric fields.

The light-emission mechanism is the same in most of these different modes of excitation. In addition to the three forms of luminescence, one speaks of *thermoluminescence* in connection with the light emitted upon heating after the crystal has been exposed to one of the preceding forms of excitation at a low temperature.

In this chapter we discuss "classical" luminescence. Injection luminescence in *pn* junctions and injection lasers are discussed in Chapter 17.

12.1. CHARACTERISTIC AND NONCHARACTERISTIC LUMINESCENCE

All phosphors have in common that certain traces of impurities are needed to produce luminescence; these impurities are called *activators*. The lattice itself is known as the *host lattice*. We distinguish between *characteristic* and *noncharacteristic* luminescence—a difference that can be understood as follows:

According to Bohr's third postulate, light quanta *hf* are emitted during a transition from a state of energy E_1 to a state of energy E_2:

$$hf = E_1 - E_2$$

In the characteristic luminescence all phenomena take place in the activator atoms; the energy levels involved are therefore those of the activator atoms, modified perhaps by the host lattice. In the noncharacteristic luminescence a charge transfer through the lattice is an essential part of the luminescence process; these luminescent materials are therefore at the same time photo-conductors. In addition the energy levels of the host lattice are involved, modified by the presence of activator atoms.

12.1a. Characteristic Luminescence

The activator atoms in crystals showing luminescence occur as positive ions, and the energy levels involved in the light emission are the energy levels of these activator ions (somewhat modified by the fact that the ions are not free, but are embedded in the host lattice). As it is usually required that the light emission be in the visible region, the energy difference $E_1 - E_2$ has to be between 1.5 and 3.0 eV. Thus only those activator ions are useful that have an excited energy level E_1 between 1.5 and 3.0 eV above the ground state E_2. It should be taken into account that the host lattice may modify the energy levels to a considerable extent; consequently, some ions, in which the energy difference between the ground state and the next excited state is 5–8 eV, may still be useful for luminescence.

The following ions are used as *activators:* Cu^+, Ag^+, Sn^{2+}, Bi^{3+}, Cr^{3+}, Mn^{4+}, Mn^{2+}, and Ce^{3+}. Some ions, such as Fe^{2+}, Ni^{2+}, and Co^{2+}, would also be suitable, but they have the property that a radiationless transition from the excited state to the ground state is much more likely than a transition with emission of light. For that reason these ions are called *killers;* in the preparation of good luminescent materials their presence should be carefully avoided.

The emission spectra of free ions consist of a number of discrete frequencies (line spectra). The spectra of luminescent activators, however, are more or less continuous, spread over a certain frequency range (band spectra). This comes about because of the vibration of the ions.

As an example of characteristic luminescence, we shall discuss the system KCl: Tl^+ in detail. This has a potassium chloride host lattice with Tl^+ ions as activators. It is not technically important, but it is one of the best understood classical examples of characteristic luminescence and shows all the essential features. Each Tl^+ ion occupies the place of a K^+ ion and is symmetrically surrounded by six Cl^- ions. The $TlCl_6$ group can thus be considered a molecule vibrating in a solid sphere formed by the surrounding lattice. For reasons of symmetry the most important mode of vibration is the one in which the Tl^+ ion is at rest, while the six Cl^- ions vibrate radially and in phase. The potential energy of the $TlCl_6$ group is then completely determined by the Tl–Cl distance. This atom group can be considered a harmonic oscillator; according to quantum theory, the energy of an oscillator vibrating with frequency f_0 can have values

$$E_v = (v + \tfrac{1}{2})hf_0 \qquad (12.1)$$

where the quantum number v can have only the values $0, 1, 2, \ldots$. At sufficiently low temperature, all vibrating ions should have $v = 0$.

During a transition of an electron from the energy state E_1 to the energy state E_2, the quantum number v may change from v_1 to v_2. The total energy involved in the transition is thus $E_1 - E_2 + E_{v1} - E_{v2}$. Many different values of v_1 and v_2 are involved, so that the spectrum for a single transition will extend over a certain frequency range and will appear as more or less continuous.

Since the luminescence is characteristic for the activator ion involved, one may associate the energy levels of the luminescence with atomic energy levels. The ground state of the Tl^+ ion is $^1S_0$, the excited state is ^3P (for this nomenclature, see Chapter 2). The excited state ^3P has $L = 1$ and $S = 1$; hence the quantum number J can have the values $J = 0$, $J = 1$, and $J = 2$. Each quantum number J corresponds to a slightly different energy; hence one has to distinguish between $^3P_0$, $^3P_1$, and $^3P_2$ states. According to Chapter 2 the following selection rules hold for electric dipole transitions between states:

For L: $\Delta L = \pm 1$.
For S: $\Delta S = 0$.
For J: $\Delta J = 0, \pm 1$; $J = 0 \rightarrow J = 0$ forbidden.

The selection rule $\Delta S = 0$ is not a very strict one for heavy atoms; the others are much less easily violated.

Applying this to the $^3P \rightarrow {}^1S$ transition of the Tl^+ ion in the KCl lattice, we see that the transition $^3P_1 \rightarrow {}^1S_0$ is only weakly forbidden (by the selection rule $\Delta S = 0$); the transition $^3P_0 \rightarrow {}^1S_0$ is strongly forbidden (by the rule that $J = 0 \rightarrow J = 0$ forbidden); and the transition $^3P_2 \rightarrow {}^1S_0$ is also strongly

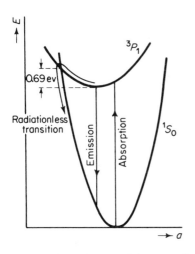

Radiationless transition *Emission* *Absorption* $^3P_1$ $^1S_0$ 0.69 ev

→ *a*

Fig. 12.1. Schematic potential energy diagram of Tl^+ ions in KCl according to Williams, showing the potential energy as a function of the Tl–Cl distance *a* for the ground state $^1S_0$ and $^3P_1$ (F. E. Williams, *Advances in Electronics*, **5**, 137–60, 1953. Courtesy Academic Press, Inc., New York).

Absorption of light occurs for atomic distances around the equilibrium distance of the ground state; *emission* of light occurs for atomic distances around the equilibrium distance of the excited state. Since the two potential energy curves cross at a point 0.69 ev above the minimum in the $^3P_1$ curve, radiationless transitions $^3P_1 \rightarrow {}^1S_0$ become possible at higher temperatures.

forbidden (by the rule $\Delta J \neq 2$). For that reason only the transition $^3P_1 \rightarrow {}^1S_0$ is important for the luminescence. If an activator ion is excited to the states $^3P_0$ or $^3P_2$, it cannot go back to the ground state with emission of light. Energy can thus be stored in these states; for that reason these states act as *traps*.

To understand the absorption and emission of these activator ions, we plot the sum of the electron energy and the potential energy of the $TlCl_6$ group in the KCl lattice as a function of the Tl–Cl distance *a*, both for the ground state $^1S_0$ and for the excited state $^3P_1$. Since the atom group acts as a harmonic oscillator, these curves are parabolas; the minima in the parabola occur at different values of *a* (Fig. 12.1).

First let us consider absorption. The Tl^+ ion is now in the ground state $^1S_0$ and the Cl^- ions vibrate around the equilibrium distance a_1. The absorption of a quantum is such a fast process that the atomic distance does not change in the absorption process.† If the $TlCl_6$ group did not vibrate, the absorption would occur at a single frequency f_1 determined by the energy difference between the ground state and the excited state at the Tl–Cl distance a_1; because of the vibration of this group, the absorption extends over a certain frequency band around f_1, the width of which depends upon the number of vibrational energy levels E_1 of the ground state that are excited thermally.

The Tl^+ ion is now in its excited state $^3P_1$ and is vibrating strongly, but it will soon lose most of its energy of vibration to the lattice. If the excited $TlCl_6$ group did not vibrate, it would emit a single frequency f_2 determined by the energy difference between the ground state and the excited state at the Tl–Cl distance a_2. Actually, this group vibrates around its equilibrium configuration; since the emission of a quantum is a very fast process, the inter-

---

† This is known as the Frank–Condon principle.

atomic distance will not change during the emission process; as a consequence, the emission will extend over a certain frequency band around f_2.

We see from the diagram that *the absorbed light must have a larger quantum energy than the emitted light;* the emission and absorption bands will coincide only (see figure) if the atomic distances a_1 and a_2 are equal. Suitably chosen luminescent materials may therefore be used to transform ultraviolet light into visible light, a process that is utilized with good success in fluorescent lamps. It is clear from the discussion that the energy difference between the absorbed quanta and the reemitted quanta is dissipated by the lattice vibrations and appears in the form of heat.

The $^3P_1$ and the $^1S_0$ curves meet at a point 0.69 eV above the minimum in the $^3P_1$ curve. If the excited configuration $TlCl_6$ had enough vibrational energy, it could pass from the $^3P_1$ state into the $^1S_0$ state in a *radiationless transition* (Fig. 12.1). At sufficiently high temperature this radiationless transition will be more probable than the radiative transition, and as a consequence the efficiency of the luminescence will decrease with increasing temperature. The energy difference of $E_a = 0.69$ eV is called the *activation energy* of the radiationless transition.

Let the probabilities at which the radiative and the radiationless transition occur be A and B, respectively; then one would expect

$$B = v_0 \exp\left(-\frac{eE_a}{kT}\right) \tag{12.2}$$

where E_a is the activation energy of the radiationless transition in electron volts and v_0 is a vibration frequency of the order of 10^{10}/s. If η_0 is the low-temperature efficiency of the luminescence process, then for the efficiency η at higher temperatures we have

$$\eta = \eta_0 \frac{A}{A + v_0 \exp\left(-eE_a/kT\right)} \tag{12.3}$$

whereas the lifetime of the excited state becomes

$$\tau = \frac{1}{A + v_0 \exp\left(-eE_a/kT\right)} = \tau_0 \frac{A}{A + v_0 \exp\left(-eE_a/kT\right)} \tag{12.3a}$$

where $\tau_0 = 1/A$. Consequently, $\eta/\eta_0\tau = A$ should be independent of temperature; this agrees with experiment (Fig. 12.2).

Therefore, for a good dependence of the luminescent efficiency upon temperature, A should be large and the activation energy E_a should be relatively large. If the activation energy is very small, radiationless transitions may predominate even at relatively low temperatures. Ions that have this property are thus *killers,* wasting the absorbed energy as heat instead of reemitting it as light.

We saw that in Tl^+ ion activators the transitions $^3P_0 \rightarrow {}^1S_0$ and $^3P_2 \rightarrow {}^1S_0$ were strongly forbidden; these energy states are therefore metastable. Energy may be stored in them at low temperatures. The energy may be

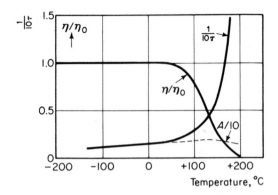

Fig. 12.2. (η/η_0) and $1/(10\tau)$ plotted against the temperature T for Mn-activated Mg_2TiO_4 phosphor according to Kröger. Note that $A = \eta/(\eta_0\tau)$ is almost independent of T. Note also the large value of τ, indicating that the transition involved is strongly forbidden. (F. A. Kröger, W. Hoogstraaten, M. Bottema, T. P. J. Botden, *Physica*, **14**, 81–96, 1948.)

released by heating; this effect is known as *thermoluminescence* (Fig. 12.3). To study this effect, the luminescent solid is excited at a low temperature. The excitation is then removed, the solid is heated at a slow linear rate, and the luminescent intensity is measured as a function of temperature; the curve thus obtained is called a "glow curve." The maxima in the glow curve correspond to the emptying of traps; the glow curve of Fig. 12.4 refers to Tl^+ ions in KCl; it shows the existence of two distinct traps that are identified with the $^3P_0$ and the $^3P_2$ states.

We can understand the mechanism of this thermoluminescence by drawing the potential-energy diagram of a trapping state and a nontrapping excited state as a function of the atomic distance a. Let these two curves meet at an energy E_t above the minimum of the curve for the trapping state (Fig. 12.3). The trapping state thus has an activation energy E_t; at sufficiently high temperature, ions in the trapping state will undergo a radiationless transition to the nontrapping state, from which a final transition to the ground state under the emission of light becomes possible. The activation energy involved can be determined from the glow curve.†

In the characteristic luminescence the excitation, the trapping, and the recombination all take place in the same ion. No electrons have to be excited into the conduction band before light emission can occur; consequently, no photoconductivity results during optical excitation or during the emptying of traps. In noncharacteristic luminescence, trapping does not occur at the same site at which the light emission occurs; hence the emptying of traps is accompanied by the flow of current.

---

† In Tl-activated KCl the nontrapping state seems to be $^1P_1$ and not $^3P_1$.

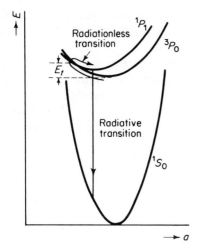

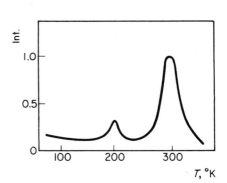

Fig. 12.3. Potential energy diagram of Tl^+ ions in KCl according to Williams, showing the emptying of trapping centers due to a radiationless transition $^3P_0 \rightarrow {}^1P_1$, followed by a radiative transition $^1P_1 \rightarrow {}^1S_0$ which is the cause of the luminescence. (F. E. Williams, *Advances in Electronics*, **5**, 137–60, 1953, courtesy Academic Press, Inc., New York.)

Fig. 12.4. Glow curve of Tl-activated KCl phosphor (0.05 % Tl); the light intensity is plotted against the temperature. Activation energies of 0.35 ev ($^3P_2$ state) and of 0.72 ev ($^3P_0$ state) can be deduced from this curve. (F. E. Williams, *Advances in Electronics*, **5**, 137–60, 1953, courtesy Academic Press, Inc., New York.)

Having discussed the example $KCl:Tl^+$ in greater detail, the discussion of presently used characteristic luminescent materials can be brief. The following are most common phosphors:

1. $Zn_3(PO_4)_2:Mn$, orthophosphate, has a red luminescent band around 6100 Å.
2. $ZnSiO_4:Mn$, willemite, has a green luminescent band around 5400 Å.
3. $YVO_4:Eu$ has a red luminescent line spectrum around 6130 Å.
4. $Y_2O_2S:Eu$ has a red luminescent line spectrum around 6050 Å.
5. Manganese-and antimony-doped halophosphates.

The first four are used in cathode-ray tubes and have a power efficiency around 6 per cent. For understanding the reason for this efficiency, see Sec. 12.2b. The last are used as fluorescent lamp phosphors; the Sb doping gives a blue band and the Mn doping gives a yellow band; together they give a reasonably white color. The power efficiency for ultraviolet radiation is quite large, much larger than the power efficiency for cathodoluminescence of the other materials.

The rare-earth phosphors (items 3 and 4) need some further discussion, since an important principle is involved here. The light is emitted due to transitions of 4 f electrons, which are located deeply in the atoms; hence the effect of the lattice is relatively small. Consequently, the observed spectra are line spectra rather than band spectra.

All atomic energy states have a symmetry property called *parity*; the parity is said to be *even* if the wave function is symmetric in r [$\psi(-x, -y, -z) = \psi(x, y, z)$], and the parity is said to be *odd* if the wave function is antisymmetric in r [$\psi(-x, -y, -z) = -\psi(x, y, z)$]. The 4$f$ energy levels of a rare-earth atom all have the same parity, either even or odd, depending on the atom involved.

Now the probability of an atomic transition between an initial state (wave function ψ_1) and a final state (wave function ψ_2) is given by an equation of the form

$$\int \psi_1(r)g(r)\psi_2^*(r)\, dr \tag{12.4}$$

where $g(r)$ characterizes the type of radiation. For electric dipole radiation $g(r)$ varies as r, so that $\psi_1(r)$ and $\psi_2(r)$ must have opposite parity; in other words, dipole transitions between 4 f levels are forbidden. For magnetic dipole radiation, $g(r)$ varies as $r\, dr/dt$; hence $\psi_1(r)$ and $\psi_2(r)$ must have the same parity. Such transitions between 4 f levels are therefore allowed. For quadrupole radiation, $\psi_1(r)$ and $\psi_2(r)$ must also have the same parity.

Nevertheless, although forbidden, electric dipole radiation can be observed because of the influence of the crystal lattice. The interaction responsible for this effect is caused by the odd crystal field terms, that is, those terms which change sign on inversion with respect to the rare-earth ion. If the rare-earth ion is located at the site that is a center of symmetry of the crystal lattice, then the odd crystal terms are absent and the parity rule applies; otherwise, they have an effect and the parity rule can be violated. In addition, the crystal field can give rise to a splitting of the atomic energy levels of a given J value.

Something should be said about the decay time of the luminescence. If the transition from the excited state to the ground state is allowed, the decay time may be as short as 10^{-7} s. For weakly forbidden transitions, as in the $^3P_1 - ^1S_0$ transition of Tl^+ activator ions, the decay time is only slightly longer ($\simeq 10^{-6}$ s). For more strongly forbidden transitions, such as the $^4G - ^6S$ transition of Mn^{2+} activator ions, the decay time may be $10^{-3} - 10^{-1}$ s, depending upon the host lattice. Rather long decay times are used in cathode-ray tubes with a persistent screen, such as in radar applications.

12.1b. Noncharacteristic Luminescence

Zinc sulfide (ZnS) phosphors are the best-known examples of noncharacteristic luminescence. Besides ZnS, mixed crystals of ZnS and CdS, ZnSe, and CdSe, and the like, are also luminescent if properly activated. Ions such as Na^+, Li^+, Ag^+, Cu^+, and Au^+ can be used as *activators*. Since Na^+ and Li^+

are ions with a closed shell structure, which do not have any energy levels
even within 10 V of the ground state, it is obvious that the luminescence
cannot be a characteristic property of the activator ion. It turns out to be a
characteristic of the host lattice.

The Zn ions occur as Zn^{2+}, and the activator ions are all singly charged.
If all that happened in the activation process were a mere substitution of
Zn^{2+} by singly charged activator ions, the charge balance would be upset and
not many activator ions would be substituted. To keep the charge balance,
one may introduce an equal number of Cl^- ions in the lattice (taking the
place of S^{2-} ions), or one may introduce an equal number of triply charged
positive ions, such as Al^{3+}, Sc^{3+}, Ga^{3+}, In^{3+} (taking the place of Zn^{2+} ions),
or one may introduce S^{2-} vacancies. In self-activated ZnS, blue-emitting
centers are formed by reduction of some Zn^{2+} ions to Zn^+ ions under the
simultaneous substitution of an equal number of S^{2-} ions by Cl^- ions or by
the formation of S^{2-} vacancies. The Cl^- ions come from halide salts that are
used as a flux in the phosphor preparation. The Cl^- ions and the triply
charged positive ions are called *coactivators;* they have nothing to do with the
luminescence itself, but only facilitate the substitution of activators into the
lattice.

Because the activator ion is only singly charged, the energy levels of
the surrounding S^{2-} ions will be slightly lifted up in comparison with normal
S^{2-} atoms. They thus give rise to *occupied* energy levels close to the top of
the valence band; these levels form the ground state of the luminescence.

In a similar way the coactivator ions give rise to *unoccupied* energy levels
close to the bottom of the conduction band; these unoccupied energy levels
act as *traps*. The depth of these traps below the conduction band can be
determined from the thermoluminescence glow curve. Since the activator
and coactivator ions are not necessarily very close together, thermally
activated electrons from traps have to *travel* before they can return to the
ground state. For that reason thermoluminescence in noncharacteristic
luminescent materials should be associated with increased conductivity;
this has indeed been observed.

The valence band in ZnS is the S^{2-} band; the next higher empty band is
the Zn^+ band (conduction band). Light absorption associated with a transition from the valence band to the conduction band can be represented as

$$S^{2-} + Zn^{2+} + hf_0 \rightleftharpoons S^- + Zn^+$$

that is, a free electron (indicated by Zn^+) and a free hole (indicated by S^-)
are generated in the process; here $\longrightarrow$ means absorption and $\longleftarrow$ means emission.

The luminescence comes about in a two-step process. First the free holes
generated by the absorbed light are trapped in the activator centers. Next
the free electrons recombine with the trapped holes under emission of light.

The empty levels near the bottom of the conduction band (such as those due to coactivator ions) would normally act as recombination centers. However, since most of the free holes are trapped in the activator levels, these centers act mainly as electron traps; nonetheless they are responsible for the radiationless transitions.

Activators	ΔE_a (Optically)	Coactivators	ΔE_L (Thermally)
Na^+	0.33 eV	Cl^-	0.37 eV
Li^+	0.66	Al^{3+}	0.37
Ag^+	0.70	Sc^{3+}	0.51
Cu^+	1.15	Ga^{3+}	0.62
Au^+	1.22	In^{3+}	0.74

The accompanying table shows the energy differences ΔE_a between the top of the valence band and the ground state caused by different activator ions (determined optically); it also shows the energy difference ΔE_L between the bottom of the conduction band and the various trapping levels caused by the coactivator ions (determined from the glow curve). The energy-level diagram for these phosphors agrees with the diagram presented in Fig. 11.5.

In mixed crystals of (Zn, Cd)S the long-wave absorption limit of unactivated crystals shifts toward the red with increasing Cd concentration, indicating a decrease in the energy gap between the valence band and the conduction band of -1.9×10^{-2} eV per mole per cent of CdS. The same shift is found in the maximum of the luminescent emission of (Zn, Cd)S activated by Ag^+; the maximum occurs at 4300 Å for ZnS, and shifts to 7400 Å for CdS. This is a clear indication that the luminescence is associated with the host lattice and not with the activator ion itself.

The advantage of the (Zn, Cd)S phosphors, besides the high efficiency that can be obtained, is that one can obtain different colors, either by using different activators or simply by using a different percentage ratio of Zn and Cd ions. By proper choice of composition one can shift the emitted band all the way from blue to red, that is, over practically the whole visible spectrum.

The following (Zn, Cd)S phosphors are commonly used in color television picture tubes:

1. ZnS: Ag^+. The material emits a blue band between 4200 and 4750 Å.
2. ZnCdS: Ag^+. The material emits a green band around 5480 Å.
3. ZnCdS: Cu: Al. The material emits a green band between 5050 and 5800 Å.
4. ZnCdS: Ag^+ with a larger Cd content. The material emits a red band around 6100 Å.
5. White phosphors consisting of a mixture of a blue phosphor (ZnS: Ag^+) and a yellow phosphor (ZnCdS: Cu: Ag with a properly chosen Cd content). Together they give a white color (Fig. 12.5).

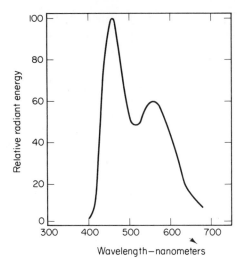

Fig. 12.5. Spectrum of white phosphor used for black and white television screens. Note the large amount of blue in the spectrum. (From S. Larach and A. E. Hardy, *Proc. IEEE*, **61**, 915, 1973).

The power efficiency of these phosphors, when used in cathodoluminescence is 13–15 per cent, but higher values have been reported in the literature. This is more than a factor of 2 better than characteristic luminescent phosphors.

Something should be said about the white color used in black and white television screens. One can describe the "whiteness" of a light source by its color temperature. For example, a 100-W incandescent lamp has a color temperature of about 2800°K, whereas sunlight has a color temperature of about 6500°K. In comparison, black and white television screens have color temperatures of about 12,000°K; apparently the viewing public prefers bluer screens.

For a survey of cathode-ray tube phosphors, compare the review paper by Larach and Hardy.†

12.1c. Efficiency of the Luminescence

If the efficiency (= light output for a given power input) η of a phosphor is plotted as a function of the activator concentration C, one finds a maximum efficiency for relatively low concentrations (1 part in 10^2–10^4) depending upon the activator (Fig. 12.6). It turns out that η can be described by an equation

$$\eta = \eta_1 \frac{\sigma'C}{\sigma'C + \sigma(1 - C)}(1 - C)^z \qquad (12.5)$$

† S. Larach and A. E. Hardy, *Proc. IEEE*, **61**, 915 (1973).

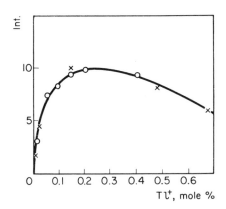

Fig. 12.6. Luminescence efficiency and thermoluminescent intensity as a function of the activator concentration in Tl-activated KCl. The solid curve is the calculated curve, the crosses represent luminescent efficiency data, the circles represent experimental glow peak intensity data. (F. E. Williams, *Advances in Electronics*, **5**, 137–60, 1953, courtesy Academic Press, Inc., New York.)

where η_1 is a constant, Z is a large number, and σ' and σ are the capture cross sections of the excitation energy for the activator and nonactivator ions, respectively. Equation (12.5) can be understood as follows: The chance that an activator absorbs the excitation energy is $\sigma'C$; the chance that a nonactivator absorbs the energy is $\sigma(1 - C)$, since $1 - C$ is the concentration of the nonactivators. If this were all, we would have

$$\eta = \eta_1 \frac{\sigma'C}{\sigma'C + \sigma(1 - C)} \quad (12.5a)$$

and we would have an optimum efficiency $\eta = \eta_1$ for $C = 1$. A maximum value of η occurs at low concentrations for two reasons:

1. In a good luminescent material σ should be very small; that is, it should be much more likely that the excitation energy is absorbed by an activator than by a nonactivator.

2. Owing to the factor $(1 - C)^Z$, where Z is large, Eq. (12.5) has a maximum for small C.

The occurrence of the latter factor can be explained as follows. The chance that the Z nearest neighbors of an activator are all nonactivators is $(1 - C)^Z$. The occurrence of this factor in (12.5) means that a transition from an excited state to the ground state under emission of light occurs only if the Z nearest neighbors of an activator ion are all nonactivators. If any one of them is another activator ion, this lowers the activation energy for a radiationless transition to such an extent that the luminescence is quenched.†

The following results have been obtained from optical studies:

Activator	T	Z
KCl : Tl$^+$	300°K	70
KCl : Mn^{2+}	300°K	22
KCl : Mn^{2+}	80°K	13
ZnS : Cu$^+$	300°K	4000

† The maximum in η is therefore caused by the competition between radiative and radiationless transitions: for large concentrations the radiationless transitions will predominate.

The first three examples indicate rather localized luminescent centers; the large value for ZnS is due to the fact that the excited state corresponds to the electron being in the conduction band. The decrease in Z with decreasing temperature can be understood from the fact that the probability $\exp(-eE_a/kT)$ of overcoming an activation energy E_a decreases with decreasing temperature.

The efficiency of phosphors can be quite high, because most of the excitation energy ends up in the activator ions, despite the small activator concentration (see Sec. 12.2a and 12.2b).

Because most luminescent materials emit in a relatively narrow wavelength range only, the light is generally colored. One can produce a white luminescence by mixing luminescent materials in an appropriate manner.

12.2. APPLICATIONS

12.2a. Fluorescent Lamps

Incandescent light bulbs have a rather poor luminous efficiency. Much higher luminous efficiencies are obtainable with fluorescent lamps. In these lamps an electric discharge is passed through a mixture of argon and mercury vapor; this by itself gives a higher luminous efficiency than an incandescent lamp, but the light has a bluish color, and an appreciable part of the power is wasted by the excitation of strong ultraviolet radiation. Both drawbacks are corrected by coating the wall of the lamp with luminescent material.

The luminous efficiency is increased by transformation of the ultraviolet light into visible radiation. For photoexcitation, efficiencies of 40–50 per cent have been reported; this makes the large luminous efficiencies (lumens per watt) of fluorescent lamps understandable. By proper choice of the luminous material, the emission bands of the luminescence may be selected such that the total light output has the desired color.

12.2b. Cathodoluminescence; Cathode-Ray Tubes

In the case of cathodoluminescence we distinguish between the excitation and the emission process. The emission process is the same as in Sec. 12.1, and the excitation process shows considerable similarity with the corresponding process in secondary emission. Owing to the differences in the applications, however, the requirements are somewhat different. In secondary emission as much of the primary energy as possible should be used to produce escaping slow secondaries; in cathodoluminescence as much of the primary energy as possible should be used to excite luminescent centers.

In the excitation process the fast primary electrons first generate a large number of secondaries from the valence band and perhaps from deeper-lying bands. As long as the energy of these secondaries, taken with respect to the bottom of the conduction band, is larger than the gap width between the valence band and the conduction band, these electrons will rapidly lose their excess energy by exciting new electrons from the valence band into the conduction band. If their energy is less than the gap width, they can lose their energy to the lattice vibrations or they may excite a luminescent center. The energy lost to the lattice vibrations is quite small, only of the order of $h\nu$ per collision, where ν is a frequency of the lattice vibrations.

The mode of excitation is different for characteristic and noncharacteristic luminescence. For characteristic luminescence the secondary electrons must excite the ground state. This is not a very efficient process, since not all secondary electrons have sufficient energy to do so. For noncharacteristic luminescence the secondary holes are trapped by the activator centers, which is a very efficient process, and the slow electrons are captured by the excited activator centers under the emission of luminescent light; this is also an efficient process. As a consequence, the overall excitation process is much more efficient than for characteristic luminescence, which explains the difference in power efficiency between the two types of luminescence.

Energy efficiencies up to 20–25 per cent have been found for ZnS and CdS phosphors at 20-kV primary energy, corresponding to a luminous efficiency of 100 1m/W. The efficiency of phosphors showing characteristic luminescence is generally lower, but luminous efficiencies of 20–40 1m/W have been reported.

Since the excitation process is essentially the same at all primary energies, one would expect the energy efficiency of cathodoluminescence to be only weakly dependent upon the primary energy. One finds this weak dependence at higher energies (above 10 kV), but for lower energies the efficiency decreases rapidly with decreasing energy (Fig. 12.7). This seems to indicate that the efficiency increases with increasing penetration depth of the primary electrons. There may be several reasons for this; it has been suggested, for example, that the ratio of radiative to nonradiative transitions increases with increasing distance to the surface. Whatever the reason, this effect indicates that a primary energy around or above 10 kV is desirable for cathode-ray screens with a high luminous efficiency.

The actual luminous efficiency in cathode-ray tubes is often considerably smaller than the value of 100 1m/W just quoted. The reason is that the light generated in the screen can travel in two directions (into and out of the tube) and that the light is scattered in the phosphor itself because of its granular structure.

If scattering did not occur, half the amount of light would emerge from each side of the screen. If the screen is provided with a thin aluminum backing, the light emitted into the tube can be reflected out of the tube; conse-

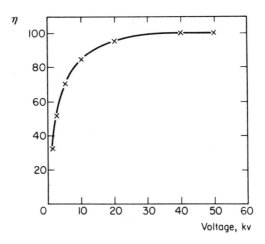

Fig. 12.7. Relative luminescent efficiency of a cathode ray screen as a function of the beam voltage for Mn-activated $ZnSO_4$ (0.4% Mn). (Bril and Klasens, *Philips Research Reports*, **7**, 401–431, 1952.)

quently, the actual luminous efficiency of the screen can be increased by a factor of almost 2 (actually it is slightly less because the reflection coefficient of the metal backing is only about 80 per cent).† The thickness of the Al layer should be about 1000 Å for electrons with primary energy between 10 and 20 kV. If the layer is too thick, it absorbs too much of the primary energy; if it is too thin, it does not reflect the light sufficiently well.

Scattering occurs because the screen consists of a large number of loosely packed small grains. Light traveling through the layer will be partially reflected at the grain boundaries. If the layer were very thin, scattering would be unimportant, and the amount of light emitted in both directions would be almost equal. The light output would be small, however, because only a small part of the primary energy would be absorbed in the layer. If the thickness increases, the part of the primary energy absorbed in the layer will become larger, and hence the total light output (sum of light energy emitted in both directions) will increase. For layers thicker than the penetration depth of the primaries the total light output will no longer increase (it will not very much decrease either, because the light absorption in the layer is quite small), but an asymmetry in the light emission is observed: the amount of light emitted *into* the tube is much larger than the amount of light traveling through the layer and coming *out* of the tube (Fig. 12.8). The reason is that light traveling through the layer and coming out of the tube is scattered many more

---

† Another advantage of the Al layer is that it absorbs any negative ions arriving at the screen, thus preventing a rapid discoloration of the screen due to the impinging negative ions (ion burn). Moreover, it prevents charging effects of the phosphor.

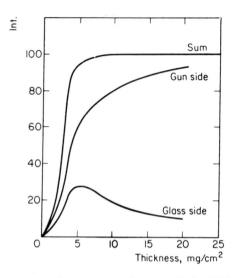

Fig. 12.8. Luminescent intensity as a function of the thickness of the luminescent layer. The intensity at the gun side is much larger than the intensity at the glass side because of light scattering. The sum of the two intensities remains constant if the thickness increases beyond a critical value, indicating that light absorption is unimportant. The critical thickness corresponds to the penetration depth of the beam electrons. (Bril and Klasens, *Philips Research Reports*, **7**, 401–431, 1952.)

times than the light emitted into the tube. Again the effect can be remedied by providing the luminescent layer with an aluminum backing that reflects the light outward.

The luminescent layer should not be made thicker than the range of the primary electrons,† since that does not add anything to the total light output.

The scattering of the light by the luminescent particles has still another effect. If very finely focused beams are used, the light spot on the screen may have a much larger diameter than the diameter of the beam. The size of the light spot has been considerably reduced by developing denser screens; because such screens have much smaller holes between the grains, the light scattering is much smaller. Such screens can, for example, be deposited by evaporation; good screens made that way are almost completely transparent.

---

† The range of the primary electrons was calculated in Eq. (10.20a), which was derived under the condition that the primary electrons are not scattered but move in straight paths. Because the primary electrons are scattered sideways, a much larger part of the primary energy will be lost close to the surface than would be expected from the calculated range of the primaries. As a consequence, the thickness of the luminescent layer may be chosen somewhat smaller than the calculated range of the primaries without much loss in output.

12.3. ELECTROLUMINESCENCE

12.3a. Direct- and Alternating-Current Electroluminescence

When a ZnS luminescent powder is inserted between two transparent electrodes and a dc voltage is applied to the electrodes (Fig. 12.9), the ZnS powder emits luminescent light, for example, of yellow color. The device is operating in a resistive mode, that is, dc current is passed through the sample, and the phenomenon is called *dc electroluminescence*.

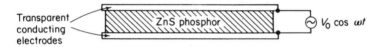

Transparent
conducting
electrodes

ZnS phosphor

$V_0 \cos \omega t$

Fig. 12.9. Electroluminescent cell, using transparent conducting electrodes.

When a ZnS luminescent powder is suspended in a transparent binder of high dielectric constant and the material is inserted between two transparent electrodes, luminescent light is emitted when an ac voltage $V_0 \sin \omega t$ is applied to the electrodes. The light emission is proportional to the frequency of the ac voltage for low frequencies and saturates at high frequencies. The device is operating in a capacitive mode, that is, no dc current is passed through the sample, and the phenomenon is called *ac electroluminescence*.

The dc electroluminescence shows a light output B that can be represented by the equation

$$B = B_0 \exp\left(-\frac{b}{V_0^{1/2}}\right) \quad (12.6)$$

where V_0 is the applied voltage and b and B_0 are constants. This exponential dependence holds for a wide range of voltages (Fig. 12.10).

The devices are quite stable. The value of B at a given voltage V_0 decreases somewhat after aging over relatively long time intervals, but this comes mostly from the decrease in the

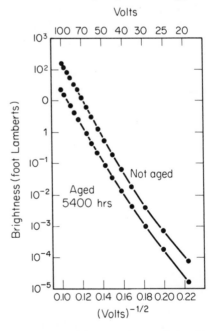

Fig. 12.10. Brightness-voltage relationship of dc electroluminescence before and after aging. (A. Vecht et al., *Proc. IEEE*, **61**, 902, 1973).

current for a given voltage V_0 due to aging. Higher light emission is obtainable by using pulse excitation with a low duty cycle.

In ac luminescence, Eq. (12.6) must be modified to

$$B = B_0(\omega) \exp\left(-\frac{b}{V_0^{1/2}}\right) \tag{12.6a}$$

where V_0 is the amplitude of the ac voltage and $B_0(\omega)$ is a constant that depends on ω. Now $B_0(\omega)$ varies linearly with frequency for low frequencies (200–1000 Hz), but saturation effects set in at high frequencies. The linear frequency dependence can be understood as follows. If the light output is observed as a function of time, it is found that a short flash of light is emitted every half-cycle. The duration τ of the flash is about 10^{-3} s, and the total amount of light emitted per flash is independent of frequency as long as the frequency is not too high. Since the number of light flashes is proportional to ω, $B_0(\omega)$ should be proportional to ω for $\omega\tau < 1$ and should saturate for $\omega\tau > 1$ (Fig. 12.11).

It was said before that the binder material should have a high dielectric constant. The reason is that the luminescence is actually caused by the high electric field in the luminescent grains. If the dielectric constant of the binder is large, the field strength in the grains may be considerably larger than the value $(V_0 \sin \omega t)/d$ expected for a homogeneous dielectric, resulting in a considerably larger light output.

The light emission in dc luminescence depends upon the polarity of the applied voltage. The largest light emission occurs if the electrode facing the observer acts as the cathode. This is caused by the fact that a luminescent

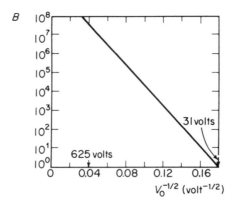

Fig. 12.11. Voltage dependence of ac electroluminescence, arbitrary units of B (see G. Diemer, *Philips Research Reports*, **11**, 353–399, 1956). In this case Eq. (12.6a) is found to hold over a range of 10^8 in light intensity. Other researchers have sometimes found deviations at the highest attainable intensities.

grain emits light at the side facing the cathode; the observed asymmetry is due to absorption and scattering of the emitted light (see Fig. 12.8). For the same reason, the two light flashes per period emitted for ac electroluminescence do not have the same intensity either.

The maximum intensity for ac luminescence occurs when the local field strength at the emitting spots is a maximum. Since this local field strength may not be in phase with the applied voltage, the light output may not be in phase with the applied voltage either.

To explain the electroluminescence, one must investigate how the luminescent centers can be emptied or how new centers may be excited. If a voltage V is applied across the grains, most of this voltage appears across a thin layer on the cathode side of the grain (Schottky barrier; see Chapter 14). There are now three possibilities.

1. The field strength in the surface layer is large enough to empty occupied luminescent centers by field emission.
2. Electrons can tunnel through the Schottky barrier on the cathode side of the grain by field emission.
3. Electrons excited by processes 1 or 2 are accelerated by the applied field and can gain enough energy to create a hole–electron pair. The process can then repeat itself, creating an avalanche of hole–electron pairs, or the holes can be trapped by the activator centers emptying them.

The probability P of each of the processes 1–3 is of the form

$$P = a \exp\left(-\frac{b}{F}\right) \tag{12.7}$$

The field F in a Schottky barrier varies as $V^{1/2}$, where V is the applied voltage, so that

$$p = a \exp\left(-\frac{b'}{V^{1/2}}\right) \tag{12.7a}$$

This explains Eqs. (12.6) and (12.6a). The three processes are pictured in Fig. 12.12.

12.3b. Applications

Direct- and alternating-current electroluminescence can be used for alphanumeric display panels. The light emission is not extremely bright, but it is sufficient for this application, especially since the emission occurs near the wavelength where the human eye has its maximum sensitivity. For details we refer the reader to recent review papers[†].

---

† E. Schlam, *Proc. IEEE*, **61**, 894 (1973). N. J. Werring, R. Ellis, and P. J. F. Smith, *Proc. IEEE*, **61**, 902 (1973). H. Kawarada and N. Ohshima, *Proc. IEEE*, **61**, 907 (1973).

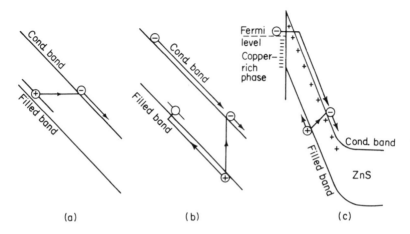

(a)   (b)   (c)

Fig. 12.12. Excitation of luminescent centers in electroluminescence. (a) Occupied luminescent ground state emptied by field emission. (b) An electron is accelerated in the conduction band and gains enough energy to create a hole-electron pair. The hole is thereupon captured by an occupied ground state, thus emptying it. (c) An electron passes through the barrier layer of a ZnS grain by tunnel effect, is accelerated in the barrier region, and gains enough energy to create a hole-electron pair. Otherwise as in (b).

Alternating-current luminescence can be used in light amplifiers. For details, see Sec. 13.3c.

REFERENCES

CURIE, D., *Luminescence in Crystals*, John Wiley & Sons, Inc., New York, 1963.

GOLDBERG, P., *Luminescence of Organic Solids*, Academic Press, Inc., New York, 1966.

IVEY, H. F., "Electroluminescence and Related Effects," in L. Marton, ed., *Advances in Electronics and Electron Physics*, supp. I. Academic Press, Inc., New York, 1963.

LARACH, S., ed., *Photoelectronic Materials and Devices*, Van Nostrand Reinhold Company, New York, 1965.

LEVERENZ, H. W., *Introduction to Luminescence of Solids*, John Wiley & Sons, Inc., New York, 1970.

Philips Technical Review, **31,** No. 10, 1970, contains several papers on characteristic luminescence.

Proc. IRE, Dec. 1955, contains many review papers on luminescence and electroluminescence, with references to the earlier literature.

Proc. IEEE, July 1973, contains many review papers on luminescence, injection luminescence, and electroluminescence, with references to the earlier literature.

PROBLEMS

1. The lifetime of the luminescent centers of a certain luminescent material at a low temperature is found to be 10^{-4} s; at 100°C its lifetime has dropped to $\frac{1}{2} \times 10^{-4}$ s because of radiationless transitions.
(a) Calculate the activation energy E of the luminescent centers, assuming $v_0 = 10^{10}$.
(b) How does the value of E change if the assumed value of v_0 was wrong by a factor 10?

Answer: (a) 0.44 eV; (b) $\pm$0.074 eV.

2. In Eq. (12.5) put $A = \sigma'/\sigma$ and $x = Z^{1/2}/(A - 1)^{1/2}$, and show that if Z is a large number and $x^2 \ll 4$, then η_{max} occurs if $C = x/Z$, and that

$$\eta_{max} = \eta_1 \frac{A}{A - 1} \frac{e^{-x}}{1 + x}$$

3. A cathode-ray tube has a control grid sensitivity of b microamperes per volt, a screen voltage V, and a luminous sensitivity of the screen of c lumens per watt. Calculate the control-grid sensitivity of the tube in lumens per volt.

Answer: $bVc \times 10^{-6}$ lm/V.

4. An insulating luminescent screen without metal backing is bombarded with electrons of energy E_p. Show that the screen will charge itself to the potential where the secondary electron current just balances the primary current arriving at the target (sticking potential). There are two cases to be considered here:
(a) The screen charges itself to a potential *lower* than the preceding electrode.
(b) The screen charges itself to about the potential of the preceding electrode. Why this distinction? Prove from the $\delta(E_p)$ curves that this is a stable operating point.

5. The fluorescent screen of a picture tube has its secondary emission factor measured in the range 1.0 to 4.0 kV to determine its sticking potential. It is found that $\delta = 5$ at $E_p = 2$ kV and that $\delta = $ const$/E_p$ above that energy. Determine the sticking potential of the screen.

Answer: 10 kV.

6. The fluorescent screen of a television picture tube has a sticking potential of 5 kV. Calculate the potential of the surface of the screen under the following conditions:

(a) The positive electrode in front of the screen has a potential of 2500 V and can collect secondary electrons from the screen.

(b) The positive electrode in front of the screen has a potential of 10 kV.

(c) To what value can we lower the positive electrode in front of the screen without changing the intensity of the picture (assuming, of course, that the beam is refocused when the electrode voltage is lowered)?

Answer: $\simeq$ 2500 V; $\simeq$ 5000 V; $\simeq$ 5000 V.

13

Applications

13.1. SECONDARY EMISSION MULTIPLICATION

13.1a. Current Amplifiers

By putting several stages of secondary emission multiplication one behind the other, we can obtain large current amplifications. This method is useful when very small electron or ion currents flowing in a vacuum must be amplified. One may then accelerate the current carriers and let them impinge upon the first electrode (dynode) of an n-stage secondary emission multiplier (Fig. 13.1). If each stage gives a secondary emission multiplication δ, then the n stages together give a current amplification δ^n. Current gains of about 10^6–10^8 can be obtained in this manner.

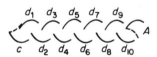

This current amplification is achieved without the use of any electrical circuit that limits the bandwidth. The only limitations are caused by the differences in electron transit times for the individual electrons passing through the multiplier. Bandwidths of about 100 MHz require no great care in the design

Fig. 13.1. Photomultiplier with a cathode c, 10 stages of secondary emission multiplication $d_1, \ldots, d_{10}$, and a final anode. Some photomultipliers used in scintillation counters have a large flat photocathode and electron-optical means for directing the photoelectrons to the first dynode. [Reproduced by permission from *Physics and Applications of Secondary Electron Emission*, Pergamon Press, New York (1954).]

of the multiplier, and by careful design much greater bandwidths can be attained.

The amplification may be so large that current pulses due to individual carriers can be counted; the multiplier then acts as a *particle counter*. Not only can a particle counter count electrons, but it may also count positive and negative ions, since the latter also cause emission of electrons when allowed to impinge upon a secondary emitter.

If the secondary emission multiplier is used in combination with a photocathode, the resulting device is called a *photomultiplier*.

13.1b. Noise and Gain Considerations

We saw in Chapter 10 that in a single stage of secondary emission multiplication the signal-to-noise power ratio deteriorated by a factor κ/δ. We shall now show that for a multistage secondary emission multiplier the signal-to-noise power ratio is deteriorated by a factor $(\kappa - 1)/(\delta - 1)$.

This easily follows if we consider the noise contributions of each dynode stage. We saw in Eq. (10.29) that for the output of a single secondary emission stage

$$\text{var } n_c = \bar{n}_p\kappa\delta = \bar{n}_p\delta^2 + \bar{n}_p(\kappa - \delta)\delta \tag{13.1}$$

Here the first term represents the multiplied shot noise of the incoming current and the second term represents the secondary emission noise of the stage. For n stages we thus have

$$\text{var } n_c = \bar{n}_p\delta^{2n} + \bar{n}_p(\kappa - \delta)\delta\cdot\delta^{2n-2} + \bar{n}_p\delta(\kappa - \delta)\delta\cdot\delta^{2n-4}$$
$$+ \ldots + \bar{n}_p\delta^{n-1}(\kappa - \delta)\delta$$
$$= \bar{n}_p\delta^{2n}\left[1 + \frac{\kappa - \delta}{\delta}\left(1 + \frac{1}{\delta} + \ldots + \frac{1}{\delta^{n-1}}\right)\right] \tag{13.2}$$
$$\simeq \bar{n}_p\delta^{2n}\left[1 + \frac{\kappa - \delta}{\delta(1 - 1/\delta)}\right] = \bar{n}_p\delta^{2n}\frac{\kappa - 1}{\delta - 1}$$

This equation can be understood as follows. The average number of particles arriving at the first dynode is $\bar{n}_p$, at the second dynode $\bar{n}_p\delta$, ..., and at the ith dynode $\bar{n}_p\delta^{i-1}$. Multiplying by $(\kappa - \delta)\delta$ gives the noise contribution of the ith dynode to the output current of that dynode. Multiplying by $\delta^{2(n-i)}$, the square of the current amplification of the remaining dynodes, gives, for the total contribution of the ith dynode,

$$\bar{n}_p\delta^{i-1}(\kappa - \delta)\delta\cdot\delta^{2(n-i)}$$

and that is exactly what the first part of Eq. (13.2) states.

Experimentally it is found that $\kappa \simeq \delta + 1$ for primary energies below a few hundred volts. The deterioration in the signal-to-noise power ratio caused by the photomultiplier is therefore approximately $\delta/(\delta - 1)$. In a good photomultiplier $\delta \simeq 5$, so the deterioration factor is about 1.25. The deterioration in signal-to-noise power ratio therefore is not very large.

Whether or not the secondary emission noise is important at all depends upon the experiment under consideration. It is important if one uses the multiplier to detect modulated light sources, since in that case one is interested in the signal-to-noise ratio. In multipliers used as *particle counters* the secondary emission noise shows up because it gives rise to a fluctuation in pulse height. Even if one only counts pulses, irrespective of their height, the secondary emission noise still shows up because the dynodes have a certain probability of producing zero secondaries per incoming primary. The secondary emission noise then determines the counting efficiency of the particle counter.

A more serious limitation is set by the dark current of the multiplier. Such current has several causes, which we discuss for a photomultiplier.

1. Leakage currents between the electrodes. This is significant only if the leakage currents enter the output circuit.

2. Ion currents. If positive ions, originating anywhere between the photocathode and the collector, arrive at the first dynodes or the photocathode, new secondary electrons are released. To prevent this the dynodes should be properly shaped.

3. Field emission from sharp edges. Sharp edges should therefore be avoided and the potential difference between the dynodes should not exceed a certain value.

4. Thermionic emission from the first dynodes. This can be improved if the thermionic work functions of the dynodes are made relatively large, for example, if the dynode material is made *p*-type rather than *n*-type.

5. Thermionic emission from the photocathode. If the work function of the photocathode is sufficiently low, it will emit some electrons, even at room temperature. The effect is most disturbing in tubes with infrared-sensitive photocathodes. If the photoemitting material is made *p*-type rather than *n*-type, the effect can be materially reduced. Other things being equal, photocathodes with a shorter cutoff wavelength should give a lower dark current; therefore, it is important to use photomultipliers with the shortest cutoff wavelength that can be tolerated in the application. Finally, cooling the photocathode with dry ice or liquid nitrogen is usually very effective in reducing the dark current.

It has been noticed that the photomultiplier gain decreases steadily upon prolonged use of the multiplier. The effect increases with increasing average collector current. There are apparently two reasons for this:

1. The secondary emission factor decreases under prolonged electron bombardment. This effect is most pronounced in the final stages of secondary emission multiplication.

2. The electron bombardment seems to release some volatile components from the last dynodes that deteriorate the emission properties of the preceding dynodes and of the photocathode. This is a relatively slow process; it takes some exposure time before the reduction in gain becomes noticeable, and it takes a long rest time before the reduction in gain has been partially or completely recovered. The effect is not insignificant and should be carefully considered in many applications.

13.1c. Required Gain

Something should be said about the multiplier gain needed in particular applications. In particle counters one wants so much gain that the individual pulses can be processed without further amplification. For example, if the multiplier has a gain of 10^8 and gives pulses of 10^{-9} s duration, then the output pulse current is 1.6×10^{-2} A, which gives a signal of 1.6 V in a 100-Ω load. This is sufficient for many applications. If a larger load resistance can be used, the gain can be correspondingly reduced.

If the output of the photomultiplier feeds into the input of an amplifier, one wants the photomultiplier gain to be so large that the amplifier noise is insignificant in comparison with the amplified shot noise.

We first consider a photodiode without multiplication. Let a modulated beam of light striking upon a photocathode give a photocurrent I_p. The output signal of a photocell device can be described by a signal current generator in parallel with the load resistance R (Fig. 13.2). To obtain large signals, one should make R large. Unfortunately, the circuit capacitance C limits the bandwidth B that can be handled by the circuit, according to the relation

$$2\pi BCR = 1$$

In addition, the thermal agitation of the conduction electrons in the resistance R produces a noise signal, which can be described by a current

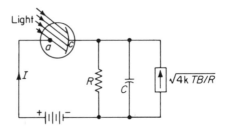

Fig. 13.2. Limitation set by thermal noise in the detection of small light signals by means of a photocell.

generator $\sqrt{4kTB/R}$ in parallel with R. For very small light signals, the signal desired may drown in this noise background. The limiting case occurs if

$$I_p = \sqrt{\frac{4kTB}{R}} = \sqrt{8\pi kTB^2C} \tag{13.3}$$

Putting $C = 20\ \mu\mu\text{F}$, $B = 4$ MHz, $T = 300°\text{K}$, and $k = 1.38 \times 10^{-23}$ J/degree yields $I_p \simeq 6 \times 10^{-9}$ A. Assuming a photocathode sensitivity of $60\ \mu\text{A/lm}$, we have a photocell that can barely detect a light flux of 10^{-4} lm. If wider bandwidths are required, the minimum detectable light flux becomes even larger.

If secondary emission multiplication is used, the multiplied shot noise of the primary current I_p must be large in comparison with the thermal noise of the load resistance R, or

$$2eI_pG^2B > \frac{4kTB}{R} \quad \text{or} \quad G^2 > \frac{2kT/e}{RI_p} \tag{13.3a}$$

Taking $I_p = 10^{-14}$ A and $R = 10,000\ \Omega$ yields $G^2 > 5 \times 10^8$, so that a gain of 3×10^4 is sufficient for making the amplified shot noise larger than the noise of the load resistance. Assuming again a photocathode sensitivity of $60\ \mu\text{A/lm}$, we have here a photomultiplier that can detect a light flux of less than 10^{-9} lm.

13.1d. Channel Multipliers

The multipliers just discussed use discrete dynodes. We now consider the channel multiplier, which uses distributed dynodes.

A channel multiplier consists of a tube of weakly conducting secondary emission material to which a potential difference of a few thousand volts is applied. Shooting in at a slight angle on the low-voltage side of the tube, the primary electrons will impinge upon the wall of the tube and produce several secondaries. These secondaries are accelerated along the tube and, because of their initial velocity, will strike the wall again, generate more secondaries, and so on. Gains of 10^6 to 10^7 can be obtained over a channel length of a few centimeters or less. The actual gain depends upon the applied voltage, the channel length, and the ratio of diameter to length.

One limitation in the channel multiplier is set by ion noise. If the multiplier is curved, these ions can be eliminated before they have traveled very far. By thus keeping the ions from arriving at the input side of the channel, we can materially reduce their effect.

Channel multipliers can be used for viewing at low light levels. The system uses a photocathode and a very large number of channel multipliers in parallel. A low-level light image is focused on the photocathode. The photoelectrons coming from the photocathode are focused upon the input side of the channel multiplier. At the output of the channel multiplier one thus obtains amplified current beams. These can be further accelerated and focused onto a cathode-ray tube screen so that an amplified replica of the light image focused on the photocathode can be displayed on the screen.

It will now be shown that the gain G of a channel multiplier of length L is given by

$$G = \exp(pL) \tag{13.4}$$

where p is the multiplication factor per unit length, and that the output noise has a spectral intensity

$$S_I(f) = 4eI_{pr}G^2 \tag{13.5}$$

if $G \gg 1$, where I_{pr} is the primary current. The output noise is thus twice the multiplied primary noise.

The proof is simple. The change in current over a length dx is

$$dI(x) = I(x)p\,dx, \qquad \frac{dI}{I} = d\ln I = p\,dx$$

Integrating over the length L, bearing in mind that $I(0) = I_{pr}$, yields

$$\frac{I(x)}{I_{pr}} = \exp(px), \qquad G = \frac{I(L)}{I_{pr}} = \exp(pL)$$

The spectral intensity of the noise is

$$S_I(f) = 2eI_{pr}\exp(2pL) + \int_0^L 2e\,dI(x)\exp[2p(L-x)]$$

since a secondary generated at the position x has a gain $\exp[p(L-x)]$, whereas the current $dI(x)$ has full shot noise. Substituting for $dI(x)$ and carrying out the integration yields

$$S_I(f) = 2eI_{pr}\exp(2pL) + 2eI_{pr}\exp(2pL)[1 - \exp(-pL)]$$

For $G \gg 1$, this reduces to Eq. (13.5).

13.1e. Scintillation Counter

Besides counting particles directly with particle counters, one can also let a fast particle impinge upon a fluorescent crystal; part of its energy is then emitted in the form of a short light flash that can be detected by a photomultiplier. The combination of the fluorescent crystal and the photomultiplier is called a *scintillation counter* (Fig. 13.3). Its wide use in nuclear physics has led to the construction of photomultipliers with a large photocathode to collect a larger part of the emitted light.

To understand the advantage of a scintillation counter, we consider an incident particle with an energy of 0.5 MeV. If the crystal produces one photon for each 50 eV of energy, then the particle produces 10,000 photons. Let 50 per cent of these photons reach the photocathode and let the cathode have a quantum efficiency of 10 per cent; then the incident particle produces 500 photoelectrons. Even if only half these photoelectrons enter the multiplier system, one obtains an appreciable gain (250 times). Moreover, charged particles, fast neutrons, and γ-quanta can all be detected by the scintillation counter; hence its great popularity.

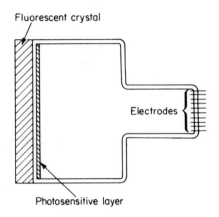

Fluorescent crystal

Electrodes

Photosensitive layer

Fig. 13.3. Scintillation counter, using a large fluorescent crystal and a photomultiplier tube with a large flat photocathode. The geometry should be chosen so that as much light as possible is received by the photocathode. The dynodes of the photomultiplier are not shown.

13.2. TELEVISION PICKUP TUBES

Television pickup tubes are used to transform light signals into electric pulses. Obviously, these tubes make use of some kind of photoelectric effect. Some use photoemission; others use photoconduction. We discuss here the commonly used tubes: the *image orthicon*, the *vidicon*, and the secondary electron conduction vidicon.

13.2a. Image Orthicon

The optical image of a scene that is to be televised is focused upon a photoemissive surface with a conductive backing. The photoemissive surface is kept at a potential of about −400 V; the ring electrode 6 of Fig. 13.4 is used to accelerate electrons away from the photocathode. A magnetic field focuses the photoelectrons upon a thin glass target ($\simeq$ 0.0002 in. thick) that is roughly at zero potential. The photoelectrons, arriving at the target with an energy of a few hundred electron volts, produce slow secondary electrons, which are collected by a screen mesh close to the glass target (distance 0.002 in. or even less); the mesh is kept at a voltage of about +2 V.

Owing to the secondary emission, the side of the target facing the photocathode is charged positively. The target potential is stabilized at zero poten-

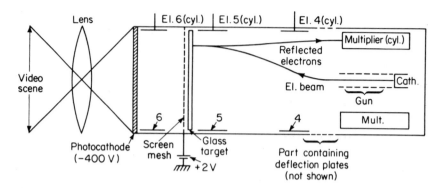

Fig. 13.4. Image orthicon. A full description of the figure is given in the text. The beam deflection section of the tube is not shown. Electrodes 4, 5, and 6 and the multiplier are cylindrical structures. [*RCA Television*, Vol. VI (1950), courtesy RCA.]

tial by an electron beam scanning the other side of the target and depositing a negative charge on that side, thereby neutralizing the positive charge on the front side at regular intervals. At sufficiently low light levels the positive charge distribution on the target thus accurately follows the light distribution at the photoemissive surface; at high light levels the positive charge is limited by the mesh potential.

Magnetic deflection methods are used to scan the thin electron beam across the other side of the target. The beam originates from the cathode of an electron gun, which is assumed to be at zero potential. Since the electrons are emitted with zero energy, they will at any point of their path have an energy equal to the potential at that point. If the electron beam arrives at the target in a direction perpendicular to the target, the electrons can land on the target if the target potential is larger than zero; if they arrive in any other direction, a certain positive target potential is needed in order to make the electrons land on the target. It will be seen that this decreases the sensitivity of the tube. To avoid the occurrence of a velocity component parallel to the target, two correcting elements are used.

First, the magnetic field at the aperture of the electron gun should be lined up with the electron beam; to that end an alignment coil is used to make the magnetic field parallel to the beam. Any alignment error would give a velocity component parallel to the target; as mentioned before, this should be avoided.

Second, the deflecting field introduces a radial velocity component† of

---

† Radial here means directed toward the axis of the tube.

the electrons, which is zero at the center of the target and increases toward the edges; this would prevent the beam electrons from arriving at the edges of the target and thus would give rise to a variable sensitivity over the target. To avoid this effect, the cylindrical decelerator electrode 5 is properly biased. If electrode 5 is operated at a positive potential between that of the target and the focusing electrode 4, a radial electrostatic field is produced, which is zero at the axis and increases toward the edges. This electrostatic field gives the electron beam a radial velocity opposite to that produced by the magnetic deflecting field.

The beam electrons arriving at the target charge the target negatively; after having neutralized the positive charge on the other side, the beam will be reflected. By scanning the electron beam over the target and collecting the reflected electrons, one thus obtains a modulation of the reflected beam that follows the positive charge distribution on the other side of the target, or, what amounts to the same thing, follows the light distribution at the photocathode. The beam current has to be such that it can neutralize the largest positive charge that can occur at the target; we saw before that this largest positive charge is determined by the mesh potential. The beam current needed is of the order of magnitude of 10^{-8} A; it depends upon the capacitance between target and screen. For example, the total charge on a target of a 2P33 image orthicon, with $+2$ V on the mesh screen, is 1.6×10^{-10} C if the photoemissive surface is highlighted. Since the capacitance is discharged in $\frac{1}{30}$ s (frame time), the signal current from highlighted areas of the target is about 5×10^{-9} A. This is also the minimum beam current. The larger the target-screen capacitance, the larger the positive charge and the larger the necessary beam current.

We can now understand why a velocity component parallel to the target will decrease the sensitivity. For since the beam electrons can then land upon the target only if the target potential is slightly positive, the negative charge deposited on the surface of the target cannot completely neutralize the positive charge on the other side. The same light input will thus produce a smaller percentage modulation of the beam.

The target must be very thin, for the positive charge on the image side and the negative charge on the scanned side of the target must combine in less than a frame time ($\frac{1}{30}$ s). The image side of the layer must have a reasonably high secondary emission factor in order to charge the target positively. The scanned side must have a very low secondary emission for low-velocity electrons (or better, a low reflection coefficient); for that reason the scanned side of the target is covered by a very thin evaporated layer of silver.

In view of the very small beam currents, it would be unwise to feed the signal obtained from the returning beam directly into the video amplifier; the order of magnitude of the video signal would not be much higher than

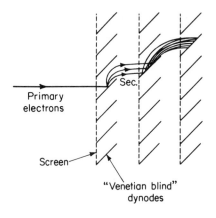

Primary electrons

Sec.

Screen

"Venetian blind" dynodes

Fig. 13.5. Part of a secondary emission multiplier with "venetian blind" type dynodes. The screen is needed in order to prevent primary electrons from passing through the venetian blind structure without hitting the dynodes.

the noise level of the amplifier. In practice, the returning electron beam is amplified by a secondary emission multiplier system. The amplification has to be such that the noise level of the pickup tube is determined by the shot noise of the electron beam. This requirement leads to a current amplification of the order of 1000. The multiplier section consists of cylindrical electrodes of the "venetian-blind" type. To obtain the most favorable conditions for drawing the secondary electrons away from a dynode, very fine screens can be interposed between the dynodes, each screen being connected to its dynode (see Fig. 13.5).

13.2b. Vidicon Pickup Tube

An interesting practical application of photoconduction is made in the vidicon tube.† It consists of a transparent conductive coating‡ on glass used as a signal plate, covered with a thin photoconductive layer of high resistivity. A scanning beam scans this photoconductive layer and deposits electrons on the scanned surface, thus charging it down to the potential of the thermionic cathode in the electron gun (Fig. 13.6). A fixed potential of 10–30 V positive, relative to the cathode, is applied to the signal plate; although a considerable field is thereby developed across the opposite faces of the photoconductor, very little dark current will flow if its dark conductivity is sufficiently low.

If a light image is focused on the target, the conductivity of the photoconductor is increased in the illuminated portions, thus permitting charge to flow. In these areas the scanned surface gradually becomes charged a volt or two positive with respect to the cathode during the $\frac{1}{30}$-s interval between successive scans. The beam deposits a sufficient number of electrons to neutralize the charge and in doing so generates the video signal in the signal plate lead. The target is sensitive to light throughout the entire frame time, thus permitting full storage of charge.

The charge–discharge cycle is identical to that of the image orthicon, but the charging of the surface is now achieved by photoconduction *through* the

† *Television*, **6**, RCA Labs, Princeton, N.J., 1949–1950.
‡ These coatings are commercially available under the trade name "Nesa."

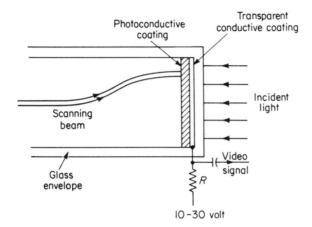

Fig. 13.6. Part of a vidicon tube, showing the glass envelope, the transparent conductive coating, the photoconductive coating, and the scanning beam. The gun structure and the beam deflection elements are not shown. (Courtesy of *Electronics*, a McGraw-Hill publication.)

target, not by secondary emission. This requires that the resistivity of the target be sufficiently high so that its time constant exceeds the $\frac{1}{30}$-s frame time between successive scans. A dark resistivity of 10^{10} ohm m or greater is satisfactory. This can be easily achieved with present photoconducting materials.

The spectral response of the target is a function of the material and of the processing. It is possible to make targets that are sensitive to the entire visible range of the spectrum. The signal-versus-light curve is linear at low light levels, with some flattening off at high light levels.

The sensitivity of photoconductors may be many orders of magnitude larger than the sensitivity of photoemitters. Unfortunately, the response of these sensitive photoconductors is far too slow to be useful in vidicon pickup tubes. As a consequence, present vidicon tubes are at least an order of magnitude less sensitive than the best image orthicons.

The advantage of the vidicon over the image orthicon is that the signal current is a minimum at zero light level and increases with increasing light level. The signals coming from the dark portions of the scene thus show the full shot effect of the beam in the image orthicon and negligible noise in the vidicon. The disadvantage of the vidicon is that no electron multiplication is possible. This requires a higher beam current.

13.2c. Secondary Electron Conduction Vidicon

In the secondary electron conduction vidicon the photoelectric signal is amplified by the processes of secondary electron conduction and secondary electron emission. The secondary electron conduction (SEC) target consists of a supporting layer of about 700 Å of Al_2O_3, a 500-Å-thick layer of Al, and a low-density layer of KCl. The Al layer serves as the signal plate and is deposited on the Al_2O_3 by vacuum evaporation. The low-density layer of KCl (about 25 μm thick) is then deposited on the Al by evaporation in an argon atmosphere at a pressure of 2 mm Hg. A schematic diagram of the tube is shown in Fig. 13.7.

A positive voltage V_T is applied to the target backplate, and the KCl side

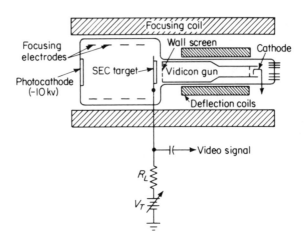

Fig. 13.7. SEC vidicon schematic according to G. W. Goetze and A. H. Boerio, *Proc. I.E.E.E.*, **52**, 1007, 1964.

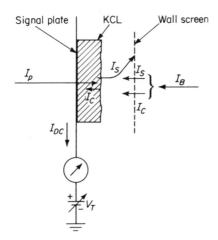

Fig. 13.8. Schematic diagram of currents involved in SEC target operation, according to G. W. Goetze and A. H. Boerio, *Proc. I.E.E.E.*, **52**, 1007, 1964.

of the target is stabilized at gun cathode potential (ground potential) by the low-energy electrons of the scanning beam. The currents involved in the operation of the target are shown schematically in Fig. 13.8. Photoelectrons with an energy of approximately 10 keV, designated by the current I_p, penetrate both the Al_2O_3 and the Al, and dissipate most of their energy in the KCl, thereby creating many low-energy secondary electrons. Owing to the electric field across the layer, established by the previous polarization, initially most of these electrons are collected by the signal plate, causing the KCl to charge to more positive values. The conduction through the layer is represented by the current I_c. This process continues until the KCl layer reaches the potential of the signal plate. Before this happens, the current I_s, due to transmitted secondary electrons, becomes significant. These emitted electrons are collected by the wall screen and continue to charge the KCl layer to even more positive values. This continues until the scanning beam (beam current I_B) returns and decreases the potential of the charged area to ground potential. This capacitive discharge results in a current pulse, which produces a video signal voltage across the load resistor R_L.

Although the process is similar to that in the standard vidicon, there is a significant difference. The conduction through the target is achieved by free electrons traveling in the interparticle volume of the layer rather than by electrons in the conduction band. With increasing voltage on the signal plate, conduction due to electrons in the conduction band will also take place and augment the observed gain. If the gain contribution of solid-state conduction is appreciable, the target shows a typical time-lag characteristic of electron-bombardment-induced conductivity. The threshold voltage for solid-state conduction lies between 30 and 40 V, depending on target thickness and structure. If operation is kept below this voltage, there is no time lag.

The total target gain, averaged over the exposure time, increases approximately linearly with the signal plate voltage V_T. At a target voltage of 30–40 V, the total gain, the sum of conduction and transmission secondary emission gains, reaches a value of 200. Owing to its transmission secondary emission gain component, the SEC target exhibits gain even at zero voltage, contrary to the behavior of a photoconductive target.

13.2d. Noise in Television Pickup Tubes

As a first example, we consider here the image orthicon. If P is the sensitivity of the photocathode in amperes per lumen, A the cathode area in square meters, and L the incident illumination in lumens per square meter, then the signal current is $PAL = I_s$ and the signal-to-noise power ratio is

$$\left(\frac{S}{N}\right)_{in} = \frac{I_s^2}{2eI_sB} = \frac{I_s}{2eB} \tag{13.6}$$

where $-e$ is the electron charge, B the system bandwidth, and $2eI_sB$ the mean square of the primary noise current in the bandwidth B.

If the target has a secondary emission factor δ_t, then the target gain $G_t = \delta_t - 1$, and the noise at the target is

$$\overline{i_t^2} = G_t^2 \cdot 2eI_sB + G_t \cdot P_t 2eI_sB \tag{13.7}$$

where $P_t = \delta_t(\kappa_t - \delta_t)/(\delta_t - 1)$, since the secondary emission noise includes again the factor $\delta_t(\kappa_t - \delta_t)$, and the net target gain is $\delta_t - 1$ rather than δ_t.

The noise current in the readout beam is

$$\overline{i_b^2} = 2eI_BB$$

where I_B is the beam current. The mean-square fluctuating return current impinging on the first dynode is therefore

$$\overline{i_d^2} = \overline{i_b^2} + \overline{i_t^2} = 2eI_BB + (G_t + P_t) \cdot 2eI_sG_tB \tag{13.8}$$

and the amplified noise coming out of the dynode system is

$$\overline{i_a^2} = G^2(\overline{i_d^2} + 2eI_dP_mB) = G^2[2eI_BB + (G_t + P_t) \cdot 2eI_sG_tB + 2eI_dP_mB] \tag{13.9}$$

where $I_d = I_B - G_tI_s$ is the current flowing to the first dynode, G is the multiplier gain, and

$$P_m = \sum_{j=1}^{n} \frac{\kappa_j - \delta_j}{\prod\limits_{k=1}^{j} \delta_k} \tag{13.9a}$$

where δ_j and κ_j are defined for the jth dynode and n is the number of dynode stages. If all dynodes were equal, this expression could be simplified, but since the first dynode must have a low gain for various practical reasons, Eq. (13.9a) is better left in the form given.

Since the signal current at the output is $I_a = GG_tI_s$, the signal-to-noise power ratio at the output is

$$\left(\frac{S}{N}\right)_{\text{out}} = \frac{I_a^2}{\overline{i_a^2}} = \left(\frac{S}{N}\right)_{\text{in}} \frac{1}{F} \tag{13.10}$$

where
$$F = 1 + \frac{P_t - P_m}{G_t} + \frac{1 + P_m}{MG_t} \tag{13.10a}$$

Here $M = G_tI_s/I_B$ is the modulation depth of the readout beam. Substituting $I_d = (1 - M)I_B$ into (13.9) and eliminating I_B with the help of the parameter M leads easily to (13.10). The factor F describes the deterioration in signal-to-noise power ratio for the image orthicon system.

For most image orthicons the factor M is relatively small,† and it becomes progressively smaller at lower light levels and lower beam currents. As a consequence the term $(1 + P_m)/MG_t$ predominates, and F can be quite large. Because of this, the limiting signal-to-noise power ratio $(S/N)_{\text{in}}$ is not reached.

One can improve the signal-to-noise ratio by making the target gain G_t larger. Or one can use a silicon transmission secondary emission multiplier stage between the photocathode and the target (Sec. 10.3). The ideal situation is obtained if the beam noise is small in comparison with the target noise.

In vidicon tubes, when the photoconductor has a small gain factor G (see Chapter 11), the main source of noise is the shot noise of the beam current. Due to the way the tube operates, this corresponds to the shot noise

---

† This comes about because the target has a large reflection coefficient for low-speed electrons.

of the photocurrent flowing through the photoconductor between successive scans. If the gain G is significant, the noise of the photoconductor also makes a contribution. It may be shown that this effect can be taken into account by multiplying the noise expression by the factor $1 + G$.

We now discuss the secondary electron conduction vidicons. If the multiplication process does not add extra noise, the noise can then be represented by the shot noise of the primary beam reaching the KCl layer. Because of the noise in secondary electron production, the noise at the output of the multiplier may be somewhat larger than the amplified shot noise of the primary beam. We may thus represent it by an equivalent noise intensity

$$S_0(f) = 2eI_sG^2F \qquad (13.11)$$

where the factor $F > 1$ describes the added noise of the multiplication processes. In good targets $F \simeq 2$.

13.3. LIGHT AMPLIFIERS

Light amplifiers are luminous devices in which the light output is controlled by the (smaller) light input. We shall discuss various examples.

The ultimate limitation to the smallest light signal that can be amplified by a light amplifier is set by the noise level of this device. In light amplifiers using electroluminescence the limit is set by the light background emitted by the device.

13.3a. Light Amplification with the Help of Photoelectric Effect

A light amplifier based upon this principle consists of a photosurface, a luminescent screen, and a high voltage (10–20 kV) applied between the two electrodes. Electron-optical focusing means should be provided to give a true electron-optical image of the photoemitting surface upon the luminescent screen. A light signal (picture) of low intensity is focused onto the photoelectric surface; the emitted photoelectrons are accelerated by the applied field, strike the luminescent screen, and give an image of the light signal focused onto the photoemitter. By proper choice of the accelerating voltage, a considerable gain in light intensity becomes feasible, as can be seen from the following calculation (Fig. 13.9).

Let B_i be the incident radiation (in lumens) and let the photosurface have a sensitivity of a amperes per lumen. Let the accelerating voltage be V and let the luminous efficiency of the screen be b lumens per watt. The photocurrent is then $B_i a$, the power arriving at the screen is $B_i aV$, and the light output B_0

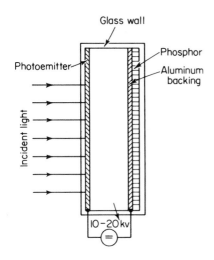

Fig. 13.9. Photoelectric light amplifier; the means of focusing the electrons (in order to form an image of the photoemitter on the phosphor) are not shown.

is, if the size of the picture is not amplified,

$$B_0 = abVB_i = GB_i \quad (13.12)$$

where

$$G = abV \quad (13.13)$$

is the gain of the light amplifier. As a numerical example, take $a = 50\,\mu\text{A/lm}$, $b = 30\,\text{lm/W}$, and $V = 20\,\text{kV}$; then $G = 30$. This shows that light amplification by this method is quite feasible.

It is essential that the luminescent screen be provided with a metal backing to prevent part of the output radiation from being fed back to the photosurface; such feedback might give rise to instability.

The electron-optical focusing means can be omitted if the distance between the photosurface and the luminescent screen is kept small. In that way it may be possible to build several stages of light amplification in cascade without much deterioration of the quality of the picture.

Another interesting application of devices of this kind is to transform light from the invisible part of the spectrum (infrared, ultraviolet, x rays) into visible radiation. Infrared converters, which convert infrared images into visible images, have been available for several years and are known by such names as snooperscope. Here the primary purpose is to obtain a visible image; obtaining a gain, although sometimes desirable, is only of secondary importance.

13.3b. Use of a Television Pickup Tube

Television pickup tubes (vidicon, image orthicon, secondary electron conduction vidicon) may be used as the input stage of a light amplifier. The signal to be amplified is focused onto the photoemitter and an electric video signal is obtained with the help of a scanning beam. The latter is amplified and fed into the control electrode of a cathode-ray tube, the beam of which scans the luminescent screen of the tube synchronously with the scanning of the pickup tube. By providing enough video amplification, we can obtain a large gain in light intensity. The advantages of this method are that the gain in light intensity can be controlled electrically, and that no optical feedback is associated with the system (Fig. 13.10).

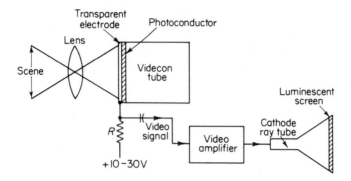

Fig. 13.10. Light amplifier using a vidicon pickup tube, a video amplifier, and a cathode ray tube. The intensified image is viewed on the cathode ray tube screen. Secondary electron conduction vidicons are preferred here.

13.3c. Basic Principles of Light Amplifiers Using AC Electroluminescence†

These light amplifiers consist of a photoconductive layer (usually CdS) on top of an electroluminescent layer (usually ZnS). A voltage $V_0 \cos \omega t$ is applied to this sandwich, thus giving rise to a voltage $V_{20} \cos (\omega t + \varphi)$ developed across the electroluminescent layer. If the photoconductive layer is illuminated locally, its conductivity will change in the illuminated regions and hence V_{20} will increase, thus giving rise to an increased light output of the electroluminescent layer at the illuminated spots. If the sandwich is properly constructed, the light output may be considerably above the light input of the layer, giving an appreciable power gain. It should be understood that the incident and the emitted radiation may have a different spectral distribution (Fig. 13.11a).

If B_1 is the incident light intensity and B_2 is the emitted light intensity, then

$$B_2 = a\omega \exp\left(-\frac{b}{V_{20}^{1/2}}\right) \qquad (13.14)$$

over a wide range of frequencies ω, where a and b are constants and V_{20} depends upon B_1. Let the electroluminescent layer have a capacitance C_2 and let the photoconductive layer have a conductance g_1 and a capacitance C_1 in parallel, where g_1 depends upon the incident light intensity; then

---

† See, for example, the December 1955 issue of *Proc. IRE* for reference papers on the subject. The discussion presented in this section is based upon G. Diemer, H. A. Klasens, and J. G. van Santen, *Philips Research Reports*, **10**, 401–424 (December 1955).

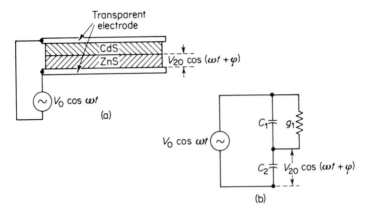

Fig. 13.11. (a) Electroluminescent light amplifier using CdS and ZnS layers sandwiched between transparent electrodes. (b) Equivalent circuit needed for the calculation of the voltage across the ZnS layer.

$$\left(\frac{V_{20}}{V_0}\right)^2 = \left|\frac{j\omega C_1 + g_1}{j\omega(C_1 + C_2) + g_1}\right|^2 = \frac{1 + (g_1/\omega C_1)^2}{(1 + C_2/C_1)^2 + (g_1/\omega C_1)^2} \quad (13.15)$$

(see Fig. 13.11b).

The conductance g_1 consists of a "dark" conductance g_0 and a conductance roughly proportional to the incident light intensity B_1. Often g_0 is negligible; assuming that g_1 is proportional to the input intensity B_1 (which is not always allowed, since the photo response may be sublinear),

$$g_1 = c_0 B_1 \quad (13.16)$$

Substituting (13.15) into (13.14), making use of (13.16), and taking the logarithm, after some arrangement of terms we get

$$\ln\frac{B_2}{a\omega} = -\frac{b}{V_{20}^{1/2}} = -\frac{b}{V_0^{1/2}}\left[1 + \frac{2p + p^2}{1 + (c_0 B_1/\omega C_1)^2}\right]^{1/4} \quad (13.17)$$

where $p = C_2/C_1$. If $\ln(B_2/a\omega)$ is plotted against $\ln(c_0 B_1/\omega C_1)$, with p as a parameter, one obtains a low-level light region where $B_2 = B_{20}$ (obtained by putting $B_1 = 0$), a high-level light region where $B_2 = B_{2\infty}$ (obtained by putting $B_1 = \infty$), and an intermediate region where

$$B_2 \simeq CB_1^\gamma \quad (13.17a)$$

Here C is a constant and γ varies slowly with B_1. Usually $\gamma > 1$, so that the image has undergone *contrast expansion*.

To prevent optical feedback, an opaque layer must be deposited between the photoconductive layer and the electroluminescent layer. Useful applications of light feedback have been described in the literature.

Electroluminescent light amplifiers have not yet lived up to the earlier expectations. The applications are interesting, however, and may become useful sooner or later.

13.4. AUGER SPECTROSCOPY

In Chapter 10 we discussed Auger effect and saw that the energy of the emitted electron was characteristic for the atom in question, so that it can be used to positively identify that atom on a surface. This application is known as Auger spectroscopy. We shall now see how this identification is done.

By sweeping out the energy distribution $N(E)$ of the emitted electrons, one obtains a small series of lines due to Auger electrons. By differentiating this signal once, one obtains sharp lines that are characteristic for the element exhibiting the Auger effect. Figure 13.12 shows $N(E)$ and $dN(E)/dE$ for an Ag target; the Auger effect displayed here is caused by an excited M shell. There are a number of lines, including two very sharp ones close together.

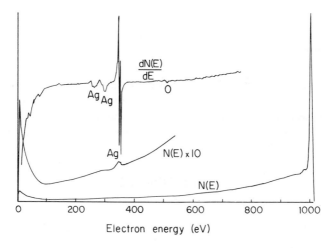

Fig. 13.12. Energy distribution and derivative of the energy distribution for a silver target with incident beam of electrons at 1000 eV.

The electron distribution can be swept out with the help of a cylindrical electron-energy analyzer; the signal thus obtained can be amplified by an electron multiplier, differentiated, and fed into an X–Y recorder or put on the screen of a cathode-ray oscillograph. The differentiation is performed to eliminate the slowly varying part of the energy distribution and to accentuate the Auger lines. A schematic layout is shown in Fig. 13.13.

Auger spectrometers based on this principle are commercially available from Physical Electronics Industries, Inc., and Varian Associates. They can be used for quantitative surface analysis. One can also sputter away part of

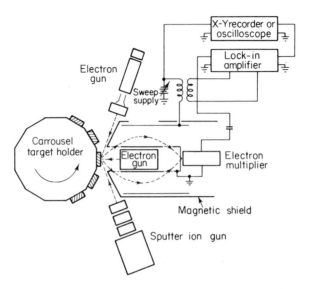

Fig. 13.13. Schematic layout of electron optics and electronics for Auger electron spectroscopy.

the surface layer and so analyze the target at various depths. By appropriate techniques one can sweep the electron beam over the target and so analyze the distribution of the various elements over the target.

Auger spectroscopy has become an indispensable tool in surface studies.

REFERENCES

BIBERMAN, L. M., and S. NUDELMAN, *Photoelectronic Imaging Devices*, vols. 1 and 2. Plenum Publishing Corporation, New York, 1971.

KAZAN, B., *Electronic Image Storage*. Academic Press, Inc., New York, 1968.

LARACH, S., ed., *Photoelectronic Materials and Devices*. Van Nostrand Reinhold Company, New York, 1965.

MCGEE, J. D., and W. L. WILCOCK, "Photoelectronic Image Devices," in L. Marton, ed., *Advances in Electronics and Electron Physics*, vol. 12. Academic Press, Inc., New York, 1960.

———, D. MCMULLAN, and E. KAHAN, "Photoelectronic Image Devices," *ibid.*, vols. 22A–22B. Academic Press, Inc., New York, 1966.

———, W. L. WILCOCK, and L. MANDEL, "Photoelectronic Image Devices," *ibid.*, vol. 16. Academic Press, Inc., New York, 1962.

PROBLEMS

1. The photocurrent I obtained from a photocathode is fed into a measuring circuit, consisting of a resistance R and a capacitance C in parallel. The bandwidth to be handled is 1 MHz, and the measuring circuit is designed so that it meets that requirement. If C is 20 $\mu\mu$F, calculate the following:
(a) The limiting photocurrent that can be measured against the thermal noise background, if the current is fed directly into the measuring circuit.
(b) The limiting photocurrent that can be measured if the signal is passed through a number of stages of secondary emission multiplication such as to make the thermal noise of R small in comparison with the amplified shot noise of the photocurrent.
(c) The secondary emission gain needed to make the shot noise of the beam predominate over the thermal noise of R in case (b).
(d) How many amplifier stages are needed in part (c) if each stage gives a gain of 4?
Hint: You may neglect the noise introduced by the secondary emission multiplication process.

Answer: (a) 1.44 $\times$ 10^{-9} A; (b) 3.2 $\times$ 10^{-13} A; (c) 4500 times; (d) 7 stages.

2. An image orthicon operates at a beam current of 10^{-8} A. The television signal obtained is first amplified in an electron multiplier and then fed into a wide-band amplifier. The wide-band amplifier has to be able to handle a bandwidth of 3 MHz, and the input circuit of the first stage consists of a capacitance C of 20 $\mu\mu$F (tube + wiring capacitance) in parallel to a resistance R which is chosen such that the input circuit has a bandwidth of 3 MHz.
(a) Calculate the secondary emission gain needed to make the amplified shot noise of the beam current of the image orthicon five times the thermal noise of the input circuit, both on an rms basis.
(b) How many stages of secondary emission amplification are needed to achieve this result, if each stage has a current amplification of 4?

Answer: 210 times; 4 stages.

3. In an image orthicon the screen mesh in front of the target is kept at $+2$ V. The surface of the glass target charges up to a maximum voltage of $+3$ V under highlight conditions and is discharged to zero potential every $\frac{1}{30}$ s by the scanning beam (that is, the frame time is $\frac{1}{30}$ s). The screen mesh has a capacitance of 30 $\mu\mu$F with respect to the glass target; the glass target has a secondary emission factor of 2, and the photocathode is irradiated by light of 5500 Å.
(a) How much charge accumulates on the target during frame time under highlight conditions?
(b) Calculate the beam current necessary to discharge the glass target under highlight conditions.
(c) Calculate the photocurrent needed to charge the glass target from 0 V to $+3$ V during frame time.

(d) Calculate the minimum amount of light (expressed in watts) needed to produce the highlight condition, if the photocathode has a yield of 10 per cent.

(e) What happens to the output signal if more than the minimum amount of light is used?

Answer: 90×10^{-12} C; 2.70×10^{-9} A; 2.70×10^{-9} A; 6.1×10^{-8} W.

4. A television pickup tube has a sensitivity of a microamperes per lumen. The output signal is fed into a video receiver with an input impedance R_i and a voltage gain g. The output signal of the amplifier is applied to the control grid of the cathode-ray tube mentioned in problem 12.3. If the illuminated areas of the photo-sensitive layer of the pickup tube and of the cathode-ray tube are equal, calculate the light amplification.

Answer: $abcR_igV \times 10^{-12}$.

5. In the light amplifier discussed in problem 4, one wants to increase the dimensions of the amplified scene by a factor p while maintaining the same light output per square meter. By what factor should the voltage gain of the video amplifier be increased to achieve this result?

Answer: Factor p^2.

6. In some discussions on light amplifiers the following expression is used for the light output of the cell:

$$B_2 = a\omega V_{20}^n$$

The value of n depends somewhat upon the operating conditions. This expression is a good approximation as long as V_{20} does not vary over too wide a range; it has the advantage that the expressions thus obtained can be more easily discussed. Derive the equivalent of (13.17) for this approximation.

Answer:
$$B_2 = a\omega V_0^n \left[\frac{1 + x^2}{(1 + p)^2 + x^2} \right]^{n/2}$$

where
$$p = \frac{C_2}{C_1} \quad \text{and} \quad x = \frac{g_1}{\omega C_1} = \frac{c_0 B_1}{\omega C_1}$$

14

Metal–Semiconductor
Diodes

14.1. CURRENT–VOLTAGE CHARACTERISTIC OF A METAL–SEMICONDUCTOR DIODE

In calculating the current–voltage characteristic of a metal–semiconductor diode, one considers two limiting cases:

1. The thickness d of the space-charge barrier layer is small in comparison with the free path length λ of the electron in the n-type semiconductor, the free path length being defined as the average distance between two collisions with the lattice. In that case the current flow is due to an *emission* of electrons across the barrier. This might apply to some silicon point-contact diodes.

2. The thickness d of the barrier is large in comparison with the free path length λ. In that case the electrons crossing the barrier suffer many collisions before arriving at the other side, and as a consequence the current flow across the barrier is a *diffusion* process. Since λ is of the order of 10^{-5}–10^{-6} cm, this diffusion model should apply to all but the thinnest barriers.

Before calculating the characteristic, we must know the potential distribution in the barrier layer. This is determined in Sec. 14.1a; the current–voltage characteristics for the emission model and for the diffusion model of the contact are calculated in Secs. 14.1b and 14.1c. Section 14.1d gives the equivalent circuit of the contact.

14.1a. Potential Distribution in the Barrier Layer;
Barrier Capacitance

We now calculate the potential distribution in the barrier layer of a metal–n-type semiconductor diode by solving Poisson's equation. Let V_{dif} be the diffusion potential of the barrier layer and let a potential difference $-V$ be applied between the semiconductor and the metal. If the surface of the semiconductor has a potential $\psi = 0$, then the potential in the interior of the semiconductor will be $V_{\text{dif}} - V$. The potential in the barrier layer rises from the value $\psi = 0$ at $x = 0$ to the value $\psi = V_{\text{dif}} - V$ for $x = d$ (Fig. 14.1a), where d is the barrier thickness. For the definition of V_{dif} see Eq. (5.23).

The voltage distribution in the layer is described by Poisson's equation:

$$\frac{d^2\psi}{dx^2} = -\frac{\rho}{\epsilon\epsilon_0} \tag{14.1}$$

where $\epsilon_0 = 8.85 \times 10^{-12}$ F/m, ϵ is the dielectric constant of the semiconductor, and ρ is the space-charge density. Let N_d be the donor density and n the electron density; then the space-charge density ρ is

$$\rho = +e(N_d - n) \tag{14.2}$$

Since the potential is constant for $x \geqq d$, we must require that the net space charge be zero for $x \geqq d$; that is, $n = N_d$ for $x \geqq d$. Going from $x = d$ to $x = 0$, the electron density decreases sharply, since the electrons have to climb a potential-energy barrier of height $\epsilon(V_{\text{dif}} - V)$. If $e(V_{\text{dif}} - V) \gg kT$,

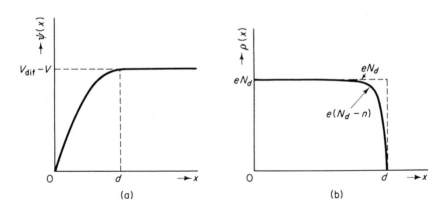

Fig. 14.1. (a) Potential distribution in the space charge layer of a metal n-type-semiconductor contact. V_{dif} is the diffusion potential, $-V$ is the applied voltage. (b) Space-charge density $e(N_d - n)$ as a function of the distance from the surface (full-drawn line). The broken line gives the approximation where the electron density n is disregarded.

it is a good approximation to put† (Fig. 14.1b)

$$\rho = eN_d, \quad \text{for} \quad 0 \leqq x \leqq d \tag{14.3}$$

and $\rho = 0$ for $x \geqq d$. The potential problem in the barrier region can now be solved easily; one only has to apply the boundary conditions

$$
\begin{aligned}
\psi &= 0, &\text{for} \quad x &= 0 \\
\psi &= V_{\text{dif}} - V, &\text{for} \quad x &\geqq d \\
\frac{d\psi}{dx} &= 0, &\text{for} \quad x &\geqq d \quad (\text{no field for } x \geqq d)‡
\end{aligned}
\tag{14.4}
$$

Substituting (14.3) into (14.1) and integrating once, using the initial condition $d\psi/dx = 0$ at $x = d$, we get

$$\frac{d\psi}{dx} = -\frac{eN_d}{\epsilon\epsilon_0}(x - d), \quad 0 \leqq x \leqq d \tag{14.5}$$

Integrating once more and applying the condition $\psi = 0$ at $x = 0$ yields

$$\psi(x) = -\frac{eN_d}{\epsilon\epsilon_0}\left(\frac{1}{2}x^2 - xd\right), \quad 0 \leqq x \leqq d \tag{14.6}$$

Applying the condition $\psi = V_{\text{dif}} - V$ at $x = d$ yields the following relation for the thickness d of the barrier layer:

$$V_{\text{dif}} - V = +\frac{eN_d}{2\epsilon\epsilon_0}d^2, \quad d = \left[\frac{2\epsilon\epsilon_0}{eN_d}(V_{\text{dif}} - V)\right]^{1/2} \tag{14.7}$$

The field strength $F = -d\psi/dx$ decreases from the maximum value

$$F_0 = -\frac{eN_d}{\epsilon\epsilon_0}d = -\left[\frac{2eN_d}{\epsilon\epsilon_0}(V_{\text{dif}} - V)\right]^{1/2} = -\frac{2(V_{\text{dif}} - V)}{d} \tag{14.8}$$

at $x = 0$ to the value $F = 0$ at $x = d$. We shall need this expression in a discussion of breakdown phenomena (Sec. 14.2).

The barrier acts as a capacitor. The total charge Q per unit area of the metal is

$$Q = -eN_d d = -[2\epsilon\epsilon_0 eN_d(V_{\text{dif}} - V)]^{1/2} \tag{14.9}$$

The small-signal capacitance C per unit area is defined as the rate of change of Q with applied voltage V§:

$$C = \frac{dQ}{dV} = \left[\frac{e\epsilon\epsilon_0 N_d}{2(V_{\text{dif}} - V)}\right]^{1/2} = \frac{\epsilon\epsilon_0}{d} \tag{14.10}$$

---

† No current would flow in the barrier region if n were rigorously zero. We require here only that $n \ll N_d$ for the major part of the barrier region; this condition is generally satisfied.

‡ Actually, this is only approximately true, since there would be no current flow in the bulk semiconductor if $d\psi/dx = 0$ in that region. However, $d\psi/dx$ is usually sufficiently small to make the boundary condition $d\psi/dx = 0$ a good approximation.

§ We use here the definition $C = dQ/dV$ because the dependence of Q upon V is non-linear. The capacitance C thus defined is the actual capacitance measured with the help of a small ac signal.

That is, the capacitance per unit area is the same as for a plane capacitor of unit area and plate distance d filled with a dielectric of dielectric constant ϵ. For an explanation of this result, see the discussion centering around Eq. (15.17).

As an example, consider the case $V_{\text{dif}} - V = 1$ V, $\epsilon = 16$, and $N_d = 5 \times 10^{22}/m^3$ (impurity of about 1 part in 10^6). In that case, we obtain $d = 1.8 \times 10^{-7}$ m $= 1800$ Å, $F_0 = 10^7$ V/m $= 10^5$ V/cm, and $C = 0.8 \times 10^{-3}$ F/m^2 $= 0.08$ μF/cm^2. The value of F_0 is not too far from the breakdown field strength in semiconductor materials.

This example indicates that one can also obtain an ohmic contact by making the surface donor density very large. The surface barrier layer is then very thin and the electric field strength in the barrier layer is very large. In that case the passage of electrons from the semiconductor into the metal is not due to thermionic emission over the barrier but to "field emission" through the barrier, resulting in a low-resistance contact (see Sec. 14.2).

If V is made negative (diode biased in the direction of difficult current flow), d increases and so does F_0, whereas C decreases. Considered as functions of N_d, F_0 and C increase with increasing N_d, whereas the barrier thickness d decreases with increasing N_d.

Equation (14.10) may be written in the form

$$\frac{1}{C^2} = \frac{2(V_{\text{dif}} - V)}{e\epsilon\epsilon_0 N_d}, \qquad d = \frac{\epsilon\epsilon_0}{C} \qquad (14.10a)$$

so that

$$-\frac{d(1/C^2)}{dV} = \frac{2}{e\epsilon\epsilon_0 N_d}$$

or

$$N_d = -\frac{2}{e\epsilon\epsilon_0} \cdot \frac{dV}{d(1/C^2)}, \qquad d = \frac{\epsilon\epsilon_0}{C} \qquad (14.10b)$$

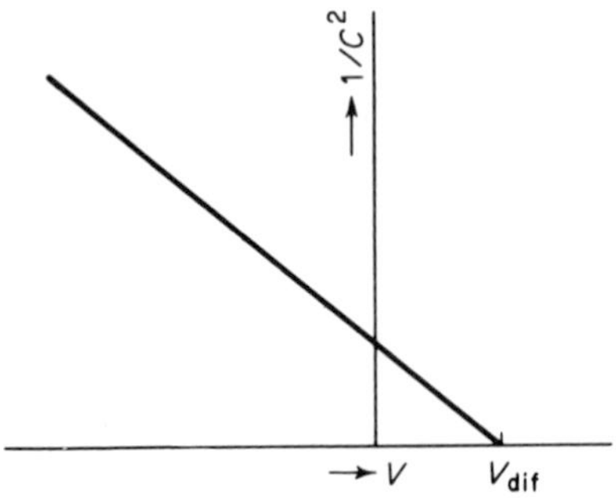

If N_d is constant throughout the barrier region, one should thus obtain a straight line by plotting $1/C^2$ against V (Fig. 14.2). Extrapolating toward the point where $1/C^2 = 0$ gives V_{dif}.

Any deviation from linearity in such a plot indicates that N_d is not constant throughout the barrier region. It can be shown that this does not alter the validity of (14.10b); these equations thus permit calculation of the donor concentration N_d as a function of the distance d below the surface from capacitance measurements.

Fig. 14.2. $1/C^2$, where C is the contact capacitance, is plotted against the voltage V applied to the metal. The curve should be a straight line, meeting the V axis at $V = V_{\text{dif}}$.

14.1b. Calculation of the Characteristic in the Emission Model

Consider a *metal–n-type semiconductor diode* of unit area. We saw in Sec. 5.4 that two equal and opposite current densities J_0 flow across the rectifying barrier for zero bias. We also saw that the current due to the flow of electrons from the metal into the semiconductor is independent of bias, but that the current due to the flow of electrons from the semiconductor into the metal depends strongly upon bias. We shall now calculate this dependence. We choose our direction of positive current flow *from the metal into the semiconductor* (positive X direction).

If $-V$ is the voltage applied to the semiconductor, the potential barrier for the electrons crossing from the semiconductor into the metal is $e(V_{\text{dif}} - V)$, where V_{dif} is the diffusion potential. The chance that an electron has the energy $\epsilon(V_{\text{dif}} - V)$ to do so is $\exp[-e(V_{\text{dif}} - V)/kT]$, and the number of electrons crossing the barrier is proportional to it. Since the flow of electrons from the semiconductor into the metal causes a current density $+J_0$ for zero bias ($V = 0$), the current density flowing when the bias $-V$ is applied is $+J_0 \exp(eV/kT)$. The current density due to the flow of electrons from the metal *into* the semiconductor is $-J_0$ for all values of V; hence the total net current density is (Fig. 14.3)

$$J = J_0\left[\exp\left(\frac{eV}{kT}\right) - 1\right] \quad (14\text{-}11)$$

This model has rectifying properties. If $V = -V_0$ and V_0 positive and $eV_0/kT \gg 1$, we have $J = -J_0$; this is called the *saturated current density*. This direction of current flow (corresponding to a flow of electrons from the metal into the semiconductor) is thus the *direction of difficult current flow* (bias in the *back direction*). If $V = +V_0$ and $eV_0/kT \gg 1$, J has opposite sign and increases strongly with increasing V_0; this direction of current flow (corresponding to a flow of electrons from the semiconductor into the metal) is the *direction of easy current flow* (bias in the *forward direction*).

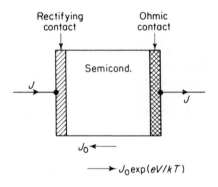

Fig. 14.3. Derivation of the characteristic of a metal-semiconductor contact in which the free path length of the electrons is large in comparison with the thickness of the barrier layer.

We finally have to calculate J_0 for this model. Let the X direction be perpendicular to the contact surface and let v_x be the X component of the velocity of the electrons in the bulk semiconductor material. The velocity distribution of these electrons is Maxwellian (compare Chapter 3); the density of electrons with a velocity between v_x and $(v_x + \Delta v_x)$ is

$$\Delta n = N_d \left(\frac{m^*}{2\pi kT}\right)^{1/2} \exp\left(-\frac{\frac{1}{2}m^* v_x^2}{kT}\right)\Delta v_x \qquad (14.12)$$

where N_d is the donor density, m^* the effective mass, and $v_x \, \Delta n$ the number of electrons in the velocity interval Δv_x arriving at the surface per unit area per second; those having $\frac{1}{2}m^* v_x^2 > eV_{\text{dif}}$ can escape. The total current density J_0 is therefore, by integrating with respect to v_x,

$$J_0 = \int_{(2eV_{\text{dif}}/m^*)^{1/2}}^{*} ev_x \, dn = eN_d \left(\frac{kT}{2\pi m^*}\right)^{1/2} \exp\left(-\frac{eV_{\text{dif}}}{kT}\right) \qquad (14.12a)$$

Introducing the average velocity component $\overline{v_x} = (2kT/\pi m^*)^{1/2}$ and a velocity v_t by the definition $v_t = \frac{1}{2}\overline{v_x} = (kT/2\pi m^*)^{1/2}$ yields

$$J = eN_d v_t \exp\left(-\frac{eV_{\text{dif}}}{kT}\right)\left[\exp\left(\frac{eV}{kT}\right) - 1\right] \qquad (14.13)$$

We need this for comparison with the diffusion case.

Since Eqs. (7.5) and (14.12a) deal with the same phenomenon, they should give identical results. According to Eq. (7.5),

$$J_0 = \frac{4\pi em^* k^2 T^2}{h^3} \exp\left[-\frac{(\phi_m - \chi_s)}{kT}\right] \qquad (14.13a)$$

since $E_b = \phi_m - \chi_s$. But $\phi_m - \chi_s = \phi_m - \phi_s + \phi_s - \chi_s = eV_{\text{dif}} - E_f$, and according to (5.21), $\exp(E_f/kT) = N_d/N_c$, where $N_c = 2(2\pi m^* kT/h^2)^{3/2}$. Substituting into (14.13a) yields (14.12a).

As a refinement of the theory, we now take into account the image force on the electron. Applying the theory of the Schottky effect of Chapter 7 yields

$$J_0 = eN_d v_t \exp\left(-\frac{eV_{\text{dif}}}{kT}\right) \exp\left[\frac{e}{2kT}\left(\frac{e|F_0|}{\pi\epsilon\epsilon_0}\right)^{1/2}\right] \qquad (14.14)$$

so that

$$J = eN_d v_t \exp\left(-\frac{eV'_{\text{dif}}}{kT}\right)\left[\exp\left(\frac{eV}{kT}\right) - 1\right] \qquad (14.14a)$$

where

$$V'_{\text{dif}} = V_{\text{dif}} - \frac{1}{2}\left(\frac{e|F_0|}{\pi\epsilon\epsilon_0}\right)^{1/2} = V_{\text{dif}} - 3.79 \times 10^{-5}\left(\frac{|F_0|}{\epsilon}\right)^{1/2} \qquad (14.14b)$$

as is found from Eq. (7.23) by replacing F_0 by $|F_0|/\epsilon$, ϕ'_0 by V_{dif}, and ϕ'' by V'_{dif}. Here $|F_0|$ follows from Eq. (14.8) and is expressed in volts per meter.

This theory has ignored the existence of minority carriers in the semiconductor. Before taking their effect into account, we must discuss the current flow in a *pn* junction (Chapter 15).

The preceding theory holds for a contact between a metal and an *n*-type semiconductor. The theory for a contact between a metal and a *p*-type semiconductor can be developed along similar lines; the only difference is that the current across the barrier is carried by holes in that case.

14.1c. Calculation of the Current–Voltage Characteristic in the Diffusion Model

Consider an n-type semiconductor having a donor concentration N_d. Let a metal–semiconductor diode be made at one side and let the thickness d of the barrier layer be large in comparison with the free path length λ of the electrons. The current flow through the barrier is then due to diffusion of the carriers; we calculate the current density by solving the diffusion equation.

Diffusion of carriers occurs if differences in carrier concentration exist; we show now that this occurs here. For if the direction perpendicular to the surface is chosen as the X direction, then the concentration $n(x)$ of the electrons depends upon x. The minimum value $n(x) = n_0$ occurs at $x = 0$; since the electron density in the metal and the height of the barrier at the metal side of the contact have constant values, n_0 will be independent of the applied voltage.† The value of $n(x)$ increases with increasing x until it reaches its maximum value $n(x) = N_d$ at $x = d$ (Sec. 14.1a, Fig. 14.1b).

According to Eq. (6.10), the current is

$$J = e\mu_n F(x) n(x) + eD_n \frac{dn}{dx} \qquad (14.15)$$

The positive direction of J is again the positive X direction. According to Chapter 6, the following relations hold for μ_n and n_0:

$$\mu_n = \frac{e}{kT} D_n \qquad (14.16)$$

$$n_0 = N_d \exp\left(-\frac{eV_{\text{dif}}}{kT}\right) \qquad (14.17)$$

We can now calculate J under the condition that a voltage $-V$ is applied to the semiconductor. Since J is independent of x, we multiply (14.15) by the factor $\exp[-e\psi(x)/kT]$, and put $F(x) = -d\psi/dx$. This yields, since $\mu_n = (e/kT)D_n$,

$$J \exp\left(-\frac{e\psi}{kT}\right) dx = -eD_n n(x) \frac{e}{kT} d\psi \exp\left(-\frac{e\psi}{kT}\right) + eD_n \, dn \exp\left(-\frac{e\psi}{kT}\right)$$

$$= eD_n \left\{ n(x) \, d\left[\exp\left(-\frac{e\psi}{kT}\right)\right] + \exp\left(-\frac{e\psi}{kT}\right) dn \right\} \qquad (14.18)$$

$$= eD_n d\left[n(x) \exp\left(-\frac{e\psi}{kT}\right) \right]$$

---

† This implies that n_0 is established by exchange of electrons between semiconductor and metal at the contact. How well this assumption applies should be investigated.

Integrating with respect to x between the limits 0 and d yields, with substitution of (14.17),

$$J \int_0^d \exp\left(-\frac{e\psi}{kT}\right) dx = eD_n\left\{-n_0 + N_d \exp\left[-\frac{e(V_{\text{dif}} - V)}{kT}\right]\right\}$$

$$= eD_nN_d \exp\left(-\frac{eV_{\text{dif}}}{kT}\right)\left[\exp\left(\frac{eV}{kT}\right) - 1\right]$$

(14.18a)

Introducing the diffusion velocity

$$v_d = \frac{D_n}{\displaystyle\int_0^d \exp\left(-\frac{e\psi}{kT}\right) dx}$$

(14.19)

yields

$$J = ev_dN_d \exp\left(-\frac{eV_{\text{dif}}}{kT}\right)\left[\exp\left(\frac{eV}{kT}\right) - 1\right]$$

(14.18b)

We must now find what v_d means. Substituting (14.6) and carrying out the integration on the left side yields†

$$\int_0^d \exp\left(-\frac{e\psi}{kT}\right) dx = \frac{\epsilon\epsilon_0 kT}{N_d e^2 d}$$

(14.19a)

so that

$$v_d = \frac{D_nN_d e^2 d}{\epsilon\epsilon_0 kT} = \mu_n|F_0|$$

(14.19b)

Hence v_d is the electron drift velocity at the contact.

To take the image effect into account, we modify Eq. (14.15). Here $-eF(x)$ is the force on the electron without image effect. But if $\varphi(x)$ is the potential energy of the electron including image effect, the *actual* force on the electron is $-d\varphi/dx$. Hence Eq. (14.15) must be written

$$J = \mu_n n(x)\frac{d\varphi(x)}{dx} + eD_n\frac{dn}{dx}$$

(14.20)

---

† Since

$$\psi(x) = -\frac{eN_d}{2\epsilon\epsilon_0}(x^2 - 2xd) = \frac{eN_d}{2\epsilon\epsilon_0}d^2 - \frac{eN_d}{2\epsilon\epsilon_0}(d - x)^2$$

the transformation

$$z^2 = \frac{e}{kT} \cdot \frac{eN_d}{2\epsilon\epsilon_0}d^2 = \frac{V_{\text{dif}} - V}{kT/e}, \qquad s^2 = \frac{e}{kT} \cdot \frac{eN_d}{2\epsilon\epsilon_0}(d - x)^2$$

transforms the integral into

$$\left(\frac{kT}{e} \cdot \frac{2\epsilon\epsilon_0}{eN_d}\right)^{1/2} I, \qquad \text{where } I = \exp(-z^2) \int_0^z \exp(s^2)\, ds$$

According to Bromwich (T.J.I'A Bromwich, *An Introduction to the Theory of Infinite Series*, Macmillan and Co., London, 1908) the integral has the asymptotic value $I_\infty = 1/(2z)$ for large z. Hence the integral (14.19a) has the asymptotic value $\epsilon\epsilon_0 kT/(N_d e^2 d)$.

This may be rewritten as

$$J \exp\left(\frac{\varphi}{kT}\right) dx = eD_n d\left[n(x) \exp\left(\frac{\varphi}{kT}\right) \right] \quad (14.20a)$$

or, by integrating between the potential energy maximum at x_m and the bulk semiconductor at $x = d$,

$$J \int_{x_m}^{d} \exp\left(\frac{\varphi}{kT}\right) dx = eD_n\left[N_d \exp\left(\frac{\varphi_s}{kT}\right) - n(x_m) \exp\left(\frac{\varphi_m}{kT}\right) \right] \quad (14.20b)$$

where φ_s is the potential energy of the electron in the bulk semiconductor and $n(x_m)$ the electron density at $x = x_m$. But in the potential energy maximum the current can also be considered as thermionic emission. That is, if $J \gg J_0$, the saturation current at the contact may be written

$$J = \int_{0}^{\infty} ev_x \, dn = en(x_m)v_t, \qquad v_t = \left(\frac{kT}{2\pi m^*}\right)^{1/2} \quad (14.21)$$

If we choose the reference point for φ at $x = x_m$, then $\varphi_m = 0$ and $\varphi_s = -e(V'_{dif} - V)$, corresponding to the negative of the barrier height seen by the electrons in the bulk semiconductor. Therefore, introducing

$$v'_d = \frac{D_n}{\displaystyle\int_{x_m}^{d} \exp\left(\frac{\varphi}{kT}\right) dx} \quad (14.20c)$$

we have

$$J = ev'_d N_d \exp\left[-\frac{e(V'_{dif} - V)}{kT} \right] - ev'_d n(x_m) \quad (14.22)$$

Equating (14.21) and (14.22), solving for $n(x_m)$, and resubstituting into (14.21) yields, for $J \gg J_0$,

$$J = e\frac{v'_d v_t}{v'_d + v_t} N_d \exp\left(-\frac{eV'_{dif}}{kT}\right) \exp\left(\frac{eV}{kT}\right) \quad (14.23)$$

so that v_d must be replaced by $v'_d v_t/(v'_d + v_t)$ and V_{dif} by V'_{dif}. This comes relatively close to the expression (14.14a) for the thermionic emission approximation.

14.1d. Equivalent Circuit of the Metal–Semiconductor Contact

Having calculated the dc response of the device, we now turn to the ac response by observing that the contact itself can be represented by a resistance R and a capacitance C connected in parallel. At not too high frequencies,

$$\frac{1}{R} = A\frac{dJ_0}{dV}, \qquad C = \frac{\epsilon\epsilon_0 A}{d} \quad (14.24)$$

In addition, one must introduce the series resistance r of the n region. Finally, one must introduce the parallel capacity C' caused by the mounting of the device. We thus obtain the equivalent circuit of Fig. 14.4.

At much higher frequencies, one must solve the continuity equation, the current equation, and Gauss's equation simultaneously and exactly. This has not yet been done. We can give the following arguments, however.

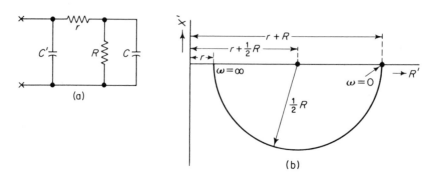

Fig. 14.4. (a) Equivalent circuit of a metal-semiconductor contact. (b) The impedance of the contact is written as $Z = R'(\omega) + jX'(\omega)$, and X' is plotted against R' with ω as a parameter. The curve should be a semicircle of radius $\frac{1}{2}R$, having its center at $R' = (\frac{1}{2}R + r)$, $X' = 0$.

1. When there are no collisions in the space-charge region, the time constant involved in the alternating current would be the dc transit time across the barrier. This time is so short that the resulting corrections of Eq. (14.24) can be neglected for all practical purposes.

2. When there are collisions, the current flow is by diffusion. One might then expect the time constant involved in the alternating current to be associated with the diffusion time through the space-charge region:

$$\tau_d = \frac{d^2}{2D_n} \qquad (14.24a)$$

In n-type silicon, $D_n \simeq 35$ cm²/s, and for $V_{\text{dif}} - V = 0.25$ V, $N_d = 10^{22}/\text{m}^3$, $\epsilon = 12$, and $\epsilon_0 = 8.85 \times 10^{-12}$ F/m, we have $d = 1830$ A; hence $\tau_d = 0.5 \times 10^{-11}$ s. This time is again so short that we may assume representation (14.24) to be satisfied for all practical frequencies.

3. The combination of the series resistance r and the RC parallel circuit yields a time constant

$$\tau = C(Rr)^{1/2} \qquad (14.24b)$$

Substituting $C = 0.5$ pF, $R = 1000\ \Omega$, and $r = 10\ \Omega$ yields $\tau = 5 \times 10^{-11}$ s. This is the longest time constant encountered so far and probably the most significant one.

We next prove Eq. (14.24b) with the help of Fig. 14.4. One would expect r, R, and C to be almost frequency independent. Writing the impedance Z of the diode as

$$Z = R' + jX' \tag{14.25}$$

we have

$$R' = \frac{R}{1 + \omega^2 C^2 R^2} + r, \qquad X' = -\frac{\omega C R^2}{1 + \omega^2 C^2 R^2} \tag{14.25a}$$

Plotting R' and X' in the complex plane with ω as a parameter, we should obtain a semicircle with origin $R' = \frac{1}{2}R + r$, $X' = 0$, and radius $\frac{1}{2}R$ (Fig. 14.4b). These plots were observed in earlier commercial rectifiers in the kilohertz range.

If an alternating current $I \cos \omega t$ is passed through the device, the power delivered to the device is $P = \frac{1}{2}I^2 R'$ and the power absorbed by the contact itself is $P_d = \frac{1}{2}I^2 R/(1 + \omega^2 C^2 R^2)$. The ratio of the two may be written

$$\frac{P_d}{P} = \frac{R}{R + r + \omega^2 C^2 R^2 r} \tag{14.26}$$

We thus see that P_d/P decreases with increasing frequency. At the frequency ω_0, where

$$\omega_0^2 C^2 R r = 1 \tag{14.27}$$

half the power fed into the device ends up in the series resistance r rather than in the contact proper, because of the shunting effect of the capacitance C. One may thus designate $f_0 = \omega_0/2\pi$ as the *cutoff frequency* of the device, and the significant time constant τ of the device as $C\sqrt{Rr}$ as stated earlier. To make τ small, one should make both C and r small.

14.1e. Practical Devices

Schottky-barrier diodes are metal–semiconductor diodes in which the contact is evaporated onto the semiconductor. The best way to make n-type Schottky-barrier diodes is to start from n^+ material, grow a thin layer of n-type material on top of it by an *epitaxial* technique, evaporate the rectifying contact, and make an ohmic contact to the n^+ region. Because the n region can be quite thin and the n^+ region has a low resistivity, the series resistance of the device can be quite small, thus ensuring good microwave operation.

Silicon point-contact diodes operate on the same principles, but here the contact is made by pressing a fine point against the semiconductor. Some types of germanium diodes closely resemble junction diodes, and they are not suitable at microwave frequencies (Sec. 15.2f).

Earlier commercial metal–semiconductor diodes were the selenium rectifier, the copper–cuprous oxide rectifier, and the magnesium–cupric oxide rectifier. They have now been replaced by silicon *pn* junctions.

14.1f. Noise in Metal–Semiconductor Diodes

Since the current flow consists of the independent random passage of carriers across a barrier, the devices should show full shot noise. If $|eV/kT|$ $\gg 1$, so that $I \gg I_0$, we have

$$\bar{i^2} = 2eI \, \Delta f = 2kTg \, \Delta f \qquad (14.28)$$

where $g = dI/dV = eI/kT$. The *equivalent noise temperature* of the diode is thus half room temperature.

In addition, the series resistance of the device shows thermal noise. In point-contact diodes the equivalent noise temperature is somewhat larger than room temperature, because the current flow heats the series resistance of the point contact, which is concentrated in the contact area. Schottky-barrier diodes hardly show this effect, and are therefore preferred.

Finally, Schottky-barrier diodes show much less flicker noise than point-contact diodes. Flicker noise has a $1/f^\alpha$ spectrum with α close to unity, so it is most pronounced at low frequencies.

14.2. BREAKDOWN PHENOMENA

At large negative bias, breakdown occurs for two main reasons:

1. Tunnel emission through the barrier (Fig. 14.5).
2. Avalanche breakdown in the space-charge region.

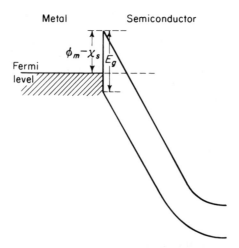

Fig. 14.5. Energy level diagram needed to explain tunnel emission.

14.2a. Tunnel Emission

i.e. (7.33) p.164

Tunnel emission was discussed in Chapter 7, and it was shown that the current was of the form

$$J = -aF_0^2 \exp\left(-\frac{b}{F_0}\right) \leftarrow \text{field strength} \quad (14.29)$$

where a and b are constants. But in the region near the contact

$$|F| = |F_0| = \left[\frac{2eN_d}{\epsilon\epsilon_0}(V_{\text{dif}} - V)\right]^{1/2} \quad (14.29a)$$

Consequently, we may write

$$J = -A(V_{\text{dif}} - V)\exp\left[-\frac{B}{(V_{\text{dif}} - V)^{1/2}}\right] \quad (14.30)$$

This can be one of the chief breakdown mechanisms in back-biased metal–semiconductor diodes. It also plays a role in electroluminescence (Chapter 12).

14.2b. Avalanche Breakdown

At very large back biases it may happen that the electrons emitted over the barrier from the metal side gain more energy during collisions than they lose in collisions. As a consequence, the electron energy will increase steadily until it is larger than the gap width E_g. Hole–electron pairs can then be generated, and these in turn can produce new hole–electron pairs, so that a single electron produces a whole avalanche of hole–electron pairs.

Let $-I_{s0}$ be the back current expected at the given back bias without multiplication, let an electron coming from the metal produce p_{1e} hole–electron pairs, let each hole–electron pair thus created produce p_2 new hole–electron pairs, and let each new hole–electron pair thus created produce p_2 new hole–electron pairs, and so on ad infinitum. The total net current is then

$$I = -I_{s0}[1 + p_{1e}(1 + p_2 + p_2^2 + p_2^3 + \ldots)]$$
$$= -I_{s0}\left(1 + \frac{p_{1e}}{1 - p_2}\right) = -I_{s0}\frac{1 + p_{1e} - p_2}{1 - p_2} = -I_{s0}M \quad (14.31)$$

where

$$M = \frac{1 + p_{1e} - p_2}{1 - p_2} \quad (14.31a)$$

is the multiplication factor of the original current $-I_{s0}$.

The parameters p_{1e} and p_2 increase with increasing field strength $|F_0|$; hence M will increase from the value of unity for small back bias to a much larger value. At the field strength F_{crit}, which is a constant for a given material, the factor p_2 reaches unity, M becomes infinite, and breakdown occurs in which the current must be limited by the external circuit. This is called *avalanche breakdown*.

For a given metal–semiconductor diode breakdown occurs at a definite voltage V_B. Empirically it is found that M can be represented as

$$M = \frac{1}{1 - [(V_{\mathrm{dif}} - V)/V_B]^n} \tag{14.32}$$

where n is a relatively large number, say 5 or 6, that depends on the type of material under investigation.

According to Eq. (14.8), F_{crit} and V_B should be related as

$$F_{\mathrm{crit}} = \left(\frac{2eN_d V_B}{\epsilon\epsilon_0}\right)^{1/2} \tag{14.33}$$

The breakdown voltage V_B should thus be inversely proportional to the donor concentration N_d. Actually, however, one finds a slower dependence of V_B on N_d.

The reason lies in the field distribution in the space-charge region. This distribution is shown in Fig. 14.6; it varies linearly from a maximum value $|F_0|$ at $x = 0$ to zero value at $x = d$. According to (14.8), $|F_0|$ varies as $[(V_{\mathrm{dif}} - V)N_d]^{1/2}$, and the width d of the space-charge region varies as $[(V_{\mathrm{dif}} - V)/N_d]^{1/2}$. With increasing value of N_d, therefore, and a given value of $V_{\mathrm{dif}} - V$, $|F_0|$ increases and d decreases. Now an electron has to travel a certain distance, equal to several free path lengths, before it has gained enough energy to generate a hole–electron pair. If N_d increases, then *for a given value of* $|F_0|$ the thickness d is inversely proportional to N_d, and hence an electron may drift out of the high-field region before it has gained enough energy to create a hole–electron pair. Consequently, at a given value of $|F_0|$, p_{1e} and p_2 will decrease with increasing N_d. But this means, since p_2 must be unity at breakdown, that the value of $|F_0|$ at breakdown must increase with increasing donor concentration N_d. Consequently, V_B will not decrease as fast as N_d increases.

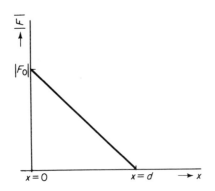

Fig. 14.6. Field distribution in the space-charge region of the metal-semiconductor contact.

In most materials, both the electrons and the holes have ionizing power. If this were not the case, avalanche multiplication could never result in avalanche breakdown. Experiments indicate that in germanium the holes have higher ionizing power than electrons, whereas the opposite is true for silicon.

The critical field strength F_{crit}, observed in the case of avalanche break-down in a uniform field, is a constant of the material. For germanium it is about 2×10^7 V/m; for silicon it is somewhat larger.

14.3. APPLICATIONS OF METAL–SEMICONDUCTOR DIODES

We discuss here the use of metal–semiconductor diodes as low-level quad-ratic detectors and as mixers.

14.3a. Low-Level Quadratic Detector

Metal–semiconductor diodes of the Schottky-barrier or the point-contact type are used as microwave low-level detectors. One condition is that the cutoff frequency ω_0, defined by

$$\omega_0^2 C^2 Rr = 1 \tag{14.34}$$

be sufficiently large. Here C is the device capacitance, r its series resistance, and R the differential resistance of the diode.

We now show that the detector responds to the *square* of the voltage amplitude. The current is

$$I = I_0 \left[\exp\left(\frac{eV}{kT}\right) - 1 \right] = I_0 \left[\frac{eV}{kT} + \frac{1}{2}\left(\frac{eV}{kT}\right)^2 + \ldots \right] \tag{14.35}$$

The input conductance at $V = 0$ is thus $G_0 = dI/dV = eI_0/kT$. If the power P_a is available at a signal source and this source is matched to the diode, then the amplitude V_0 of the signal applied to the device terminals follows from

$$P_a = \frac{1}{2} V_0^2 G_0 \quad \text{or} \quad V_0^2 = \frac{2kTP_a}{eI_0} \tag{14.36}$$

Substituting $V = V_0 \cos \omega t$ into (14.35), the rectified current I_d is

$$I_d = \frac{1}{4} I_0 \left(\frac{eV_0}{kT}\right)^2 \quad \text{or} \quad \frac{I_d}{P_a} = \frac{1}{2}\frac{e}{kT} \tag{14.37}$$

for relatively small V_0. Note that I_d/P_a is independent of diode parameters.

14.3b. Diode Mixers

The first stage in a microwave receiver is usually a mixer stage. These mixer stages use point-contact or Schottky-barrier diodes as mixing ele-ments because of their favorable frequency characteristics,† small power loss, and favorable noise characteristics.

To understand the operation of the diode mixer, we assume that a local

---

† The favorable frequency characteristics are due to the small transit time through the barrier region, the small input capacitance, and so on.

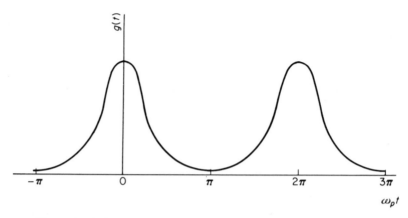

Fig. 14.7. Diode conductance $g(t)$ in a mixer circuit as a function of $\omega_p t$, where ω_p is the local oscillator frequency.

oscillator voltage $V_p \cos \omega_p t$ is applied to the diode, and plot the diode conductance $g(t)$ as a function of $\omega_p t$ (Fig. 14.7). Since $g(t)$ is periodic in $\omega_p t$ with period 2π, we may make a Fourier analysis of $g(t)$:

$$g(t) = a_0 + a_1 \cos \omega_p t + a_2 \cos 2\omega_p t + \dots \qquad (14.38)$$

which yields

$$a_0 = \frac{1}{2\pi} \int_{-\pi}^{\pi} g(t)\, d(\omega_p t) = g_0 \qquad (14.38a)$$

$$a_n = \frac{2}{2\pi} \int_{-\pi}^{\pi} g(t) \cos n\omega_p t\, d(\omega_p t) = 2g_{cn} \qquad (14.38b)$$

so that (14.38) may be written, if $g_0 = a_0$ and $g_{cn} = \frac{1}{2} a_n$,

$$g(t) = g_0 + 2g_{c1} \cos \omega_p t + 2g_{c2} \cos 2\omega_p t + \dots \qquad (14.39)$$

If a small input signal $V_{i0} \cos \omega_i t$ is applied, the diode current becomes

$$i(t) = g(t) V_{i0} \cos \omega_i t = g_0 V_{i0} \cos \omega_i t + 2g_{c1} V_{i0} \cos \omega_i t \cos \omega_p t + \dots$$

or

$$i(t) = g_0 V_{i0} \cos \omega_i t + g_{c1} V_{i0} \cos (\omega_i - \omega_p)t + \text{other terms without interest}$$
$$(14.40)$$

Let now the output be tuned to the frequency $\omega_0 = \omega_i - \omega_p$ (assuming $\omega_i > \omega_p$, of course). We then see that the mixer diode has an *input conductance* g_0 and that the *conversion conductance* (or transfer conductance) from input to output is g_{c1}.

If a signal $V_{o0} \cos \omega_o t$ is applied to the output, we find in the same way that the mixer diode has an *output conductance* g_0, too, and that the *conversion conductance* (or transfer conductance) from output to input is again g_{c1} (Fig. 14.8a).

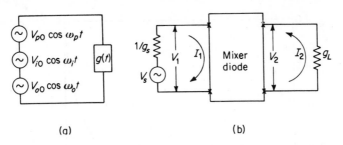

(a)   (b)

Fig. 14.8. (a) The local oscillator voltage, the input voltage, and the output voltage are applied to a mixer diode. (b) Mixer diode considered as a four-terminal network.

We now consider the diode mixer as a four-terminal network having an input signal of frequency ω_i and an output signal of frequency ω_o. Since in the standard complex notation the symbols for voltages and currents refer only to the amplitude and phase of the signals, we can write the equations for the mixer circuit in complex notation as (Fig. 14.8b)

$$I_1 = g_0 V_1 - g_{c1} V_2$$
$$I_2 = -g_{c1} V_1 + g_0 V_2 \tag{14.41}$$

Connecting the mixer to an input source of emf V_s and internal resistance $1/g_s$ and the output load to a load conductance g_L, we have

$$I_1 = (V_s - V_1)g_s, \qquad I_2 = -V_2 g_L \tag{14.42}$$

Substituting into (14.41) and solving for V_2, we obtain

$$V_2 = \frac{V_s g_s g_{c1}}{(g_0 + g_s)(g_0 + g_L) - g_{c1}^2} \tag{14.43}$$

The power fed into the load is $\frac{1}{2}|V_2|^2 g_L$, and the power available at the source is $\frac{1}{8}|V_s|^2 g_s$. The power gain G is now defined as the ratio

$$G = \frac{\text{power fed into load}}{\text{power available at source}} = 4\left|\frac{V_2}{V_s}\right|^2 \frac{g_L}{g_s}$$
$$= \frac{4 g_s g_L g_{c1}^2}{[(g_0 + g_s)(g_0 + g_L) - g_{c1}^2]^2} \tag{14.44}$$

This expression has a maximum value G_{max} if g_s and g_L are chosen such that

$$g_s = g_L = (g_0^2 - g_{c1}^2)^{1/2} \tag{14.45}$$

in which case (Fig. 14.9)

$$G_{max} = \frac{g_{c1}^2}{[g_0 + (g_0^2 - g_{c1}^2)^{1/2}]^2} \tag{14.46}$$

The mixer stage thus gives some power loss, but it converts the high-frequency signal of frequency ω_i into an intermediate-frequency (IF) signal

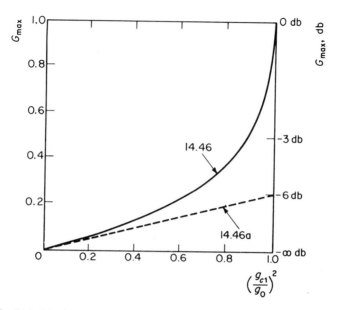

Fig. 14.9. Maximum power gain $G_{\max}$ of a mixer diode as a function of $(g_{c1}/g_0)^2$, where g_{c1} is the conversion conductance and g_0 the input (or output) conductance of the mixer diode.

that can be more easily amplified. A good mixer gives a power loss of less than 6 dB.

Another interesting feature of the diode mixer is its relatively low noise figure. According to van der Ziel,[†] the minimum noise figure of the mixer is

$$F_{\min} = 1 + \frac{n(1 - G_{\max})}{G_{\max}} \quad \text{for} \quad g_s = (g_0^2 - g_{c1}^2)^{1/2} \qquad (14.47)$$

where n is the noise ratio of the diode (= total noise over thermal noise). For a Schottky-barrier diode $n = \frac{1}{2}$, according to Eq. (14.28).

The mixer is followed by an IF amplifier. If F_2 is the noise figure of the IF amplifier, then the noise figure of the combination is

$$F = F_{\min} + \frac{F_2 - 1}{G_{\max}} = 1 + \frac{n(1 - G_{\max}) + F_2 - 1}{G_{\max}} \qquad (14.48)$$

This can be understood as follows. The noise of the IF amplifier itself can be represented by a current generator $[(F_2 - 1) \cdot 4kTg_{\text{out}} \, \Delta f]^{1/2}$ in parallel with the output conductance g_{out} of the mixer. This corresponds to a current generator $[4kTg_s \, \Delta f \cdot (F_2 - 1)/G_{\max}]^{1/2}$ in parallel with the source conductance g_s, so that its contribution to F is $(F_2 - 1)/G_{\max}$.

[†] A. van der Ziel, *Noise: Sources, Characterization, Measurement.* Prentice-Hall, Inc., Englewood Cliffs, N.J., 1970.

In Doppler radar receivers one usually starts with a mixer and uses zero IF frequency. One then wants to avoid that the beat frequencies drown in the noise background of the receiver. To that end one must require the mixer diode to have very low flicker noise. Again, it seems that well-constructed Schottky-barrier diodes meet this requirement best.

14.3c. Nonlinear Capacitance Mixing, Parametric Amplifiers, Parametrons

In nonlinear capacitance mixing one starts with a device having a non-linear charge characteristic $Q = f(V)$ and applies a dc signal V_0, a pump signal ΔV_p, and a *small* input signal ΔV_i. One thus has

$$Q = f(V_0 + \Delta V_p + \Delta V_i) \tag{14.49}$$

One can now make a Taylor expansion of Q with respect to ΔV_i and write

$$Q = f(V_0 + \Delta V_p) + \frac{dQ}{dV}\bigg|_{\Delta V_i = 0} \Delta V_i = Q(V_0 + \Delta V_p) + C(V_0 + \Delta V_p)\, \Delta V_i \tag{14.50}$$

since higher-order terms in ΔV_i can be neglected. The mixing is thus accomplished by the nonlinear, small-signal capacitance

$$C(t) = C(V_0 + \Delta V_p) = \frac{dQ}{dV}\bigg|_{V = V_0 + \Delta V_p} \tag{14.51}$$

If $\Delta V_p = v_{p0} \cos \omega_p t$, $C(t)$ can be written as a Fourier series

$$C(t) = C_0 + 2C_1 \cos \omega_p t + 2C_2 \cos 2\omega_p t + \dots \tag{14.52}$$

If $\Delta V_i = v_{i0} \cos (\omega_p t + \varphi_i)$, and the output is short-circuited, the significant signals are contained in the charge

$$\Delta Q = C_0 v_{i0} \cos (\omega_i t + \varphi_i) + \sum_{n=1}^{\infty} C_n v_{i0} \{\cos [(n\omega_p + \omega_i)t + \varphi_i]$$
$$+ \cos [n(\omega_p - \omega_i)t - \varphi_i]\} \tag{14.53}$$

and the significant currents are obtained by differentiation with respect to time:

$$i(t) = -\omega_i C_0 v_{i0} \sin (\omega_i t + \varphi_i)$$
$$-\sum_{n=1}^{\infty} \{(n\omega_p + \omega_i)C_n v_{i0} \sin [(n\omega_p + \omega_i)t + \varphi_i] \tag{14.54}$$
$$+ (n\omega_p - \omega_i)C_n v_{i0} \sin [(n\omega_p - \omega_i)t - \varphi_i]\}$$

The nonlinear device thus represents a capacitance C_0 to the input, and the signal transfer from the frequency ω_i to the frequencies $n\omega_p + \omega_i$ and $|n\omega_p - \omega_i|$ is represented by the capacitance C_n.

There are now three possibilities:

1. $n\omega_p + \omega_i = \omega_o$, or $n\omega_p = \omega_o - \omega_i$. In that case the significant output current is $-\omega_o C_n \sin(\omega_o t + \varphi_i)$, so that there is no phase reversal involved in the mixing process.

2. $n\omega_p - \omega_i = -\omega_o$, or $n\omega_p = \omega_i - \omega_o$. In that case the significant output current is $-\omega_o C_n \sin(\omega_o t + \varphi_i)$, so that there is again no phase reversal in the mixing process.

3. $n\omega_p - \omega_i = \omega_o$ or $n\omega_p = \omega_i + \omega_o$. In that case the significant output current is $-\omega_o C_n \sin(\omega_o t - \varphi_i)$, so that there is a phase reversal in the mixing process.

We now short-circuit the input and apply a small signal $V_0 = v_{00} \cos(\omega_o t + \varphi_o)$ to the output. One then obtains an output current flowing *into* the device

$$-\omega_o C_0 v_{00} \sin(\omega_o t + \varphi_o) \tag{14.55}$$

and a significant input current

$$-\omega_i C_n v_{00} \sin(\omega_i t \pm \varphi_o) \tag{14.56}$$

flowing *out* of the device. Here the plus sign holds if $|\omega_o - \omega_i| = n\omega_p$, whereas the minus sign holds if $\omega_o + \omega_i = n\omega_p$. There is thus a phase reversal in the input current in the latter case, just as before.

Changing over to complex notation, we have

$$\left. \begin{aligned} i_i &= j\omega_i C_0 v_i - j\omega_i C_n v_o \\ i_o &= -j\omega_o C_n v_i + j\omega_o C_0 v_o \end{aligned} \right\} \quad \text{if } |\omega_o - \omega_i| = n\omega_p \tag{14.57}$$

and, if the asterisk denotes the complex conjugate,

$$\left. \begin{aligned} i_i &= j\omega_i C_0 v_i - j\omega_i C_n v_o^* \\ i_o &= -j\omega_o C_n v_i^* + j\omega_o C_0 v_o \end{aligned} \right\} \quad \text{if } \omega_o + \omega_i = n\omega_p \tag{14.57a}$$

These equations take into account the possibility of a phase reversal in the mixing process. Complex notation is again possible, because a complex signal refers only to the amplitude and the phase of the signal. Moreover, in each line of Eqs. (14.57) and (14.57a) all complex signals refer to signals of the same frequency.

Applying this to practical mixers, we consider the circuit of Fig. 14.10. The input circuit plus the capacitance C_0 is tuned at the frequency ω_i, whereas the output circuit plus the capacitance C_0 is tuned at the frequency ω_o. We first consider the case $|\omega_o - \omega_i| = n\omega_p$ and substitute into (14.57):

$$\begin{aligned} i_i &= i_s - v_i\left(g_s + \frac{1}{j\omega_i L_i} + j\omega_i C_i\right) \\ i_o &= -v_o\left(g_L + \frac{1}{j\omega_o L_2} + j\omega_o C_2\right) \end{aligned} \tag{14.58}$$

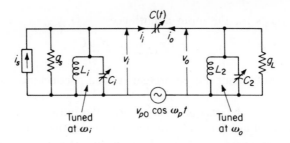

Fig. 14.10. Capacitance mixer with tuned input and tuned output (from A. van der Ziel, *Electronics*, Allyn and Bacon, Inc., Boston, 1966).

This yields

$$i_s = v_i\left(g_s + \frac{1}{j\omega_i L_i} + j\omega_i C_i + j\omega_i C_0\right) - j\omega_i C_1 v_o$$

$$= g_s v_i - j\omega_i C_1 v_o$$

$$0 = -j\omega_o C_1 v_i + v_o\left(g_L + \frac{1}{j\omega_o L_2} + j\omega_o C_2 + j\omega_o C_0\right) \tag{14.59}$$

$$= -j\omega_o C_1 v_i + g_L v_o$$

because of the tuning conditions $\omega_o^2 L_2(C_2 + C_0) = \omega_i^2 L_i(C_i + C_0) = 1$. Solving for v_o yields

$$v_o = \frac{j\omega_o C_1}{g_s g_L + \omega_o \omega_i C_1^2} i_s \tag{14.60}$$

The power gain is therefore

$$G = 4 g_s g_L \left|\frac{v_o}{i_s}\right|^2 = \frac{4 g_s g_L \omega_o^2 C_1^2}{(g_s g_L + \omega_o \omega_i C_1^2)^2} \tag{14.61}$$

which has the optimum value

$$G_{\max} = \frac{\omega_o}{\omega_i}, \qquad \text{if } g_s g_L = \omega_o \omega_i C_1^2 \tag{14.62}$$

The circuit thus has a considerable power gain if $\omega_o \gg \omega_i$ (up converter), whereas there is a considerable power loss if $\omega_o \ll \omega_i$ (down converter). The down converter is thus not very useful.

It will now be shown that the output circuit presents a load conductance $\omega_o \omega_i C_1^2/g_L$ to the input circuit. To that end we substitute

$$i_o = -v_o\left(g_L + \frac{1}{j\omega_o L_2} + j\omega_o C_2\right)$$

into Eq. (14.57) and apply the tuning condition $\omega_o^2 L_2(C_2 + C_0) = 1$. This yields the equations

$$i_i = j\omega_i C_0 v_i - j\omega_i C_1 v_o$$

$$0 = -j\omega_o C_1 v_i + g_L v_o \tag{14.63}$$

Solving i_i/v_i yields for the input admittance

$$Y_i = \frac{i_i}{v_i} = j\omega_i C_0 + \frac{\omega_o \omega_i C_1^2}{g_L} \tag{14.64}$$

so that the output circuit indeed presents a load conductance $\omega_o \omega_i C_1^2/g_L$ to the input.

We next turn to the case $\omega_o + \omega_i = n\omega_p$. We must now start from Eq. (14.57a). Substituting Eq. (14.58) and applying the tuning conditions $\omega_o^2 L_2(C_2 + C_0) = \omega_i^2 L_i(C_i + C_0) = 1$ yields the equations

$$\begin{aligned} i_s &= g_s v_i - j\omega_i C_1 v_o^* \\ 0 &= -j\omega_o C_1 v_i^* + g_L v_o \end{aligned} \tag{14.65}$$

Consequently,

$$v_o = \frac{j\omega_o C_1 v_i^*}{g_L}, \qquad v_o^* = -\frac{j\omega_o C_1 v_i}{g_L}$$

and hence

$$i_s = v_i \left(g_s - \frac{\omega_o \omega_i C_1^2}{g_L} \right) \tag{14.66}$$

The output circuit thus presents a *negative* load conductance $-\omega_o \omega_i C_1^2/g_L$ to the input circuit. This is a consequence of the phase-reversal effect.

Solving for v_o yields

$$v_o = \frac{j\omega_o C_1}{g_s g_L - \omega_o \omega_i C_1^2} i^* \tag{14.67}$$

and hence the power gain is

$$G = 4g_s g_L \left| \frac{v_o}{i_s} \right|^2 = \frac{4g_s g_L \omega_o^2 C_1^2}{(g_s g_L - \omega_o \omega_i C_1^2)^2} \tag{14.68}$$

which becomes infinite if $g_s g_L = \omega_o \omega_i C_1^2$. We see that this condition corresponds to zero admittance of the total input circuit. If $g_s g_L < \omega_o \omega_i C_1^2$, oscillations will occur; that is, the frequencies ω_i and ω_o will be generated simultaneously. Circuits operating in that manner are called *parametric oscillators* or *parametrons*.

The fact that the mixer circuit of Fig. 14.10 presents a negative conductance to the input circuit is used in the *parametric amplifier*. It consists of a mixer circuit with an idler circuit tuned at the frequency $\omega_2 = \omega_p - \omega_i$, thus presenting a negative conductance $-\omega_2 \omega_i C_1^2/g_2$ to the input circuit. The inductance L_i is used to tune the input capacitance C_0 of the mixer circuit ($\omega_i^2 L_i C_0 = 1$). The negative conductance $-\omega_2 \omega_i C_1^2/g_2$ makes it possible that the power delivered to the load conductance g_L is larger than the available power $P_{av} = \frac{1}{8}|i_s|^2/g_s$ of the source (Fig. 14.11).

The power fed into the load is $P_{out} = \frac{1}{2}|v_i|^2 g_L$, and hence the power gain is

$$G = \frac{P_{\text{out}}}{P_{\text{av}}} = 4g_s g_L \left| \frac{v_i}{i_s} \right|^2 = \frac{4g_s g_L}{(g_s - \omega_2 \omega_i C_1^2 / g_2 + g_L)^2} \qquad (14.69)$$

If $g_s > \omega_2 \omega_i C_1^2 / g_2$, the maximum gain is obtained if $g_L = g_s - \omega_2 \omega_i C_1^2 / g_2$. This gain is the *available gain* G_{av} of the circuit:

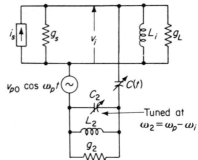

$$G_{\text{av}} = \frac{g_s}{g_s - \omega_2 \omega_i C_1^2 / g_2} \qquad (14.69a)$$

The gain goes to infinity if g_s approaches $\omega_2 \omega_i C_1^2 / g_2$.

If $g_s < \omega_2 \omega_i C_1^2 / g_2$, an infinite power gain is obtained if $g_s + g_L = \omega_2 \omega_i C_1^2 / g_2$, whereas oscillations occur if $g_s + g_L < \omega_2 \omega_i C_1^2 / g_2$. This condition should therefore be avoided.

We now evaluate the noise figure of the mixer. Since an ideal capacitor has no losses, and hence no noise, the

Fig. 14.11. Parametric amplifier with tuned input and tuned idler circuit (from A. van der Ziel, *Electronics*, Allyn and Bacon, Inc., Boston, 1966).

only noise in the ideal circuit is due to the source conductance g_s and the load conductance g_L of the mixer. But the noise of g_L is counted as belonging to the next stage; hence the noise figure of the ideal mixer is unity. In the actual mixer there are some circuit losses; hence its noise figure is somewhat larger than unity.

We finally turn to the parametric amplifier. The equivalent circuit of the amplifier is shown in Fig. 14.12. Here g_L is the load conductance, ω_i and ω_2 are the input and the idler frequencies, g_i represents the losses of the input circuit, L_i and L_2 are so chosen that the input and idler circuits are tuned, and g_2 is the conductance of the tuned idler circuit. As is common practice, we shall count the noise of g_L as belonging to the next stage.

We short-circuit the input; the noise current i in the short-circuiting lead is then due to three sources: (1) the noise of g_s, (2) the noise of g_i, and (3) the converted noise of g_2. Its mean-square value is

$$\overline{i^2} = 4kTg_s \, \Delta f + 4kTg_i \, \Delta f + \left(\frac{4kT \, \Delta f}{g_2} \right) \omega_i^2 C_1^2 \qquad (14.70)$$

Dividing $\overline{i_2}$ by its first term gives the noise figure F:

$$F = 1 + \frac{g_i}{g_s} + \frac{\omega_i^2 C_1^2}{g_s g_2} = 1 + \frac{g_i}{g_s} + \frac{g_{c2}}{g_s} \frac{\omega_i}{\omega_2} \qquad (14.71)$$

where $-g_{c2} = -\omega_2 \omega_i C_1^2 / g_2$ is the negative conductance appearing at the input owing to the tuned idler conductance g_2. For good performance of the amplifier, g_s should be comparable with g_{c2}, as shown by our previous dis-

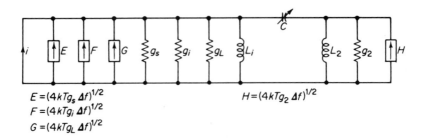

$$E = (4kTg_s\,\Delta f)^{1/2}$$
$$F = (4kTg_i\,\Delta f)^{1/2}$$
$$G = (4kTg_L\,\Delta f)^{1/2}$$

$$H = (4kTg_2\,\Delta f)^{1/2}$$

Fig. 14.12. Schematic circuit diagram for the parametric amplier.

cussion of the amplifier circuit; hence

$$F \simeq 1 + \frac{\omega_i}{\omega_2} \tag{14.71a}$$

if the input losses are small.† It is thus advantageous to use a high idler frequency ω_2. For $\omega_i \simeq \omega_2$, $F \simeq 2$ (3 dB).

14.3d. Capacitance Frequency Multipliers

Since all nonlinear elements can be used for frequency multiplication, it is not surprising that nonlinear capacitive elements can be used for that purpose. The merit of the nonlinear capacitance frequency multiplier is that the nonlinear element does not dissipate power but transforms input power into harmonic power. Ideally the transformation would be completely loss-less and all the high-frequency power fed into the device would appear at the output as harmonic power. Because of diode and circuit losses, and because power is generated at other harmonics, this ideal condition is not achieved. The harmonic power generation with nonlinear capacitive elements can be much more efficient than harmonic power generation with diodes, however.

The circuit is illustrated in Fig. 14.13. An input signal $v_i \cos \omega_i t$ is applied

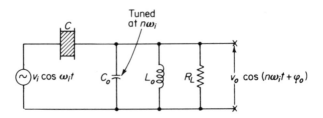

Fig. 14.13. Varactor diode multiplier circuit (from A. van der Ziel, *Electronics*, Allyn and Bacon, Inc., Boston, 1966).

---

† According to (14.71), we could make F arbitrarily close to unity by making g_s large. Since G_{av} also approaches unity in that case, this is not useful.

to a nonlinear capacitance $C(V)$. In series with the nonlinear device is a tuned circuit tuned to the harmonic frequency $n\omega_i$ and loaded by a load resistance R_L. Because of the nonlinear operation of the device, a voltage $v_o \cos (n\omega_i t + \varphi_o)$ will appear across this circuit and a harmonic power $\frac{1}{2}v_o^2/R_L$ is dissipated in the load resistance R_L. This power comes from the input signal.

The transfer of power from the input to the output comes about in a complicated mixing process already discussed. Because of this mixing process the output load resistance R_L presents an input conductance g_i to the input. If the nonlinear capacitor had no losses, the power fed into the input would be equal to the power delivered to the load:

$$\frac{1}{2}v_i^2 g_i = \frac{1}{2}\frac{v_o^2}{R_L} \qquad (14.72)$$

and the efficiency would be 100 per cent. Because of circuit losses and generation of power at other harmonics, the actual efficiency will be less. Nevertheless, appreciable efficiencies can be obtained.

The maximum amount of power that can be transformed in this manner depends upon the input signal amplitude v_i and the input conductance g_i. The first is limited by avalanche breakdown and the latter is determined by the mixing process and by R_L. By proper design of the circuit the power transfer can be optimized. The difference between the input power P_{in} and the harmonic output power P_{out} is dissipated in the diode and appears as heat. The detailed theoretical treatment of this problem is too involved to be presented in this book, and we refer the reader to the original literature on the subject.†

REFERENCES

BLACKWELL, L. A., and K. L. KOTZEBUE, *Semiconductor Diode Parametric Amplifiers*. Prentice-Hall, Inc., Englewood Cliffs, N.J., 1961.

CHANG, K. K., *Parametric and Tunnel Diodes*. Prentice-Hall, Inc., Englewood Cliffs, N.J., 1964.

HENISH, K., *Metal Rectifiers*. Oxford University Press, Inc., New York, 1949.

LEVINE, S. N., and R. R. KURZROK, *Selected Papers on Semiconductor Microwave Electronics*. Dover Publications, Inc., New York, 1964.

PENFIELD, P., JR., and P. RAFUSE, *Varactor Applications*. The M.I.T. Press, Cambridge, Mass., 1962.

---

† Some early papers have been reprinted in Levine and Kurzrok's book (see reference list).

SZE, S. M., *Physics of Semiconductor Devices*. John Wiley & Sons, Inc. (Interscience Division), New York, 1969.

VAN DER ZIEL, A., *Electronics*, Chap. 20. Allyn and Bacon, Inc., Boston, 1966.

PROBLEMS

1. The capacitance C of a contact rectifier of 1-cm^2 area is measured for back bias as a function of the applied voltage V, and $1/C^2$ is plotted against V. A linear relationship is found that can be expressed as $1/C^2 = A - BV$. Calculate the diffusion potential V_{dif} and the number of donors per cubic centimeter from this graph, if the dielectric constant of the material is known.

Answer: $V_{dif} = A/B$; $N_d = 2 \times 10^8/(e\epsilon\epsilon_0 B)$.

2. A contact rectifier of 10^{-4}-cm^2 area is operated at a back bias of 10 V. The measured capacitance at that operating point is 1.2 $\mu\mu$F. Dielectric constant $\epsilon = 9$, and $V_{dif} = 1$ V. Calculate the following:
(a) The thickness of the barrier layer.
(b) The (uniform) donor concentration N_d.

Answer: 0.66×10^{-4} cm; 2.5×10^{16}/cm^3.

3. A contact rectifier has a contact area of 10^{-4} cm^2; the donor concentration in the semiconductor is 10^{16}/cm^3 and the dielectric constant is 9. If $V_{dif} = 1$ V and the back bias voltage is 10 V, calculate the following:
(a) The thickness of the barrier layer.
(b) The barrier capacitance.
(c) The field strength at the contact.

Answer: 1.05 μm; $C = 0.76$ $\mu\mu$F; 2.1×10^5 V/cm.

4. The impedance of a low-level detector is measured as a function of frequency, and its resistance R' and series reactance X' are plotted in the complex plane. The plot is a semicircle with its origin at 200 Ω and a radius of 190 Ω. Calculate the diode resistance R and the series resistance r. If $R' = 200$ Ω at $\omega = 10^{10}$ per second, calculate the capacitance C.

Answer: $R = 380$ Ω; $r = 10$ Ω; $C = 0.26$ $\mu\mu$F.

5. A diode detector has a capacitance of 1.0 $\mu\mu$F, a series resistance of 10 Ω, and a diode resistance of 250 Ω. Calculate the cutoff frequency of this diode as a low-level detector.

Answer: 3200 MHz.

6.   The maximum power gain of a good diode mixer is $\frac{1}{4}$. Calculate the ratio g_{c1}/g_0 needed to obtain it. Find the source and load impedance needed for maximum power gain.

Answer: $g_{c1}/g_0 = 0.80; g_L = g_s = 0.60g_0$.

7.   Repeat problem 6 for a power gain of $\frac{1}{2}$.

Answer: $g_{c1}/g_0 = (2\sqrt{2})/3; g_L = g_s = 0.33g_0$.

8.   Calculate the minimum noise figure for problem 7, if $n = \frac{1}{2}$ and $F_2 = 1.40$.

Answer: $F_{min} = 2.30$.

9.   A parametric amplifier has an input frequency of 1000 MHz, an idler frequency of 10,000 MHz, $C_1/C_0 = \frac{1}{4}$, $C_0 = 0.30 \, \mu\mu F$, and the idler circuit has a Q factor of 10. If the total circuit capacitance of the idler circuit is equal to C_0, determine the negative conductance at the input.

Answer: -1.16×10^{-3} S.

10.   Calculate the minimum noise figure of the amplifier in problem 9.

Answer: $F_{min} = 1.10$.

15

PN Junction Diodes

15.1. POTENTIAL DISTRIBUTION IN THE *PN* JUNCTION DIODE

15.1a. Potential Distribution and Transition-Region Capacitance of a Junction

We calculate the potential distribution in the transition region of a junction diode by solving Poisson's equation. The result obtained also allows the determination of the capacitance of the transition region and of the field strength in that region.

We consider a *pn* junction consisting of a *p*-type and an *n*-type section of the same semiconducting material in intimate contact. Such a junction is known as a *step junction*. We use the model because it allows a simple calculation of the properties of the junction. Let the two sections of material make contact at $x = x_0$; let the material for $x < x_0$ be *p* type with N_a acceptors per unit volume, and let the material for $x > x_0$ be *n* type with N_d donors per unit volume.

Before contact, the Fermi levels of the *p* and *n* sections are at different heights; the Fermi level in the *p* region is close to the top of the filled band and the Fermi level in the *n* region is close to the bottom of the conduction band. After contact has been established, electrons from the contact area of the *n* section flow into the *p* region and recombine with free holes until a new equilibrium condition is attained in which the Fermi levels are at equal height. The energy levels in the n region are thereby lowered by an amount equal to the difference in work function for the p-type and n-type material, and as a consequence a potential difference V_{dif} (diffusion potential) will be maintained between the n- and p-type regions.

The potential difference V_{dif} comes about because of a distributed dipole layer at the contact; the contact area of the p-type section has a negative space charge due to ionized acceptors, whereas the contact area of the n-type section has a positive space charge due to ionized donors. The space-charge region around the contact at $x = x_0$ is called the transition region; it extends from $x = x_1$ to $x = x_2$ (see Fig. 15.1a).

At equilibrium the hole concentrations at $x = x_1$ and at $x = x_2$ are N_a and p_n, respectively, whereas V_{dif} is the potential difference across the junction. But p_n and N_a should be related by the Boltzmann factor exp $(-eV_{\text{dif}}/kT)$, so that (compare Sec. 15.1c)

$$p_n = N_a \exp\left(\frac{-eV_{\text{dif}}}{kT}\right) \quad \text{or} \quad V_{\text{dif}} = \frac{kT}{e} \ln \frac{N_a}{p_n}$$

$$= \frac{kT}{e} \ln \frac{N_a N_d}{n_i^2} \tag{15.1}$$

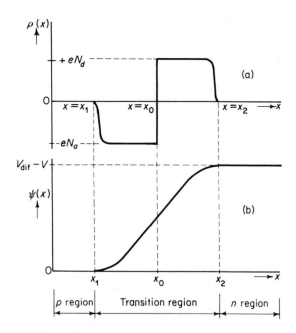

Fig. 15.1. (a) Space-charge distribution in a *p-n* step junction. (b) Voltage distribution in a *p-n* junction; a voltage $-V$ is applied to the n region. V_{dif} is the diffusion potential.

since $p_n = n_i^2/N_d$. Substituting (5.19a), we obtain

$$V_{\text{dif}} = E_g + \frac{kT}{e} \ln\left[\frac{N_d N_a}{6.3 \times 10^{38}} \cdot \left(\frac{m^2}{m_n^* m_p^*}\right)^{3/2}\left(\frac{300}{T}\right)^3\right] \tag{15.1a}$$

where E_g is in electron volts and N_d and N_a are numbers per cubic centimer. Take $E_g = 1.1$ eV, $N_a = N_d = 10/^{16}$cm^3, $m_n^* = m_p^* = m$, and $T = 300°$K; then $V_{\text{dif}} = 0.70$ V. Only if N_d and N_a are of the order of 10^{19}/cm^3 will V_{dif} be nearly equal to E_g.

The potential $\psi(x)$ in the contact region is related to the space-charge density $\rho(x)$ in that region by Poisson's equation,

$$\frac{d^2\psi}{dx^2} = -\frac{\rho(x)}{\epsilon\epsilon_0} \tag{15.2}$$

If a voltage $-V$ is applied to the n-type section, the initial conditions are $\psi(x) = 0$ and $d\psi/dx = 0$ for $x \leq x_1$ (we disregard the smaller voltage drop for $x \leq x_1$ due to the flow of direct current) and $\psi(x) = V_{\text{dif}} - V$ and $d\psi/dx = 0$ for $x \geq x_2$ (making the same approximation as for $x \leq x_1$) (Fig. 15.1b).

The hole density p and the electron density n in the space-charge region are approximately given by the equations (see Sec. 15.1c for their derivation)

$$p = p_p \exp\left(-\frac{e\psi}{kT}\right), \qquad n = n_n \exp\left[-\frac{e(V_{\text{dif}} - V - \psi)}{kT}\right] \tag{15.3}$$

where p_p is the hole density in the p region for $x = x_1$, n_n is the electron density in the n region for $x = x_2$, whereas $\exp(-e\psi/kT)$ and $\exp[-e(V_{\text{dif}} - V - \psi)/kT]$ are the appropriate Boltzmann factors. In our case $p_p \simeq N_a$ and $n_n \simeq N_d$.

Since $\rho(x) = -e(N_a - p + n)$ for $x < x_0$, where n is very small for $x < x_0$ and p is very small if $\psi > kT/e$, it is a good approximation to put $\rho(x) = -eN_a$ for $x_1 \leq x \leq x_0$. Similarly, it is a good approximation to put $\rho(x) = +eN_d$ for $x_0 \leq x \leq x_2$; but $\rho(x) = 0$ outside the two regions. Figure 15.1 shows $\rho(x)$ and $\psi(x)$ for the case $N_a = N_d$.

Poisson's equation may be readily solved for the two regions $x_1 \leq x \leq x_0$ and $x_0 \leq x \leq x_2$. Calling these solutions $\psi = \psi_1$ for $x_1 \leq x \leq x_0$ and $\psi = \psi_2$ for $x_0 \leq x \leq x_2$, we require

$$\psi_1(x) = \psi_2(x), \qquad \frac{d\psi_1}{dx} = \frac{d\psi_2}{dx}, \qquad \text{at } x = x_0 \tag{15.4}$$

The solutions are

$$\frac{d\psi_1}{dx} = \frac{eN_a}{\epsilon\epsilon_0}(x - x_1), \qquad \psi_1(x) = \frac{eN_a}{2\epsilon\epsilon_0}(x - x_1)^2 \tag{15.5}$$

for $x_1 \leq x \leq x_0$, satisfying the conditions $d\psi_1/dx = 0$ and $\psi_1(x) = 0$ at $x = x_1$; and

$$\frac{d\psi_2}{dx} = \frac{eN_d}{\epsilon\epsilon_0}(x_2 - x), \qquad \psi_2(x) = (V_{\text{dif}} - V) - \frac{eN_d}{2\epsilon\epsilon_0}(x_2 - x)^2 \qquad (15.6)$$

for $x_0 \leq x \leq x_2$, satisfying the conditions $d\psi_2/dx = 0$ and $\psi_2(x) = V_{\text{dif}} - V$ at $x = x_2$. Applying the condition (15.4) at $x = x_0$ yields

$$N_a(x_0 - x_1) = N_d(x_2 - x_0) \qquad (15.7)$$

$$\frac{eN_a}{2\epsilon\epsilon_0}(x_0 - x_1)^2 = V_{\text{dif}} - V - \frac{eN_d}{2\epsilon\epsilon_0}(x_2 - x_0)^2 \qquad (15.8)$$

Substituting (15.7) into (15.8) and solving for $x_0 - x_1$ yields

$$
\begin{aligned}
x_0 - x_1 &= \left[\frac{2\epsilon\epsilon_0(V_{\text{dif}} - V)}{eN_a(1 + N_a/N_d)}\right]^{1/2} \\
x_2 - x_0 &= \left[\frac{2\epsilon\epsilon_0(V_{\text{dif}} - V)}{eN_d(1 + N_d/N_a)}\right]^{1/2}
\end{aligned}
\qquad (15.9)
$$

and the thickness d of the transition region becomes

$$
\begin{aligned}
d = x_2 - x_1 &= (x_0 - x_1) + (x_2 - x_0) \\
&= \left[\frac{2\epsilon\epsilon_0(V_{\text{dif}} - V)(N_a + N_d)}{eN_aN_d}\right]^{1/2}
\end{aligned}
\qquad (15.10)
$$

The transition region contains a positive charge $+Q$ per unit area on the n side of the contact and a negative charge $-Q$ per unit area on the p side:

$$Q = eN_d(x_2 - x_0) = \left[2e\epsilon\epsilon_0(V_{\text{dif}} - V)\frac{N_aN_d}{N_a + N_d}\right]^{1/2} \qquad (15.11)$$

The small-signal transition-region capacitance per unit area is therefore

$$C = -\frac{dQ}{dV} = \left[\frac{e\epsilon\epsilon_0}{2(V_{\text{dif}} - V)} \cdot \frac{N_aN_d}{N_a + N_d}\right]^{1/2} = \frac{\epsilon\epsilon_0}{d} \qquad (15.12)$$

The space-charge layer thus acts as a parallel planar capacitor with plate distance d.

The maximum field strength occurs at $x = x_0$; we have

$$
\begin{aligned}
F = -F_{\max} = -\frac{d\psi_1}{dx}\bigg|_{x=x_0} &= -\frac{eN_a}{\epsilon\epsilon_0}(x_0 - x_1) \\
&= -\left[\frac{2eN_aN_d(V_{\text{dif}} - V)}{\epsilon\epsilon_0(N_a + N_d)}\right]^{1/2} = -\frac{2(V_{\text{dif}} - V)}{d}
\end{aligned}
\qquad (15.13)
$$

These formulas hold for the case in which a sudden step occurs from p-type to n-type material. In other junctions the transition is more gradual. It is then a better approximation to assume that N_d and N_a are linear functions of x (linearly graded junctions). If $N_d = N_a = N_0$ at $x = x_0$, then (Fig. 15.2)

$$N_d = N_0 + a_1(x - x_0), \qquad N_a = N_0 - a_2(x - x^0) \qquad (15.14)$$

Solving again Poisson's equation as before and calculating the width d of the space charge region and the transition-region capacitance C, one obtains, if $a = a_1 + a_2$,

$$d = \left[\frac{12\epsilon\epsilon_0(V_{\text{dif}} - V)}{ea} \right]^{1/3} \qquad (15.15)$$

$$C = \left[\frac{ea\epsilon^2\epsilon_0^2}{12(V_{\text{dif}} - V)} \right]^{1/3} = \frac{\epsilon\epsilon_0}{d} \qquad (15.16)$$

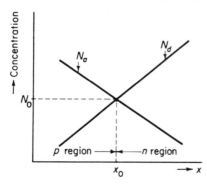

Fig. 15.2. Impurity distribution in a graded junction where N_a and N_d vary linearly with distance.

The transition capacitance thus varies as $(V_{\text{dif}} - V)^{-1/3}$, whereas it varies as $(V_{\text{dif}} - V)^{-1/2}$ in the other model. The capacitance C and the field strength F_{max} at $x = x_0$ may be much smaller than in the previous case if $a_1 + a_2$ is sufficiently small, because of the larger width d of the transition region. It is interesting to note that the equation $C = \epsilon\epsilon_0/d$ is valid under both conditions.

That this is the case is not so surprising. Suppose that we have a voltage V_0 applied to the p region. Let the width of the space-charge region for that bias be d and let $-Q_0$ be the charge per unit area in the p side of the space-charge region. We now apply a small change ΔV in the voltage. Then the charge per unit area in the p side of the contact changes to $-Q_0 + \Delta Q$, and the charge per unit area in the n side of the contact changes to $Q_0 - \Delta Q$. We thus have two charges, $+\Delta Q$ and $-\Delta Q$, a distance d apart, caused by a change in voltage ΔV. The small-signal capacitance per unit area is therefore

$$C = \frac{\Delta Q}{\Delta V} = \frac{\epsilon\epsilon_0}{d} \qquad (15.17)$$

independent of how the space charge is distributed.

For very large back bias the field strength F_{max} can become so large that electrical breakdown occurs. This problem is discussed in the next section.

15.1b. Breakdown Phenomena

In a *pn* junction, Zener breakdown and avalanche breakdown can occur. We first discuss the avalanche process in a way similar to Section 14.2b. One difference occurs in the avalanche process, and it comes about because both electrons and holes take part in the conduction process. We shall see in the next section that the saturation current I_0 consists of a part I_{n0} due to electrons and a part I_{p0} due to holes. Both currents are multiplied differently in the avalanche multiplication process. The multiplication factor M_n for electrons is, in analogy with Eq. (14.31a),

$$M_n = \frac{1 + p_{1e} - p_2}{1 - p_2} \qquad (15.18)$$

where p_{1e} is the number of secondary hole–electron pairs produced by a primary electron and p_2 the number of hole–electron pairs produced per secondary pair. Hence if p_{1h} is the number of secondary hole–electron pairs produced by a primary hole, the multiplication factor M_p for holes will be

$$M_p = \frac{1 + p_{1h} - p_2}{1 - p_2} \qquad (15.18a)$$

and the saturated current will be

$$I_s = M_n I_{n0} + M_p I_{p0} \qquad (15.19)$$

Breakdown again occurs when $p_2 = 1$. This is the case at a well-defined value of the maximum field strength in the space-charge region, corresponding to a well-defined value of the applied voltage.

Next we discuss Zener breakdown. It is caused by tunneling from the valence band to the conduction band under the influence of a large applied field; in silicon diodes it occurs in devices with a breakdown voltage smaller than 6 V. Since it is a tunnel effect, the theory of Sec. 7.3 is qualitatively valid, the only differences being that the density of states on the left side of the barrier is somewhat different and that the exact shape of the barrier in the forbidden gap is uncertain. Under the assumption of a parabolic barrier shape, Moll[†] has obtained

$$J = \frac{2m^{*1/2}e^3 F |V|}{4\pi^3 \hbar^2 E_g^{1/2}} \exp\left[-\frac{\pi}{4} \left(\frac{2m^*}{\hbar^2} \right)^{1/2} \frac{E_g^{3/2}}{eF} \right] \qquad (15.20)$$

where E_g is the gap width and V the applied voltage. This shows considerable resemblance to Eq. (7.30).

Since the effect is strongest where the field in the space-charge region is largest, we substitute into (15.20)

$$F = F_0 = \left[\frac{2eN_dN_a}{\epsilon\epsilon_0(N_d + N_a)}(V_{\text{dif}} - V) \right]^{1/2}$$

This gives again a characteristic of the form $\exp[-b/(V_{\text{dif}} - V)^{1/2}]$ as in the Schottky-barrier-diode case.

[†] J. M. Moll, *Physics of Semiconductors*, McGraw-Hill Book Company, New York, 1964.

Zener breakdown occurs only at relatively low breakdown voltages, whereas avalanche breakdown occurs at higher breakdown voltages. The reason is that in homogeneous semiconductors such as silicon and germanium avalanche multiplication occurs at a lower critical field strength, F_{crit}, than Zener breakdown. If the space-charge region at breakdown is sufficiently wide, avalanche breakdown will occur before Zener breakdown can become significant. For narrower space-charge regions—that is, for higher donor and acceptor concentrations—an electron or hole may drift out of the high-field region before it has gained enough energy to generate a hole–electron pair. As a consequence, the maximum field strength at which breakdown occurs will increase with increasing N_a and N_d until finally the breakdown field strength for avalanche breakdown is larger than the field strength at which Zener emission sets in. In silicon diodes Zener breakdown occurs at breakdown voltages below 6 V, whereas avalanche breakdown occurs at higher breakdown voltages.

If N_a and N_d are very large, the Zener breakdown condition occurs for a slight back bias. The saturated part of the characteristic then occurs for a limited range of back bias only. To avoid this, N_a and N_d should be properly chosen.

15.1c Carrier Distribution and Quasi-Fermi Levels in the Space-Charge Region

In the space-charge region of a *pn* junction there is a large field strength F and a large carrier density gradient. In the current density equation for holes

$$J_p = e\mu_p p F - e D_p \frac{dp}{dx}$$

the current density J_p is the difference between two large opposing currents, so that $|J_p| \ll e\mu_p p |F|$ and $|J_p| \ll e D_p |dp/dx|$ for most of the space-charge region. Hence in good approximation

$$-e\mu_p p \frac{d\psi}{dx} - e D_p \frac{dp}{dx} \simeq 0 \quad \text{or} \quad \frac{dp}{p} \simeq -\frac{\mu_p}{D_p} d\psi$$

or
$$p(x) = p_p \exp\left[-\frac{e\psi(x)}{kT}\right] \tag{15.21}$$

for $x_1 \leq x \leq x_2$, where p_p is the hole concentration at $x = x_1$. In the same way, for electrons we have

$$n(x) = n_n \exp\left\{-\frac{e[V_{dif} - V_j - \psi(x)]}{kT}\right\} \tag{15.21a}$$

for $x_1 \leq x \leq x_2$, where n_n is the electron concentration at $x = x_2$, $(V_{dif} - V_j)$ the potential difference across the space-charge region, and V_j the potential applied to the junction itself.

Applying this to the points $x = x_1$ and $x = x_2$, respectively, we have†

$$p(x_2) = p_p \exp\left[-\frac{e(V_{\text{dif}} - V_j)}{kT}\right], \quad n(x_1) = n_n \exp\left[-\frac{e(V_{\text{dif}} - V_j)}{kT}\right]$$
$$(15.22)$$

These boundary conditions are called the *Fletcher* boundary conditions, or *Boltzmann* boundary conditions.

Since there is approximate space-charge neutrality outside the space-charge region, $p_p = N_a + n(x_1)$ and $n_n = N_d + p(x_2)$. Substituting into (15.22) and solving for $p(x_2)$ and $n(x_1)$ yields

$$p(x_2) = \frac{N_d B^2 + N_a B}{1 - B^2}, \quad n_n = N_d + p(x_2) = \frac{N_d + N_a B}{1 - B^2}$$

$$n(x_1) = \frac{N_a B^2 + N_d B}{1 - B^2}, \quad p_p = N_a + n(x_1) = \frac{N_a + N_d B}{1 - B^2}$$
$$(15.23)$$

where $B = \exp\left[-e(V_{\text{dif}} - V_j)/kT\right]$. Usually $B^2 \ll 1$.

If $B^2 \ll 1$ and $N_a \gg N_d$, these equations may be written

$$p(x_2) = N_a B = p_n \exp\left(\frac{eV_j}{kT}\right), \quad n_n = N_d + N_a B$$

$$n(x_1) = N_a B^2 + N_d B, \quad p_p = N_a$$
$$(15.23a)$$

whereas a similar set of equations holds for $N_d \gg N_a$. We need these equations for the discussion of the *pin* diode (Sec. 15.2c).

In Eqs. (15.23a) one can distinguish between two cases: (1) $B < N_d/N_a$. In that case $n_n \simeq N_d$, and $n(x_1) \simeq N_d B$. This is called the case of *low injection*. (2) $B > N_d/N_a$. In that case $n_n \simeq N_a B$ and $n(x_1) \simeq N_a B^2$. This is called the case of *high injection*.

According to the definition of the quasi-Fermi levels $\overline{\mu}_n$ and $\overline{\mu}_p$,

$$\overline{\mu}_p = \mu_{cp} - e\psi = kT \ln\left(\frac{n_i}{p}\right) - e\psi = kT \ln\left(\frac{n_i}{p_p}\right)$$
$$(15.24)$$

$$\overline{\mu}_n = \mu_{cn} - e\psi = kT \ln\left(\frac{n}{n_i}\right) - e\psi = kT \ln\left(\frac{n_n}{n_i}\right) - e(V_{\text{dif}} - V_j)$$

Hence

$$\overline{\mu}_n - \overline{\mu}_p = kT \ln\left(\frac{n_n p_p}{n_i^2}\right) - e(V_{\text{dif}} - V_j) \qquad (15.25)$$

At low injection $p_p = N_a$ and $n_n = N_d$, or

$$\overline{\mu}_n - \overline{\mu}_p = kT \ln\left(\frac{N_d N_a}{n_i^2}\right) - e(V_{\text{dif}} - V_j) = eV_j \qquad (15.25a)$$

because of definition (15.1). At high injection either $p_p > N_a$ or $n_n > N_d$ or

---

† For a verification of these boundary conditions, compare A. van der Ziel, *Solid State Electronics*, **16**, 1509 (1973).

both, and hence

$$\overline{\mu}_n - \overline{\mu}_p > kT \ln \left(\frac{N_d N_a}{n_i^2}\right) - e(V_{\text{dif}} - V_j) = eV_j \qquad (15.25b)$$

This is caused by the voltage drops in the bulk n and p regions.

15.2. CURRENT FLOW IN A *PN* JUNCTION

The accurate calculation of the characteristic of a *pn* diode under arbitrary conditions is very difficult, since one must solve five simultaneous equations:

1. Two current equations for the current densities J_p and J_n.
2. Two continuity equations for the carrier concentrations p and n.
3. Gauss law for the field F.

Even if we reduce this to a one-dimensional problem it remains difficult because the drift terms in J_p and J_n are nonlinear.

It will be shown in Sec. 15.2a that for low-level injection, that is, for $p(x_2) \ll N_d$ and $n(x_1) \ll N_a$, one can ignore the effect of the drift terms in J_p and J_n altogether. The problem then becomes linear, and we obtain the diffusion equations (6.59a) and (6.60a); for direct current they reduce to Eqs. (6.61) and (6.62), which are solved easily. The high-level injection case is discussed in Sec. 15.2b.

15.2a. Direct-Current Characteristics

We consider two limiting cases: (1) the *very long diode* in which nearly all the injected minority carriers recombine before reaching the ohmic contact, and (2) the *very short diode* in which nearly all the injected minority carriers reach the ohmic contacts.†

We first solve the problem for the long diode (Fig. 15.3). Let the space-charge region again extend from $x = x_1$ to $x = x_2$, and let the p and n regions have widths w_p and w_n, respectively; then the ohmic contact to the p region is at $x = x_1 - w_p$ and the contact to the n region at $x = x_2 + w_n$. Let p_n and n_p be the equilibrium hole and electron concentrations in the bulk p and n regions, respectively. The hole concentration $p(x_2)$ at x_2 and the electron concentration $n(x_1)$ at x_1 are then given by

$$p(x_2) = p_n \exp\left(\frac{eV}{kT}\right) \qquad n(x_1) = n_p \exp\left(\frac{eV}{kT}\right) \qquad (15.26)$$

where V is the voltage applied to the p region.

---

† A contact is *ohmic* if a free exchange of *majority* carriers is possible across the contact; that does not specify what happens to the minority carriers, which is clarified at the end of this section.

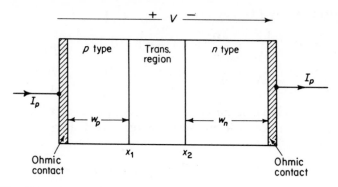

Fig. 15.3. A long *p-n* junction with two ohmic contacts and a space-charge region. For forward bias V must be positive.

The hole diffusion equation is now

$$\frac{\partial p}{\partial t} = -\frac{p - p_n}{\tau_p} + D_p \frac{\partial^2 p}{\partial x^2} \tag{15.27}$$

and the steady-state equation is $(\partial p/\partial t = 0)$

$$\frac{d^2 p}{dx^2} = \frac{p - p_n}{D_p \tau_p} = \frac{p - p_n}{L_p^2} \tag{15.27a}$$

where $L_p = (D_p \tau_p)^{1/2}$. This must now be solved under the conditions $p = p(x_2)$ at $x = x_2$, and $p = p_n$ for sufficiently large x. Equation (15.27a) has solutions $\exp\left[-(x - x_2)/L_p\right]$ and $\exp\left[(x - x_2)/L_p\right]$, but the second solution violates the boundary condition for large x; so the full solution of (15.27a) is

$$p(x) = p_n + [p(x_2) - p_n]\exp\left(-\frac{x - x_2}{L_p}\right) \tag{15.28}$$

We thus see that practically no holes reach the ohmic contact if $w_n > 5L_p$.

We shall now show that the name "diffusion length" for L_p is quite appropriate, because it represents the average distance traveled by the injected holes. According to (15.28), the normalized probability that a hole recombination occurs between x and $x + dx$ is

$$dP = \exp\left(-\frac{x - x_2}{L_p}\right)\frac{dx}{L_p}, \qquad \text{so that} \quad \int_{x_2}^{\infty} dP = 1 \tag{15.29}$$

Hence the average distance traveled is

$$\overline{x - x_2} = \int_{x_2}^{\infty} (x - x_2)\, dP = L_p \tag{15.30}$$

Since the drift term in the hole currrent density could be neglected, the hole current density $J_p(x_2)$ at x_2 is

$$J_p(x_2) = -eD_p \frac{dp}{dx}\bigg|_{x=x_2} = \frac{eD_p}{L_p}[p(x_2) - p_n] = J_{p0}\left[\exp\left(\frac{eV}{kT}\right) - 1\right] \quad (15.31)$$

$$J_{p0} = ep_n\frac{D_p}{L_p} = eN_a\frac{D_p}{L_p}\exp\left(-\frac{eV_{\text{dif}}}{kT}\right) = \frac{en_i^2}{N_d}\frac{D_p}{L_p} = ep_n\left(\frac{D_p}{\tau_p}\right)^{1/2} \quad (15.31a)$$

For $x_1 < x < x_2$ the hole current density J_p is equal to $J_p(x_2)$, provided that no recombination occurs in the space-charge region. For $x > x_2$ the hole current density is

$$J_p(x) = -eD_p\frac{dp}{dx} = \frac{eD_p}{L_p}[p(x_2) - p_n]\exp\left(-\frac{x - x_2}{L_p}\right)$$

$$= J_p(x_2)\exp\left(-\frac{x - x_2}{L_p}\right) \quad (15.31b)$$

This gives rise to an electron current density

$$J_n(x) = J_p(x_2)\left[1 - \exp\left(-\frac{x - x_2}{L_p}\right)\right] \quad (15.31c)$$

The same theory can be applied to the p region. If $w_p > 5L_n$ we have instead

$$J_n(x_1) = e\frac{D_n}{L_n}[n(x_1) - n_p] = J_{n0}\left[\exp\left(\frac{eV}{kT}\right) - 1\right] \quad (15.32)$$

where $L_n = (D_n\tau_n)^{1/2}$ is the diffusion length for electrons and

$$J_{n0} = en_p\frac{D_n}{L_n} = eN_d\frac{D_n}{L_n}\exp\left(-\frac{eV_{\text{dif}}}{kT}\right) = \frac{en_i^2}{N_a}\frac{D_n}{L_n} = en_p\left(\frac{D_n}{\tau_n}\right)^{1/2} \quad (15.32a)$$

For $x_1 < x < x_2$ the electron current density J_n is also equal to $J_n(x_1)$; but for $x < x_1$ we have, in analogy with (15.31b),

$$J_n(x) = J_n(x_1)\exp\left(\frac{x - x_1}{L_n}\right) \quad (15.32b)$$

This gives rise to a hole current density

$$J_p(x) = J_n(x_1)\left[1 - \exp\left(\frac{x - x_1}{L_p}\right)\right] \quad (15.32c)$$

Since for $x_1 < x < x_2$ the hole current is $J_p(x_2)$ and the electron current is $J_n(x_1)$, the total current density J is everywhere

$$J = J_p(x_2) + J_n(x_1) = (J_{p0} + J_{n0})\left[\exp\left(\frac{eV}{kT}\right) - 1\right] \quad (15.33)$$

because J is continuous.

We have now the whole picture of the hole and electron current densities in the device. It is shown in Fig. 15.4, which gives $J(x)$, $J_p(x)$, and $J_n(x)$ as functions of x.

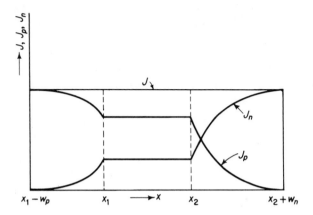

Fig. 15.4. Current density distribution in a long *p-n* junction (from A. van der Ziel, *Electronics*, Allyn and Bacon Co., Boston, 1966).

We can now draw the following conclusions:

1. The diode is forward biased for $V > 0$ and back biased for $V < 0$. For $V < 0$ and $e\,|V|/kT \gg 1$, the current densities are $J_n = -J_{n0}$ and $J_p = -J_{p0}$. They are called *saturation* currents and they are independent of V.

2. If $N_a \gg N_d$, $J_{p0} \gg J_{n0}$, or nearly all current is carried by holes. If $N_d \gg N_a$, $J_{n0} \gg J_{p0}$, or nearly all current is carried by electrons.

3. For forward bias ($V > 0$) the *p* region injects holes into the *n* region and the *n* region injects electrons into the *p* region. For back bias ($V < 0$) the *p* region extracts holes from the *n* region and the *n* region extracts electrons from the *p* region. This is important for understanding transistor operation.

4. The larger V_{dif}, the smaller the saturation currents. Because of the larger gap width, the saturation current in a silicon diode is many orders of magnitude smaller than in a germanium diode.

For a discussion of recombination at contacts, we rewrite Eqs. (15.31) and (15.32) as

$$J_p = e s_{pj}[p(x_2) - p_n], \qquad s_{pj} = \frac{D_p}{L_p} \tag{15.34}$$

$$J_n = e s_{nj}[n(x_1) - n_p], \qquad s_{nj} = \frac{D_n}{L_n} \tag{15.35}$$

where s_{pj} and s_{nj} are called the *junction recombination velocities*, respectively. This can now be applied to surfaces. We write at the surface

$$J_{ps} = e s_p(p_s - p_n), \qquad J_{ns} = e s_n(n_s - n_p) \tag{15.36}$$

where s_p and s_n are the *surface recombination velocities* and p_s and n_s are the minority carrier concentrations at the surface. For contacts we write

$$J_p = es_{cp}(p_c - p_n), \qquad J_n = es_{cn}(n_c - n_p) \tag{15.37}$$

where s_{cp} and s_{cn} are the *contact recombination velocities* and p_c and n_c are the minority carrier concentrations at the contact. Often s_{cp} and s_{cn} are very large; the contacts are then said to be *perfect sinks* for minority carriers.

We now apply this to a very short diode. Since practically no holes recombine in the n region, $dp/dx = -[p(x_2) - p_c]/w_n$, where p_c is the hole concentration at the contact. Consequently,

$$J_p = eD_p \frac{p(x_2) - p_c}{w_n} = es_{cp}(p_c - p_n) \tag{15.38}$$

since all holes recombine at the contact. Hence

$$p_c = \frac{p(x_2)D_n/w_n + p_n s_{cp}}{D_n/w_n + s_{cp}} \tag{15.39}$$

and

$$J_p = e\frac{s_{cp} \cdot D_p/w_n}{s_{cp} + D_p/w_n}[p(x_2) - p_n] = e\frac{s_{cp} \cdot D_p/w_n}{s_{cp} + D_p/w_n}p_n\left[\exp\left(\frac{eV}{kT}\right) - 1\right] \tag{15.40}$$

Usually, $s_{cp} \gg D_p/w_n$ and then

$$J_p = \frac{eD_p}{w_n}p_n\left[\exp\left(\frac{eV}{kT}\right) - 1\right] \tag{15.40a}$$

We next discuss the case of an n region of arbitrary length in which the contact is a perfect sink for holes. We then have

$$p(x) = p(x_2), \quad \text{at } x = x_2; \qquad p(x) = p_n, \quad \text{at } x = x_2 + w_n$$

Although $\exp[(x - x_2)/L_p]$ and $\exp[-(x - x_2)/L_p]$ are again independent solutions of the diffusion equation, it is more convenient to use as equivalent independent solutions $\sinh[(w_n + x_2 - x)/L_p]$ and $\sinh[(x - x_2)/L_p]$. Then

$$p(x) - p_n = A \sinh\left(\frac{w_n + x_2 - x}{L_p}\right) + B \sinh\left(\frac{x - x_2}{L_p}\right) \tag{15.41}$$

Introducing the boundary conditions at $x = x_2$ and $x = x_2 + w_n$ yields $A = [p(x_2) - p_n]/[\sinh(w_n/L_p)]$ and $B = 0$. Hence

$$p(x) - p_n = [p(x_2) - p_n]\frac{\sinh(w_n + x_2 - x)/L_p}{\sinh(w_n/L_p)} \tag{15.42}$$

$$J_p(x_2) = -eD_p\left.\frac{dp}{dx}\right|_{x=x_2} = \frac{eD_p}{L_p}\frac{p(x_2) - p_n}{\tanh(w_n/L_p)} \tag{15.43}$$

For $w_n/L_p \ll 1$, this reduces to (15.40a); for $w_n/L_p > 2$, it reduces to (15.31).

In the more general case where the contact is not a perfect sink for holes, $J_p(x_2)$ becomes

$$J_p(x_2) = e \frac{D_p}{L_p} \frac{[p(x_2) - p_n]}{\tanh(w_n/L_p)}$$

$$\cdot \left\{ 1 - \frac{D_p/L_p}{[s_{cp} \tanh(w_n/L_p) + D_p/L_p] \cosh^2(w_n/L_p)} \right\} \tag{15.44}$$

which reduces to (15.43) in most cases.

15.2b. Current Flow in the High-Injection Case for a P^+N Diode

To start the discussion we use the full current equations

$$J_p = e\mu_p pF - eD_p \frac{dp}{dx}, \qquad J_n = e\mu_n nF + eD_n \frac{dn}{dx} \tag{15.45}$$

We assume a p^+n diode, so that $p_p = N_a$ and in the n region $n = N_d + p$, where p is the injected hole density (approximate space-charge neutrality). Since practically no electrons cross the barrier, $J_n = 0$ at $x = 0$; hence

$$F(x_2) = -\frac{D_n}{\mu_n} \frac{1}{n} \frac{dn}{dx}\bigg|_{x_2} \tag{15.46}$$

Substituting in the expression for J_p and taking into account $D_n/\mu_n = D_p/\mu_p$ and $dn/dx \simeq dp/dx$ (space-charge neutrality),

$$J_p = -eD_p \frac{p}{n} \frac{dp}{dx}\bigg|_{x=x_2} - eD_p \frac{dp}{dx}\bigg|_{x=x_2} = -eD_p \left(1 + \frac{p}{n}\right) \frac{dp}{dx}\bigg|_{x=x_2} \tag{15.47}$$

For $p(x_2) \ll N_d$, $p/n \ll 1$ and we obtain the equation used for the low-injection case; for $p(x_2) \gg N_d$, $p/n \simeq 1$ and J_p is twice as large as expected for low injection. The boundary between low and high injection is properly put at $p(x_2) \simeq N_d$ or

$$N_a \exp\left[-\frac{e(V_{dif} - V)}{kT} \right] \simeq N_d \quad \text{or} \quad V_{dif} - V \simeq \frac{kT}{e} \ln \frac{N_a}{N_d} \tag{15.48}$$

Consequently, since $p(x_2) \gg p_n$, we have for that case

$$J_p \simeq e \left(\frac{D_p}{\tau_p}\right)^{1/2} N_d \tag{15.49}$$

For silicon $D_p = 12 \text{ cm}^2/\text{s}$; taking $\tau_p = 10^{-6} \text{ s}$ and $N_d = 10^{16}/\text{cm}^3$ yields $J_p \simeq 5 \text{ A/cm}^2$. Since junction areas of 10^{-3}–10^{-4} cm^2 are feasible, the high-injection case is easily reached.

We may now write

$$J_p = -eD_{p\,eff} \frac{dp}{dx}\bigg|_{x_2}, \qquad \frac{dp}{dx}\bigg|_{x_2} = -\frac{p(x_2)}{L_{peff}} \tag{15.50}$$

where $D_{p\,\text{eff}}$ and $L_{p\,\text{eff}}$ are effective values, not too far different from the low-injection values. Hence

$$J_p = e \frac{D_{p\,\text{eff}}}{L_{p\,\text{eff}}} p_n \exp\left(\frac{eV_j}{kT}\right) \tag{15.50a}$$

since $p(x_2) = p_n \exp(eV_j/kT)$. Since there should be no drastic difference between $D_{p\,\text{eff}}/L_{p\,\text{eff}}$ and D_p/L_p, the *low-level theory remains essentially correct if the voltage V is replaced by the junction voltage V_j*. However, the equation

$$J_p = e \frac{D_p}{L_p} p_n \exp\left(\frac{eV}{kT}\right) \tag{15.50b}$$

becomes quite incorrect, because $V = V_j + V_n$, where V_n is the voltage drop in the n region.

This agrees quite well with Fulkerson's computer solution of the problem. When J_p was expressed in terms of the junction voltage V_j, the increase of J_p with V_j at high injection was only slightly larger than expected for the low level theory. But a large deviation was found when J_p was expressed in terms of the applied voltage V (Fig. 15.5).

Naïvely, one would expect an ohmic region of the characteristic for $V_j \simeq V_{\text{dif}}$; that is, one should have

$$J_p \simeq \frac{V - V_{\text{dif}}}{R_s} \tag{15.51}$$

where R_s is the sum of the series resistances of the bulk n and p regions. Actually, this does not occur; because of majority carrier injection at the contacts, the current flow in the p and (or) n region becomes space-charge-limited.

15.2c. Voltage Drop in the N Region of a P⁺N Diode

We now calculate the voltage drop V_n in a p^+n junction with arbitrary length w_n of the n region under the assumption that the ohmic contact to the n region is a good sink for holes, so that $p(w_n) \ll N_d$.

Adding the two current equations (15.45), for the total current $J_0 = J_p + J_n$, we obtain

$$J_0 = e[(\mu_p + \mu_n)p_0 + \mu_n N_d]F_0 + e(D_n - D_p)\frac{dp_0}{dx} \tag{15.52}$$

where we have put $n_0 = N_d + p_0$ and $dn_0/dx = dp_0/dx$. Solving for the field F_0, we obtain

$$F_0 = \frac{J_0}{e[(\mu_p + \mu_n)p_0 + \mu_n N_d]} - \frac{(D_n - D_p)dp_0/dx}{[(\mu_p + \mu_n)p_0 + \mu_n N_d]} \tag{15.52a}$$

so that, putting $x_2 = 0$, the voltage V_n between the boundary to the space-

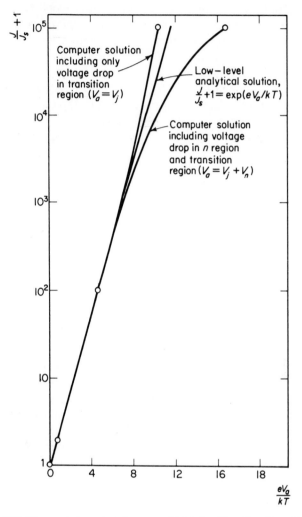

Fig. 15.5. Forward characteristic curve of high-level junction (from D. E. Fulkerson and A. Nussbaum, *Solid State Electronics*, **9**, 709–719, 1966).

charge region and the contact is

$$V_n = \int_0^{w_n} F_0 \, dx = J_0 \int_0^{w_n} \frac{dx}{e[(\mu_p + \mu_n)p_0 + \mu_n N_d]} \\ + \frac{kT}{e}\left(\frac{\mu_n - \mu_p}{\mu_n + \mu_p}\right) \ln\left[\frac{(\mu_p + \mu_n)p_0(0) + \mu_n N_d}{\mu_n N_d}\right]$$

(15.53)

where use has been made of the condition $p(w_n) \ll N_d$.

We note that the integral in the first term is simply equal to $J_0 R_{dc}$, where

R_{dc} is the dc resistance of the n region. Since the second term exists only because D_n and D_p are different, we call it the *diffusion voltage drop* V_{nd}.

We now apply this calculation to a p^+n junction with a very short n region; that is, we assume that $w_n \ll L_p$ at all injection levels. We furthermore assume that $J_n \ll J_p$, so that $J_0 \simeq J_p$.

For very low injection $[p_0(0) \ll N_d]$, the voltage V_n is very small. For very high injection $[p_0(0) \gg N_d]$, we have, according to Eq. (15.47),

$$J_0\, dx \simeq J_p\, dx \simeq -eD_p\left(1 + \frac{p_0}{n_0}\right)\frac{dp_0}{dx}\, dx \simeq -2eD_p\, dp_0 \qquad (15.53\text{a})$$

and hence

$$J_0 R_{dc} \simeq \int_0^{w_n} \frac{-2eD_p\, dp_0}{e[(\mu_p + \mu_n)p_0 + \mu_n N_d]}$$

$$\simeq \frac{kT}{e}\frac{2\mu_p}{\mu_p + \mu_n}\cdot \ln\left[\frac{(\mu_p + \mu_n)p_0(0) + \mu_n N_d}{\mu_n N_d}\right] \qquad (15.53\text{b})$$

so that

$$V_n \simeq \frac{kT}{e}\ln\frac{(\mu_p + \mu_n)p_0(0)}{\mu_n N_d}\bigg] \qquad (15.53\text{c})$$

since $(\mu_p + \mu_n)p_0(0) \gg \mu_n N_d$. According to the Fletcher condition (15.22),

$$p_0(0) = \frac{n_i}{N_d}\exp\left(\frac{eV_j}{kT}\right) \qquad (15.53\text{d})$$

so that

$$V_n = V_j - \frac{kT}{e}\ln\left[\frac{N_d^2}{n_i^2}\frac{\mu_n}{\mu_p + \mu_n}\right] = V_j - V_i \qquad (15.53\text{e})$$

where V_i is a voltage caused by the impurity concentration N_d.

At low-injection levels the voltage V_D across the terminals is equal to V_j, whereas at high injection $V_D = V_j + V_n = 2V_j - V_i$. Since the characteristic varies as $\exp (eV_j/kT)$ in both regions, we have a characteristic of the form $\exp (eV_D/kT)$ at very low injection and of the form $\exp [e(V_D + V_i)/2kT]$ at very high injection, when expressed in terms of the terminal voltage V_D.

We now repeat the calculation for a short n^+p junction. Here Eq. (15.53) must be replaced by

$$V_p = J_0 \int_0^{w_p} \frac{dx}{e[(\mu_p + \mu_n)n_0 + \mu_p N_a]}$$

$$+ \frac{kT}{e}\left(\frac{\mu_n - \mu_p}{\mu_n + \mu_p}\right)\ln\left[\frac{(\mu_p + \mu_n)n_0(0) + \mu_p N_a}{\mu_p N_a}\right] \qquad (15.54)$$

where V_p is the voltage across the p region.

For very low injection $[n_0(0) \ll N_a]$, the voltage V_p is quite small. For very high injection $[n_0(0) \gg N_a]$, we have

$$J_0\, dx \simeq J_n\, dx \simeq eD_n\left(1 + \frac{n_0}{p_0}\right)\frac{dn_0}{dx}\, dx \simeq 2eD_n\, dn_0 \qquad (15.54\text{a})$$

so that

$$J_0 R_{dc} \simeq -\frac{kT}{e} \frac{2\mu_n}{\mu_p + \mu_n} \ln\left[\frac{(\mu_p + \mu_n)n_0(0) + \mu_p N_a}{\mu_p N_a}\right] \qquad (15.54\text{b})$$

and

$$V_p \simeq -\frac{kT}{e} \ln\left[\frac{(\mu_p + \mu_n)n_0(0)}{\mu_p N_a}\right] \qquad (15.54\text{c})$$

since $(\mu_p + \mu_n)n_0(0) \gg \mu_p N_a$. According to the Fletcher condition,

$$n_0(0) = \frac{n_i^2}{N_a} \exp\left(-\frac{eV_j}{kT}\right) \qquad (15.54\text{d})$$

so that

$$V_p = V_j + \frac{kT}{e} \ln\left[\frac{N_a^2}{n_i^2} \frac{\mu_p}{\mu_n + \mu_p}\right] = V_j + V_i' \qquad (15.54\text{e})$$

where V_i' is a voltage caused by the impurity concentration N_a.

At low injection levels the voltage V_D across the terminals is equal to V_j, whereas at high injection $V_D = V_j + V_n = 2V_j + V_i'$. Since the characteristic varies as $\exp(-eV_j/kT)$ in both regions, we have a characteristic of the form $\exp(-eV_D/kT)$ at very low injection and of the form $\exp[-e(V_D - V_i')/2kT]$ at very high injection, when expressed in terms of the terminal voltage V_D.

It should be noted in view of (15.54d) that the last term in (15.54) corresponds to $-V_j(\mu_n - \mu_p)/(\mu_n + \mu_p) +$ const at high injection, whereas the last term in (15.53) corresponds to $+V_j(\mu_n - \mu_p)/(\mu_n + \mu_p) +$ const at high injection. Adding the junction voltage drop and this diffusion voltage drop, we obtain a combined voltage $V_j(1 + m)$ for the p^+n diode and $V_j(1 - m)$ for the n^+p diode, where $m = (\mu_n - \mu_p)/(\mu_n + \mu_p)$. This distinction is important for the evaluation of the high-frequency impedance of p^+n and n^+p junctions (Sec. 15.3b).

15.2d. Current Flow in the PIN Diode

Fletcher† has given a simplified discussion of the *pin* diode. Such devices are diodes in which the end contacts are separated by a region of relatively high resistivity. They are important technically, in that they are able to withstand a large back bias (compare Sec. 15.1b). For that reason they are useful as power rectifiers.

The simplification consists in assuming that the diode shown in Fig. 15.6 is completely symmetric. That is, the distribution of the hole concentration in the p region is the same as the distribution of the electron concentration in the n region. Furthermore, $\mu_p = \mu_n$ and $D_p = D_n$ in each of the regions.‡ Finally $p \simeq n$ in the intrinsic region. Furthermore, it is assumed that the width

---

† N. Fletcher, *Proc. IRE*, **45**, 862–872 (June 1957).

‡ This neglects the small voltage drop in the *i* region caused by the fact that $D_n \neq D_p$ (see Sec. 15.2c).

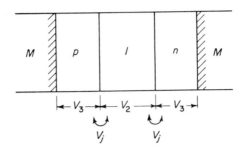

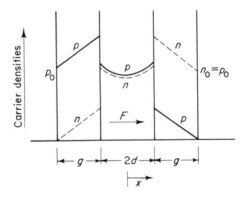

Fig. 15.6. (a) The *p-i-n* diode; *M* are ohmic metal contacts. (b) Carrier density distribution in the diode.†

$2d$ of the *i* region is large in comparison with the width g of the *p* and the *n* regions; as a matter of fact, g is assumed to be so small that the electron concentration in the *p* region and the hole concentration in the *n* region vary linearly with distance. The contacts are assumed to have a large recombination velocity for minority carriers, so that the minority carrier concentrations at the contacts are extremely small.

Since the device is symmetric, the voltages V_j applied to the *pi* and the *in* junctions are equal. The holes going from *p* to *i* have to climb a potential-energy barrier $e(V_{dif} - V_j)$ and the same holds for the holes going from *i* to *n*, where V_{dif} is the diffusion potential of each junction. Consequently if $p_0 = N_a$ is the equilibrium hole concentration in the *p* region, $p(P)$ and $n(P)$ the hole and electron concentrations on the *p* side of the junction, respectively, and $p(I) = n(I)$ the hole and electron concentrations on the *i* side of the junction, then, according to Eqs. (15.23a), for $B^2 \ll 1$, we have

$$p(P) = p_0, \qquad n(P) = p_0 B^2, \qquad p(I) = p_0 B \qquad (15.55)$$

where
$$B = \exp\left[-\frac{e(V_{dif} - V_j)}{kT}\right] \qquad (15.55a)$$

This is allowed until $V_{dif} - V_j$ becomes comparable with kT/e.

Fletcher now assumes a diffusion constant αD and a mobility $\alpha \mu$ in the i region and a diffusion constant βD and a mobility $\beta \mu$ in the p region. Here D and μ are the values found for lattice scattering, so that $D/\mu = kT/e$. The current equations in the i region are then

$$J_p = e\alpha\mu pF - e\alpha D \frac{\partial p}{\partial x}, \qquad J_n = e\alpha\mu nF + e\alpha D \frac{\partial n}{\partial x} \qquad (15.56)$$

and, since $n \simeq p$,

$$J = J_p + J_n = 2e\alpha\mu pF \qquad (15.56a)$$

is independent of x over the i region. Consequently, pF is independent of x; hence by substitution of (15.56) into the continuity equations

$$\frac{\partial p}{\partial t} = -\frac{p}{\tau} - \frac{1}{e}\frac{\partial J_p}{\partial x}, \qquad \frac{\partial n}{\partial t} = -\frac{n}{\tau} + \frac{1}{e}\frac{\partial J_n}{\partial x} \qquad (15.57)$$

the effect of the drift terms disappears completely. Hence

$$\frac{\partial p}{\partial t} = -\frac{p}{\tau} + \alpha D \frac{\partial^2 p}{\partial x^2} \qquad (15.58)$$

and a corresponding equation holds for n. Under steady-state conditions this yields

$$\frac{d^2 p}{dx^2} = \frac{p}{L^2}, \qquad L = (\alpha D\tau)^{1/2} \qquad (15.58a)$$

with the initial condition $p = p(I)$ at $x = \pm d$. The solution is

$$p(x) = p(I) \frac{\cosh x/L}{\cosh d/L} \qquad (15.59)$$

Substituting into (15.56a), we obtain

$$F = \frac{J}{2e\alpha\mu p(I)} \frac{\cosh d/L}{\cosh x/L} \qquad (15.60)$$

and hence the voltage drop across the p region is

$$V_2 = \int_{-d}^{d} \frac{J\cosh d/L}{2e\alpha\mu p(I)} \frac{dx}{\cosh x/L}$$
$$= \frac{2JL}{e\alpha\mu p(I)} \cosh \frac{d}{L}\left[\tan^{-1}\left(\exp\frac{d}{L}\right) - \frac{\pi}{4}\right] \qquad (15.61)$$

In the same way we calculate the voltage drops in the p and n regions. The current equations are

$$J_p = e\beta\mu pF - e\beta D \frac{dp}{dx}, \qquad J_n = e\beta\mu nF + e\beta D \frac{dn}{dx} \qquad (15.62)$$

and, since $dn/dx \simeq dp/dx$ (space-charge neutrality),

$$J = J_p + J_n = e\beta\mu(n + p)F \simeq e\beta\mu p_0 F \qquad (15.63)$$

since $n \ll p$ if $B^2 \ll 1$. Consequently,

$$F = \frac{J}{e\beta\mu p_0} \tag{15.64}$$

and

$$V_3 = \int_0^g F \, dx = \frac{Jg}{e\beta\mu p_0} \tag{15.64a}$$

The total applied voltage is

$$V = 2V_j + V_2 + 2V_3 \simeq 2V_j + V_2 \tag{15.65}$$

since usually $2V_3 \ll 2V_j + V_2$.

We finally calculate J by observing that J_p is continuous at the junction. If $J_p|_P$ and $J_p|_I$ are the current densities on the two sides of the junction, then from (15.56), (15.56a), (15,62), and (15.63), we have

$$J_q|_I = \frac{1}{2}J - e\alpha D \frac{dp}{dx}\bigg|_I = J_p|_P = J - e\beta D \frac{dp}{dx}\bigg|_P \tag{15.66}$$

so that

$$J = 2\left(e\beta D \frac{dp}{dx}\bigg|_P - e\alpha D \frac{dp}{dx}\bigg|_I\right) \tag{15.67}$$

We now bear in mind that $p(I) = n(I)$, and

$$\frac{dp}{dx}\bigg|_P = \frac{dn}{dx}\bigg|_P = \frac{n(P)}{g}, \quad \frac{dp}{dx}\bigg|_I = \frac{dn}{dx}\bigg|_I = -\frac{p(I)}{L}\tanh\frac{d}{L} \tag{15.68}$$

whereas $p(P) = p_0$ and $n(P) = p_0 B^2$. Substituting into (15.67) yields

$$J = 2eD\left[\alpha\frac{p(I)}{L}\tanh\frac{d}{L} + \frac{\beta}{g}n(P)\right] \tag{15.69}$$

We now bear in mind that $B = \exp[-\epsilon(V_{\text{dif}} - V_j)/kT]$; in addition $p_0 \exp(-eV_{\text{dif}}/kT) = n_i$, and

$$n(P) = p_0 B^2 = \frac{n_i^2}{p_0}\exp\left(\frac{2eV_j}{kT}\right) \tag{15.70}$$

Consequently,

$$J = \frac{2e\alpha D}{L}n_i\left[\tanh\frac{d}{L}\exp\left(\frac{eV_j}{kT}\right) + \frac{\beta}{\alpha}\frac{L}{g}\frac{n_i}{p_0}\exp\left(\frac{2eV_j}{kT}\right)\right] \tag{15.71}$$

Depending on conditions, either the first or the second term predominates. We shall now investigate under what conditions these situations occur.

First we take d/L quite small (very short i region). Then the first term in (15.71) is negligible, and the characteristic varies as $\exp(2eV_j/kT)$ or as $\exp(eV/kT)$, since $V_2 \ll 2V_j$ and hence $V \simeq 2V_j$. Next take d/L somewhat larger. Then the first term has some value and predominates for small V_j when

$$\tanh\frac{d}{L} > \frac{\beta}{\alpha}\frac{L}{g}\frac{n_i}{p_0} \tag{15.72}$$

The characteristic then varies as $\exp(eV_j/mkT)$, where $m \simeq 2$ for small V_j

and $m \simeq 1$ for large V_j. The region $m \simeq 2$ may extend over a large voltage range if d/L is large so that tanh $d/L \simeq 1$.

If we now express J in terms of the *applied* voltage V, we can also write $I = \text{const} \exp(eV/mkT)$, but now m first decreases from 2 to 1 at relatively small currents and then gradually increases with increasing current for large currents, because the voltage drop V_2 becomes comparable to or larger than $2V_j$. For very large values of d/L, the region $m \simeq 1$ may not be reached; in that case m increases more or less steadily from the value $m = 2$ at small currents to a value $m > 2$ at large currents. In commercial *pin* diodes with a relatively small value of d/L, the change from $m \simeq 2$ to $m \simeq 1$ lies at such low currents that it is not observed; all one sees is that m has a value of about unity a very small currents and increases steadily at larger currents.

Figure 15.7 shows Fletcher's results for germanium with $g \simeq 0.01$ cm, $\tau = 50 \ \mu s$, $n_0 = p_0 = 10^{18}/\text{cm}^3$, and $T = 300°\text{K}$. The width $2d$ of the i region is a parameter in the figure. Fletcher found good agreement between his predicted results and his experiments.

For very large voltages V, the voltage drop $V_2 + 2V_3$ becomes large in comparison with $2V_j$ and then the characteristic levels off in the manner shown in Fletcher's figure. We now try to find the limiting characteristic. In well-designed units $V_2 \gg 2V_3$. We see from Eq. (15.61) that V_2 varies as

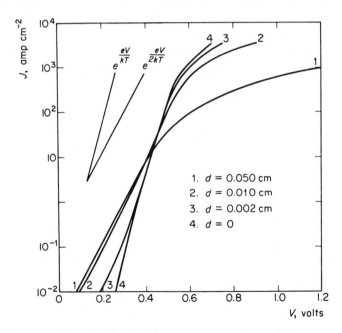

Fig. 15.7. Theoretical *p-i-n* diode characteristic (from N. Fletcher, *Proc. I.R.E.*, **45**, 862–872, June 1957).

$J/p(I)$. Now J varies as exp $(2eV_j/kT)$ at high currents and $p(I)$ varies as exp (eV_j/kT) as long as $B^2 \ll 1$. Hence V_2 is proportional to $\exp(eV_j/kT)$, or proportional to $J^{1/2}$. Consequently, $V_2 = \text{const } J^{1/2}$ at high currents. Hence for $V_2 > 2V_j$ the current density J varies approximately as V^2 at high currents. This is about what the space-charge-limited theory of Sec. 19.1 predicts. Fletcher's analysis thus leads naturally to something like space-charge-limited flow at high injection.

15.2e. Recombination in the Space-Charge Region

Fletcher's theory predicts characteristics of the form exp (eV/mkT) for some *pin* diodes with m going from the value 2 at rather low currents to the value $m \simeq 1$ at larger currents. This is actually found in some devices that are not of the *pin* variety. In such cases the effect is caused by generation recombination in the space-charge region due to Shockley–Read–Hall recombination centers.

For back bias the number of carriers in most parts of the space-charge region is very small, and as a consequence the centers alternately generate an electron and a hole. The electrons are collected by the *n* region and the holes by the *p* region; the generation process thus adds to the back current. For forward bias the number of carriers all through the space-charge region is much larger, and as a consequence the centers alternately capture an electron and a hole and thus add to the forward current.

The calculation of the characteristic is simple in principle. Let $R(p, n)$ be the rate at which hole–electron pairs are captured by the centers; then

$$J = e \int_{x_1}^{x_2} R(p, n) \, dx \tag{15.73}$$

where the integration is carried out over the whole space-charge region.

The question is now to find the correct expressions for R, p, and n. For $x_1 \leq x \leq x_2$ and relatively low injection $p(x_2) \ll N_d$, $n(x_1) \ll N_a$, one would expect $p(x)$ and $n(x)$ to be governed by the appropriate Boltzmann factors. That is [see Eqs. (15.21) and (15.21a)],

$$p(x) = N_a \exp\left(-\frac{e\psi}{kT}\right) \tag{15.74}$$

$$n(x) = N_d \exp\left[-\frac{e(V_{\text{dif}} - V - \psi)}{kT}\right] \tag{15.75}$$

so that, since $N_a N_d = n_i^2 \exp(eV_{\text{dif}}/kT)$,

$$p(x)n(x) = n_i^2 \exp\left(\frac{eV}{kT}\right) = N_a n(x_1) \tag{15.76}$$

all through the space-charge region. The recombination rate for Shockley–Read–Hall centers is [Eq. (6.53)]

$$R = \frac{pn - n_i^2}{(p + p_1)\tau_{n0} + (n + n_1)\tau_{p0}} \tag{15.77}$$

where p_1 and n_1 are the densities of free carriers when the Fermi level equals the trap level E_T, τ_{n0} is the lifetime of electrons in highly doped p-type material, and τ_{p0} the lifetime of holes in highly doped n-type material.

For zero bias $R(p, n) = 0$ and hence $J = 0$. For back-biased diodes

$$R = -\frac{n_i^2}{p_1\tau_{n0} + n_1\tau_{p0}} \tag{15.77a}$$

and the current density becomes

$$J \simeq -e\frac{n_i^2}{p_1\tau_{n0} + n_1\tau_{p0}}(x_2 - x_1) = -\text{const}\,(V_{\text{dif}} - V)^{1/2} \tag{15.78}$$

since $x_2 - x_1 = \text{const}\,(V_{\text{dif}} - V)^{1/2}$. The current density thus increases for increasing back bias and does not saturate.

For forward-biased diodes $p \gg p_1$ and $n \gg n_1$ for most of the space-charge region; hence, since $np \gg n_i^2$,

$$\begin{aligned} J &= en_i^2 \exp\left(\frac{eV}{kT}\right) \int_{x_1}^{x_2} \frac{dx}{p\tau_{n0} + n_i^2 \exp(eV/kT)\tau_{p0}/p} \\ &= en_i^2 \exp\left(\frac{eV}{kT}\right) \int_{x_1}^{x_2} \frac{dx \exp(e\psi/kT)}{N_a\tau_{n0} + n(x_1)\tau_{p0}[\exp(e\psi/kT)]^2} \end{aligned} \tag{15.79}$$

as is found by substituting for p and putting $n_i^2 \exp(eV/kT) = N_a n(x_1)$.

We calculate the integral for a symmetrical junction, in which the recombination centers lie at midband, by taking $N_a = N_d$ and $\tau_{n0} = \tau_{p0} = \tau$. The integrand then has its maximum value at the center of the space-charge-limited region where $p = n = n_i \exp(eV/2kT)$ and $d\psi/dx$ has its maximum value F_{max}. We now introduce $u = \exp(e\psi/kT)$, and put $du = u(e/kT)(d\psi/dx)\,dx \simeq u(eF_{\text{max}}/kT)dx$, since only the region where $d\psi/dx \simeq F_{\text{max}}$ gives a major contribution to the integral. We then observe that $\exp[e(V_{\text{dif}} - V)\,kT]$ is a very large number, so that the upper limit of integration may be replaced by ∞, and that $n(x_1)/N_a$ is a very small number at relatively low injection, so that the lower limit of integration may be replaced by zero. Hence

$$\begin{aligned} J &= \frac{n_i^2 \exp(eV/kT)}{\tau F_{\text{max}}/kT} \int_0^\infty \frac{du}{N_a + n(x_1)u^2} \\ &= \frac{\pi}{2}\frac{kT}{\tau F_{\text{max}}}n_i \exp\left(\frac{eV}{2kT}\right) \end{aligned} \tag{15.80}$$

since $n(x_1) = n_p \exp(eV/kT)$ and $N_a n_p = n_i^2$. We thus see that the characteristic varies as $\exp(eV/2kT)$ rather than as $\exp(eV/kT)$.

The example just discussed is a rather artificial one that does not conform to practical situations. Almost all diodes now in operation are of the p^+n

or n^+p varieties. Moreover, it is by no means certain that the recombination centers lie at midband.

Nussbaum† has made an exact calculation for a wide range of conditions and has found that the characteristic can always be represented as $I = $ const $\exp(eV/mkT)$ over a surprisingly wide range of currents. The factor m depends on the doping levels and on the position of the recombination centers in the forbidden band. It always lies between 1.0 and 2.0, but is typically about 1.5.

15.2f. Application to Metal–Semiconductor Diodes

Now that we have calculated the current flow and the admittance in a *pn* junction, it is only a small matter to calculate the effect of minority carrier flow on the characteristic and on the admittance of a metal–semiconductor diode. We shall do so for a metal–*n*–type semiconductor configuration.

Deep in the *n* region, the hole concentration $p_n = n_i^2/N_d$, where n_i is the intrinsic carrier concentration and N_d the donor concentration. If the diode has a diffusion potential V_{dif}, then the hole concentration at the contact at equilibrium is

$$p(0) = p_n \exp\left(\frac{eV_{\text{dif}}}{kT}\right) \tag{15.81}$$

This is independent of bias, since $p(0)$ comes about by an exchange of holes between the metal and the semiconductor side of the contact. According to Eq. (15.31a), the hole-current density is therefore

$$J_p = e\left(\frac{D_p}{\tau_p}\right)^{1/2} p_n\left[\exp\left(\frac{eV}{kT}\right) - 1\right] \tag{15.82}$$

which must be compared with the electron-current density (14.18b):

$$J_n = eN_d v_d \exp\left(-\frac{eV_{\text{dif}}}{kT}\right)\left[\exp\left(\frac{eV}{kT}\right) - 1\right] \tag{15.83}$$

Consequently, at relatively large forward current the low-frequency conductances per unit area are

$$G_{n0} = \frac{eJ_n}{kT}, \qquad G_{p0} = \frac{eJ_p}{kT} \tag{15.84}$$

and the high-frequency admittance due to the hole-current flow is

$$Y_p = G_{p0}(1 + j\omega\tau_p)^{1/2} \simeq \frac{e^2}{kT}(j\omega D_p)^{1/2}\frac{n_i^2}{N_d}\exp\left(\frac{eV}{kT}\right) \tag{15.85}$$

if $\omega\tau_p \gg 1$, which must be compared with the high-frequency conductance due to electron-current flow:

---

† A. Nussbaum, *Phys. Status Solidi A*, **19**, 441 (1973).

$$G_n = G_{n0} = \frac{e^2}{kT} N_d v_d \exp\left(-\frac{eV_{\text{dif}}}{kT}\right) \exp\left(\frac{eV}{kT}\right) \qquad (15.85a)$$

We have put here $p_n = n_i^2/N_d$.

We now have the following requirements for diode mixers, if they are to operate satisfactorily at microwave frequencies. We see that Y_p increases at $\omega^{1/2}$, so that it constitutes a serious input loss mechanism in the diode at microwave frequencies. To make the device operable at microwave frequencies, one must thus require $|Y_p| \ll G_{n0}$. This means that one should make the donor concentration N_d large and the diffusion voltage V_{dif} not too large —at least considerably smaller than the gap width E_g. These conditions become more and more stringent at higher frequencies. The condition $|Y_p| < G_{n0}$ leads to

$$V_{\text{dif}} < \frac{kT}{e} \ln\left[\frac{N_d^2}{n_i^2} \frac{v_d}{(\omega D_p)^{1/2}}\right] \qquad (15.85b)$$

15.2g. Practical PN Junction Diodes

In the early days of device technology, one made grown junctions, alloy junctions, and gold-bonded diodes. These devices have now been replaced by diffused junctions. We discuss here the case of a diffused junction made on n-type material.

First, an oxide is grown or deposited on the semiconducting material; then well-defined holes are etched into the oxide by a *photoresist process*, and p-type impurities are diffused in through these holes. The material is then cut into units, and contacts are made to the p and n regions. Since many units can be processed simultaneously, this is a cheap process; the main expense is in the testing, cutting, bonding, and mounting of the individual units.

The bulk semiconductor gives the device a series resistance. The magnitude of this resistance can be decreased by starting with a heavily doped (n^+) wafer, growing a thin, weakly doped, n layer onto this material by an *epitaxial* technique, and then applying the diffusion process just described. The more weakly doped n-type material gives the diode its characteristic and its breakdown voltage, whereas the n^+ material provides the required low series resistance.

15.3. ALTERNATING-CURRENT ADMITTANCE OF A JUNCTION DIODE

15.3a. Alternating-Current Admittance of a Junction Diode at Low Injection

We now apply a direct voltage $-V$ to the n region of the junction diode and superimpose a small alternating voltage $-v_1 \exp(j\omega t)$, with $|ev_1/kT| \ll 1$, to determine the ac response to the diode. For the sake of simplicity it will be first assumed that practically all the current is carried by holes. We also assume again that practically no holes reach the ohmic contact of the n region ($w \gg L_p$, Fig. 15.3). We put the origin of the coordinate system at the n side of the space-charge region ($x_2 = 0$).

Since the rearrangement of charge on both sides of the transition region is a very fast process, it may be assumed that the hole concentration at $x = 0$ follows the applied voltage instantaneously. We now calculate the ac impedance of the junction in three stages.

1. We calculate the hole concentration at $x = 0$. Instead of (15.23a),† we have

$$p(x)|_{x=0} = p_n \exp \left\{ \frac{e[V + v_1 \exp(j\omega t)]}{kT} \right\}$$

$$= p_0(0) \exp \left[\left(\frac{ev_1}{kT} \right) \exp(j\omega t) \right] \qquad (15.86)$$

$$= p_0(0) \left[1 + \frac{ev_1}{kT} \exp(j\omega t) \right] = p_0(0) + p_1(0) \exp(j\omega t)$$

where
$$p_0(0) = p_n \exp \left(\frac{eV}{kT} \right), \qquad p_1(0) = p_0(0) \frac{ev_1}{kT} \qquad (15.86a)$$

2. We calculate the ac hole concentration for $x > 0$. According to (15.28), we may write

$$p = p_n + [p_0(0) - p_n] \exp \left(-\frac{x}{L_p} \right) + p_1(x) \exp(j\omega t) \qquad (15.87)$$

which reduces to (15.28) for $p_1(x) = 0$. By substituting (15.87) into (15.27) we obtain a differential equation for p_1:

$$j\omega p_1(x) = -\frac{p_1(x)}{\tau_p} + D_p \frac{d^2 p_1}{dx^2}$$

or
$$\frac{d^2 p_1}{dx^2} = \frac{p_1(x)(1 + j\omega \tau_p)}{L_p^2} \qquad (15.88)$$

where $L_p = (D_p \tau_p)^{1/2}$ as before. This equation must be solved under the initial conditions

$$p_1(x) = p_1(0) \quad \text{at } x = 0; \quad p_1(x) = 0, \quad \text{at } x = \infty$$

Comparing with the solution of (15.28), we obtain

$$p_1(x) = p_1(0) \exp \left[-\frac{x(1 + j\omega \tau_p)^{1/2}}{L_p} \right] \qquad (15.89)$$

3. We calculate the alternating hole-current density $j_p \exp(j\omega t)$ at $x = 0$, neglecting the influence of the field term in (6.9), which is allowed at low

---

† We apply here the formula $e^{-x} = 1 - x$ for small x where $x = ev_1 \exp(j\omega t)/kT$; this is allowed since $|ev_1/kT| \ll 1$.

injection. If the direction of positive current flow is again the positive x direction, this yields

$$j_p \exp(j\omega t) = -eD_p \frac{dp_1}{dx}\bigg|_{x=0} \exp(j\omega t)$$

$$= \frac{eD_p p(0)}{L_p}(1 + j\omega\tau_p)^{1/2}\frac{ev_1}{kT}\exp(j\omega t) \tag{15.90}$$

Putting

$$j_p = Y_p v_1 \tag{15.91}$$

we obtain the junction admittance per unit area Y_p.† Substituting (15.31) and (15.31a), we obtain

$$Y = \frac{e(J_p + J_{p0})}{kT}(1 + j\omega\tau_p)^{1/2} \tag{15.91a}$$

where $J_{p0} = ep_n D_p/L_p$ is the saturated hole-current density for back bias.

This calculation can be easily extended to the case where the current density is carried by electrons. By analogy with (15.91a), for the junction admittance per unit area Y_n caused by electron storage in the p region, we obtain

$$Y_n = \frac{e(J_n + J_{n0})}{kT}(1 + j\omega\tau_n)^{1/2} \tag{15.91b}$$

where $J_{n0} = en_p D_n/L_n$ is the saturated electron-current density for back bias.

In conclusion, therefore, for the total junction admittance we have

$$Y = (Y_p + Y_n) = G_{p0}(1 + j\omega\tau_p)^{1/2} + G_{n0}(1 + j\omega\tau_n)^{1/2} \tag{15.92}$$

where

$$G_{p0} = \frac{e(J_{p0} + J_p)}{kT} = \frac{dJ_p}{dV}, \qquad G_{n0} = \frac{e(J_{n0} + J_n)}{kT} = \frac{dJ_n}{dV} \tag{15.92a}$$

are the low-frequency junction conductances due to holes and electrons, respectively.

Putting

$$Y = G + jB \tag{15.93}$$

we calculate the ac conductance G and the ac susceptance B of the junction. For the sake of simplicity we do so for the case in which practically all current is carried by holes. In that case $G = G_p$ and $B = B_p$, and hence

$$G_p = G_{p0}[\tfrac{1}{2}(1 + \omega^2\tau_p^2)^{1/2} + \tfrac{1}{2}]^{1/2} \tag{15.94}$$

$$B_p = G_{p0}[\tfrac{1}{2}(1 + \omega^2\tau_p^2)^{1/2} - \tfrac{1}{2}]^{1/2} \tag{15.94a}$$

Figure 15.8 shows G_p and B_p as functions of $\omega\tau_p$.

---

† Of this admittance the part $G_{p0} = e(J_p + J_{p0})/kT$ is caused by injection of holes into the n region, and the part $Y_p - G_{p0}$ is caused by *storage* of holes in the n region. This follows from corpuscular arguments.

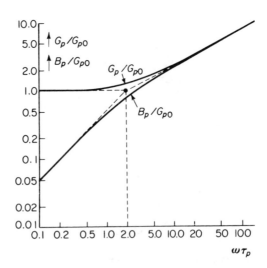

Fig. 15.8. G_p/G_{p0} and B_p/G_{p0} as a function of $\omega\tau_p$. The following approximations are reasonably good: (a) $G_p/G_{p0} \simeq 1.00$ and $B_p/G_{p0} \simeq \frac{1}{2}\omega\tau_p$ for $\omega\tau_p < 2$, indicating that the junction may be represented by a frequency independent conductance and capacitance in parallel. (b) $G_p/G_{p0} \simeq B_p/G_{p0} \simeq (\frac{1}{2}\omega\tau_p)^{1/2}$ for $\omega\tau_p > 2$.

If practically all current is carried by holes, the hole charge stored per unit area in the n region of a long junction in our present notation is

$$Q = \int_0^{\infty} e[p(0) - p_n] \exp\left(-\frac{x}{L_p}\right) dx$$

$$= ep_n\left[\exp\left(\frac{eV}{kT}\right) - 1\right]L_p = J_p\tau_p \tag{15.95}$$

so that the hole storage capacitance per unit area of the device can be defined as

$$C_s = \frac{dQ}{dV} = G_{p0}\tau_p \tag{15.95a}$$

where $G_{p0} = dJ_p/dV$, as before.

This capacitance C_s is not measured directly, however. What is measured is the admittance Y_p. Making a Taylor expansion in $\omega\tau_p$ yields

$$Y_p = G_p + j\omega C_p = G_{p0}(1 + \tfrac{1}{2}j\omega\tau_p) = G_{p0} + \tfrac{1}{2}j\omega C_s \tag{15.96}$$

The low-frequency capacitance C_p is thus equal to $\frac{1}{2}C_s$.

To understand Eq. (15.96), we rewrite it as

$$Y_p = G_{p0} - \tfrac{1}{2}j\omega C_s + j\omega C_s \tag{15.96a}$$

We now bear in mind that the ac junction current consists of a convection current and a displacement current. The displacement current gives, of course, a contribution $j\omega C_s$ to Y_p, but the ac convection current has in first approximation a phase shift at high frequencies so that it varies as $\exp\left(-\tfrac{1}{2}j\omega\tau_p\right)$; hence it gives the following contribution to Y_p:

$$G_{p0}\exp\left(-\tfrac{1}{2}j\omega\tau_p\right) \simeq G_{p0}\left(1 - \tfrac{1}{2}j\omega\tau_p\right) = G_{p0} - \tfrac{1}{2}j\omega C_s$$

As an example, put $J_{p0} = 0.10$ A/m^2 and $J_p = 0$ (zero bias); this yields $G_{p0} = 3.84$ S/m^2. Putting $\tau_p = 100\ \mu$s gives $C_s = 384\ \mu$F/m^2. The capacitance C_s may be considerably larger than the capacitance C_T of the transition region, especially if J_p is large. For silicon diodes J_{p0}, and hence C_s at zero bias, is much smaller.

Note that the capacitance C_s is most important for bias in the forward direction ($V > 0$) and disappears for back bias ($V < 0$) as soon as $|V|$ exceeds a value larger than a few times kT/e.

Up to now we have assumed that the length w of the n region is much larger than the diffusion length L_p of the holes. In this case practically all holes injected into the n region recombine before they reach the ohmic contact. The case $w \ll L_p$ is left as a problem for the reader (problem 15.8). The theory can also be developed for the case that w and L_p are comparable.

The equivalent circuit of a pn junction diode is essentially identical to that of a metal–semiconductor diode, but in order to accommodate all sources of conductance we extend it here. Instead of the series resistance r of the contact diode, we introduce the bulk resistance r_p of the p region and the bulk resistance r_n of the n region. Instead of the diode resistance R and the capacitance C in parallel, we introduce the hole conductance G_p and the electron conductance G_n in parallel to the three parallel capacitances $C_p = B_p/\omega$, $C_n = B_n/\omega$, and C_T, representing, respectively, the hole capacitance, the electron capacitance, and the transition-region capacitance. Sometimes it is necessary to introduce also the leakage conductance G_l of the junction and the capacitance C between the electrodes (Fig. 15.9).

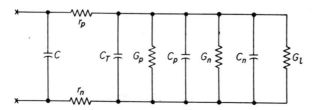

Fig. 15.9. Equivalent circuit of a *p-n* junction, taking into account the electrode capacitance C, the contact resistances r_p and r_n, the admittances due to the injected electrons and holes, the capacitance of the transition region C_T, and the leakage conductance G_l.

15.3b. Alternating-Current Impedance of a Diode at High Injection

At high injection one must not only take into account the effect of the junction itself, but also the effect of the n region. To that end we add again the current equations (15.45), and for the total current of a p^+n diode, we obtain

$$J = J_p + J_n = e[(\mu_p + \mu_n)p + \mu_n N_d]F + e(D_n - D_p)\frac{dp}{dx} \qquad (15.97)$$

where we have made use of the space-charge neutrality condition.

We now substitute for the junction voltage $V_j + v_j \exp(j\omega t)$, and $J = J_0 + j_1 \exp(j\omega t)$; $p(x) = p_0(x) + p_1(x)\exp(j\omega t)$; $F = F_0 + f_1 \exp(j\omega t)$. Collecting the dc terms (subscript 0) and ac terms (subscript 1) yields

$$J_0 = e[(\mu_p + \mu_n)p_0 + \mu_n N_d]F_0 + e(D_n - D_p)\frac{dp_0}{dx} \qquad (15.97a)$$

$$j_1 = e[(\mu_p + \mu_n)p_0 + \mu_n N_d]f_1 + e(\mu_n + \mu_p)F_0 p_1 + e(D_n - D_p)\frac{dp_1}{px} \qquad (15.97b)$$

so that

$$f_1 = \frac{j_1}{e[(\mu_p + \mu_n)p_0 + \mu_n N_d]} - \frac{e(\mu_p + \mu_n)F_0}{e[(\mu_p + \mu_n)p_0 + \mu_n N_d]}p_1$$
$$- \frac{e(D_n - D_p)}{e[(\mu_p + \mu_n)p_0 + \mu_n N_d]}\frac{dp_1}{dx} \qquad (15.98)$$

and

$$F_0 = \frac{J_0}{e[(\mu_p + \mu_n)p_0 + \mu_n N_d]} - \frac{e(D_n - D_p)}{e[(\mu_p + \mu_n)p_0 + \mu_n N_d]}\frac{dp_0}{dx} \qquad (15.98a)$$

Substituting (15.98a) into (15.98) and collecting terms yields

$$f_1 = \frac{j_1}{e[(\mu_p + \mu_n)p_0 + \mu_n N_d]} - J_0\frac{(\mu_p + \mu_n)p_1(x)}{e[(\mu_p + \mu_n)p_0 + \mu_n N_d]^2}$$
$$- (D_n - D_p)\frac{d}{dx}\left[\frac{p_1(x)}{(\mu_p + \mu_n)p_0 + \mu_n N_d}\right] \qquad (15.99)$$

Therefore the ac voltage across the n region is

$$v_n = \int_0^{w_n} f_1\,dx = j_1\int_0^{w_n}\frac{dx}{e[(\mu_p + \mu_n)p_0 + \mu_n N_d]}$$
$$- J_0\int_0^{w_n}\frac{(\mu_p + \mu_n)p_1(x)}{e[(\mu_p + \mu_n)p_0 + \mu_n N_d]^2}\,dx \qquad (15.100)$$
$$+ (D_n - D_p)\frac{p_1(0)}{[(\mu_p + \mu_n)p_0(0) + \mu_n N_d]}$$

since $p_1(w_n) \simeq 0$ and $p_0(w_n) \ll N_d$. Since $p_1(0) = p_0(0)\,ev_j/kT$ and $Z_j = v_j/j_1$ is the junction impedance proper, we have if $m = (\mu_n - \mu_p)/(\mu_n + \mu_p)$

$$Z_{tot} = \frac{v_1}{j_1} + \frac{v_n}{j_1} = Z_j + R_{dc} + Z_{mod} + m\frac{(\mu_p + \mu_n)p_0(0)}{[(\mu_p + \mu_n)p_0(0) + \mu_n N_d]}Z_j \qquad (15.101)$$

so that Z_{tot} consists of the junction impedance Z_j, the dc resistance R_{dc}, the modulation impedance Z_{mod}, and the diffusion impedance Z_d, the last term in (15.101), in series. Here

$$R_{dc} = \int_0^{w_n} \frac{dx}{e[(\mu_p + \mu_n)p_0 + \mu_n N_d]}$$

$$Z_{mod} = -\frac{J_0}{j_1} \int_0^{w_n} \frac{(\mu_p + \mu_n)p_1(x)\,dx}{e[(\mu_p + \mu_n)p_0 + \mu_n N_d]^2}$$

Since

$$p_1(0) = \frac{ev_1}{kT} p_0(0),$$

$$\frac{J_0 p_1(x)}{j_1} = J_0 \frac{p_1(0)}{j_1} \frac{p_1(x)}{p_1(0)} = \frac{eJ_0}{kT} Z_j \cdot p_0(0) \frac{p_1(x)}{p_1(0)},$$ (15.101a)

Z_{mod} may be written

$$Z_{mod} = -\frac{eJ_0}{kT} Z_j \int_0^{w_n} \frac{(\mu_p + \mu_n)p_0(0)}{e[(\mu_p + \mu_n)p_0 + \mu_n N_d]^2} \frac{p_1(x)}{p_1(0)}\,dx$$ (15.101b)

To evaluate Z_j and Z_{mod}, we assume that $|J_n| \ll J_p$ at the junction $x_2 = 0$. Equation (15.47) is then valid for both the dc current density J_{p0} and the ac current density j_{p1}, so that $J_0 \simeq J_{p0}$ and $j_1 \simeq j_{p1}$ at $x = 0$. Consequently, at high injection ($p \simeq n$),

$$J_0 = -2eD_p \left(\frac{dp_0}{dx}\right)_{x=0}, \qquad j_1 = -2eD_p \left(\frac{dp_1}{dx}\right)_{x=0}$$ (15.102)

We must now evaluate dp_0/dx and dp_1/dx at $x = 0$. We then use the ambipolar diffusion equation

$$\frac{\partial p}{\partial t} = -\frac{p}{\tau} - \mu_a F \frac{\partial p}{\partial x} + D_a \frac{\partial^2 p}{\partial x^2}$$ (15.103)

where

$$\mu_a = \frac{\mu_n \mu_p (n - p)}{\mu_n n + \mu_p p} = \frac{\mu_n \mu_p N_d}{[(\mu_p + \mu_n)p_0 + \mu_n N_d]} \simeq 0$$ (15.103a)

$$D_a = \frac{D_n D_p (2p_0 + N_d)}{(D_p + D_n)p_0 + D_n N_d} \simeq \frac{2D_n D_p}{D_n + D_p}$$ (15.103b)

This approximation holds if $p_0 > N_d$ for most of the n region. Strictly speaking, these equations are not valid near the ohmic contact; but if $p_0(0) \gg N_d$, this is only a small region. The ambipolar drift term in (15.103) then disappears, and the equation becomes linear. Substituting $p(x) = p_0(x) + p_1(x)$ exp $(j\omega t)$, we then have two linear equations for $p_0(x)$ and $p_1(x)$:

$$\frac{d^2 p_0}{dx^2} = \frac{p_0}{D_a \tau} = \frac{p_0}{L_a^2}, \qquad \frac{d^2 p_1}{dx^2} = \frac{p_1(1 + j\omega\tau)}{D_a \tau} = \frac{p_1}{L_a'^2}$$ (15.104)

with the initial conditions

$$p_0(0) \text{ and } p_1(0) \quad \text{given;} \quad p_0(w_n) \simeq 0, \quad p_1(w_n) \simeq 0 \quad (15.104a)$$

Here L_a is the ambipolar diffusion length

$$L_a = (D_a \tau)^{1/2} \quad \text{and} \quad L'_a = \left(\frac{D_a \tau}{1 + j\omega\tau}\right)^{1/2} = \frac{L_a}{(1 + j\omega\tau)^{1/2}} \quad (15.105)$$

The solutions of these equations are therefore

$$p_0(x) = p_0(0)\frac{\sinh\left[(w_n - x)/L_a\right]}{\sinh w_n/L_a}, \quad p_1(x) = p_1(0)\frac{\sinh\left[(w_n - x)/L'_a\right]}{\sinh w_n/L'_a} \quad (15.106)$$

Substituting into Eq. (15.102) yields

$$J_0 = \frac{2eD_p}{L_a \tanh w_n/L_a}p_0(0), \quad j_1 = \frac{2eD_p}{L'_a \tanh w_n/L'_a}p_1(0) \quad (15.107)$$

We now observe that

$$p_0(0) = p_n \exp\left(\frac{eV_j}{kT}\right), \quad p_1(0) = \frac{ev_j}{kT}p_0(0) \quad (15.107a)$$

Consequently, the low-frequency junction conductance is

$$G_0 = \frac{dJ_0}{dV_j} = \frac{eJ_0}{kT} \quad (15.108)$$

and the ac junction impedance is

$$Z_j = \frac{v_j}{j_1} = \frac{kT}{eJ_0}\frac{L'_a \tanh w_n/L'_a}{L_a \tanh w_n/L_a} \quad (15.108a)$$

All these expressions can now be substituted into (15.101)–(15.101b) and Z_{mod} can be evaluated:

$$Z_{\text{mod}} = -\frac{L'_a \tanh w_n/L'_a}{L_a \tanh w_n/L_a}\int_0^{w_n} \frac{(\mu_p + \mu_n)p_0(0)}{e[(\mu_p + \mu_n)p_0 + \mu_n N_d]^2}\frac{\sinh\left[(w_n - x)/L'_a\right]}{\sinh w_n/L'_a}dx \quad (15.109)$$

as is found by substituting for Z_j and for $p_1(x)/p_1(0)$ in Eq. (15.101b).

We now apply this for a very short *pn* junction, that is, a junction for which $w_n \ll L_a$. Then, according to (15.106),

$$p_0 = p_0(0)\left(\frac{w_n - x}{w_n}\right) = p_0(0)y, \quad y = \frac{w_n - x}{w_n}$$

We furthermore expand

$$\sinh\frac{w_n - x}{L'_a} \simeq \frac{w_n - x}{L'_a} + \frac{1}{6}\frac{(w_n - x)^3}{L'^3_a} = \frac{w_n}{L'_a}y + \frac{1}{6}\left(\frac{w_n}{L'_a}\right)^3 y^3$$

and write, since $\tanh w_n/L_a \simeq w_n/L_a$,

$$\frac{L'_a \tanh w_n/L'_a}{L_a \tanh w_n/L_a \cdot \sinh w_n/L'_a} \simeq \frac{L'_a/w_n}{\cosh w_n/L'_a} = \frac{L'_a/w_n}{1 + \frac{1}{2}(w_n^2/L'^2_a)}$$

Consequently,

$$Z_{\text{mod}} = -\frac{w_n}{e(\mu_p + \mu_n)p_0(0)}\left[\int_0^1 \frac{y\,dy}{(y+b)^2} + \frac{1}{6}\frac{w_n^2}{L_a'^2}\int_0^1 \frac{y^3\,dy}{(y+b)^2}\right]$$

$$\times \frac{1}{1+\frac{1}{2}(w_n^2/L_a'^2)} \tag{15.110}$$

where $b = \mu_n N_d/[(\mu_p + \mu_n)p_0(0)]$. Carrying out the integrations yields, at high injection ($b \ll 1$),

$$\int_0^1 \frac{y\,dy}{(y+b)^2} = \ln\left(\frac{b+1}{b}\right) - \frac{1}{b+1} \simeq \ln\left[1 + \frac{(\mu_n + \mu_p)p_0(0)}{\mu_n N_d}\right] - 1 \tag{15.110a}$$

$$\int_0^1 \frac{y^3\,dy}{(y+b)^2} = \frac{1}{2} - 2b + 3b^2\ln\left(\frac{1+b}{b}\right) - \frac{b^2}{1+b} \simeq \frac{1}{2} \tag{15.110b}$$

For a wide frequency range,

$$\int_0^1 \frac{y\,dy}{(y+b)^2} \gg \frac{1}{6}\left|\frac{w_n^2}{L_a'^2}\right|\left|\int_0^1 \frac{y^3\,dy}{(y+b)^2}\right|$$

so that in reasonable approximation

$$Z_{\text{mod}} = -\frac{w_n}{e(\mu_p + \mu_n)p_0(0)}\left\{\ln\left[1 + \frac{(\mu_p + \mu_n)p_0(0)}{\mu_n N_d}\right] - 1\right\}\frac{1}{1+\frac{1}{2}(w_n^2/L_a'^2)} \tag{15.110c}$$

This has a low-frequency value (we assumed $w_n^2 \ll L_a^2$) of

$$R_{\text{mod}} = -\frac{w_n}{e(\mu_p + \mu_n)p_0(0)}\left\{\ln\left[1 + \frac{(\mu_p + \mu_n)p_0(0)}{\mu_n N_d}\right] - 1\right\} \tag{15.110d}$$

so that

$$Z_{\text{mod}} = R_{\text{mod}} - R_{\text{mod}}\frac{\frac{1}{2}w_n^2/L_a'^2}{1+\frac{1}{2}w_n^2/L_a'^2} = R_{\text{mod}} + (-R_{\text{mod}})\frac{\frac{1}{2}j\omega(w_n^2/D_a)}{1+\frac{1}{2}j\omega(w_n^2/D_a)}$$

$$= R_{\text{mod}} + \frac{R\cdot j\omega L}{R + j\omega L} \tag{15.110e}$$

Hence Z_{mod} consists of a negative resistance R_{mod} in series with the parallel connection of a positive resistance $R = -R_{\text{mod}}$ and an inductance $L = \frac{1}{2}(w_n^2/D_a)R$; thus the time constant of the parallel connection is $w_n^2/(2D_a)$.

We further observe that the dc resistance is

$$R_{dc} = \int_0^{w_n} \frac{dx}{e[(\mu_p + \mu_n)p_0 + \mu_n N_d]} = \frac{w_n}{e(\mu_p + \mu_n)p_0(0)}\ln\left[1 + \frac{(\mu_p + \mu_n)p_0(0)}{\mu_n N_d}\right] \tag{15.111}$$

so that the low-frequency series resistance is

$$r_s = R_{dc} + R_{\text{mod}} = \frac{w_n}{e(\mu_p + \mu_n)p_0(0)} \tag{15.111a}$$

Finally, the junction impedance and the diffusion impedance of Eq. (15.111) may be lumped together as

$$Z'_j = Z_j + Z_d = Z_j(1 + m) = \frac{kT}{eJ_0}(1 + m)\left[1 - \frac{1}{3}j\omega\frac{w_n^2}{D_a}\right] \quad (15.112)$$

where $m = (\mu_n - \mu_p)/(\mu_n + \mu_p)$, since

$$\frac{L'_a \tanh w_n/L'_a}{L_a \tanh w_n/L_a} \simeq \frac{\tanh w_n/L'_a}{w_n/L'_a} \simeq 1 - \frac{1}{3}\left(\frac{w_n^2}{D_a}\right)j\omega$$

This can be considered as a resistance R'_j and a capacitance C'_j connected in parallel, where

$$R'_j = \frac{kT(1 + m)}{eJ_0}, \qquad C'_j = \frac{w_n^2}{3D_aR'_j} \quad (15.112a)$$

so that the time constant of the parallel connection is $w_n^2/(3D_a)$.

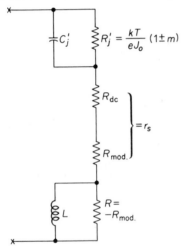

The calculation presented so far holds for a p^+n diode. A corresponding calculation for an n^+p diode gives similar results, but with the following modifications:

1. The expressions for R_{mod}, R_{dc}, and r_s must be appropriately modified; $p(_00)$ becomes $n_0(0)$; μ_nN_d becomes μ_pN_a; and so on.
2. $Z'_j = Z_j(1 - m)$, so that $R'_j = kT(1 - m)/eJ_0$.

The equivalent circuit for the two cases is thus as shown in Fig. 15.10; here the plus sign holds for a p^+n junction and the minus sign for an n^+p junction.

Fig. 15.10. H. F. equivalent circuit of a *pn* diode at high injection; plus sign for p^+n, minus sign for n^+p. $R'_jC'_j = \frac{1}{3}w_n^2/D_a$; $L/R = \frac{1}{2}w_n^2/D_a$.

15.3c. Noise in Junction Diodes

We can evaluate the noise of junction diodes by considering the possible groups of carriers taking part in the current flow. We do so for a *pn* junction in which practically all current is carried by holes. In that case there are three groups of minority carriers (Fig. 15.11):

1. Holes going from *p* to *n* and recombining there.
2. Holes generated in the *n* region and going from *n* to *p*.
3. Holes injected into the *n* region but returning to the *p* region by back diffusion before recombining.

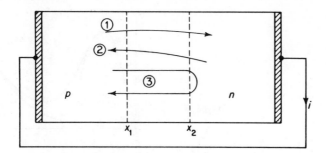

Fig. 15.11. Three groups of holes in a *p-n* junction.

The first two processes constitute series of independent events occurring at random. Since the first group gives a contribution $I_p + I_{p0}$ to the total current, its contribution to the noise spectrum is $2e(I_p + I_{p0})$. Since the second group gives a contribution $-I_{p0}$ to the total current, its contribution to the noise spectrum is $2eI_{p0}$. The third group gives a contribution $G_p - G_{p0}$ to the ac conductance of the device and, since diffusion is a thermal process, one would expect full thermal noise of it. Its contribution to the noise spectrum is therefore $4kT(G_p - G_{p0})$. Consequently, the total spectrum of the current fluctuations is

$$S_i(f) = 2e(I_p + 2I_{p0}) + 4kT(G_p - G_{p0}) \qquad (15.113)$$

The case of high injection is still being investigated.

15.4. TUNNEL DIODES

15.4a. Qualitative Description

Tunnel diodes are diodes in which the *p* and *n* regions are so heavily doped that they are degenerate. At equilibrium, part of the valence band in the *p* region is empty and part of the conduction band in the *n* region is filled (Fig. 15.12a).

A slight back bias brings some filled energy levels of the valence band of the *p* region opposite to empty energy levels of the conduction band of the *n* region, and as a consequence electrons will flow from the *p* region to the *n* region by quantum-mechanical tunnel effect. Since the number of available valence-band energy levels increases with back bias, and since the distance over which the electrons must tunnel decreases with increasing back bias, the back current increases very rapidly with increasing back bias (Zener effect). Since the direction of positive current flow is from *p* to *n*, the current is negative (Fig. 15.12b).

A slight forward bias brings some levels of the filled part of the conduction band of the *n* region opposite to empty levels of the valence band of the *p* region. Quantum-mechanical tunneling occurs, and electrons flow from the

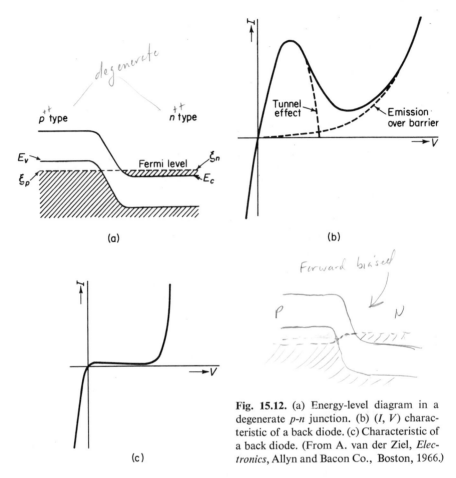

degenerate

p type   *n* type

E_v Fermi level ξ_n
ξ_p E_c

(a)

Tunnel effect   Emission over barrier

(b)

Forward biased

P   *N*

(c)

Fig. 15.12. (a) Energy-level diagram in a degenerate *p-n* junction. (b) (I, V) characteristic of a back diode. (c) Characteristic of a back diode. (From A. van der Ziel, *Electronics*, Allyn and Bacon Co., Boston, 1966.)

n region to the *p* region, giving a positive current that first increases with increasing back bias. When the filled part of the conduction band of the *n* region is opposite to the empty part of the valence band of the *p* region, the current goes through a maximum. After that the current decreases with increasing forward bias and should become almost zero if the filled part of the conduction band of the *n* region lies opposite the forbidden energy gap of the *p* region. At still larger forward bias electrons and holes are injected "over the barrier" into the *p* and *n* regions, respectively, resulting in a rapid increase in current for increasing forward bias (Fig. 15.12b).

The *I, V* characteristic thus has a negative conductance part in the forward region of the characteristic. Since the current flow in that part of the characteristic is by quantum-mechanical tunnel effect, the name *tunnel diode* is quite appropriate. We shall see in Chapter 17 how such a negative conductance characteristic can be put to good use.

The current near the valley of the characteristic (see Fig. 15.12b) is larger than expected theoretically. Direct tunneling is impossible, since the filled part of the conduction band of the n region is opposite to the forbidden energy gap of the p region. Minority carrier injection "over the barrier" is still a rather uncommon process. The most likely explanation is that of tunneling via (impurity) energy states in the forbidden energy gap of the space-charge region.

If the doping levels of the p and n regions are now decreased, the filled part of the conduction band of the n region and the empty part of the valence band of the p region become narrower. As a consequence, there will be fewer energy levels from which tunneling under forward bias conditions can occur. Hence the maximum in the forward I, V characteristic becomes lower and lower until it practically disappears at a given doping level. One then has also a rectifying characteristic, but it is the exact opposite of that of a normal diode: it has a large current in the back direction and a low current in the forward direction. Such a diode is therefore called a *back diode* (Fig. 15.12c).

Finally, if the doping levels are lowered further, one obtains a normal diode characteristic with "injection over the barrier" at forward bias and Zener breakdown at a critical back-bias voltage. A continuous transition from normal diodes via back diodes to tunnel diodes can thus be achieved by gradually increasing the impurity concentrations.

We now turn to the noise in the negative resistance region. Since the probability that an electron will tunnel through the barrier region of the junction is quite small, the passage of electrons through the barrier can be considered as a series of independent, random events. Consequently, one would expect full shot noise of the diode. Hence if I_d is the diode current, then the spectral intensity of the noise is

$$S_1(f) = 2eI_d \qquad (15.113a)$$

15.4b. Calculation of the I, V Characteristic of the Tunnel Diode

If a voltage V is applied, there is a current I_{vc} due to electrons tunneling from the valence band to the conduction band and a current I_{cv} due to electrons tunneling from the conduction band to the valence band. If Z_{vc} and Z_{cv} are the tunneling rates in the two directions, respectively, then the currents are given by the expressions

$$I_{vc} = A \int_{E_c}^{E_v} Z_{vc} f_v(E) \rho_v(E) [1 - f_c(E)] \rho_c(E) \, dE \qquad (15.114)$$

$$I_{cv} = A \int_{E_c}^{E_v} Z_{cv} f_c(E) \rho_c(E) [1 - f_v(E)] \rho_v(E) \, dE \qquad (15.115)$$

Here A is the junction area and $f_v(E)$ and $f_c(E)$ are the probabilities that the energy levels E in the valence band or in the conduction band are filled and

$1 - f_v(E)$ and $1 - f_c(E)$ the probabilities that they are empty. $\rho_v(E)$ and $\rho_c(E)$ are the densities of states in the valence band and the conduction band, respectively; E_v and E_c represent the energies at the top of the valence band and the bottom of the conduction band, respectively. The net electron current I_d flowing across the junction is the difference between I_{cv} and I_{vc}. Since it may be assumed that $Z_{cv} = Z_{vc} = Z$, the expression for I_d becomes

$$I_d = I_{cv} - I_{vc} = A \int_{E_c}^{E_v} Z[f_c(E) - f_v(E)]\rho_c(E)\rho_v(E)dE \qquad (15.116)$$

This integral is difficult to evaluate. We shall use here an approximation† that may not be too accurate but that has the merit of giving the forward characteristic in closed form.

Let us denote the (quasi) Fermi levels in the valence band and in the conduction band by $\bar{\mu}_p$ and $\bar{\mu}_n$, respectively. If we assume that

$$E_1 = \bar{\mu}_n - E_c \lesssim 2kT, \qquad E_2 = E_v - \bar{\mu}_p \lesssim 2kT \qquad (15.117)$$

we are allowed to approximate the Fermi distribution by a straight line

$$f_c = \frac{1}{2} - \frac{E - \bar{\mu}_n}{4kT}, \qquad f_v = \frac{1}{2} + \frac{\bar{\mu}_p - E}{4kT} \qquad (15.118)$$

We further assume that Z is practically independent of voltage for the voltage range considered here. We also take into account that $\rho_v(E)$ and $\rho_c(E)$ vary as $(E - E_c)^{1/2}$ and $(E_v - E)^{1/2}$, respectively. Finally, we bear in mind that $\bar{\mu}_n - \bar{\mu}_p = eV$ and $E_v - E_c = E_1 + E_2 - eV$. If we take the zero level of the energy at the bottom of the conduction band, so that $E_c = 0$, we have

$$I_d = A' \int_{E_c}^{E_v} \frac{\bar{\mu}_n - \bar{\mu}_p}{4kT}(E - E_c)^{1/2}(E_v - E)^{1/2} \, dE$$

$$= A' \int_0^{E_1 + E_2 - eV} \frac{eV}{4kT}[u(E_1 + E_2 - eV - u)]^{1/2} \, du \qquad (15.119)$$

$$= \frac{A''V}{kT/e}\left(\frac{E_1 + E_2}{e} - V\right)^2 = \frac{A''V(V_1 - V)^2}{kT/e}$$

where $u = E - E_c$, A' and A'' are constants, and $V_1 = (E_1 + E_2)/e$. It is interesting to note that in the early development of tunnel-diode circuitry the characteristic was often represented by a cubic characteristic of the preceding form. This finds here its theoretical basis.

The characteristic has zeros at $V = 0$ and at $V = V_1$, but (15.119) becomes invalid for $V > V_1$. The conductance

$$g = \frac{dI}{dV} = \frac{A''}{kT/e}(V_1 - V)(V_1 - 3V) \qquad (15.120)$$

---

† J. Karlovsky, *Phys. Rev.*, **127**, 419 (1962).

has zeros at $V = \frac{1}{3}V_1$ and at $V = V_1$ and is negative for $\frac{1}{3}V_1 < V < V_1$. We shall call this negative conductance $-g_d$; it plays an important role in the operation of the device as an amplifier.

REFERENCES

FRANKL, D. R., *Electrical Properties of Semiconductor Surfaces*. Pergamon Press, Inc., Elmsford, N.Y., 1967.

GROVE, A. S., *Physics and Technology of Semiconductor Devices*. John Wiley & Sons, Inc., New York, 1967.

HUNTER, L. P., ed., *Handbook of Semiconductor Electronics*, 3rd ed. McGraw-Hill Book Company, New York, 1970.

LINDMAYER, J., and C. Y. WRIGLEY, *Fundamentals of Semiconductor Devices*. Van Nostrand Reinhold Company, New York, 1965.

NUSSBAUM, A., *Semiconductor Device Physics*. Prentice-Hall, Inc., Englewood Cliffs, N.J., 1962.

RUNYAN, W., *Silicon Semiconductor Technology*. McGraw-Hill Book Company, New York, 1965.

SHIVE, J. N., *The Properties, Physics and Design of Semiconductor Devices*. Van Nostrand Reinhold Company, New York, 1959.

SHOCKLEY, W., *Electrons and Holes in Semiconductors*. Van Nostrand Reinhold Company, Inc., New York, 1950.

STRUTT, M. J. O., *Semiconductor Devices*. Academic Press, Inc., New York, 1966.

SZE, S. M., *Physics of Semiconductor Devices*. John Wiley & Sons, Inc. (Interscience Division), New York, 1969.

PROBLEMS

1. (a) Calculate the avalanche breakdown voltage of a germanium pn step junction in which $N_a = N_d = 10^{15}/\text{cm}^3$, if $\epsilon = 16$ and if the critical field strength is 200,000 V/cm. (b) Repeat the calculation if N_a is increased by a factor of 10. (c) Calculate this voltage for the limiting case $N_d \gg N_a$; $N_a = 10^{15}/\text{cm}^3$.

Answer: (a) 354 V; (b) 195 V; (c) 177 V.

2. (a) Calculate the donor concentration in n-type germanium having a resistivity of 1 Ω cm. $\mu_n = 3600 \text{ cm}^2/\text{V/s}$. (b) Repeat the calculation for p-type germanium of 1 Ω cm resistivity. $\mu_p = 1700 \text{ cm}^2/\text{V/s}$.

Answer: (a) $N_d = 1.74 \times 10^{15}/\text{cm}^3$; (b) $N_a = 3.68 \times 10^{15}/\text{cm}.^3$ Use these values in later calculations where applicable.

3. (a) Calculate the thickness of the barrier region of an n^+-p step junction on p-type germanium of $1\ \Omega$ cm resistivity, as a function of the applied voltage. (b) Repeat the calculation on a p^+-n step junction on n-type germanium. In both case you may assume that the heavily doped side of the junction has a very small resistivity, and that $\epsilon = 16$.

Answer: (a) $d = 0.694 \times 10^{-6}(V_{\text{dif}} - V)^{1/2}$ m; (b) $1.01 \times 10^{-6} \times (V_{\text{dif}} - V)^{1/2}$ m.

4. (a) Using the expressions for d in problem 3, calculate the capacitance per unit area for a pn step junction (n^+-p type) on p-type material of $1\ \Omega$ cm resistivity. (b) Repeat the calculation for a p^+-n step junction on n-type material of $1\ \Omega$ cm resistivity. (c) Apply these two cases to a barrier junction with a circular contact having a diameter of 6 mils (either case), in particular for $V = -3$ V, $V_{\text{dif}} = 0.6$ V.

Answer:

(a) $C_{pT} = \dfrac{20{,}400}{(V_{\text{dif}} - V)^{1/2}} \mu\mu\text{F}/\text{cm}^2.$

(b) $C_{nT} = \dfrac{14{,}000}{(V_{\text{dif}} - V)^{1/2}} \mu\mu\text{F}/\text{cm}^2.$

(c) $C_{pT} = 3.7/(V_{\text{dif}} - V)^{1/2}\ (= 1.9\ \mu\mu\text{F})$,

$C_{nT} = 2.56/(V_{\text{dif}} - V)^{1/2}\ (= 1.3\ \mu\mu\text{F}).$

5. (a) Calculate the avalanche breakdown voltage for an n^+-p junction on p-type germanium of $1\ \Omega$ cm resistivity. (b) Repeat the calculation for a step junction on n-type material of the same resistivity. Assume a critical field strength of 200,000 V/cm and a dielectric constant of 16 in either case.

Answer: (a) 48 V; (b) 102 V.

6. An n^+-p step junction is biased in the forward direction. The junction is made on p-type germanium of $1\ \Omega$ cm resistivity. (a) Calculate the value of J_{n0}, if $\tau_n = 100\ \mu$s and $D_n = 93\ \text{cm}^2/\text{s}$. (b) Apply this to a surface barrier junction with a contact area of 6 mil diameter. *Hint:* The resistivity of intrinsic germanium at 300°K is $47\ \Omega$ cm. Use this value to calculate the carrier concentration p_i in intrinsic material; then calculate the equilibrium electron concentration n_p in the p-type material, using the value for N_a that can be obtained from problem 2. You may assume $\mu_n = 3600$ and $\mu_p = 1700\ \text{cm}^2/\text{V/s}$.

Answer: $p_i = 2.51 \times 10^{13}/\text{cm}^3$, $n_p = 1.71 \times 10^{11}/\text{cm}^3$, $J_{n0} = 26.4\ \mu\text{A}/\text{cm}^2$, $I_{n0} = 0.0047\ \mu\text{A}$.

7. Show that in Fletcher's model of the *pin* diode half the current is carried by electrons if the second term in Eq. (15.69) strongly predominates over the first one.

8. Show that the ac admittance of the junction of a short diode is

$$Y_p = g_{p0} \frac{(1 + j\omega\tau_p)^{1/2} \tanh w/L_p}{\tanh [(1 + j\omega\tau_p)^{1/2}w/L_p]}, \quad \text{where } g_{p0} = \frac{dJ_p}{dV}$$

9. Show that in Fletcher's model of the *pin* diode the ac admittance may be written

$$Y = \frac{J_1}{2v_1} = \frac{e}{kT} \left[\frac{2e\beta Dn_0(P)}{g} \right] \frac{g/L_n'}{\tanh g/L_n'} + \frac{e}{kT} \left[\frac{e\alpha Dp_0(I) \tanh d/L_i}{L_i} \right] \frac{L_i \tanh d/L_i'}{L_i' \tanh d/L_i}$$

where

$$L_n' = \left(\frac{\beta D\tau_n}{1 + j\omega\tau_n} \right)^{1/2} \quad \text{and} \quad L_i' = \left(\frac{\alpha D\tau}{1 + j\omega\tau} \right)^{1/2}$$

Here τ_n is the electron lifetime in the *p* region, and J_1 and $2v_1$ are the amplitudes of the total current and the total voltage, respectively. *Hints:* (a) Solve (15.58) and the corresponding equations for electrons in the *p* region by assuming that $p(I) = p_0(I) + p_1(I) \exp (j\omega t)$; $n(P) = n_0(P) + n_1(P) \exp (j\omega t)$. (b) Show that $p_1(I) = p_0(I)ev_1/kT$; $n_1(P) = n_0(P)2ev_1/kT$, where $V_j = V_{j0} + v_1 \exp (j\omega t)$. (c) Show that Eq. (15.66) must be replaced by

$$J_{p1}|_I = \frac{1}{2}J_1 - e\alpha D \frac{dp_1}{dx}\bigg|_I = J_{p1}|_P = J_1 - e\beta D \frac{dn_1}{dx}\bigg|_P$$

Using the results from parts (a) and (b), solve for J_1.

10. A *pn* junction diode with a heavily doped *p* region and a very narrow *n* region is biased in the forward direction at relatively low injection. Neglecting recombination in the *n* region and assuming that the *n* contact is a perfect sink for holes, calculate the field in the *n* region. Calculate also the voltage drop in the *n* region, and show that it is equal to $(\mu_n/\mu_p)IR$, where I is the current and R the resistance of the *n* region without injection. Explain this result in terms of ambipolar drift effects.

11. A silicon p^+n junction has an n^- region consisting of 5 μm of a carrier concentration of $10^{14}/\text{cm}^3$, followed by a 20-μm n^+ region of carrier concentration $10^{19}/\text{cm}^3$. The first region has a hole lifetime of 10^{-7} s and the second a hole lifetime of 10^{-8} s.
(a) Find the excess hole distribution in the *n* region for forward bias.
(b) Find the characteristic of the junction.
(c) Determine numerical values for the various parameters.

Hint: You may assume that you have relatively low injection, so that the applied voltage appears across the p^+n junction. Put for the n^- region, $p(x) = A \sinh (x/L_{p1}) + B \sinh [(w - x)/L_{p1}]$, whereas $p(x) = C \exp (-x/L_{p2})$ for the n^+ region. Match the boundary conditions at $x = 0$ and $x = w$. Here w is the width of the n^- region, and L_{p1} and L_{p2} are the appropriate diffusion lengths. $D_p = 12.4 \text{ cm}^2/\text{s}$. You may neglect p_n everywhere.

Answer:

$$A = \frac{p(0)}{\sinh w/L_{p1} \cosh w/L_{p1}[1 + (L_{p1}/L_{p2}) \tanh w/L_{p1}]}$$

$$B = \frac{p(0)}{\sinh w/L_{p1}}$$

$$C = \frac{p(0) \exp (w/L_{p2})}{\cosh w/L_{p1}[1 + (L_{p1}/L_{p2}) \tanh w/L_{p1}]}$$

$$J = \frac{eD_p p(0)}{L_{p1}} \left(\frac{L_{p2} \sinh w/L_{p1} + L_{p1} \cosh w/L_{p1}}{L_{p2} \cosh w/L_{p1} + L_{p1} \sinh w/L_{p1}} \right)$$

$$L_{p1} = 1.11 \times 10^{-3} \text{ cm}; \quad L_{p2} = 3.52 \times 10^{-4} \text{ cm}$$

16

Transistors

A transistor is a three-layer device, consisting either of two p layers separated by a thin n layer (*pnp* transistor) or of two n-layers separated by a thin p-layer (*npn* transistor). We shall discuss the operation of the *pnp* transistor, that of the *npn* transistor being analogous. In the *pnp* transistor the one pn junction is forward biased. If $N_A \gg N_D$, the p region *injects* holes into the n region. The second pn junction is back biased so that this p region *extracts* holes from the n region. For these reasons the first p region is called the *emitter* and the second p region the *collector*. The middle region is called the *base*, for a reason that relates to the original method of manufacturing transistors.

Since the emitter junction is biased in the forward direction, it is operating at a low impedance level. The collector junction is biased in the back direction; hence it is operating at a high impedance level. If almost all the holes injected by the emitter are collected at the collector, then the same current that is injected at a low impedance level is collected at a high impedance level, so that the output power is much larger than the input power. In other words, the device can amplify ac signals.

We can obtain a better circuit by connecting the small ac signal between base and emitter, inserting a load resistance between the collector and the emitter, and taking the amplified signal from this load. The reason is that in a good transistor almost all the current injected by the emitter is collected by the collector so that hardly any current flows to the base. As a consequence, the base provides a much higher input impedance to the signal source than the emitter. Moreover, since the small base current now controls the much larger collector current, a considerable current amplification is achieved.

The first circuit is called the *common-base circuit* and the second circuit the *common-emitter circuit*. For the reason just mentioned, the common-emitter circuit is used almost exclusively.

16.1. CURRENT FLOW IN TRANSISTORS

16.1a. Ebers–Moll Equations

A *pnp* transistor consists of two diodes back to back; for a forward-biased emitter and a back-biased collector the emitter thus injects carriers into the base and the collector extracts carriers from the base. For reverse bias, the opposite is true. With both electrodes forward biased, the emitter current I_E and the collector current I_C may be written

$$I_E = I_{Es} \exp\left(\frac{eV_{EB}}{kT}\right) - \alpha_R I_{Cs} \exp\left(\frac{eV_{CB}}{kT}\right) - I_{BE}$$

$$I_C = -\alpha_F I_{Es} \exp\left(\frac{eV_{EB}}{kT}\right) + I_{Cs} \exp\left(\frac{eV_{CB}}{kT}\right) - I_{BC}$$

(16.1)

Here $I_{Es} \exp(eV_{EB}/kT)$ represents the current due to holes injected from the emitter into the base and the small current due to electrons injected from the base into the emitter. The part α_F of this current is due to holes collected by the collector; the minus sign occurs because I_C is counted positive if the current flows into the collector. Moreover, $I_{Cs} \exp(eV_{CB}/kT)$ represents the current due to holes injected from the collector into the base and the current due to electrons injected from the base into the collector. The part α_R of this current is due to holes collected by the emitter. Most transistors are of the p^+np^- type; hence the electron current injected from the base into the emitter is quite small, and for a good device geometry, α_F is very close to unity; typically, $\alpha_F = 0.980$–0.995. However, the electron current injected from the base into the collector can be quite large, and hence α_R is often quite small; this is accentuated by the fact that the collector area is usually much larger than the emitter area. α_F and α_R are appropriately called the *forward* and *reverse current amplification factors*, respectively.

The small current I_{BE} is due to holes generated in the base and collected by the emitter and to electrons generated in the emitter and collected by the base. The small current I_{BC} is due to holes generated in the base and collected by the collector and to electrons generated in the collector and collected by the base. The situation is pictured in Fig. 16.1.

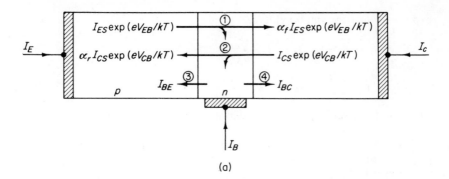

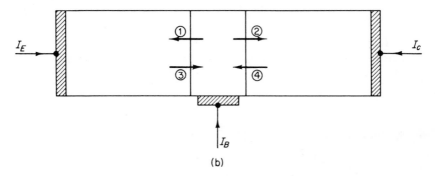

Fig. 16.1. (a) Current flow in a *p-n-p* transistor due to four groups of holes. (b) Current flow in a *p-n-p* transistor due to four groups of electrons. (From A. van der Ziel, *Electronics*, Allyn and Bacon Co., Boston, 1966.)

We observe that $I_E = I_C = 0$ if $V_{EB} = V_{CB} = 0$. Hence

$$I_{BE} = I_{Es} - \alpha_R I_{Cs}, \qquad I_{BC} = I_{Cs} - \alpha_F I_{Es} \qquad (16.1a)$$

Substituting back into Eq. (16.1) yields

$$I_E = I_{Es}\left[\exp\left(\frac{eV_{EB}}{kT}\right) - 1\right] - \alpha_R I_{Cs}\left[\exp\left(\frac{eV_{CB}}{kT}\right) - 1\right]$$

$$I_C = -\alpha_F I_{Es}\left[\exp\left(\frac{eV_{EB}}{kT}\right) - 1\right] + I_{Cs}\left[\exp\left(\frac{eV_{CB}}{kT}\right) - 1\right] \qquad (16.2)$$

These equations are called the *Ebers–Moll equations*.

The four parameters I_{Es}, I_{Cs}, α_F, and α_R are related as

$$\alpha_F I_{Es} = \alpha_R I_{Cs} \qquad (16.3)$$

This can be called the *Onsager relation* for transistors. The proof follows from the observation that for $|eV_{EB}/kT| \ll 1$ and $|eV_{CB}/kT| \ll 1$ the circuit is a passive, linear resistive circuit that should have reciprocity. Since

$\exp x - 1 = x$ for small $|x|$, (16.2) yields

$$I_E = I_{Es}\frac{eV_{EB}}{kT} - \alpha_R I_{Cs}\frac{eV_{CB}}{kT} \tag{16.2a}$$

$$I_C = -\alpha_F I_{Es}\frac{eV_{EB}}{kT} + I_{Cs}\frac{eV_{CB}}{kT} \tag{16.2b}$$

This is a linear set of equations; reciprocity yields Eq. (16.3).

We evaluate I_{Es}, I_{Cs}, α_F, and α_R for a one-dimensional model of a p^+np^+ transistor at low injection; in this case practically all current is carried by holes. The diffusion equation in the region may then be written

$$\frac{d^2p}{dx^2} = \frac{p - p_n}{L_p^2} \tag{16.4}$$

where $L_p = (D_p\tau_p)^{1/2}$ is the diffusion length of holes in the base region.

Let the base region extend from $x = 0$ (emitter side of base) to $x = w_B$ (collector side of base); then the initial conditions for $p(x)$ are

$$p(0) = p_n \exp\left(\frac{eV_{EB}}{kT}\right), \qquad p(w_B) = p_n \exp\left(\frac{eV_{CB}}{kT}\right) \tag{16.4a}$$

Since $\sinh x/L_p$ and $\sinh (w_B - x)/L_p$ are solutions for $p - p_n$, we put

$$p - p_n = A \sinh\left(\frac{w_B - x}{L_p}\right) + B \sinh\frac{x}{L_p}$$

where A and B are constants. Applying the boundary conditions (16.4a) yields

$$p(x) - p_n = [p(0) - p_n]\frac{\sinh (w_B - x)/L_p}{\sinh w_B/L_p} + [p(w_B) - p_n]\frac{\sinh x/L_p}{\sinh w_B/L_p} \tag{16.5}$$

Consequently, the hole current densities $J_p(0)$ and $J_p(w_B)$ are

$$J_p(0) = -eD_p\frac{dp}{dx}\bigg|_{x=0}$$

$$= \frac{eD_p}{L_p \tanh w_B/L_p}[p(0) - p_n] + \frac{eD_p}{L_p \sinh w_n/L_p}[p(w_B) - p_n] \tag{16.6}$$

$$J_p(w_B) = eD_p\frac{dp}{dx}\bigg|_{x=w_B}$$

$$= -\frac{eD_p}{L_p \sinh w_B/L_p}[p(0) - p_n] + \frac{eD_p}{L_p \tanh w_B/L_p}[p(w_B) - p_n] \tag{16.7}$$

Multiplying by the junction area A and comparing with Eq. (16.2) yields

$$I_{Es} = I_{Cs} = \frac{eD_pAp_n}{L_p \tanh w_B/L_p}, \qquad \alpha_F = \alpha_R = \frac{1}{\cosh w_B/L_p} \tag{16.8}$$

Usually, w_B/L_p is quite small. Making Taylor expansions yields

$$I_{Es} = I_{Cs} = \frac{eD_pAp_n}{w_B}, \qquad \alpha_F = \alpha_R = 1 - \frac{1}{2}\frac{w_B^2}{L_p^2} \qquad (16.9)$$

It is obvious that I_{Es} and I_{Cs} are both proportional to the equilibrium hole concentration $p_n = n_i^2/N_d$, so that they are strongly temperature dependent.

In this one-dimensional model, where all current is carried by holes, $I_{Es} = I_{Cs}$ and $\alpha_F = \alpha_R$. These equalities disappear for a three-dimensional model in which the junctions have arbitrary shape and in which part of the current is carried by electrons. What remains true, however, is that for a good transistor $\alpha_F \simeq 1$, and if the geometry is planar

$$I_{Es} \simeq \frac{eD_pAp_n}{w_B} \qquad (16.10)$$

It is not always true, however, that

$$\alpha_F \simeq 1 - \frac{1}{2}\frac{w_B^2}{L_p^2} \qquad (16.11)$$

For in modern transistors w_B is so small that the right side of (16.11) is very nearly unity, whereas α_F is not much larger than 0.990–0.995. This discrepancy comes from the fact that a small fraction of $I_{Es} \exp(eV_{EB}/kT)$ comes from electrons injected from the base into the n region. We come back to this problem in Sec. 16.2b.

We notice that, for a strongly back biased collector, $p(w_B) \simeq 0$. This can never be fully true, for beyond the point $x = w_B$ is the collector space-charge region, and here the current flow is by drift. The minimum hole density will occur if the holes drift with their critical velocity $u_c(u_c \simeq 10^7 \text{ cm/s})$. In that case

$$I_C = ep(w_B)u_c A \qquad (16.12)$$

so that $p(w_B)$ must satisfy the inequality

$$p(w_B) > \frac{I_C}{eu_c A} \qquad (16.12a)$$

For example, if $I_C = 16$ mA, $A = 10^{-6}$ cm², and $u_c = 10^7$ cm/s, then $p(w_B) > 10^{16}$ cm³. It is usually true, however, that for such large current densities $p(0) \gg p(w_B)$.

16.1b. Transistors with Back-Biased Collector

In most transistor applications the emitter is forward biased and the collector is back biased. In that case

$$I_E = I_{Es} \exp\left(\frac{eV_{EB}}{kT}\right) - I_{BE} = I_{Es}\left[\exp\left(\frac{eV_{EB}}{kT}\right) - 1\right] + \alpha_R I_{Cs} \qquad (16.13)$$

Hence the small-signal emitter–base conductance g_{eb0} is

$$g_{eb0} = \frac{dI_E}{dV_{EB}} = \frac{e}{kT} I_{Es} \exp\left(\frac{eV_{EB}}{kT}\right) = \frac{e}{kT}(I_E + I_{BE}) \qquad (16.13a)$$

The collector current is

$$
\begin{aligned}
I_C &= -\alpha_F I_{Es}\left[\exp\left(\frac{eV_{EB}}{kT}\right) - 1\right] - I_{Cs} \\
&= -\alpha_F I_E + I_{Cs}(1 - \alpha_F \alpha_R) = -\alpha_F I_E - I_{C0}
\end{aligned}
\qquad (16.14)
$$

where $I_{C0} = I_{Cs}(1 - \alpha_R \alpha_F)$ is the collector current for open emitter; it is called the *collector saturated current*. The transconductance g_m is defined as

$$g_{m0} = -\frac{dI_C}{dV_{EB}} = \alpha_F g_{eb0} = \frac{e}{kT}(-I_C - I_{BC}) \qquad (16.14a)$$

For silicon transistors, I_{BE}, I_{C0} and I_{BC} can be neglected and $g_{m0} \simeq -eI_C/kT$, unless one operates at very small currents.

The current I_B is (positive current flows into base)

$$
\begin{aligned}
I_B &= -I_E - I_C = -(1 - \alpha_F)I_{Es}\left[\exp\left(\frac{eV_{EB}}{kT}\right) - 1\right] + (1 - \alpha_R)I_{Cs} \\
&= -(1 - \alpha_F)I_E + I_{C0}
\end{aligned}
\qquad (16.15)
$$

In common-emitter connection the dc bias $V_{BE} = -V_{EB}$; hence the small-signal base conductance is

$$g_{be0} = \frac{dI_B}{dV_{BE}} = -\frac{dI_B}{dV_{EB}} = (1 - \alpha_F)g_{eb0} = \frac{e}{kT}(-I_B + I_{BE} + I_{BC}) \quad (16.15a)$$

If α_F is close to unity, then the base–emitter conductance g_{be0} is much smaller than g_{eb0}, a great advantage for the common-emitter circuit. For silicon transistors $g_{be0} \simeq -eI_B/kT$.

It is now convenient to express I_C in terms of I_B. We find with the help of (16.15) and (16.14) that

$$I_C = \frac{\alpha_F}{1 - \alpha_F} I_B - \frac{I_{C0}}{1 - \alpha_F} = \beta_F I_B - I_{C0}(1 + \beta_F) \qquad (16.16)$$

where β_F is the forward current amplification factor in the common-emitter connection. It is very large if α_F is close to unity. In silicon transistors $I_{C0}(1 + \beta_F)$ is negligible and $\beta_F = I_C/I_B$. At high temperature the term $I_{C0}(1 + \beta_F)$ is important; it is much larger than the collector saturated current I_{C0}.

Figure 16.2a shows $|I_C|$ plotted as a function of $-V_{CB}$ for the common-emitter connection with the emitter current I_E as a parameter. Figure 16.2b shows $|I_C|$ plotted versus $-V_{CE}$ for the common-emitter connection with the base current $|I_B|$ as a parameter.

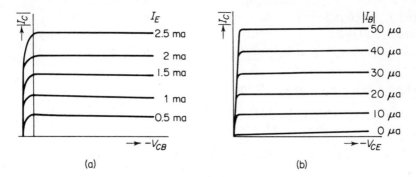

Fig. 16.2. (a) $(|I_C|, -V_{CB})$ characteristic of a *p-n-p* transistor with I_E as a parameter. (b) $(|I_C|, -V_{CE})$ characteristic of a *p-n-p* transistor with $|I_B|$ as a parameter.

16.1c. High-Frequency Theory of the Transistor

We now calculate the high-frequency input admittance Y_{eb}, the high-frequency current amplification factor α, the high-frequency transconductance g_m, and the high-frequency common-emitter input conductance Y_{be}. We do so for a one-dimensional model with a base length w_B and a back-biased collector. Later we shall correct the result for other geometries.

In this model let a voltage

$$V(t) = V_{EB} + v_{eb} \exp(j\omega t) \tag{16.17}$$

be applied to the emitter, where $|ev_{eb}/kT| \ll 1$ (small-signal condition). Then the hole concentration at $x = 0$ is

$$
\begin{aligned}
p(0) &= p_n \exp\left(\frac{eV}{kT}\right) = p_n \exp\left(\frac{eV_{EB}}{kT}\right)\left[1 + \frac{ev_{eb}}{kT}\exp(j\omega t)\right] \\
&= p_0(0) + p_1(0)\exp(j\omega t)
\end{aligned} \tag{16.18}
$$

where $\qquad p_0(0) = p_n \exp\left(\dfrac{eV_{EB}}{kT}\right), \qquad p_1(0) = p_0(0)\dfrac{ev_{eb}}{kT} \qquad$ (16.18a)

Here $p_0(0)$ and $p_1(0)$ are the dc and ac hole concentrations at $x = 0$, respectively. We have substituted $\exp(x) \simeq 1 + x$ for $x = (ev_{eb}/kT)\exp(j\omega t)$.

The hole concentration in the base region satisfies the time-dependent diffusion equation

$$\frac{\partial p}{\partial t} = -\frac{p - p_n}{\tau_p} + D_p \frac{\partial^2 p}{\partial x^2} \tag{16.19}$$

with $p(x) = p(0)$ at $x = 0$ and $p(x) \simeq 0$ at $x = w_B$. To solve Eq. (16.19), we substitute for $p(x)$

$$p(x) = p_0(x) + p_1(x)\exp(j\omega t) \tag{16.20}$$

where $p_0(x)$ satisfies the dc diffusion equation

$$D_p \frac{d^2 p_0}{dx^2} - \frac{p_0 - p_n}{\tau_p} = 0 \quad \text{or} \quad \frac{d^2 p_0}{dx^2} = \frac{p_0 - p_n}{L_p^2} \tag{16.21}$$

and $L_p = (D_p\tau_p)^{1/2}$. Substituting (16.20) into (16.19) and collecting the ac terms yields

$$D_p\frac{d^2p_1}{dx^2} - \frac{p_1}{\tau_p} = j\omega p_1 \quad \text{or} \quad \frac{d^2p_1}{dx^2} = \frac{p_1}{L_p'^2} \qquad (16.22)$$

where $p_1(x) = p_1(0)$ at $x = 0$ and $p_1(x) = 0$ at $x = w_B$, and

$$L_p' = \frac{(D_p\tau_p)^{1/2}}{(1 + j\omega\tau_p)^{1/2}} = \frac{L_p}{(1 + j\omega\tau_p)^{1/2}} \qquad (16.22a)$$

It is easily seen by substitution that (16.22) has the solution

$$p_1(x) = p_1(0)\frac{\sinh[(1 + j\omega\tau_p)^{1/2}(w_B - x)/L_p]}{\sinh[(1 + j\omega\tau_p)^{1/2}w_B/L_p]} \qquad (16.23)$$

If A is the junction area, the ac currents at $x = 0$ and $x = w_B$ are

$$i_e = -eD_pA\frac{dp_1(x)}{dx}\bigg|_{x=0} = \frac{eD_pAp_0(0)(ev_{eb}/kT)(1 + j\omega\tau_p)^{1/2}}{L_p\tanh[(1 + j\omega\tau_p)^{1/2}w/L_p]} \qquad (16.24)$$

$$i_c = eD_pA\frac{dp_1}{dx}\bigg|_{x=w_B} = -\frac{eD_pAp_0(0)(ev_{eb}/kT)(1 + j\omega\tau_p)^{1/2}}{L_p\sinh[(1 + j\omega\tau_p)^{1/2}w/L_p]} \qquad (16.25)$$

Hence the ac current amplification factor is

$$\alpha = -\frac{i_c}{i_e} = \alpha_F\frac{\cosh w_B/L_p}{\cosh[(1 + j\omega\tau_p)^{1/2}w_B/L_p]} \qquad (16.26)$$

$$g_m = -\frac{i_c}{v_{eb}} = g_{m0}\frac{(1 + j\omega\tau_p)^{1/2}\sinh w_B/L_p}{\sinh[(1 + j\omega\tau_p)^{1/2}w_B/L_p]} \qquad (16.27)$$

$$Y_{eb} = \frac{i_e}{v_{eb}} = g_{eb0}\frac{(1 + j\omega\tau_p)^{1/2}\tanh w_B/L_p}{\tanh[(1 + j\omega\tau_p)^{1/2}w_B/L_p]} \qquad (16.28)$$

where, for $I_E \gg I_{BE}$,

$$\alpha_F = \frac{1}{\cosh w_B/L_p}, \quad g_{m0} = \frac{eD_pAp_0(0)e/kT}{L_p\sinh w_B/L_p} = -\frac{eI_C}{kT},$$

$$g_{eb0} = \frac{eD_pAp_0(0)e/kT}{L_p\tanh w_B/L_p} = \frac{eI_E}{kT} \qquad (16.29)$$

according to the preceding theory. The input admittance Y_{be} for the common-emitter connection follows from

$$Y_{be} = -\frac{i_b}{v_{eb}} = \frac{i_e + i_c}{v_{eb}}$$

$$= g_{eb0}\frac{(1 + j\omega\tau_p)^{1/2}\tanh w_B/L_p}{\tanh[(1 + j\omega\tau_p)^{1/2}w_B/L_p]} - g_{m0}\frac{(1 + j\omega\tau_p)^{1/2}\sinh w_B/L_p}{\sinh[(1 + j\omega\tau_p)^{1/2}w_B/L_p]}$$

$$\simeq g_{be0} + g_{m0}\left\{\frac{(1 + j\omega\tau_p)^{1/2}\tanh w_B/L_p}{\tanh[(1 + j\omega\tau_p)^{1/2}w_B/L_p]} - \frac{(1 + j\omega\tau_p)^{1/2}\sinh w_B/L_p}{\sinh[(1 + j\omega\tau_p)^{1/2}w_B/L_p]}\right\}$$

$$(16.30)$$

where $g_{be0} = g_{eb0} - g_{m0}$ is the low-frequency input conductance of the common-emitter connection.

We now observe that w/L_p is a small number. Hence $\sinh w/L_p \simeq w/L_p$, $\tanh w/L_p \simeq w/L_p$, and

$$z = (1 + j\omega\tau_p)^{1/2}\frac{w_B}{L_p} \simeq \left(j\omega\tau_p\frac{w_B^2}{L_p^2}\right)^{1/2} = (2j\omega\tau_d)^{1/2} \qquad (16.31)$$

since $L_p^2 = D_p\tau_p$, and

$$\tau_d = \frac{w_B^2}{2D_p} \qquad (16.32)$$

is the diffusion time of the carriers through the base region.

We may thus write in good approximation

$$\frac{\alpha}{\alpha_F} \simeq \frac{1}{\cosh\left[(2j\omega\tau_d)^{1/2}\right]} \qquad (16.26a)$$

$$\frac{g_m}{g_{m0}} \simeq \frac{(2j\omega\tau_d)^{1/2}}{\sinh\left[(2j\omega\tau_d)^{1/2}\right]} \qquad (16.27a)$$

$$\frac{Y_{eb}}{g_{eb0}} \simeq \frac{(2j\omega\tau_d)^{1/2}}{\tanh\left[(2j\omega\tau_d)^{1/2}\right]} \qquad (16.28a)$$

$$Y_{be} \simeq g_{be0} + g_{m0}\frac{(2j\omega\tau_d)^{1/2}}{\sinh[(2j\omega\tau_d)^{1/2}]}\{\cosh\left[(2j\omega\tau_d)^{1/2}\right] - 1\} \quad (16.30a)$$

Note that in these expressions all references to the particular model and to the lifetime τ_p have disappeared, and only the diffusion time τ_d through the base region is retained. The results should thus be valid for any model for which the diffusion time τ_d can be defined, that is, for any planar geometry.

We now make Taylor expansions of the right side of these equations and introduce

$$f_\alpha = \frac{1}{2\pi\tau_d} = \frac{D_p}{\pi w_B^2} \quad \text{or} \quad (2j\omega\tau_d)^{1/2} = (1 + j)\left(\frac{f}{f_\alpha}\right)^{1/2}$$

Then for small values of f/f_α we obtain

$$\frac{\alpha}{\alpha_F} \simeq \frac{1}{1 + jf/f_\alpha} \qquad (16.26b)$$

$$\frac{g_m}{g_{m0}} \simeq \frac{1}{1 + \frac{1}{3}(jf/f_\alpha)} \simeq 1 - \frac{1}{3}j\omega\tau_d \qquad (16.27b)$$

$$\frac{Y_{eb}}{g_{eb0}} \simeq \frac{1}{1 - \frac{2}{3}(jf/f_\alpha)} \simeq 1 + \frac{2}{3}j\omega\tau_d \qquad (16.28b)$$

$$Y_{be} = Y_{eb} - g_m \simeq g_{be0} + j\omega\tau_d g_{m0} \qquad (16.30b)$$

The input admittance of the common-base connection thus consists of a conductance g_{eb0} and a capacitance $C_d' = \frac{2}{3}g_{eb0}\tau_d$ in parallel, and the input admittance for the common-emitter connection consists of a conductance $g_{be0} = g_{eb0} - g_{m0}$ and a capacitance $C_d = g_{m0}\tau_d$ in parallel. Moreover,

$|\alpha| = \frac{1}{2}\sqrt{2}\,\alpha_F$ at $f = f_\alpha$, whereas $|g_m|$ is practically equal to g_{m0} up to $f = f_\alpha$, with only a small phase angle for g_m.

We now evaluate these expressions more accurately. In the first place we write

$$\alpha = |\alpha| \exp(-j\varphi_\alpha) \tag{16.33}$$

Figure 16.3a shows $|\alpha|/\alpha_F$ as a function of f/f_α, and Fig. 16.3b shows φ_α as a function of f/f_α. Note that φ_α can be larger than $\pi/2$, in contrast with Eq. (16.26b). We also read from the graph that $|\alpha|/\alpha_F = \frac{1}{2}\sqrt{2}$ at $f = f_\alpha' = 1.215$ f_α. If we plot $(1 + f^2/f_\alpha^2)^{-1/2}$ and $(1 + f^2/f_\alpha'^2)^{1/2}$ versus f/f_α, we see that the actual value of $|\alpha|/\alpha_F$ lies in between. If we look at Fig. 16.3b, where the function $\tan^{-1}(f/f_\alpha)$ has been plotted versus f/f_α, we see a large discrepancy with φ_α for large f/f_α; Eq. (16.26b) is therefore only approximately correct.

Next we write

$$g_m = |g_m| \exp(-j\varphi_m) \tag{16.34}$$

Figure 16.4a shows $|g_m|/g_{m0}$ as a function of f/f_α, and Fig. 16.4b shows φ_m as a function of f/f_α. Note that $|g_m|$ decreases only slowly with increasing frequency and that φ_m is much smaller than φ_α.

Figure 16.5 shows the real and imaginary parts of $(Y_{be} - g_{be0})/g_{m0}$ as a function of f/f_α. Also shown is the line $\omega C_d/g_{m0}$. It is seen that the imaginary part of our expression differs from $\omega C_d/g_{m0}$ for $f > f_\alpha$, and that the real part of $(Y_{be} - g_{be0})/g_{m0}$ has an appreciable value when f is of the order of f_α. The expression

$$Y_{be} = g_{be0} + j\omega C_d \tag{16.35}$$

is therefore only approximately correct. Nevertheless, it is useful for $f < f_\alpha$, since the imaginary part of $Y_{be} - g_{be0}$ effectively shunts its real part so that the deviations from (16.35), although reasonably large, have little final effect.

We shall now show that the capacitance C_d can be interpreted as the hole storage capacitance of the base region. If the base length is small ($w_B \ll L_p$), the hole distribution in the base region is linear:

$$p(x) = p_n + [p_0(0) - p_n]\frac{w_B - x}{w_B} \tag{16.36}$$

The hole charge Q stored in the base region is

$$Q = eA \int_0^{w_B} [p(x) - p_n]\,dx = \frac{1}{2}e[p_0(0) - p_n]w_B A \tag{16.36a}$$

where A is the junction area. Hence the storage capacitance C_s is

$$C_s = \frac{dQ}{dV_{EB}} = \frac{1}{2}ep_0(0)Aw_B\frac{e}{kT} = g_{m0}\tau_d = C_d \tag{16.37}$$

since $p_0(0) = p_n \exp(eV_{EB}/kT)$, $g_{m0} \simeq eD_p p_0(0)A(e/kT)/w_B$ according to (16.29), and $\tau_d = w_B^2/2D_p$.

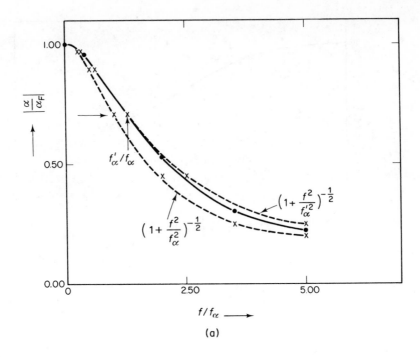

(a)

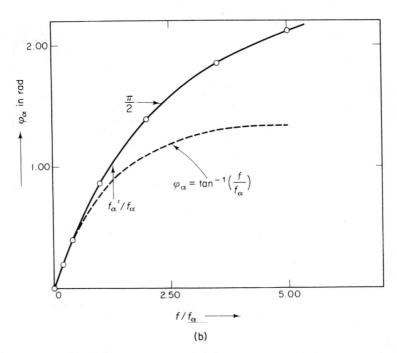

(b)

Fig. 16.3. (a) $|\alpha|/\alpha_F$ as a function of f/f_α. (b) Phase angle φ of α versus f/f_α.

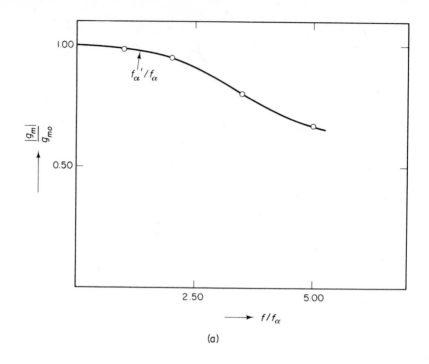

(a)

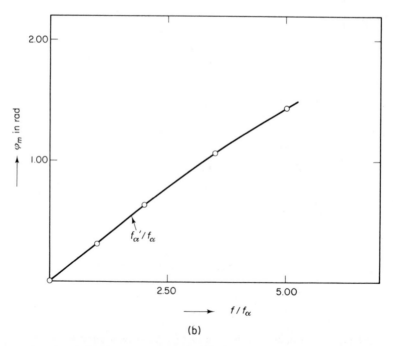

(b)

Fig. 16.4. (a) $|g_m|/g_{m0}$ versus f/f_α. (b) Phase angle φ_m of g_m versus f/f_α.

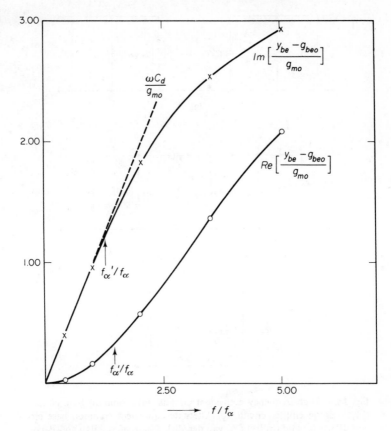

Fig. 16.5. Real and imaginary parts of $(Y_{be} - g_{be0})/g_{m0}$ versus f/f_α.

Figure 16.6a shows the equivalent circuit for the common-base connection and Fig. 16.6b the equivalent circuit for the common-emitter connection. Also shown are the capacitances C_{je} and C_{jc} of the emitter–base and the collector–base space-charge regions. The difference between C_d and C'_d comes from the phase angle of the transconductance g_m.

For small currents $C'_d \ll C_{je}$, and this affects f_α. This is easily seen as follows. Now

$$\alpha = -\frac{i_c}{i_e} = -\frac{i_c/v_{eb}}{i_e/v_{eb}} = \frac{g_m}{Y_{eb}} = \frac{\alpha_F}{1 + jf/f_\alpha} \tag{16.38}$$

or

$$1 + j\frac{f}{f_\alpha} \simeq \left[1 + \frac{j\omega(C'_d + C_{je})}{g_{eb0}}\right]\left(1 + \frac{1}{3}j\omega\tau_d\right) \simeq 1 + j\omega\left(\tau_d + \frac{C_{je}}{g_{eb0}}\right) \tag{16.38a}$$

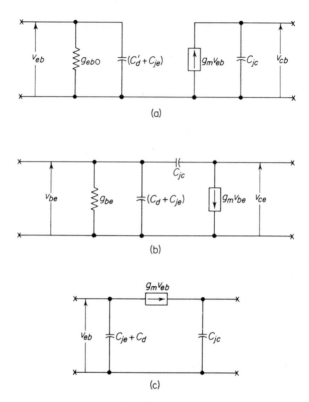

Fig. 16.6. High-frequency equivalent circuits. (a) Common base circuit. (b) Common emitter circuit. (c) Alternate equivalent common base cir-ciut. [Parts (a) and (b) from A. van der Ziel, *Electronics*, Allyn and Bacon Co., Boston, 1966.]

since $C'_d = \frac{2}{3} g_{eb0} \tau_d$ and $\alpha_F = g_{m0}/g_{eb0}$, so that

$$\frac{1}{f_\alpha} = 2\pi\left(\tau_d + \frac{C_{Je}}{g_{eb0}}\right) \tag{16.39}$$

This means that f_α can be orders of magnitude smaller than $1/(2\pi\tau_d)$ at very small currents. This is important in many low-current transistor circuits.

We finally introduce the high-frequency current amplification factor β for the common-emitter connection. Since $i_c = -\alpha i_e$ and $i_b = -i_e - i_c$, we have

$$\beta = \frac{i_c}{i_b} = \frac{\alpha}{1-\alpha} = \frac{1}{\cosh\left[(1 + j\omega\tau_p)^{1/2} w/L_p\right] - 1} \tag{16.40}$$

Substituting $\alpha = \alpha_F/(1 + jf/f_\alpha)$ yields, if $\beta_F = \alpha_F/(1 - \alpha_F)$,

$$\beta = \frac{\beta_F}{1 + jf/f_\beta} = \frac{\beta_F}{1 + j\omega\tau_\beta} \tag{16.41}$$

where $f_\beta = f_\alpha/\beta_F$ and $\tau_\beta = \tau_d\beta_F$. We need this equation for the pulse response of transistors (Chapter 17).

We finally need to know when $|\beta| = 1$. Let this be at $f = f_T$, then

$$|\cosh (2j\omega_T\tau_d)^{1/2} - 1| = 1 \tag{16.41a}$$

A graphical solution of the equation yields $f_T \simeq 1.00 f_\alpha$.

16.1d. Manufacturing Techniques for Transistors

In the early days of transistor technology, transistors were either grown from the melt or were made by alloying techniques. These techniques have now been replaced by diffusion techniques.

In the double-diffusion technique for *pnp* transistors, one starts with a thin layer of *p*-type material epitaxially deposited on a p^+ substrate, which becomes the collector, and subsequently an *n*-type base region and a *p*-type emitter region are diffused in. Finally, one provides ohmic contacts to the two regions. The p^+ region is used to reduce the collector series resistance.

In integrated circuits one usually makes *npn* transistors. The reason is that the *p*-diffusion is made by boron doping and the *n*-diffusion by phosphorus doping. The latter gives the highest dopant concentration and hence makes better emitters.

In integrated circuits one often uses a triple-diffusion technique. Here one starts with a thin layer of *n*-type material epitaxially deposited on a *p*-type substrate. To isolate the individual transistors, one first diffuses *p*-type sections all the way through the *n*-type region; this cuts out *n*-type islands. Then a *p*-type base and an *n*-type emitter are diffused in and the original *n*-type section is used as a collector.

16.2. MISCELLANEOUS TRANSISTOR PROBLEMS

First we discuss various contributions to the base current:

1. Recombination at the surface of the base region.
2. Injection of carriers from the base into the emitter.
3. Recombination in the emitter–base space-charge region.

In addition, a few other transistor effects will be discussed.

16.2a. Recombination at the Surface of the Base Region

Up to now we assumed that the geometry was one-dimensional and that the recombination was volume recombination. Actually, the geometry is not fully one dimensional and most recombination occurs at the surface. That is, we may often put $\tau_p = \infty$ and $L_p = (D_p\tau_p)^{1/2} = \infty$. The diffusion equation then becomes

$$\nabla^2 p' = \frac{\partial^2 p'}{\partial x^2} + \frac{\partial^2 p'}{\partial y^2} + \frac{\partial^2 p'}{\partial z^2} = 0 \tag{16.42}$$

where $p' = p - p_n$, and p_n is the equilibrium hole concentration. The boundary conditions at emitter and collector are (Fig. 16.7)

$$p' \simeq p_n\left[\exp\left(\frac{eV_{EB}}{kT}\right) - 1\right] \quad \text{at emitter,} \qquad p' \simeq 0 \quad \text{at collector} \tag{16.42a}$$

The current density in any point is, in vector notation,

$$\mathbf{J} = -eD_p\nabla p' \tag{16.43}$$

where $\nabla p'$ has components $\partial p'/\partial x$, $\partial p'/\partial y$, and $\partial p'/\partial z$. At the surface the normal component J_n of the hole current density $\mathbf{J}$ is

$$J_n = ep's \tag{16.44}$$

where s is the surface recombination velocity. Hence

$$\frac{1}{p'}\nabla_n p' = \frac{1}{p'}\frac{\partial p'}{\partial n}\bigg|_s = -\frac{s}{D_p} \tag{16.45}$$

where $\nabla_n p'$ is the normal component of the gradient of p' at the surface. Equation (16.42) can now be solved for the boundary conditions (16.42a) and (16.45). Usually this must be done on a computer.

Usually, the emitter–base junction and the collector–base junction are closely parallel, and the theory of Sec. 16.1 holds for I_C, so that, if w_B is the

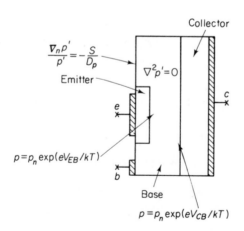

Fig. 16.7. Boundary conditions at the surface of the base region if the flow of holes is not one-dimensional and if all recombination occurs at the surface (surface recombination velocity s).

effective base length,

$$I_C = -\frac{eD_p A}{w_B} p_n \exp\left(\frac{eV_{EB}}{kT}\right) \tag{16.46}$$

The base current I_B is then

$$I_B = -\int_S J_n \, dS = -es \int_S p' \, dS = -es A_{\text{eff}} p_n \exp\left(\frac{eV_{EB}}{kT}\right) \tag{16.47}$$

where A_{eff} is an "effective" base area that depends only very slightly on the bias at the emitter and at the collector.

If this base current mechanism predominates, we have

$$\beta_F = \frac{I_C}{I_B} = \frac{D_p A}{s w_B A_{\text{eff}}} \tag{16.48}$$

Usually, however, other effects are more important.

16.2b. Electron Current in PNP Transistors

The theory of Sec. 16.2a holds for the hole currents I_{Cp} and I_{Bp}:

$$I_{Cp} = -\frac{eD_p A}{w_B} p_n \exp\left(\frac{eV_{EB}}{kT}\right), \qquad I_{Bp} = -es A_{\text{eff}} p_n \exp\left(\frac{eV_{EB}}{kT}\right) \tag{16.49}$$

To evaluate the contribution I_{Bn} of the electron current injected into the emitter, we assume that the emitter is very shallow; that is, the emitter length w_E is small in comparison with the diffusion length L_{ne} of the electrons in the emitter. In that case (Chapter 15), I_{Bn} may be written

$$I_{Bn} = -\frac{eD_n A}{w_E} n_p \exp\left(\frac{eV_{EB}}{kT}\right) \tag{16.50}$$

Here D_n is the electron diffusion constant and n_p the equilibrium electron concentration in the emitter. This does not depend on collector bias.

The dc current amplification factor β_F is therefore

$$\beta_F = \frac{I_C}{I_B} = \frac{I_{Cp}}{I_{Bp} + I_{Bn}} = \frac{(D_p/w_B)p_n}{s A_{\text{eff}} p_n/A + (D_n/w_E)n_p} \tag{16.51}$$

Hence for $I_{Bn} < I_{Bp}$, β_F is determined by the hole flow,

$$\beta_F \simeq \frac{D_p/w_B}{s A_{\text{eff}}/A} \tag{16.52}$$

and for $I_{Bn} > I_{Bp}$, β_F is determined by the electron flow,

$$\beta_F \simeq \frac{(D_p/w_B)p_n}{(D_n/w_E)n_p} = \frac{(D_p/w_B)N_a}{(D_n/w_E)N_d} \tag{16.53}$$

Almost always the emitter junction is a p^+n junction, and hence one would expect n_p to be extremely small. But this is erroneous; in most high-β transistors the electron injection effect is significant and Eq. (16.53) prevails.

To understand this, we must bear in mind that in the heavily doped emitter region the energy gap E_g decreases with increasing doping level. Therefore, since the square of the intrinsic carrier concentration is proportional to $\exp(-eE_g/kT)$, we have

$$p_n = \frac{n_{ib}^2}{N_d}, \qquad n_p = \frac{n_{ie}^2}{N_a} = \frac{n_{ib}^2}{N_a}\exp\left(\frac{e\,\Delta E_g}{kT}\right) \tag{16.54}$$

where n_{ie} and n_{ib} are the intrinsic carrier concentrations in emitter and base, respectively, and ΔE_g is the difference in gap width between base and emitter. Hence (16.53) becomes

$$\beta_F \simeq \frac{D_p/w_B}{D_n/w_E}\frac{N_a}{N_d}\exp\left(-\frac{e\,\Delta E_g}{kT}\right) \tag{16.55}$$

and this is considerably smaller than (16.53). As a consequence, the limit (16.55) for β_F is often reached.†

For the ac currents, we have

$$i_e = i_{pe} + i_{ne}, \qquad i_c = i_{pc} \tag{16.56}$$

Consequently, the transconductance of the device is not affected by the flow of electron current, but the input admittance Y_{ce} and the ac current amplification factor are. For example,

$$\alpha = -\frac{i_c}{i_e} = -\frac{i_{ep}}{i_{ep} + i_{en}}\cdot\frac{i_{cp}}{i_{ep}} = \gamma\alpha_c \tag{16.57}$$

where $\gamma = i_{ep}/(i_{ep} + i_{en})$ is called the ac emitter efficiency and $\alpha_c = -i_{cp}/i_{ep}$ is called the ac collector efficiency. Obviously, α_c corresponds to (16.26); by evaluating i_{ep} and i_{en} as functions of frequency, the frequency dependence of γ can also be evaluated.

In transistors operating in the reverse manner, that is, with the collector forward biased and the emitter back biased, the reverse current amplification factor α_R can be an order of magnitude smaller than unity. Most transistors are of the p^+np^- variety; that is, the collector contains a region that is only lightly doped. As a consequence, the base–collector electron current I_{Bn} can be considerably larger than the collector hole current I_{Cp}. Moreover, the emitter usually has a much smaller area than the collector, so that the emitter hole current I_{Ep} may be considerably smaller than I_{Cp} in magnitude. Hence

$$\alpha_R = -\frac{I_{Ep}}{I_{Cp} + I_{Bn}}$$

can be an order of magnitude smaller than unity. This is important in the transient response of saturated transistors (Sec. 17.1).

---

† R. P. Mertens, H. J. de Man, and R. J. van Overstraeten, *Trans. IEEE*, **ED 20**, 772 (1973).

16.2c. Recombination in the Emitter–Base Space-Charge Region

According to Eq. (15.80) if $eV_{EB}/kT \gg 1$, the recombination current in the space-charge region may be written†

$$I_R = \frac{\pi}{2} \frac{kTA}{\tau F_{max}} n_i \exp\left(\frac{eV_{EB}}{2kT}\right) = I_{R0} \exp\left(\frac{eV_{EB}}{2kT}\right) \quad (16.58)$$

The base current I_B contains I_R, whereas, if $eV_{EB}/kT \gg 1$, the hole currents may be written

$$I_{Ep} = \frac{eD_pA}{w_B} p_n \exp\left(\frac{eV_{EB}}{kT}\right) = I_{Ep0} \exp\left(\frac{eV_{EB}}{kT}\right), \qquad I_{Cp} = -\alpha_{CF} I_{Ep} \quad (16.59)$$

where $\alpha_{CF} = -I_{Cp}/I_{Ep}$ is the dc collector efficiency.

The dc emitter efficiency γ_{dc} is defined as

$$\gamma_{dc} = \frac{I_{Ep}}{I_{Ep} + I_R} = \left[1 + \frac{I_{R0}}{I_{Ep0}} \exp\left(-\frac{eV_{EB}}{2kT}\right)\right]^{-1} \quad (16.60)$$

For large eV_{EB}/kT this is close to unity, but for small eV_{EB}/kT γ_{dc} becomes small since I_{R0}/I_{Ep0} contains the large factor $n_i/p_n = N_d/n_i$. Hence

$$\alpha_F = \gamma_{dc}\alpha_{CF}; \quad \beta_F \simeq \frac{\gamma_{dc}}{1 - \gamma_{dc}} \quad (16.61)$$

if α_{CF} is close to unity.

The low frequency ac emitter efficiency is defined as

$$\gamma_0 = \frac{i_{ep}}{i_{ep} + i_r} = \frac{dI_{Ep}/dV_{EB}}{dI_{Ep}/dV_{EB} + dI_R/dV_{EB}}$$

$$= \left[1 + \frac{1}{2} \frac{I_{R0}}{I_{Ep0}} \exp\left(-\frac{eV_{EB}}{2kT}\right)\right]^{-1} \quad (16.62)$$

whereas the low-frequency ac current amplification α_0 is

$$\alpha_0 = \gamma_0\alpha_{c0}; \quad \beta_0 \simeq \frac{\gamma_0}{1 - \gamma_0} \quad (16.63)$$

Usually $\alpha_{c0} = i_{cp}/i_{ep}$ is equal to α_{CF} and is close to unity.

We now turn to the effect of recombination on g_{m0} and g_{eb0}. The effect of g_{m0} is non-existent, since $I_C = I_{Cp}$ and I_{Cp} is not affected. We have for g_{eb0}

$$g_{eb0} = \frac{dI_{EB}}{dV_{EB}} + \frac{dI_R}{dV_{EB}} = \frac{e}{kT}\left[I_{Ep0} \exp\left(\frac{eV_{EB}}{kT}\right) + \frac{1}{2} I_{R0} \exp\left(\frac{eV_{EB}}{2kT}\right)\right]$$

$$= \frac{eI_E}{kT} \frac{\gamma_{dc}}{\gamma_0} \simeq \frac{eI_E}{kT} \frac{\alpha_F}{\alpha_0} \quad (16.64)$$

---

† It is probably somewhat better to write the exponential as $\exp(eV_{EB}/mkT)$, with m differing slightly from 2 (Sec. 15.2e).

It is found that the effect becomes more pronounced at lower temperatures because I_{R0}/I_{Ep0} contains the factor N_d/n_i. In well-designed silicon units operating at room temperature, α_F remains practically constant down to emitter currents of 1 μA.

For high-frequency operation, one must bear in mind that the ac hole currents i_{ep} and i_{cp} are frequency dependent in the usual manner, whereas the ac recombination current i_r is not. The frequency dependence can be determined from the relationships

$$\gamma = \frac{i_{ep}}{i_{ep} + i_r}, \quad \alpha = -\frac{i_{cp}}{i_{ep} + i_r} = -\frac{i_{cp}}{i_{ep}}\gamma, \quad Y_{eb} = \frac{i_{ep} + i_r}{v_{eb}} = \frac{i_{ep}}{v_{eb}}\frac{1}{\gamma} \quad (16.65)$$

where v_{eb} exp $(j\omega t)$ is the ac emitter–base voltage. The effect is not important enough to warrant a detailed discussion.

16.2d. Effect of Built-In Fields

We saw in Sec. 16.1c that if we put $\alpha = |\alpha| \exp(-j\varphi_\alpha)$ then $|\alpha| \simeq \alpha_0/(1 + f^2/f_\alpha'^2)^{1/2}$ for properly chosen f_α', whereas φ_α is considerably larger than $\tan^{-1}(f/f_\alpha')$. We could remedy the situation by writing

$$\alpha = \frac{\alpha_0 \exp(-j\omega\tau)}{1 + jf/f_\alpha'} \quad (16.66)$$

with τ and f_α' properly chosen.

In many *pnp* transistors the donor concentration tapers off toward the collector. According to Sec. 6.5c, this results in an electric field so directed that it speeds up the flow of minority carriers. If the current flow in the base region were completely due to drift, we might write

$$\alpha = \alpha_0 \exp(-j\omega\tau_{dr}) \quad (16.66a)$$

where τ_{dr} is the drift time through the base region. Consequently, it is not surprising that the electric field in the base region increases τ in Eq. (16.66). This has no effect on $|\alpha|$, but it decreases $|\beta|$ at high frequencies. The effect of the exp $(-j\omega\tau)$ term is therefore much more significant for the common-emitter than for the common-base transistor.

16.2e. Avalanche Breakdown in the Collector Space-Charge Region

What happens if current multiplication occurs in the collector junction space-charge region? According to the theory of Sec. 14.2b, the collector current in a *pnp* transistor, and hence α_F, is multiplied by the factor

$$M = \frac{1}{[1 - |V_{CB}|/V_B]^n} \quad (16.67)$$

where $|V_{CB}|$ is the collector–base voltage, V_B the breakdown voltage of the junction, and n an experimental factor; usually $n \simeq 6$. It is thus possible to make α_F larger than unity. It can be shown that this leads to a negative input resistance in the common-base connection, which can be used for switching purposes. Since avalanche multiplication is a very fast process, very fast switching circuits can be obtained in this manner. However, increases in speed can also be obtained by making the base length smaller, and that is the procedure usually followed. Avalanche multiplication is therefore a nuisance effect that should be avoided. It sets an upper limit to $|V_{CB}|$ in power amplifiers.

16.2f. Punch-Through

In a transistor the base length becomes smaller if $|V_{CB}|$ is increased, because the collector space-charge region extends farther into the base region. It may thus happen that the base length becomes zero for a particular value of $|V_{CB}|$. This effect is called *punch-through* and should be avoided, since it makes the transistor inoperative because the base voltage no longer controls the (now large) collector current. By designing the transistor such that avalanche breakdown in the collector sets in before punch-through occurs, this can be avoided.

16.3. HIGH-LEVEL INJECTION EFFECTS IN TRANSISTORS

In this section we discuss the following high-injection effects:

1. At high injection an electric field develops in the base region that speeds up the flow of minority carriers.
2. At high injection a voltage drop develops in the base region, and this affects the external characteristics (Ebner–Gray effect).
3. At high currents the current flow in the collector space-charge region becomes space-charge-limited. This leads to the Kirk effect and the van der Ziel–Agouridis effect.
4. At high injection the emitter efficiency γ decreases.

16.3a. Electric Field in the Base Region

We consider here the base region of a *pnp* transitor extending from $x = 0$ (emitter side) to $x = w_B$ (collector side). The full current equations are now

$$J_p = e\mu_p pF - eD_p\frac{dp}{dx}, \qquad J_n = e\mu_n nF + eD_n\frac{dn}{dx} \simeq 0 \qquad (16.68)$$

where $J_n \simeq 0$ means $J_n \ll J_p$. Since $n \simeq N_d + p$ and $dn/dx \simeq dp/dx$ because of space-charge neutrality, we find that

$$F \simeq -\frac{D_n}{\mu_n}\frac{1}{n}\frac{dn}{dx} = -\frac{D_p}{\mu_p}\frac{1}{N_d + p}\frac{dp}{dx} \qquad (16.69)$$

so that J_p may be written

$$J_p = -eD_p\left(1 + \frac{p}{N_d + p}\right)\frac{dp}{dx} \qquad (16.68a)$$

Consequently,

$$-\frac{J_p}{eD_p}x = \int_0^x \left(1 + \frac{p}{N_d + p}\right) dp = -2[p(0) - p(x)] + N_d \ln \frac{N_d + p(0)}{N_d + p(x)}$$

$$(16.70)$$

Putting $x = w_B$ and assuming that $p(w_B) \ll p(0)$, we obtain

$$J_p = \frac{eD_p p(0)}{w_B}\left\{2 - \frac{N_d}{p(0)}\ln\left[1 + \frac{p(0)}{N_d}\right]\right\} \qquad (16.70a)$$

For low-injection levels $[p(0) \ll N_d]$ this is equal to $eD_p p(0)/w_B$, since $\ln(1 + x) \simeq x$ for small x. For high-injection levels this is twice as large and equal to $2eD_p p(0)/w_B$. This is completely analogous to the diode case.

16.3b. Effect of the Voltage Drop in the Base Region

We consider the problem again for a *pnp* transitor. Because of the complicated shape of the base region, the problem is three dimensional and

$$J_p = e\mu_p pF - eD_p\nabla p, \qquad J_n = e\mu_n nF + eD_n\nabla n \qquad (16.71)$$

so that the total current density may be written

$$J = J_p + J_n = e(\mu_p p + \mu_n n)F + e\left(\frac{kT}{e}\right)(\mu_n\nabla n - \mu_p\nabla p) \qquad (16.72)$$

Because of space-charge neutrality, $\nabla p \simeq \nabla n$ and $n \simeq N_d + p$.

We now observe that the potential difference ψ_{ab} between the emitter side of the base and the base contact is independent of the path taken (see Fig. 16.8). Hence

$$\psi_{ab} = \int_b^a \nabla\psi \cdot ds = \int_c^a \nabla\psi \cdot ds + \int_d^c \nabla\psi \cdot ds + \int_b^d \nabla\psi \cdot ds \qquad (16.73)$$

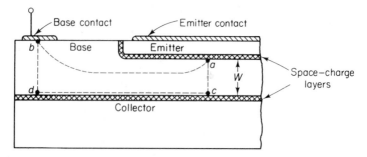

Fig. 16.8. Illustrative structural arrangement (not to scale) of a transistor.

But at the collector side of the base region $\nabla \psi$ is perpendicular to the path, unless there is a significant electron current returning from the collector space-charge region owing to avalanche multiplication, so that the second integral is zero. Moreover, there is no field along the path db to speak of; hence the third integral is negligible. Consequently,

$$\psi_{ab} = \int_{c}^{a} \nabla \psi \cdot \mathbf{ds} = \int_{0}^{w_B} F \, ds \qquad (16.73a)$$

Solving for F from Eq. (16.72) yields

$$F = \frac{J}{e[(\mu_p + \mu_n)p + \mu_n N_d]} - \frac{e(D_n - D_p) \, dp/dx}{e[(\mu_p + \mu_n)p + \mu_n N_d]} \qquad (16.74)$$

Substituting into Eq. (16.73a) and carrying out the integration yields, in analogy with the diode case [Eq. (15.53)],

$$\psi_{ab} = JR_{dc} + \frac{kT}{e}\left(\frac{\mu_n - \mu_p}{\mu_n + \mu_p}\right)\ln\left[1 + \frac{(\mu_n + \mu_p)p(0)}{\mu_n N_d}\right] \qquad (16.75)$$

where R_{dc} is the lateral resistance across the base region:

$$R_{dc} = \int_{0}^{w_B} \frac{dx}{e[(\mu_p + \mu_n)p + \mu_n N_d]} \qquad (16.75a)$$

Assuming, in analogy with (15.53a), that

$$J \, dx = -2eD_p \, dp$$

and carrying out the integration yields, as in the diode case [Eq. (15.53b)],

$$JR_{dc} \simeq \frac{kT}{e}\frac{2\mu_p}{\mu_p + \mu_n}\ln\left[1 + \frac{(\mu_p + \mu_n)p(0)}{\mu_n N_d}\right] \qquad (16.75b)$$

so that, for $(\mu_p + \mu_n)p(0) \gg \mu_n N_d$,[†]

$$\psi_{ab} \simeq \frac{kT}{e}\ln\frac{(\mu_p + \mu_n)p(0)}{\mu_n N_d} \qquad (16.76)$$

But according to the Fletcher condition (15.22)

$$p(0) = \frac{n_i^2}{N_d}\exp\left(\frac{eV_j}{kT}\right) \qquad (16.76a)$$

so that

$$\psi_{ab} = V_j - \frac{kT}{e}\ln\left[\frac{N_d^2}{n_i^2}\frac{\mu_n}{(\mu_p + \mu_n)}\right] = V_j - V_i \qquad (16.76b)$$

where V_i is a voltage caused by the impurity concentration N_d.

---

† If we had substituted F from (16.68), we would have obtained

$$\psi_{ab} = \frac{kT}{e}\ln\left[1 + \frac{p(0)}{N_d}\right] \simeq \frac{kT}{e}\ln\frac{p(0)}{N_d}$$

which differs from (16.76) by the small, constant amount $(kT/e)\ln(1 + \mu_p/\mu_n)$. This indicates that the condition $J_n \simeq 0$ is not satisfied, or that Eq. (15.53a) is not accurate.

At low-injection levels the voltage V_{EB} across the terminals is equal to V_j, whereas at high injection $V_{EB} = V_j + \psi_{ab} = 2V_j - V_i$. Since the characteristic varies as exp (eV_j/kT) in both regions, we have a characteristic of the form exp (eV_{EB}/kT) at very low injection, and of the form exp $[e(V_{EB} + V_i)/2kT]$ at very high injection.

An analogous situation occurs in *npn* transistors. Carrying out the same calculation as before, we find for the voltage drop ψ_{ab} at high injection [see Eq. (15.54e)]

$$\psi_{ab} = V_j + \frac{kT}{e} \ln\left[\frac{N_a^2}{n_i^2}\frac{\mu_p}{(\mu_n + \mu_p)}\right] = V_j + V_i' \qquad (16.77)$$

where V_i' is a voltage caused by the impurity concentration N_a.

At low-injection levels the voltage V_{EB} across the terminals is equal to V_j, whereas at high injection $V_{EB} = V_j + \psi_{ab} = 2V_j + V_i'$. Since the characteristic varies as exp $(-eV_j/kT)$ in both regions, we have a characteristic of the form exp (eV_{BE}/kT) at very low injection and of the form exp $[e(V_{BE} + V_i')/2kT]$ at very high injection, when expressed in terms of the terminal voltage $V_{BE} = -V_{EB}$.

Therefore, in both cases the characteristic should vary from exp (eV_{EB}/kT) to exp $(eV_{EB}/2kT)$ for a *pnp* transistor or from exp (eV_{BE}/kT) to exp $(eV_{BE}/2kT)$ for an *npn* transistor when going from low to high injection. This does not agree with what Ebner and Gray found.[†] The difference is due to the fact that they made the base current I_B exactly zero by a small amount of avalanche multiplication in the collector space-charge region. In that case the preceding calculation is not valid. According to Ebner and Gray's calculation, the expected characteristic should now be of the form exp $eV_{EB}/[(1 + m)kT]$ for a *pnp* transistor and exp $eV_{BE}/[(1 - m)kT]$ for an *npn* transistor. This is called the Ebner–Gray effect, and for $I_B = 0$ this was found experimentally. As our calculation indicates, however, the effect should not occur if there is no avalanche multiplication in the base–collector region.

An Ebner–Gray effect at high injection should occur, however, for the emitter impedance at high frequencies. In that case the emitter junction impedance Z_{ej} must be replaced by $Z_{ej}(1 \pm m)$, where the plus sign holds for a *pnp* transistor and the minus sign for an *npn* transistor. The reason is the same as in the diode case: one must combine the junction impedance Z_{ej} and the diffusion impedance $\pm m Z_{ej}$. To prove this, one simply repeats the calculation of the diode impedance given in Sec. 15.3b.

The full equivalent circuit of the common-base connection is shown in Fig. 16.9. Here there is an impedance $Z_{ej}(1 \pm m)$, a dc resistance R_{dc} (lateral dc resistance of the base region), a negative modulation resistance R_{mod} (caused by the current modulation of R_{dc}), and a high frequency-addition

---

† G. C. Ebner and P. E. Gray, *Trans. IEEE*, **ED-13**, 692–700 (Oct. 1966).

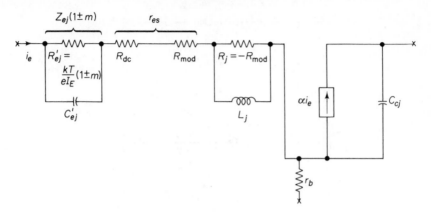

Fig. 16.9. H.F. equivalent circuit of a transistor at high injection. r_{es} is the a-c lateral resistance of the base region, r_b is the base resistance proper, C_{jc} is the capacitance of the collector junction, α is the h.f. current amplification factor.

consisting of a resistance $R_j = -R_{mod}$ and an inductance L_j in parallel. $Z_{ej}(1 \pm m)$ is represented by a resistance R'_{ej} and a capacitance C'_{ej} in parallel. Here $R'_{ej}C'_{ej} = \frac{1}{3}w_B^2/D_a$, and $L_j/R_j = \frac{1}{2}w_B^2/D_a$, where D_a is the ambipolar diffusion constant.

To make the equivalent circuit complete, we have also added the resistance r_b of the base region proper (see Sec. 16.4d) and the collector junction capacitance C_{jc}. The only effects not taken into account are the Early effects discussed in Section 16.4a.

16.3c. High-Current Effects in the Collector Space-Charge Region

We consider the problem for a *pnp* device. *Pnp* transistors are usually of the $p^+np^-p^+$ variety; that is, between the base and the bulk collector region (which is heavily doped) there is a thin region of weakly *p*-type material. The reason for this structure is that it makes the space-charge region wider, since the space-charge region now extends mainly into the p^- region, and hence makes the collector junction capacitance C_{jc} smaller. The width of the p^- region is such that it can accommodate the space-charge region for normal bias.

If there is a voltage V_{CB} across the collector space-charge region, the field in that region follows from

$$\frac{dF}{dx} = -\frac{e(N_a - p)}{\epsilon \epsilon_0} \tag{16.78}$$

with $F = F_0$ at the boundary between the n and p^- region. We assume that V_{CB} is kept constant and that the hole current density J_p increases. For small

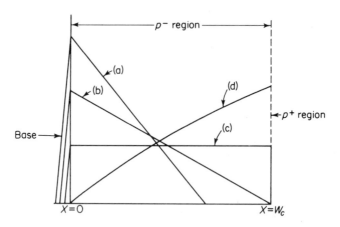

Fig. 16.10. Field distribution in the p^- region.

J_p the field distribution is as in curve a (Fig. 16.10). With increasing J_p, and hence decreasing $N_a - p$, the value of $|F_0|$ decreases and the space-charge region expands until it fills the p^- region (curve b). Finally, the point is reached where $N_a - p = 0$; then $|F|$ is constant in the p^- region (curve c). With further increases in J_p we finally come to the point where $F_0 = 0$; the current flow in the p region then becomes space-charge-limited, and J_p attains its value $J_{\max}$ (curve d). Any further increase in collector current must either result in a narrowing of the space-charge region (Kirk effect), or in a widening of the region in which the collector current flows (van der Ziel–Agouridis effect). Actually, both effects occur simultaneously.

First, let us consider the Kirk effect. According to the theory of space-charge-limited flow (Sec. 19.1), the current in a device of length d to which a voltage V_a is supplied is $I_0 = CV_a^2/d^3$, where C is a constant. If we apply this to our p^- region of width w_c, and assume that the region of current flow does not widen, then the length of the space-charge region must shrink, as soon as I_C exceeds a critical value I_0, to a value (Fig. 16.11a)

$$d' = \left(\frac{CV_{CB}^2}{I_C}\right)^{1/3} = w_c\left(\frac{I_0}{I_C}\right)^{1/3} \qquad (16.79)$$

for $I_C > I_0$. The carriers thus diffuse over the width

$$w = w_B + w_c - d' = w_B + w_c\left[1 - \left(\frac{I_0}{I_C}\right)^{1/3}\right] \qquad (16.80)$$

so that the diffusion time is

$$\tau_D = \frac{w^2}{2D_p} = \frac{w_B^2}{2D_p}\left\{1 + \frac{w_c}{w_B}\left[1 - \left(\frac{I_0}{I_C}\right)^{1/3}\right]\right\}^2 \qquad (16.81)$$

where $\tau_B = w_B^2/2D_p$ is the diffusion time at lower currents. Hence

$$f_\alpha = \frac{1}{2\pi\tau_D} = f_{\alpha 0}\left\{1 + \frac{w_c}{w_B}\left[1 - \left(\frac{I_0}{I_C}\right)^{1/3}\right]\right\}^{-2} \qquad (16.81a)$$

where $f_{\alpha 0} = (2\pi\tau_B)^{-1}$ is the cutoff frequency at lower currents. The alpha cutoff frequency f_α can thus decrease considerably with increasing I_C, since w_c/w_B is a large number.

We now discuss the van der Ziel–Agouridis effect. A cross section of the *pnp* device is shown in Fig. 16.11b. The emitter has a length L and a width $w_e \ll L$. In the p^- region let $J_p = J_{\max}$ for a width $w' > w_e$ and zero otherwise. Space-charge-limited flow sets in at a collector current $I_0 = J_{\max}Lw_e$, and for $I_C > I_0$ we have $I_C = J_{\max}Lw'$, so that $I_C/I_0 = w'/w_e$.

If a small alternating-current ripple is superimposed upon the direct current, then this can only happen because the width w' is modulated in the rhythm of the alternating current. If w_B is the length of the base region, the holes must diffuse over a length

$$w = \left[w_B^2 + \frac{(w' - w_e)^2}{4} \right]^{1/2} \tag{16.82}$$

Hence if $\tau_B = w_B^2/2D_p$, the effective diffusion time τ_d is

$$\tau_d = \frac{w^2}{2D_p} = \frac{w_B^2 + \frac{1}{4}(w' - w_e)^2}{2D_p} = \tau_B \left[1 + \left(\frac{w_e}{2w_B}\right)^2 \left(\frac{I_C}{I_0} - 1\right)^2 \right] \tag{16.83}$$

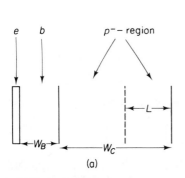

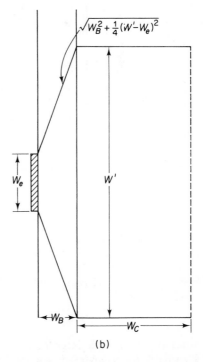

Fig. 16.11a. Widening of the base region (Kirk effect).

Fig. 16.11b. Spreading of the beam in the base and p^- region (van der Ziel-Agouridis effect).

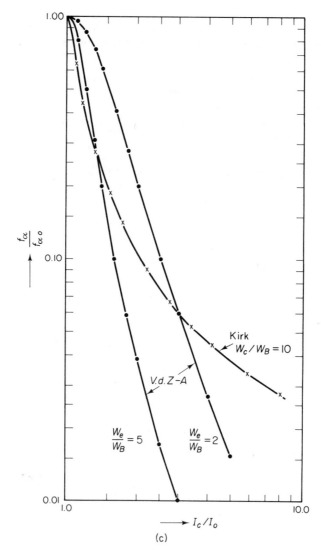

Fig.16.11c. Comparison of the Kirk effect for $w_c/w_B = 10$ and the van der Ziel–Agouridis effect for $w_E/w_B = 2$ and 5.

Hence, if $f_{\alpha 0} = 1/(2\pi\tau_B)$ is the cutoff frequency for small I_C,

$$f_\alpha = \frac{1}{2\pi\tau_d} = f_{\alpha 0}\left[1 + \left(\frac{w_e}{2w_B}\right)^2\left(\frac{I_C}{I_0} - 1\right)^2\right]^{-1} \qquad (16.83a)$$

Since $(w_e/2w_B)^2$ is a large quantity, the decrease in f_α can be significant. Figure 16.11c compares the Kirk and the van der Ziel–Agouridis effects by plotting $f_\alpha/f_{\alpha 0}$ versus I_C/I_0. There is some similarity between the two cases.

Slotboom† has given a two-dimensional computer-aided analysis of the two effects. He finds that both effects are present but that the van der Ziel–Agouridis approach considerably overestimates the spreading of the carrier paths in the base region; usually the Kirk effect is more important.

His analysis also shows an enhanced injection near the emitter rim; this effect is known as *emitter crowding*, and is also a high-level injection effect.

16.3d. Decrease in Emitter Efficiency at High Injection

We discuss the effect again for a *pnp* transistor and define γ as

$$\gamma = \frac{J_p}{J_p + J_n} = \frac{1}{1 + J_n/J_p} \tag{16.84}$$

At high injection, for a shallow emitter of length w_E,

$$J_p = \frac{2eD_{pb}p(x_2)}{w_B}, \qquad J_n = \frac{eD_{ne}n(x_1)}{w_E} \tag{16.85}$$

where D_{pb} is the hole diffusion constant in the base and D_{ne} the electron diffusion constant in the emitter; J_p follows from (16.70a) and J_n from (16.50). Now, according to (15.23a), for $N_a \gg N_d$ and $B^2 = \exp{[-2e(V_{dif} - V)/kT]} \ll 1$, we have $p(x_2) = N_aB$ and $n(x_1) = N_aB^2$ at high injection, or

$$J_p = \frac{2eD_{pb}N_aB}{w_B}, \qquad J_n = \frac{eD_{ne}N_aB^2}{w_E} = \frac{1}{4}\frac{D_{ne}}{D_{pb}^2}\frac{w_B^2}{w_E}\frac{J_p^2}{eN_a} \tag{16.85a}$$

Consequently, since $J_p \simeq J$ in our approximation,

$$\gamma = \left[1 + \frac{1}{4}\frac{D_{ne}}{D_{pb}^2}\frac{w_B^2}{w_E}\frac{J}{eN_a}\right]^{-1} \tag{16.86}$$

which should be reasonably accurate for $0.9 < \gamma < 1$. For smaller value of γ the device would be less useful.

As a numerical example, take $\gamma = 0.90$, $D_{ne} = D_{np} = 10 \text{ cm}^2/\text{s}$, $w_B = 10^{-3}$ cm, $w_E = 0.4 \times 10^{-4}$ cm, and $N_a = 10^{19}/\text{cm}^3$; neglecting the narrowing of the emitter gap width for large N_a (Sec. 16.2b), then $J \simeq 250$ A/cm². This seems to be the observed magnitude of the effect. For microwave transistors, w_B is very small ($< 10^{-4}$ cm) and the effect is negligible.

16.4. SECONDARY EFFECTS (EARLY EFFECTS)

Early effects refer to the effects caused by the modulation of the base width w_B by the collector base voltage V_{CB}. If in a *pnp* transistor the voltage V_{CB} is made more negative, the collector space-charge region expands into the base region and the base length w_B decreases. An ac collector voltage ΔV_{CB} thus

---

† J. W. Slotboom, *Trans. IEEE*, **ED-20**, 669 (1973).

modulates the base width w_B; the resulting effects were first described by Early, hence the name *Early effects*.

16.4a. Equivalent Π Circuit of the Common-Emitter Connection

To develop the equivalent Π network of a common-emitter transistor, we express the currents I_B and I_C as functions of the input voltage V_{BE} and the output voltage V_{CE} (Fig. 16.12a):

$$I_B = f_1(V_{BE}, V_{CE}), \qquad I_C = f_2(V_{BE}, V_{CE}) \tag{16.87}$$

The small-signal low-frequency response is thus described by

$$\Delta I_B = \frac{\partial I_B}{\partial V_{BE}} \Delta V_{BE} + \frac{\partial I_B}{\partial V_{CE}} \Delta V_{CE}, \qquad \Delta I_C = \frac{\partial I_C}{\partial V_{BE}} \Delta V_{BE} + \frac{\partial I_C}{\partial V_{CE}} \Delta V_{CE}$$

$$\tag{16.88}$$

where ΔV_{BE}, ΔV_{CE}, ΔI_B, and ΔI_C represent small increments. But if we represent the active Π network by three conductances g_{be}, g_{bc}, g_{ce} and the transconductance g_m, and assume that g_{bc} is small in comparison with the other parameters, we have (Fig. 16.12b)

$$\Delta I_B = g_{be} \Delta V_{BE} - g_{bc} \Delta V_{CE}, \qquad \Delta I_C = g_m \Delta V_{BE} + g_{ce} \Delta V_{CE} \tag{16.89}$$

so that

$$g_{be} = \frac{\partial I_B}{\partial V_{BE}}, \quad g_m = \frac{\partial I_C}{\partial V_{BE}}, \quad g_{bc} = -\frac{\partial I_B}{\partial V_{CE}}, \quad g_{ce} = \frac{\partial I_C}{\partial V_{CE}} \tag{16.90}$$

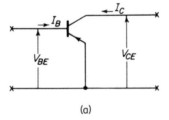

(a)

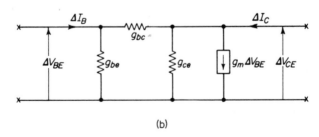

(b)

Fig. 16.12. Equivalent network of the common emitter circuit. (a) Sign convention of the circuit. (b) Equivalent π network. (From A. van der Ziel, *Electronics*, Allyn and Bacon Co., Boston, 1966.)

But we know already from the preceding discussion that

$$g_m = -\frac{eI_C}{kT}, \qquad g_{be} = \frac{g_m}{\beta_F} \qquad (16.91)$$

In the expressions for g_{bc} and g_{ce}, V_{BE} is kept constant. But

$$\left.\frac{\partial}{\partial V_{CE}}\right|_{V_{BE}=\text{const}} = \left.\frac{\partial}{\partial V_{CB}}\right|_{V_{BE}=\text{const}}$$

since $V_{CE} = V_{CB} + V_{BE}$. Hence

$$g_{bc} = -\left.\frac{\partial I_B}{\partial V_{CB}}\right|_{V_{BE}=\text{const}}, \qquad g_{ce} = \left.\frac{\partial I_C}{\partial V_{CB}}\right|_{V_{BE}=\text{const}} \qquad (16.92)$$

We now assume that g_{bc} and g_{ce} are caused by the dependence of w_B on V_{CB}. In that case

$$g_{bc} = -\frac{\partial I_B}{\partial w_B}\frac{\partial w_B}{\partial V_{CB}} = -\frac{e}{kT}w_B\frac{\partial I_B}{\partial w_B}\cdot\frac{kT}{e}\frac{1}{w_B}\frac{\partial w_B}{\partial V_{CB}} = -\frac{e}{kT}w_B\frac{\partial I_B}{\partial w_B}\mu \qquad (16.93)$$

$$g_{ce} = \frac{\partial I_C}{\partial w_B}\frac{\partial w_B}{\partial V_{CB}} = \frac{e}{kT}w_B\frac{\partial I_C}{\partial w_B}\cdot\frac{kT}{e}\frac{1}{w_B}\frac{\partial w_B}{\partial V_{CB}} = \frac{e}{kT}w_B\frac{\partial I_C}{\partial w_B}\mu \qquad (16.94)$$

where

$$\mu = \frac{kT}{e}\frac{1}{w_B}\frac{\partial w_B}{\partial V_{CB}} \qquad (16.95)$$

is a quantity of the order of 10^{-3} or less (see below).

We now apply this to various cases:

1. I_B is due to recombination in the base region between emitter and collector.

$$I_C = -\frac{eAD_p}{w_B}p_n\exp\left(-\frac{eV_{BE}}{kT}\right), \qquad \frac{e}{kT}w_B\frac{\partial I_C}{\partial w_B} = -\frac{eI_C}{kT} = g_m$$

or

$$g_{ce} = g_m\mu \qquad (16.96)$$

Furthermore, since $I_B = -(1-\alpha_F)I_E$ and $\alpha_F = 1 - \frac{1}{2}w_B^2/L_p^2$,

$$I_B = -\frac{1}{2}\frac{eAD_pw_B}{L_p^2}p_n\exp\left(-\frac{eV_{BE}}{kT}\right), \qquad \frac{e}{kT}w_B\frac{\partial I_B}{\partial w_B} = -\frac{1}{2}\frac{w_B^2}{L_p^2}\frac{eI_C}{kT}$$

or

$$g_{bc} = (1-\alpha_F)g_m\mu = g_{be}\mu \qquad (16.97)$$

According to Fig. 16.12b, for open-circuited input we have

$$\Delta V_{BE} \simeq \Delta V_{CE}\frac{g_{bc}}{g_{be}} = \Delta V_{CE}\mu \qquad (16.98)$$

when $g_{bc} \ll g_{be}$. Hence in this model μ is the reverse feedback factor of the common-emitter connection.

2. I_B is caused by recombination at the base surface. In that case g_{ce} is

again given by (16.96), but I_B does not depend on w_B. According to Eq. (16.47),

$$g_{bc} = -\frac{\partial I_B}{\partial V_{CB}}\bigg|_{V_{BE}=\text{const}} = \frac{1}{A_{\text{eff}}} \frac{\partial A_{\text{eff}}}{\partial V_{CB}} |I_B|$$ (16.99)

Since $\partial A_{\text{eff}}/\partial V_{CB}$ is small, g_{bc} is small.

3. I_B is due to electron injection into the emitter. Here I_B is given by (16.50); since it does not seem to depend on V_{CB}, we expect $g_{bc} \simeq 0$. The expression for g_{ce} is again given by (16.96).

We shall now show that μ always corresponds to the reverse feedback factor in a common-base connection. To prove this we turn to Fig. 16.13.

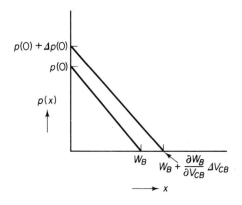

Fig. 16.13. Diagram for the derivation of Eq. 16.100. From A. van der Ziel, *Electronics*, Allyn and Bacon Co., Boston (1966).

Open-circuited input means here that I_E, and hence dp/dx, is kept constant when V_{CB} changes to V_{CB} to $(V_{CB} + \Delta V_{CB})$. Hence, from the properties of similar triangles, the change $\Delta p(0)$ in the hole concentration $p(0)$ at $x = 0$ is

$$\Delta p(0) = p(0)\frac{1}{w_B} \frac{\partial w_B}{\partial V_{CB}} \Delta V_{CB} = \frac{ep(0)}{kT} \mu \, \Delta V_{CB}$$

where μ has the same meaning as before. But the change $\Delta p(0)$ implies a change ΔV_{EB} in open-circuited voltage. Since

$$p(0) = p_n \exp\left(\frac{eV_{EB}}{kT}\right), \qquad \Delta p(0) = \frac{e}{kT}p(0)\,\Delta V_{EB}$$

or
$$\Delta V_{EB} = \mu \, \Delta V_{CB}$$ (16.100)

16.4b. Early Voltage

If the collector current I_C of a transistor is plotted versus the collector–emitter voltage V_{CE} at constant base current I_B, and tangents are drawn to these characteristics at the operating point V_{CE0}, it is found experimentally

that all tangents intersect the V_{CE} axis at the same voltage V_{CE1}, independent
of I_B. This voltage V_{CE1} is called the *Early voltage.*

To investigate this more closely, we write

$$I_C = \beta_F I_B$$

and make a Taylor expansion of β_F around the operating point V_{CE0}, where
β_F has the value β_{F0}. At constant I_B, the tangents to the characteristic satisfy
the equation

$$I_C = \left[\beta_{F0} + \left(\frac{\partial \beta_F}{\partial V_{CE}}\right)_0 (V_{CE} - V_{CE0}) \right] I_B$$

This straight line intersects the V_{CE} axis at V_{CE1}, where

$$V_{CE1} = V_{CE0} - \frac{\beta_{F0}}{(\partial \beta_F / \partial V_{CE})_0} \qquad (16.101)$$

Hence an Early voltage exists if $\beta_{F0}/(\partial \beta_F/\partial V_{CE})_0$ is independent of I_B.

A sufficient condition for this to happen is that $\beta_F = f(V_{CE})g(V_{EB})$, where
f and g are functions of the stated variables only. For in that case

$$\frac{\beta_{F0}}{(\partial \beta_F / \partial V_{CE})_0} = \frac{f(V_{CE})}{df/dV_{CE}} \bigg|_0 \qquad (16.101a)$$

and this is independent of V_{EB}. But at $V_{CE} = V_{CE0}$ any change in base current
I_B must come from a change in V_{EB}. Thus V_{CE1} is independent of V_{EB}, and
hence of I_B.

We now investigate this for various cases:

1. Recombination in the base region between emitter and collector. Here
I_B and I_C both depend on V_{EB} as $\exp(eV_{EB}/kT)$; hence, according to Sec.
16.4a,

$$\beta_F = \frac{I_C}{I_B} = \frac{2L_p^2}{w_B^2} \qquad (16.102)$$

This is practically independent of V_{EB} and depends on V_{CE} only. Hence the
existence of an Early voltage is verified.

2, 3. Recombination at the base surface, or electron injection into the
emitter. In that case I_B is practically independent of V_{CE}, and I_B and I_C
depend again on V_{EB} in the same manner. Hence

$$\beta_F = \frac{\text{const}}{w_B} \qquad (16.103)$$

This is independent of V_{EB} and depends on V_{CE} only. Again an Early voltage
exists.

16.4c. Generalized Expressions for I_C and I_B

We shall now write generalized expressions for I_C and I_B. From Eq.
(16.46), for a step junction transistor of small base width w_B, since $p_n =$

n_i^2/N_d, we have

$$I_C = -\frac{eD_p n_i^2 A}{w_B N_d} \exp\left(\frac{eV_{EB}}{kT}\right) = -\frac{C_{ES}}{Q_B} \exp\left(\frac{eV_{EB}}{kT}\right) \qquad (16.104)$$

where $-Q_B$ represents the electron charge in the base region:

$$Q_B = eN_d w_B A, \qquad C_{ES} = e^2 D_p n_i^2 A^2 \qquad (16.104a)$$

For nonuniform donor concentration, Q_B must be written†

$$Q_B = eA \int_B N_d(x)\, dx \qquad (16.105)$$

where the integral is extended over the length w_B. At high injection, $N_d(x)$ must be replaced by $n(x)$, the electron concentration.

Next we write a generalized expression for the base current I_B when I_B is due to electron injection from the base into the emitter. Then, according to Eq. (16.50), since $n_p = n_{ie}^2/N_a$, where n_{ie} is the intrinsic carrier concentration in the emitter, which is larger than the concentration n_i in more lightly doped material, we have for a step junction model

$$I_B = -\frac{eD_n n_{ie}^2 A}{w_E N_a} \exp\left(\frac{eV_{EB}}{kT}\right) = -\frac{C_{BS}}{Q_E} \exp\left(\frac{eV_{EB}}{kT}\right) \qquad (16.106)$$

where w_E is the emitter length, and Q_E is a kind of "effective hole charge" in the emitter:

$$Q_E = eN_a w_E A \frac{n_i^2}{n_{ie}^2}, \qquad C_{BS} = e^2 D_n n_i^2 A^2 \qquad (16.106a)$$

For a nonuniform acceptor concentration in the emitter, Q_E must be replaced by the integral‡

$$Q_E = eA \int_E N_a(x) \frac{n_i^2}{n_{ie}^2(x)}\, dx \qquad (16.107)$$

where the integral is extended over the emitter length w_E.

16.4d. Base Resistance

We must now introduce the resistance of the base region. If recombination occurs in the base region, electrons must flow from the base contact to the spots where recombination occurs. If the base current is due to electron injection into the emitter, electrons must flow from the base contact to the periphery of the emitter. This electron flow is restricted to the narrow base region and causes a voltage drop in the base layer that will be different for different parts of the base region. We may thus approximate the distributed

---

† H. K. Gummel and H. C. Poon, *Bell System Tech. J.*, **49**, 801–852 (1970); H. K. Gummel, *Bell System Tech. J.*, **49**, 115–120 (1970).

‡ R. P. Mertens, H. J. de Man, and R. J. van Overstraeten, *Trans. IEEE*, **ED-20**, 772, (1973).

resistance of the base by a lumped resistance r_b, which is a kind of weighted average.

If A is the junction area and w_B the base length, then

$$r_b = \text{const} \frac{A}{\mu_n N_d w_B} \qquad (16.108)$$

where N_d is the donor concentration and μ_n the electron mobility. The tendency is to make the base length w_B smaller and smaller. But one wants to keep r_b constant or smaller. This means the junction area must be made smaller simultaneously. Further improvement can be obtained by making N_d larger.

Figure 16.14 shows the resulting equivalent circuit of the common-emitter configuration. It is approximately correct up to the cutoff frequency $f_T = f_\alpha$.

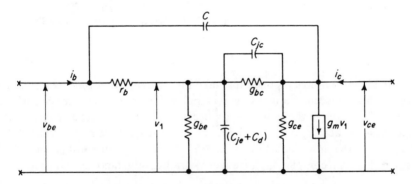

Fig. 16.14. Full equivalent circuit of the common emitter circuit. (From A. van der Ziel, *Electronics*, Allyn and Bacon Co., Boston, 1966.)

In the preceding sections the conductances g_{ce} and g_{bc} were calculated for low frequencies only. At high frequencies g_{bc} is shunted by the capacitance C_{jc}, and hence its effect can be ignored. Most circuits are operated at a relatively low impedance level to achieve bandwidth; hence the effect of g_{ce} can often be neglected. The equivalent circuit of Fig. 16.14 can then be appropriately simplified.

16.5. NOISE

16.5a. Derivation of the Basic Equations

The passage of carriers across space-charge regions constitutes a series of independent random events. For that reason one would expect full shot noise in transistors. Let us now see what this means:

The current I_E is due to electrons crossing the emitter junction. One neglects the noise due to the very small current I_{BE} flowing from emitter to base. According to the section on diode noise, one should represent the noise by a current generator i_1 at the input, so that at all frequencies of practical interest

$$\overline{i_1^2} = 2eI_E\,\Delta f + 4kT(g_{eb} - g_{eb0})\Delta f = 4kTg_{eb}\Delta f - 2eI_E\Delta f \quad (16.109)$$

The current I_C is due to electrons crossing the collector junction. Since the noise should be again full shot noise, one should represent it by a current generator i_2 at the output, so that at all frequencies of practical interest

$$\overline{i_2^2} = 2eI_C\,\Delta f \quad (16.110)$$

where I_C is taken positive.

The current generators i_1 and i_2 are very strongly correlated. The correlation comes about because the noise in the current I_C flowing from emitter to collector is common to both i_1 and i_2. At low frequency one would therefore expect

$$\overline{i_1 i_2} = 2eI_C\,\Delta f = 2kTg_{m0}\Delta f \quad (16.111)$$

At higher frequencies one must take into account that there is a random time delay between the input current pulse due to a given carrier and the output pulse due to the same carrier, because of diffusion. As a consequence, one must replace $\overline{i_1 i_2}$ by $\overline{i_1^* i_2}$ and determine the magnitude of the latter. This gives that the low-frequency transconductance g_{m0} in (16.111) must be replaced by the complex high-frequency transconductance g_m, so that

$$\overline{i_1^* i_2} = 2kTg_m\Delta f = 2kT\alpha Y_{eb}\Delta f \quad (16.111a)$$

which reduces to (16.111) at low frequencies. We thus obtain the equivalent circuit of Fig. 16.15a.

For the common-base circuit it is more convenient to replace the equivalent circuit by an alternate one, consisting of an emf $e_e = i_1 Z_e$ in series with the emitter junction and a current generator $i = i_2 - \alpha i_1$ in parallel with the collector junction (Fig. 16.15b). The advantage of this representation is that e_e and i are practically uncorrelated, which greatly simplifies the noise figure calculations. Since $R_{e0} \simeq kT/eI_E$, we thus have,

$$\overline{e_e^2} = \overline{i_1^2}\,|Z_e|^2 \simeq \overline{i_1^2}R_{e0}^2 \simeq 2kTR_{e0}\,\Delta f \quad (16.112)$$

$$\overline{i^2} = \overline{(i_2 - \alpha i_1)(i_2^* - \alpha^* i_1^*)}$$

$$= 2e[I_C - |\alpha|^2 I_E]\Delta f \quad (16.113)$$

$$\simeq 2e\left[\frac{1 - \alpha_F + f^2/f_\alpha^2}{1 + f^2/f_\alpha^2}\right]\alpha_F I_E\,\Delta f$$

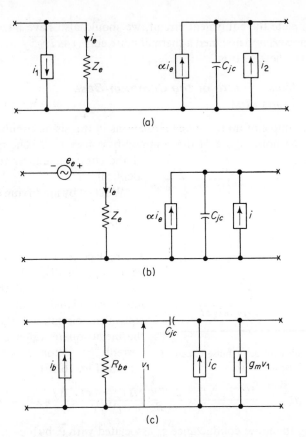

Fig. 16.15. (a) Equivalent noise circuit of the transistor. (b) Alternate circuit of the common base transistor. (c) Alternate circuit for the common emitter transistor at relatively low frequencies.

if we replace α_0 by α_F. It is here taken into account that $\alpha_F = I_C/I_E$ and $|\alpha|^2 = \alpha_0^2/(1 + f^2/f_\alpha^2)$. We thus see that $\overline{e_e^2}$ is practically frequency independent, but that $\overline{i^2}$ increases from the low-frequency value $\overline{i^2} = 2e\alpha_F I_B\Delta f$ for $f < f_\alpha\sqrt{1 - \alpha_0}$ to the high-frequency value $\overline{i^2} = 2eI_C\Delta f$ for $f > f_\alpha$.

For the common-emitter circuit it is more convenient to replace the equivalent circuit by an alternate one consisting of a current generator $i_b = i_1 - i_2$ in parallel with $R_{be} = 1/g_{be}$ and the current generator $i_c = i_2$ in parallel with the collector junction (Fig. 16.15c). The advantage is again that i_b and i_c are practically independent at low frequencies. In that case

$$\overline{i_b^2} = \overline{(i_1 - i_2)^2} = \overline{i_1^2} + \overline{i_2^2} - 2\overline{i_1 i_2}$$
$$\simeq 2e(I_E - I_C)\Delta f = 2eI_B\Delta f \tag{16.114}$$

To complete the equivalent circuit, we should also have introduced the base resistance r_b and assigned a thermal noise emf $\sqrt{4kTr_b\Delta f}$ to it. We shall do this in the next section.

16.5b. Noise Figure of the Common-Base Transistor

As an example of noise-figure calculations in transistor circuits we determine here the noise figure of the common-base transistor. The noise figure of the common-emitter transistor is identical.

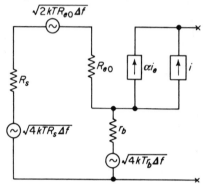

Fig. 16.16. Full equivalent noise circuit of the common base transistor.

We start by modifying the equivalent circuit of Fig. 16.15b. We add a source resistance R_s with a thermal noise emf $\sqrt{4kTR_s\Delta f}$ to the circuit. We replace Z_e by its low-frequency value R_{e0}. Finally, we introduce the base resistance r_b, and the thermal noise associated with it, in the equivalent circuit. We then find for the mean-square value of the total current generator $i + \alpha i_e$ at the output (Fig. 16.16)

$$\overline{(i + \alpha i_e)^2} = \frac{4kTR_s\,\Delta f + 2kTR_{e0}\,\Delta f + 4kTr_b\Delta f}{(R_s + R_{e0} + r_b)^2}\,|\alpha|^2 + \overline{i^2} \quad (16.115)$$

Introducing the noise conductance g_n associated with $\overline{i^2}$ by

$$4kTg_n\,\Delta f = \frac{\overline{i^2}}{|\alpha|^2}, \qquad g_n = \frac{\alpha_F}{\alpha_0^2}\frac{1}{2R_{e0}}\left(1 - \alpha_F + \frac{f^2}{f_\alpha^2}\right) \quad (16.116)$$

we may write the noise figure

$$
\begin{aligned}
F &= \frac{\overline{(i + \alpha i_e)^2}}{4kTR_s\,\Delta f\,|\alpha|^2/(R_s + R_{e0} + r_b)^2} \\
&= 1 + \frac{\tfrac{1}{2}R_{e0} + r_b}{R_s} + \frac{g_n}{R_s}(R_s + R_{e0} + r_b)^2
\end{aligned}
\quad (16.117)
$$

which has a minimum value

$$F_{\min} = 1 + 2g_n(R_{e0} + r_b) + 2[g_n(\tfrac{1}{2}R_{e0} + r_b) + g_n^2(R_{e0} + r_b)^2]^{1/2} \quad (16.118)$$

for

$$R_s = \left[\frac{\tfrac{1}{2}R_{e0} + r_b}{g_n} + (R_{e0} + r_b)^2\right]^{1/2} \quad (16.118a)$$

For low frequencies, replacing α_F/α_0^2 by unity, and bearing in mind that

$g_n(R_{e0} + r_b)$ is a small quantity, we obtain

$$F_{\min} \simeq 1 + 2[g_n(\tfrac{1}{2}R_{e0} + r_b)]^{1/2}$$
$$= 1 + (1 - \alpha_F)^{1/2}\left(1 + \frac{2r_b}{R_{e0}}\right)^{1/2} \tag{16.119}$$

at
$$R_s \simeq \left[\frac{1 + 2r_b/R_{e0}}{1 - \alpha_F}\right]^{1/2} R_{e0} \tag{16.119a}$$

We can obtain very low noise figures by making $1 - \alpha_F$ and r_b/R_{e0} small. This low noise figure extends beyond the frequency $f_1 = f_\alpha(1 - \alpha_F)^{1/2}$. At higher frequencies $F_{\min}$ increases, but acceptable noise figures can be obtained up to about the α-cutoff frequency f_α. Since transistors with higher and higher cutoff frequencies are becoming available, low-noise operation now extends well into the microwave range.

REFERENCES

GRAY, P. E., D. DE WITT, A. R. BOOTHROYD, and J. F. GIBBONS, *Electronics and Circuit Models of Transistors*, S.E.E.C. Notes, vol. 2. John Wiley & Sons, Inc., New York, 1965.

GROVE, A. S., *Physics and Technology of Semiconductor Devices*. John Wiley & Sons, Inc., New York, 1967.

HUNTER, L. P., ed., *Handbook of Semiconductor Electronics*, 3rd ed. McGraw-Hill Book Company, New York, 1969.

LINDMAYER, J., and C. Y. WRIGLEY, *Fundamentals of Semiconductor Devices*. Van Nostrand Reinhold Company, New York, 1965.

STRUTT, M.J.O., *Semiconductor Devices*. Academic Press, Inc., New York, 1966.

SZE, S. M., *Physics of Semiconductor Devices*. John Wiley & Sons, Inc. (Interscience Division), New York, 1969.

VALDES, L. B., *The Physical Theory of Transistors*. McGraw-Hill Book Company, New York, 1961.

van der ZIEL, A., *Noise: Sources, Characterization, Measurement*. Prentice-Hall Inc., Englewood Cliffs, N.J., 1970.

WARNER, R. M., et al., *Integrated Circuits, Design Principles and Fabrication*. McGraw-Hill Book Company, New York, 1965.

PROBLEMS

1. Find the current in a *pnp* transistor (germanium) at $T = 300°K$, if all current is carried by holes, the junction area is 10^{-4} cm^2, $N_d = 10^{16}$/cm^3, and the base width is 10^{-4} cm, which is small in comparison with the diffusion length of holes in the base region. $D_p = 49$ cm^2/s, $n_i = 2.4 \times 10^{13}$/cm^3.

Answer: $I_E = 0.45 \times 10^{-6} [\exp(38.6 V_{EB}) - 1]$A.

2. The transistor of problem 1 is operated in common-base connection at an emitter current of 1 mA. Find the input conductance g_{eb}, the diffusion time in the base region, the hole storage capacitance C_d, and the alpha-cutoff frequency, the latter under the assumption that $C_d \gg C_{je}$.

Answer: 3.86×10^{-2} S; 1.02×10^{-10} s; $3.9 \ \mu\mu$F; 1560 MHz.

3. The transistor of problem 1 is operated in common-emitter connection at a current of 1 mA. If $\beta_0 = 100$, find the conductance g_{be}, the transconductance g_m, and the β-cutoff frequency.

Answer: 3.83×10^{-4} S; 3.83×10^{-2} S; 15.4 MHz.

4. A transistor has a high-current cutoff frequency of 1000 MHz. If the emitter junction has a space-charge capacitance of 1.0 $\mu\mu$F, find the following at $T = 300°$K:
(a) The current at which $f_\alpha = 100$ MHz.
(b) The current dependence of f_α at low emitter currents.
You may assume that $I_{BE} \ll I_E$.

Answer: 18.0 μA; $6.2 \times 10^6 I_E$ with I_E in μA.

5. In a transistor with recombination in the emitter space-charge region

$$I_E = I_{ES} \exp\left(\frac{eV_{EB}}{kT}\right) + I_{R0} \exp\left(\frac{eV_{EB}}{2kT}\right)$$

and 98 per cent of the injected holes are collected by the collector:
(a) Calculate the dc current amplification factor $\alpha_{dc} = I_C/I_E$ in common-base connection and express it in terms of I_E.
(b) Calculate the ac current amplification factor $\alpha_0 = \partial I_C/\partial I_E$ in common-base connection and express it in terms of I_E.
(c) Calculate the ac current amplification factor

$$\beta_0 = \frac{\alpha_0}{1 - \alpha_0}$$

Hint: Use the expression for I_E as an equation with the help of which one can express $\exp(eV_{EB}/kT)$ in terms of I_E.

Answer:

$$\alpha_{dc} = 0.98\left\{1 - 2\frac{I_{R0}^2}{4 I_{ES} I_E}\left[\left(1 + \frac{4 I_{ES} I_E}{I_{R0}^2}\right)^{1/2} - 1\right]\right\}$$

$$\alpha_0 = 0.98\left[1 - \left(1 + \frac{4 I_{ES} I_E}{I_{R0}^2}\right)^{-1/2}\right]$$

$$\beta_0 = 49\frac{(1 + 4 I_{ES} I_E/I_{R0}^2)^{1/2} - 1}{(1 + 4 I_{ES} I_E/I_{R0}^2)^{1/2} + 49}$$

6. Find the value of I_E for which β_0 has half the limiting value of 49.

Answer: $I_E = 650 I_{RO}^2/I_{ES}$.

7. According to Eq. (16.71), $J_p = eD_p p(0)/w$ at low injection and $J_p = 2eD_p p(0)/w$ at high injection. Find the value of $p(0)/N_d$ at which $J_p = 1.5eD_p p(0)/w$. This leads to an equation that must be solved by trial and error.

Answer: $p(0)/N_d \simeq 2.5$.

8. A microwave *pnp* transistor has a p^- region 2 μm thick on the collector side of the base. The acceptor concentration in the p^- region is $10^{15}/cm^3$. Find the hole-current density at which the space charge in the p^- region is zero. You may assume that the holes move through the p^- region with a limiting velocity of 10^7 cm/s.

Answer: $J_p = 1600$ A/cm^2.

9. In Eq. (16.86) the emitter efficiency γ is calculated for a "short" emitter region. Show that w_E must be replaced by the diffusion length L_{ne} if $w_E \gg L_{ne}$.

10. A *pnp* germanium transistor has a planar geometry, and a junction area of 10^{-4} cm^2. The emitter has $N_a = 10^{18}/cm^3$, the base has $N_d = 10^{16}/cm^3$, and the collector has $N_a = 10^{16}/cm^3$. The junctions are abrupt, and the length of the base region is 2 μm. The actual length of the base region is less, since the emitter and the collector space-charge regions extend into the base. $\epsilon = 16$, $n_i = 2.4 \times 10^{13}/cm^3$ at $T = 300°$K, $D_p = 49$ cm^2/s; $E_g = 0.720$ eV, $m_n^* = m_p^* = m$.
(a) Calculate the diffusion potentials of the two junctions at $T = 300°$K.
(b) Calculate the emitter bias V_{EB} needed to produce a current of 1 mA, assuming a net base length w_B of 1 μm.
(c) Calculate the part of the emitter space-charge region that extends into the base region under the same conditions as in part (b).
(d) Calculate the collector bias V_{CB} needed to produce a net base length w_B of 1 μm at a current of 1 mA.
(e) Express the net base length w_B in terms of V_{CB} at $I_E = 1$ mA, assuming that the width of the emitter space-charge region does not depend on V_{CB}.
(f) Calculate g_m, μ, and g_{ce} under the condition of part (d).

Answer: (a) 0.430 V and 0.311 V; (b) 0.198 V; (c) 0.202 μm; (d) $V_{CB} = -6.9$ V; (e) $w_B = 1.798 - 0.298(V_{dif} - V_{CB})^{1/2}$ μm; (f) $\mu = 1.44 \times 10^{-3}$; $g_{ce} = 5.6 \times 10^{-5}$ S; $g_m = 3.86 \times 10^{-2}$ S $(g_m \simeq g_{eb})$.

11. Assume in problem 1 that the junction cross section is a square of side length $d = 10^{-2}$ cm. Show the following:
(a) The base resistance can be written $r_b = L_{eff}/(e\mu_n N_D w_B d)$, where L_{eff} is the effective height of the base region as seen from the base contact.
(b) $L_{eff} = d/2$ if the base contact is made to one side only.
(c) $L_{eff} = d/6$ if the base contact is made to two opposing sides.

Hint: In case (b) the resistance between a section at a distance x from the contact and the contact itself is $x/(e\mu_n N_D w_B d)$, and this must be averaged over x. In case (c) the resistance between a section at distance x from the one contact and the two contacts is $(e\mu_n N_D w_B d)^{-1}[x^{-1} + (d - x)^{-1}]^{-1}$, and this must be averaged over x.

12. Find the numerical values for cases (b) and (c) of problem 11 if $\mu_n = 3900$ cm^2/V s.

Answer: (b) 800 Ω; (c) 270 Ω.

17

Applications

17.1. PULSE RESPONSE IN DIODES AND TRANSISTORS

There are several methods of approach to the problem of pulse response in diodes and transistors. The most rigorous approach consists in solving the diffusion equation; we shall do so for the "long" *pn* diode. A second method is the "charge-control approach," by which one looks for the hole charge stored in the *n* region of a *pn* junction diode or for the hole charge stored in the base region of an *npn* transistor. For the long diode this leads to inaccurate results, but for the transistor it is often a very reasonable approximation.

17.1a. Charge Control Approximation for a Long Diode

In the charge control approximation the behavior of a device is expressed in terms of its stored charge. We do so for a long p^+n diode. Let V_D be the diode voltage and $L_p = (D_p\tau_p)^{1/2}$ be the diffusion length; then the excess hole distribution in the *n* region is

$$p'(x) = p_n\left[\exp\left(\frac{eV_D}{kT}\right) - 1\right]\exp\left(-\frac{x}{L_p}\right) \tag{17.1}$$

Consequently, if A is the junction area, the stored hole charge is

$$Q_s = \int_0^\infty eAp'(x)\,dx = eAL_p p_n\left[\exp\left(\frac{eV_D}{kT}\right) - 1\right] = I_D\tau_p \qquad (17.2)$$

where I_D is the dc diode current:

$$I_D = \frac{eD_p A p_n}{L_p}\left[\exp\left(\frac{eV_D}{kT}\right) - 1\right] = I_{D0}\left[\exp\left(\frac{eV_D}{kT}\right) - 1\right] \qquad (17.2a)$$

If one expresses transients in terms of voltage, the response is very nonlinear. However, in terms of stored charge we can try to represent the transient by the differential equation

$$i_D(t) = \frac{dq_s}{dt} + \frac{q_s(t)}{\tau_p} = au(t) \qquad (17.3)$$

where a is a constant and $u(t)$ is the unit step function. This equation expresses the fact that the part q_s/τ_p of the diode current i_D is used to supply the recombination current, whereas the part dq_s/dt is used to change the stored charge. The initial condition is $q_s = q_{s0}$ at $t = 0$.

For the turn-on transient, $q_{s0} = 0$; hence

$$q_s(t) = a\tau_p\left[1 - \exp\left(-\frac{t}{\tau_p}\right)\right] \qquad (17.3a)$$

so that $q_s(\infty) = a\tau_p$. For the turn-off transient, $q_{s0} = a\tau_p$; hence

$$q_s(t) = a\tau_p \exp\left(-\frac{t}{\tau_p}\right) \qquad (17.3b)$$

We shall now replace Eq. (17.3) by a more accurate approximation. In Eq. (15.96) we approximated the junction admittance as

$$Y_p = G_{p0} + \frac{1}{2}j\omega G_{p0}\tau_p = G_{p0} + \frac{1}{2}j\omega C_s$$

where $C_s = dQ_s/dV_D = \tau_p\,dI_D/dV_D = G_{p0}\tau_p$ is the hole storage capacitance. The small-signal equation is therefore

$$i_D(t) = \frac{1}{2}C_s\frac{dv_D}{dt} + G_{p0}v_D = au(t)$$

Introducing the stored charge $q_s = C_s v_D$, we obtain

$$i_D(t) = \frac{1}{2}\frac{dq_s}{dt} + \frac{q_s}{\tau_p} = au(t) \qquad (17.4)$$

Since all reference to small signals has disappeared, this equation should be valid for arbitrary signals. The only difference with Eq. (17.3) is that in the solution $\exp(-t/\tau_p)$ must be replaced by $\exp(-2t/\tau_p)$ so that the response is speeded up by a factor of 2.

We are finally interested in determining the diode voltage $V_D(t)$. According to Eq. (17.2a),

$$V_D = \frac{kT}{e} \ln\left(1 + \frac{I_D}{I_{D0}}\right) = \frac{kT}{e} \ln\left(1 + \frac{Q_s}{Q_{s0}}\right)$$

where $Q_{s0} = I_{D0}\tau_p$. Consequently, we write

$$V_D(t) = \frac{kT}{e} \ln\left[1 + \frac{q_s(t)}{Q_{s0}}\right] \qquad (17.5)$$

The charge control method thus solves the transient problem.

Obviously, this is only an approximation. One cannot rigorously transform dc concepts into transient concepts; for example, the instantaneous recombination current $q_s(t)/\tau_p$ does not represent a current flowing across the junction at the instant t. A more accurate approach appears in Sec. 17.1c.

17.1b. Charge Control Equations for Transistors

In transistors we start again with the basic dc equations of the device and then modify them for the transient case. We do so for a step junction *npn* transistor in which all current is carried by electrons. In analogy with the *pnp* transistor, for the excess electron distribution in the base (base length w_B) we have

$$n'(x) = n_p\left[\exp\left(\frac{eV_{BE}}{kT}\right) - 1\right]\frac{w_B - x}{w_B} \qquad (17.6)$$

Therefore, if A is the junction area, the collector current is

$$I_C = \frac{eD_n A}{w_B} n_p\left[\exp\left(\frac{eV_{BE}}{kT}\right) - 1\right]$$

Hence, if $\tau_d = w_B^2/2D_p$, the stored electron charge stored in the base is

$$Q_n = -\int_0^{w_B} eAn'(x)\,dx = -\frac{1}{2}eAw_B n_p\left[\exp\left(\frac{eV_{BE}}{kT}\right) - 1\right] = -I_C\tau_d \qquad (17.7)$$

and the corresponding hole charge in the base region is

$$Q_B = -Q_n = I_C\tau_d = I_B\tau_{BF} \qquad (17.8)$$

where $\tau_{BF} = \beta_F\tau_d$. Consequently,

$$I_B = \frac{Q_B}{\tau_{BF}}, \qquad I_C = \beta_F I_B = \beta_F\frac{Q_B}{\tau_{BF}} \qquad (17.8a)$$

whereas the emitter current is

$$I_E = -I_C - I_B = -\frac{1 + \beta_F}{\tau_{BF}}Q_B \qquad (17.8b)$$

(the minus sign is used because I_E flows into the emitter). These equations are of general validity since all references to the model involved have disappeared. The only requirement is that the time constant τ_{BF} can be defined.

We shall now show that

$$\tau_{BF} = C_{be}R_{be} \qquad (17.8c)$$

where $R_{be} = 1/g_{be}$ is the input resistance of the common-base transistor. According to Eq. (16.30b), $C_{be} = C_d = g_m\tau_d$ and $R_{be} = 1/g_{be} = \beta_F/g_m$. Hence $C_{be}R_{be} = \beta_F\tau_d = \tau_{BF}$.

If we neglect feedback through the capacitance C_{jc}, we have, in analogy with (17.3) and (17.8a), for the stored charge $q_F(t)$

$$i_b(t) = \frac{q_F(t)}{\tau_{BF}} + \frac{dq_F}{dt} \qquad (17.9)$$

$$i_c(t) = \frac{\beta_F}{\tau_{BF}}q_F(t) \qquad (17.10)$$

Consequently,

$$i_e(t) = -i_b(t) - i_c(t) = -\frac{1 + \beta_F}{\tau_{BF}}q_F(t) - \frac{dq_F}{dt} \qquad (17.11)$$

We need Eq. (17.11) when discussing saturated transistors.

To prove these equations, we write for small signals (Fig. 17.1)

$$i_b(t) = C_{be}\frac{dv_{be}}{dt} + \frac{v_{be}}{R_{be}}$$

Substituting $q_F = C_{be}v_{be}$, and bearing in mind that $C_{be}R_{be} = \tau_{BF}$, yields (17.9). In addition, since $\beta_F = g_mR_{be}$,

$$i_c(t) = g_mv_{be}(t) = g_mR_{be}\frac{v_{be}(t)}{R_{be}} = \beta_F\frac{q_F(t)}{\tau_{BF}}$$

corresponding to (17.10). Since all references to the small-signal model have disappeared, the equation should hold for arbitrary signals.

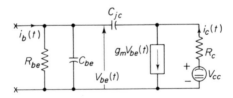

Fig. 17.1. Equivalent circuit of transistor, used for derivation of charge control equations.

The effect of the collector capacitance C_{jc} can be represented by a Miller effect capacitance $C_{jc}(1 + g_mR_c)$ in parallel to C_{be}. The total capacitance C_T is thus nearly equal to $C_{be} + C_{jc}g_mR_c$, and the time constant τ_T becomes

$$\tau_T = R_{be}C_T = C_{be}R_{be} + g_mR_{be}C_{jc}R_c = \tau_{BF} + \beta_FC_{jc}R_c \qquad (17.12)$$

so that

$$i_b(t) = C_T \frac{dv_{be}}{dt} + \frac{v_{be}}{R_{be}} = \frac{C_T}{C_{be}}\frac{dq_F}{dt} + \frac{q_F}{\tau_{BF}} = \frac{\tau_T}{\tau_{BF}}\frac{dq_F}{dt} + \frac{q_F}{\tau_{BF}}$$

or

$$\frac{dq_F}{dt} + \frac{q_F}{\tau_T} = \frac{\tau_{BF}}{\tau_T}i_b(t) \tag{17.13}$$

whereas it is still true that

$$i_c(t) = \frac{\beta_F}{\tau_{BF}}q_F \tag{17.13a}$$

It should be borne in mind that C_{jc} varies with $V_{CE}(t)$ so that a suitably chosen intermediate value must be used.

For a saturated transistor the collector is effectively clamped to about zero volt and hence the Miller effect disappears. But the base–collector junction is also forward biased and injects electrons into the base. The hole charge stored in the base is now $q_F + q_R$, where q_R is caused by electrons injected from the collector. A reverse current amplification factor β_R and a reverse time constant τ_{BR} are associated with q_R. Equations (17.9) and (17.10) now become

$$i_b(t) = \frac{q_F}{\tau_{BF}} + \frac{dq_F}{dt} + \frac{q_R}{\tau_{BR}} + \frac{dq_R}{dt} \tag{17.14}$$

$$i_c(t) = \beta_F \frac{q_F}{\tau_{BF}} - q_R\left(\frac{1 + \beta_R}{\tau_{BR}}\right) - \frac{dq_R}{dt} \tag{17.14a}$$

These are the charge control equations of the saturated transistor. The last two terms in Eq. (17.14a) come about because the collector now emits electrons into the base, and for this process Eq. (17.11) must be applied.

17.1c. Transient Response in Long Diodes

We first use the charge control equation (17.4). The circuit is shown in Fig. 17.2a, in the turn-on transient, $V(t)$ goes from 0 to V_1 at $t = 0$ (Fig. 17.2b); in the turn-off transient, $V(t)$ goes from V_1 to $-V_2$ at $t = 0$ (Fig. 17.2c).

If $V_1 > V_{D0}$, for the turn-on transient we have

$$i_D(t) = I_1 = \frac{V_1 - V_{D0}}{R_s} \tag{17.15}$$

and for the turn-off transient

$$i_D(t) = -I_2 = -\frac{V_2 + V_{D0}}{R_s}, \qquad I_2 = \frac{V_2 + V_{D0}}{R_s} \tag{17.15a}$$

where V_{D0} ($\simeq 0.70$ V for a silicon diode) is the turn-on voltage of the diode, that is, the voltage at which $I_D = 1$ mA.

At first it takes a certain time before the diode is turned on. During that

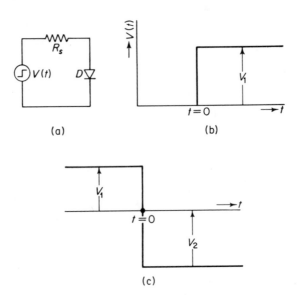

Fig. 17.2. (a) Diode circuit for large pulse response. (b) Step-up V_1 applied at $t = 0$. (c) Step-down from $+ V_1$ to $- V_2$ at $t = 0$.

period it acts as a capacitor of capacitance C_0; hence the transient is

$$V_D(t) = V_1\left[1 - \exp\left(-\frac{t}{C_0 R_s}\right)\right] \qquad (17.16)$$

This part of the transient stops if $V_D(t) = V_{D1}$ ($\simeq 0.55$ V), at which point current begins to flow. Let this occur at $t = t_1$; then

$$V_{D1} = V_1\left[1 - \exp\left(-\frac{t_1}{C_0 R_s}\right)\right], \qquad t_1 = -R_s C_0 \ln\left(1 - \frac{V_{D1}}{V_1}\right) \simeq R_s C_0 \frac{V_{D1}}{V_1} \qquad (17.16a)$$

This is called the *turn-on delay*. At this point the stored charge is still small. For $t > t_1$, Eq. (17.4) must be applied with $q_s(t_1) \simeq 0$ and $i_D(t) = I_{D1}$. For $q_s(t)$, this yields

$$q_s(t) = I_1 \tau_p \left\{1 - \exp\left[-\frac{2(t - t_1)}{\tau_p}\right]\right\} \qquad (17.17)$$

so that, according to Eq. (17.5), for $t - t_1$ not too small,

$$V_D(t) = \frac{kT}{e}\ln\left(\frac{I_1}{I_{D0}}\right) + \frac{kT}{e}\ln\left\{1 - \exp\left[-\frac{2(t - t_1)}{\tau_p}\right]\right\} \qquad (17.17a)$$

Now $kT/e \ln (I_1/I_{dc}) \simeq 0.70$ V, so that the transient is 90 per cent completed if the second term in (17.17a) is -0.07 V or

$$1 - \exp\left[-\frac{2(t - t_1)}{\tau_p}\right] = 0.066 \quad \text{or} \quad t - t_1 = t_2 \simeq 0.034\,\tau_p$$

This is the *turn-on time*. The total duration of the turn-on transient for 90 per cent completion is

$$t_1 + t_2 \simeq R_s C_0 \frac{V_{D1}}{V_1} + 0.034\tau_p \qquad (17.17b)$$

Next we look at the turn-off transient. The stored charge disappears according to Eq. (17.4) with $I_D(t) = -I_2$, $q_s(0) = I_1\tau_p$, and $q_s(\infty) = -I_2\tau_p$. Hence

$$q_s(t) = I_1\tau_p \exp\left(-\frac{2t}{\tau_p}\right) - I_2\tau_p\left[1 - \exp\left(-\frac{2t}{\tau_p}\right)\right] \qquad (17.18)$$

This transient stops if $q_s(t) \simeq 0$. Let this occur at $t = t_3$; then

$$t_3 = \frac{1}{2}\tau_p \ln \frac{I_1 + I_2}{I_2} = \frac{1}{2}\tau_p \ln \frac{V_1 + V_2}{V_2 + V_{D0}} \qquad (17.18a)$$

For example, if $V_1 = 5.0$ V, $V_{D0} = 0.70$ V, $V_2 = 0$ V, and $\tau_p = 10^{-6}$ s, then $t_3 \simeq 1.0 \times 10^{-6}$ s.

After this the diode acts again as a capacitor and the diode voltage continues to drop until the value $-V_2$ is reached. Since the diode capacitance is now voltage dependent, a suitable intermediate value must be chosen for C_0. When this has been done, the transient will be 90 per cent completed after a time interval t_4, where

$$t_4 = 2.3C_0R_s \qquad (17.19)$$

To calculate the transient response more accurately, we must solve the diffusion equation for the excess minority carriers. We shall see that $1 - \exp\left(-2t/\tau_p\right)$ must be replaced by $\operatorname{erf}(t/\tau_p)^{1/2}$ and $\exp\left(-2t/\tau_p\right)$ by $\operatorname{erfc}(t/\tau_p)^{1/2}$. These functions, which are called the *error function* and the *complementary error function*, respectively, are defined as

$$
\begin{aligned}
\operatorname{erf} u &= \frac{2}{\sqrt{\pi}} \int_0^u \exp\left(-x^2\right) dx, \\
\operatorname{erfc} u &= 1 - \operatorname{erf} u = \frac{2}{\sqrt{\pi}} \int_u^\infty \exp\left(-x^2\right) dx
\end{aligned} \qquad (17.20)
$$

Figure 17.3 compares $1 - \exp\left(-2x\right)$ and $\operatorname{erf}(x^{1/2})$. The two functions behave quite similarly, but for small x $\operatorname{erf}(x^{1/2})$ increases faster with x than $1 - \exp\left(-2x\right)$. Figure 17.4 shows the normalized admittances $Y/g_0 = (1 + j\omega\tau_p)^{1/2}$ and $(1 + \frac{1}{2}j\omega\tau_p)$ plotted in the complex plane with $\omega\tau_p$ as a parameter. The agreement is quite reasonable for relatively small $\omega\tau_p$ but rather poor for large $\omega\tau_p$. This means that the approximate expression $1 + \frac{1}{2}j\omega\tau_p$ gives a rather poor approximation for very short times, but much better for large times, in agreement with Fig. 17.3.

We bear in mind that $V_D(t)$ is related to the excess minority carrier concentration at the junction. If $x = 0$ at the junction, and practically all the

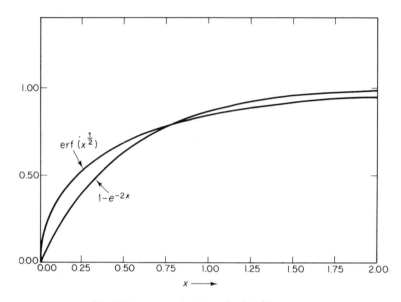

Fig. 17.3. $1 - \exp(-2x)$ and $\operatorname{erf}(x^{1/2})$ versus x.

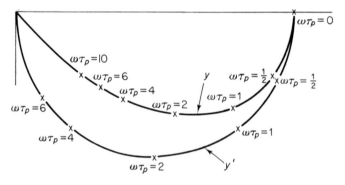

Fig. 17.4. $Y = g_0(1 + j\omega\tau_p)^{1/2}$ and $Y' = g_0(1 + \frac{1}{2}j\omega\tau_p)$ plotted in the complex plane with $\omega\tau_p$ as a parameter.

current at the junction is carried by holes, then the excess hole concentration $p'(0, t)$ at the junction is

$$p'(0, t) = p_{n0}\left\{\exp\left[\frac{eV_D(t)}{kT}\right] - 1\right\} \quad \text{or} \quad V_D(t) = \frac{kT}{e}\ln\left[\frac{p'(0, t)}{p_{n0}} + 1\right]$$
(17.21)

The problem thus reduces to the calculation of $p'(0, t)$.

The accurate calculation of $p'(0, t)$ is performed by solving the diffusion equation

$$\frac{\partial p'}{\partial t} = -\frac{p'}{\tau_p} + D_p\frac{\partial^2 p'}{\partial x^2}$$
(17.22)

under the appropriate initial conditions:

(a) $$p'(\infty, t) = 0 \qquad (17.22a)$$

This expresses the fact that the junction is a long junction ($w_n \gg L_p$).

(b) $$p'(x, 0) = \phi(x) \qquad \text{is given} \qquad (17.23)$$

This expresses the stored hole concentration in the n region. For the turn-on transient $\phi(x) = 0$; but for the turn-off transient

$$\phi(x) = p_{n0}\left[\exp\left(\frac{eV}{kT}\right) - 1\right]\exp\left(-\frac{x}{L_p}\right), \qquad \text{where } L_p = (D_p\tau_p)^{1/2} \qquad (17.23a)$$

(c) $$-eD_p\frac{\partial p'(x, t)}{\partial x}\bigg|_{x=0} = Ju(t) \quad \text{or} \quad \frac{\partial p'(x, t)}{\partial x}\bigg|_{x=0} = -\alpha u(t) \qquad (17.24)$$

where $\alpha = J/eD_p$ and $u(t)$ is the unit step function.

J and hence α have different values for the turn-on and the turn-off transient. We have

$$J = J_f = \frac{V_1 - V_{D0}}{R_sA}, \qquad \alpha = \alpha_f = \frac{V_1 - V_{D0}}{eD_pR_sA} \qquad (17.25)$$

for the turn-on transient, whereas

$$J = -J_r = -\frac{V_2 + V_{D0}}{R_sA}, \qquad \alpha = -\alpha_r = -\frac{V_2 + V_{D0}}{eD_pR_sA} \qquad (17.25a)$$

for the turn-off transient. Consequently, $\phi(x)$ for the turn-off transient may be written

$$\phi(x) = \frac{J_fL_p}{eD_p}\exp\left(-\frac{x}{L_p}\right) = \alpha_fL_p\exp\left(-\frac{x}{L_p}\right) \qquad (17.26)$$

Equation (17.22) is best solved by the Fourier cosine transform method. The Fourier cosine transform of $p'(x, t)$ is defined as

$$P(k, t) = \int_0^\infty p'(x, t)\cos kx\, dx \qquad (17.27)$$

Since

$$\int_0^\infty \frac{\partial p'}{\partial t}\cos kx\, dx = \frac{d}{dt}P(k, t) \qquad (17.27a)$$

$$\int_0^\infty -\frac{p'}{\tau_p}\cos kx\, dx = -\frac{P(k, t)}{\tau_p} \qquad (17.27b)$$

$$\int_0^\infty D_p\frac{\partial^2 p'}{\partial x^2}\cos kx\, dx$$

$$= D_p\frac{\partial p'}{\partial x}\cos kx\bigg|_0^\infty + kD_p\int_0^\infty \frac{\partial p'}{\partial x}\sin kx\, dx$$

$$= D_p\alpha u(t) + D_pkp'(x, t)\sin kx\bigg|_0^\infty - D_pk^2\int_0^\infty p'(x, t)\cos kx\, dx$$

$$= D_p\alpha u(t) - D_pk^2P(k, t) \qquad (17.27c)$$

the Fourier cosine transform of Eq. (17.22) must satisfy the equation

$$\frac{dP(k, t)}{dt} = -P(k, t)\left(\frac{1}{\tau_p} + D_p k^2\right) + \alpha D_p u(t) \tag{17.28}$$

with $\quad P(k, 0) = \int_0^\infty p'(x, 0) \cos kx \, dx = \int_0^\infty \phi(v) \cos kv \, dv \tag{17.28a}$

(we have changed here to the new variable v to avoid confusion later).

The solution of Eq. (17.28) for $t > 0$ is

$$P(k, t) = \frac{\alpha D_p}{1/\tau_p + D_p k^2} + \left[P(k, 0) - \frac{\alpha D_p}{1/\tau_p + D_p k^2}\right] \exp\left[-t\left(\frac{1}{\tau_p} + D_p k^2\right)\right] \tag{17.29}$$

as we find by substituting into Eq. (17.28) and by observing that the initial condition is satisfied. By transforming back, we obtain $p'(x, t)$. According to textbooks on Fourier cosine transforms,

$$p'(x, t) = \frac{2}{\pi} \int_0^\infty P(k, t) \cos kx \, dk \tag{17.30}$$

Fortunately, we only need to know

$$p'(0, t) = \frac{2}{\pi} \int_0^\infty P(k, t) \, dk$$

$$= \frac{2}{\pi} \alpha D_p \int_0^\infty \frac{dk}{1/\tau_p + D_p k^2} - \frac{2}{\pi} \alpha D_p \int_0^\infty \frac{\exp\left[-t(1/\tau_p + D_p k^2)\right]}{1/\tau_p + D_p k^2} dk$$

$$+ \frac{2}{\pi} \int_0^\infty \exp\left[-t\left(\frac{1}{\tau_p} + D_p k^2\right)\right] dk \int_0^\infty \phi(v) \cos kv \, dv \tag{17.31}$$

The first integral is αL_p. The second one has the value $\alpha L_p[\text{erf}\,(t/\tau_p)^{1/2} - 1]$. The third integral is zero for the turn-on transient, since $\phi(v) = 0$, whereas it is equal to $\alpha_f L_p \,\text{erfc}\,(t/\tau_p)^{1/2}$ for the turn-off transient.

Consequently, for the turn-on transient $\alpha = \alpha_f$ and

$$p'(0, t) = \alpha_f L_p \,\text{erf}\left(\frac{t}{\tau_p}\right)^{1/2} \tag{17.32}$$

and for the turn-off transient

$$p'(0, t) = -\alpha_r L_p \,\text{erf}\left(\frac{t}{\tau_p}\right)^{1/2} + \alpha_f L_p \,\text{erfc}\left(\frac{t}{\tau_p}\right)^{1/2} \tag{17.33}$$

which holds until $p'(0, t)$ becomes equal to zero.

We now substitute $p'(0, t)$ into (17.21) to obtain $V_D(t)$ for the turn-on transient. For $p'(0, t)/p_{n0} \gg 1$, this yields

$$V_D(t) \simeq \frac{kT}{e} \ln\left(\frac{\alpha_F L_p}{p_{n0}}\right) + \frac{kT}{e} \ln\left[\text{erf}\left(\frac{t}{\tau_p}\right)^{1/2}\right] \tag{17.33a}$$

Since the first term is of the order of 0.70 V, $V_D(t)$ has risen to about 90 per cent of the full value if

$$\text{erf}\left(\frac{t}{\tau_p}\right)^{1/2} \simeq 0.066, \quad \text{or} \quad \frac{t}{\tau_p} \simeq 0.004 \tag{17.33b}$$

This is much faster than in the first approximation.

We now discuss the turn-off transient. The turn-off time is obtained by substituting $p'(0, t) = 0$ in Eq. (17.33). This yields

$$(\alpha_f + \alpha_r)L_p \, \text{erf}\left(\frac{t}{\tau_p}\right)^{1/2} = \alpha_f L_p, \quad \text{erf}\left(\frac{t}{\tau_p}\right)^{1/2} = \frac{V_1 - V_{D0}}{V_1 + V_2} \tag{17.33c}$$

For $V_1 = 5.0$ V, $V_{D0} = 0.70$ V, and $V_2 = 0$, this yields $t/\tau_p = 1.09$. The shape of the transient is

$$V_D(t) \simeq \frac{kT}{e} \ln \frac{p'(0, t)}{p_{n0}} \simeq \frac{kT}{e} \ln \left(\frac{\alpha_f L_p}{p_{n0}}\right) - \frac{kT}{e} \ln \left[1 - \frac{\alpha_f + \alpha_r}{\alpha_f} \text{erf}\left(\frac{t}{\tau_p}\right)^{1/2}\right] \tag{17.33d}$$

This depends very slowly on t up to where the transient stops.

17.1d. Pulse Response for NPN Transistors

We consider transients in an *npn* transistor caused by switching the input voltage $V(t)$ from $-V_2 \, (V_2 > 0)$ to $V_1 \, (V_1 > 0)$ and back (Fig. 17.5a, b).

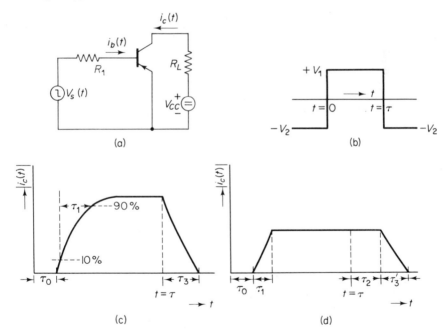

Fig. 17.5. (a) Transistor circuit in which the transient is studied. (b) Input pulse used in the transient study. (c) Collector current transient for the unsaturated case, showing the turn-on delay, the turn-on time τ_1, and the turn-off time τ'_3. (d) Same for the saturated case, showing also the storage delay time τ_2.

If $V(t)$ has been at $-V_2$ for a long time, the emitter–base junction is forward biased and current is flowing. There are two possibilities:

1. The collector is not saturated in the "on" condition.
2. The collector is saturated in the on condition and V_{ce} is effectively clamped to about zero volt.

In the first case there is a turn-on delay τ_0, because it takes time for the emitter junction to become forward biased, a turn-on time τ_1, and a turn-off time τ_3; τ_1 is measured by the rise in collector current from 10–90 per cent of the final value, whereas τ_3 can be measured as the total time needed to let $i_c(t)$ go to zero (Fig. 17.5c).

In the second case there is a turn-on delay τ_0, which is the same as in the first case, a turn-on time τ_1', a storage delay time τ_2, defined as the time it takes the transistor to become unsaturated, and the turn-off time τ_3'. All switching times can now be defined as the total time needed to complete the operation (Fig. 17.5d).

First, we calculate the time it takes to charge the input capacitance by means of the current flowing through the resistor R_1. The input capacitance of the transistor has the value $C_{in} = C_{je} + C_{jc}$, where C_{je} and C_{jc} are the space charge capacitances of the base–emitter and base–collector junctions, respectively; because both depend on voltage, suitable intermediate values must be chosen. Since $v(t) = -V_2$ for $t < 0$ and $v(t) = V_1$ for $t > 0$, the solution is

$$v_b(t) = V_1\left[1 - \exp\left(-\frac{t}{\tau_b}\right)\right] - V_2\exp\left(-\frac{t}{\tau_b}\right) \tag{17.34}$$

where $\tau_b = RC_{in}$. This part of the transient stops if current just begins to flow, or $v_b(t) = V_{BE1} \simeq 0.55$ V for a silicon transistor. Let this be at $t = \tau_0$; then

$$V_1\left[1 - \exp\left(-\frac{\tau_0}{\tau_b}\right)\right] - V_2\exp\left(-\frac{\tau_0}{\tau_b}\right) = V_{bE1},$$

$$\tau_0 = \tau_b\ln\frac{V_2 + V_1}{V_1 - V_{BE1}} \tag{17.35}$$

To find the turn-on time τ_1, we write Eq. (17.13) as (Fig. 17.6)

$$\frac{dq_F}{dt} + \frac{q_F}{\tau_T} = \frac{\tau_{BF}}{\tau_T}I_{B1}, \qquad I_{B1} = \frac{V_1 - V_{BE0}}{R_1} \tag{17.36}$$

where $V_{BE0} \simeq 0.70$ V. Since $q_F(0) \simeq 0$, the solution is,

$$q_F(t) = I_{B1}\tau_{BF}\left[1 - \exp\left(-\frac{t}{\tau_T}\right)\right] \tag{17.37}$$

The turn-on time for $i_c(t)$ is therefore

$$\tau_1 = 2.2\tau_T = 2.2(\tau_{BF} + \beta_F C_{jc}R_c) \tag{17.38}$$

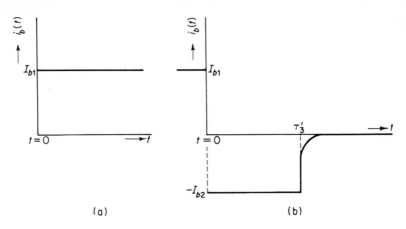

Fig. 17.6. (a) $i_b(t)$ as a function of time for the turn-on transient. (b) $i_b(t)$ as a function of time for the turn-off transient; $i_b(t)$ drops to zero for $t \simeq \tau'_3$.

We next find τ_3. As long as the transistor is not turned off, a constant current $-I_{B2}$ flows into the transistor (Fig. 17.6):

$$-I_{B2} = -\frac{V_2 + V_{BE0}}{R} \tag{17.39}$$

and Eq. (17.13) becomes

$$\frac{dq_F}{dt} + \frac{q_F}{\tau_T} = -\frac{\tau_{BF}}{\tau_T} I_{B2} \tag{17.40}$$

with the initial condition $q_F = I_{B1}\tau_{BF}$, so that

$$q_F(t) = -I_{B2}\tau_{BF}\left[1 - \exp\left(-\frac{t}{\tau_T}\right)\right] + I_{B1}\tau_{BF}\exp\left(-\frac{t}{\tau_T}\right) \tag{17.41}$$

The transient now stops if $i_c(t)$, and hence $q_F(t)$, is zero, or

$$-I_{B2}\left[1 - \exp\left(\frac{\tau_3}{\tau_T}\right)\right] + I_{B1}\exp\left(-\frac{\tau_3}{\tau_T}\right) = 0 \quad \text{or} \quad \tau_3 = \tau_T \ln\frac{I_{B2} + I_{B1}}{I_{B2}} \tag{17.42}$$

If the transistor is saturated part of the time, nothing changes as far as τ_0 is concerned. The solution (17.37) also remains correct for the turn-on transient. But the largest value that $i_c(t)$ can attain is $I_{Cs} = V_{CC}/R_c$. The corresponding value of $q_F(t)$ is $I_{Cs}\tau_{BF}/\beta_F = I_{Bs}\tau_{BF}$, where $I_{Bs} = I_{Cs}/\beta_F = V_{CC}/(\beta_F R_c)$. Hence the unsaturated part of the transient (17.37) stops when $q_F(t)$ reaches the value $q_{Fs} = I_{Bs}\tau_{BF}$. Let this be at τ'_1; then

$$I_{Bs}\tau_{BF} = I_{B1}\tau_{BF}\left[1 - \exp\left(-\frac{\tau'_1}{\tau_T}\right)\right] \quad \text{or} \quad \tau'_1 = \tau_T \ln\frac{I_{B1}}{I_{B1} - I_{Bs}} \tag{17.43}$$

For $I_{Bs} \ll I_{B1}$, that is, for strongly overdriving the base,

$$\tau_1' \simeq \tau_T \frac{I_{Bs}}{I_{B1}} \tag{17.43a}$$

which is small in comparison with $\tau_1 = 2.2\tau_T$. Note that the stored charge at the limit of saturation is

$$q_{Fs} = I_{Bs}\tau_{BF} \tag{17.44}$$

For $t > \tau_1'$, $i_c(t)$ remains clamped at I_{Cs}, but the stored charge continues to rise until it reaches an equilibrium situation described by the steady-state solution of Eqs. (17.14) and (17.14a).

Next let $i_b(t)$ change from I_{B1} to $-I_{B2}$ at $t = 0$. At first nothing happens to $i_c(t)$; it remains clamped to I_{Cs} as long as the stored charge is larger than q_{Fs}. Let the stored charge reach the value q_{Fs} at $t = \tau_2$; then the transistor becomes unsaturated for $t > \tau_2$ and the solution of Eq. (17.40) is

$$q_F(t) = I_{Bs}\tau_{BF} \exp\left(-\frac{t-\tau_2}{\tau_T}\right) - I_{B2}\tau_{BF}\left[1 - \exp\left(-\frac{t-\tau_2}{\tau_T}\right)\right] \tag{17.45}$$

This part of the transient stops when $q_F(t) = 0$. Let this be at $t - t_2 = \tau_3'$; then

$$\tau_3' = \tau_T \ln \frac{I_{B2} + I_{Bs}}{I_{B2}} \simeq \tau_T \frac{I_{Bs}}{I_{B2}} \tag{17.46}$$

if $I_{B2} \gg I_{Bs}$, resulting in a considerable speed up of the response.

We finally evaluate the storage delay time τ_2. To do so rigorously means solving Eqs. (17.14) and (17.14a); but this involves solving two simultaneous differential equations that give two time constants. Usually, however, there is a predominant mode of decay, and then the problem can be simplified. One thereby introduces the excess stored charge $q_e = q_F + q_R - I_{Bs}\tau_{BF}$, and a predominant lifetime τ_e, which will be evaluated in a moment. The differential equation for q_e is

$$\frac{dq_e}{dt} + \frac{q_e}{\tau_e} = i_b(t) - I_{Bs} \tag{17.47}$$

For the turn-on part of q_e we had $i_b(t) = I_{B1}$; hence $q_e(\infty) = (I_{B1} - I_{Bs})\tau_e$. Hence, for the turn-off transient, $q_e(0) = (I_{B1} - I_{Bs})\tau_e$ and $q_e(\infty) = -(I_{B2} + I_{Bs})\tau_e$, since $i_b(t) = -I_{B2}$, so that

$$q_e(t) = (I_{B1} - I_{Bs})\tau_e \exp\left(-\frac{t}{\tau_e}\right) - (I_{B2} + I_{Bs})\tau_e\left[1 - \exp\left(-\frac{t}{\tau_e}\right)\right] \tag{17.48}$$

This transient stops when $q_e(t) = 0$. Let this be at $t = \tau_2$; then

$$\tau_2 = \tau_e \ln \frac{I_{B2} + I_{B1}}{I_{B2} + I_{Bs}} \tag{17.49}$$

We now evaluate τ_e as the predominant lifetime of the process described by Eqs. (17.14) and (17.14a). Substituting

$$q_F = A \exp(-st), \qquad q_R = B \exp(-st)$$

into the homogeneous equations of (17.14) and (17.14a), we obtain

$$A\left(\frac{1}{\tau_{BF}} - s\right) + B\left(\frac{1}{\tau_{BR}} - s\right) = 0$$

$$A\frac{\beta_F}{\tau_{BF}} - B\left(\frac{1 + \beta_R}{\tau_{BR}} - s\right) = 0$$

These equations have a solution only if

$$\begin{vmatrix} \dfrac{1}{\tau_{BF}} - s & \dfrac{1}{\tau_{BR}} - s \\[2mm] \dfrac{\beta_F}{\tau_{BF}} & -\left(\dfrac{1 + \beta_R}{\tau_{BR}} - s\right) \end{vmatrix} = 0$$

or

$$s^2 - s\frac{(1 + \beta_F)\tau_{BR} + (1 + \beta_R)\tau_{BF}}{\tau_{BF}\tau_{BR}} + \frac{1 + \beta_F + \beta_R}{\tau_{BF}\tau_{BR}} = 0 \qquad (17.50)$$

Now any equation $x^2 - px + q = 0$ with $p^2 \gg 4q$ has a large root $x = x_1 \simeq p$ and a small root $x = x_2 \simeq q/p$. Therefore, the predominant lifetime of the excess carriers is

$$\tau_e = \frac{1}{s_2} = \frac{(1 + \beta_F)\tau_{BR} + (1 + \beta_R)\tau_{BF}}{1 + \beta_F + \beta_R} \qquad (17.51)$$

Usually, $\beta_F \gg \beta_R$; in that case $\tau_e \simeq \tau_{BR}$.

We finally evaluate τ'_1 and τ'_3 in some greater detail. Since $\tau_{BF} = \beta_F \tau_d$, $\tau_d = w_B^2/2D_n = (2\pi f_\alpha)^{-1}$, and $I_{Bs} = I_{Cs}/\beta_F$, we may write

$$\tau'_1 = (\tau_{BF} + \beta_F C_{jc}R_L)\frac{I_{Bs}}{I_{B1}} = (\tau_d + C_{jc}R_c)\frac{I_{Cs}}{I_{B1}} \qquad (17.52)$$

$$\tau'_3 = (\tau_{BF} + \beta_F C_{jc}R_L)\frac{I_{Bs}}{I_{B2}} = (\tau_d + C_{jc}R_c)\frac{I_{Cs}}{I_{B2}} \qquad (17.52a)$$

The conditions for small values of τ'_1 and τ'_3 are therefore a large alpha cutoff frequency f_α, a small product $C_{jc}R_c$, and relatively small ratios I_{Cs}/I_{B1} and I_{Cs}/I_{B2}. However, the latter occur at the expense of the turn-off delay τ_2.

17.2. PHOTODIODE

Let a *pn* junction diode have its junction close to, and parallel with, the surface. If the surface is irradiated by light of quantum energy *hf* larger than the gap width, hole–electron pairs will be generated. The holes will move to the *p* region and the electrons to the *n* region, resulting in a photocurrent I_s flowing from *n* to *p*.

There are two ways of operating this diode.

1. The junction is biased in the back direction and the photocurrent is fully collected. The diode is then said to be operated in the *photodiode mode*.

2. The junction electrodes are left open. The diode then biases itself in the forward direction, so that the net current is zero and an open-circuit voltage is developed (photovoltaic effect). The diode is then said to be operated as a *photovoltaic cell.*

17.2a. Operation as a Photodiode and as a Photovoltaic Cell

Let the diode have a characteristic†

$$I = -I_0\left[\exp\left(\frac{eV}{kT}\right) - 1\right] \qquad (17.53)$$

when kept in the dark; then the characteristic for the illuminated diode is

$$I = I_s - I_0\left[\exp\left(\frac{eV}{kT}\right) - 1\right] \qquad (17.54)$$

where I_s is the photocurrent. We now calculate I_s.

Let E be the incident radiant power and let R be the power reflection coefficient of the surface. Then $E(1 - R)$ is the energy entering into the photodiode, and the number of hole–electron pairs that is generated per second is $E(1 - R)/hf$. Let η be the probability that the hole–electron pair is collected; then the photocurrent is

$$I_s = \frac{eE(1 - R)\eta}{hf} \qquad (17.55)$$

If the diode is operated as a photodiode, the device current is

$$I = I_0 + I_s \qquad (17.56)$$

where I_0 is called the *dark* current. In good diodes one should keep the dark current small.

If the diode is operated as a photovoltaic cell, the open-circuit voltage is

$$V = V_{oc} = \frac{kT}{e}\ln\left(1 + \frac{I_s}{I_0}\right) \qquad (17.57)$$

as we see by putting $I = 0$ in (17.54) and solving for V. For a good response one should again keep the dark current I_0 of the diode small.

For very low light levels $I_s \ll I_0$ and

$$V_{oc} \simeq \frac{kT}{e}\frac{I_s}{I_0} \simeq I_s R_0 \qquad (17.57a)$$

where $R_0 = kT/eI_0$ is the internal diode resistance for zero bias. The device then operates in its *linear* range. For higher levels, $I_s \gg I_0$ and the response

---

† We choose here the direction of positive current flow from n to p. V is the voltage applied to the p region of the junction.

is *logarithmic*:

$$V_{oc} \simeq \frac{kT}{e} \ln \frac{I_s}{I_0} \tag{17.57b}$$

The light does not penetrate very deeply into the material. Light absorption generally follows an exponential law:

$$E = E_0 \exp(-\alpha x) \tag{17.58}$$

where E_0 is the incident light intensity, E the intensity at a depth x, and α the absorption coefficient of the light. The penetration depth of the light is thus of the order of $1/\alpha$. Since $\alpha \simeq 10^5/\text{cm}$ for direct transitions, the penetration depth of the light is of the order of 1000 Å. This means that the junction must lie very close to the surface and that the surface region must be very thin. This condition is satisfied best for a Schottky-barrier photodiode with a transparent metal electrode.

A thin surface region thus allows one to obtain a higher collection efficiency η, since few hole–electron pairs will recombine before collection. But not only is the collection efficiency η larger, the response is also faster. This can be seen as follows. If the surface region is much narrower than the penetration depth of the light, then the hole–electron pairs are generated in the space-charge region, and the longest travel time of the carriers is the travel time across the space-charge region. That is,

$$\tau = \frac{w}{\bar{u}_d} = \frac{w}{\mu \bar{E}} = \frac{w^2}{\mu |V_{\text{dif}} - V|} \tag{17.59}$$

where w is the width of the space-charge region, $\bar{E} = |V_{\text{dif}} - V|/w$, and $|V_{\text{dif}} - V|$ is the voltage developed across the junction. Taking $w = 10^{-5}$ cm, $\mu = 1000 \text{ cm}^2/\text{V s}$, and $|V_{\text{dif}} - V| = 1$ Volt yields $\tau = 10^{-13}$ s.

But if the surface region is much wider than the penetration depth of the light, then the hole–electron pairs are generated in the surface region. The minority carriers must first diffuse toward the space-charge region and then be collected. If w_s is the width of the surface region and D the diffusion coefficient of the minority carriers, then the diffusion time τ_d of the minority carriers is

$$\tau_d \simeq \frac{w_s^2}{2D} \tag{17.59a}$$

Taking $w_s = 10^{-4}$ cm and $D = 25 \text{ cm}^2/\text{s}$ gives $\tau_d = 2 \times 10^{-10}$ s, which is several orders of magnitude larger. We see from this example that to ensure good microwave operation of the device one should keep the surface region very thin. This is especially pertinent for diodes used to detect microwave-modulated laser light.

Figure 17.7a gives the equivalent circuit of the photodiode. If $I_s(t)$ is the received photocurrent, one may write

$$I_s(t) = I_{s0} + I_{s1} \cos \omega_p t \tag{17.60}$$

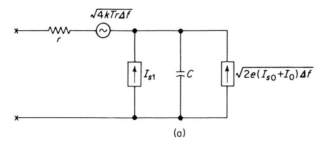

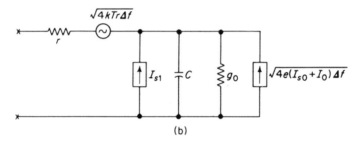

Fig. 17.7. (a) Equivalent circuit of a solid state photodiode. (b) Equivalent circuit of a photovoltaic cell.

where ω_p is the modulation frequency. For good microwave operation both the capacitance C (small junction area) and the series resistance r should be small. The effect of the capacitance C can always be tuned out, even at microwave frequencies, but the series resistance r has a detrimental effect.

Figure 17.7a also shows the noise properties of the diode. The series resistance r shows, of course, thermal noise, and the current $I_{s0} + I_0$ shows shot noise. Figure 17.7b gives the equivalent circuit of the photovoltaic cell. It is similar to that of the photodiode, but the capacitance C is larger (since the space-charge capacitance is voltage dependent and since there may be some storage capacitance). Also, there is now a conductance $g_0 = e(I_{s0} + I_0)/kT$ in parallel to it. In addition there are now two equal and opposite currents $I_{s0} + I_0$, and hence the shot noise is twice as large. The photodiode thus has a slight advantage over the photovoltaic cell, but the latter has more stable operation.

We now evaluate the noise properties of a photovoltaic cell used as a low-level detector, so that $I_s \ll I_0$. Then from Eq. (17.55), if $hf = eV_{ph}$, where V_{ph} is the photon energy in electron volts, we have

$$I_s = \frac{E(1 - R)\eta}{V_{ph}}$$

The *minimum detectable power* or *noise equivalent power is now* defined as the power E_{min}, for which I_s equals the rms noise current $(4eI_0)^{1/2}$ per unit bandwidth. Hence

$$\frac{E_{min}(1-R)\eta}{V_{ph}} = (4eI_0)^{1/2} \quad \text{or} \quad E_{min} = \frac{V_{ph}}{(1-R)\eta}(4eI_0)^{1/2} \quad (17.61)$$

As an example, we put $V_{ph} = 1.5$ V, $R = 0$, $\eta = 1.0$, and $I_0 = 10^{-8}$ A, and obtain $E_{min} = 1.2 \times 10^{-13}$ W/Hz$^{1/2}$. A photodiode is at best a factor $\sqrt{2}$ better.

17.2b. Photovoltaic Diode as an Energy Converter

If a photovoltaic cell is connected to a load resistance R, dc power is delivered to resistance R. By appropriate choice of R, we can optimize this dc power. The cell thus converts radiant power into dc power. The most likely power source is the sun; a photovoltaic cell used to convert solar power into dc power is called a *solar battery* or a *solar cell*.

We start with Eq. (17.54) but change over to current density, since it is important to know the power generated per square centimeter of irradiated surface. Hence

$$J = J_s - J_0\left[\exp\left(\frac{eV}{kT}\right) - 1\right] \quad (17.62)$$

The open-circuit voltage V_{oc} is

$$V_{oc} = \frac{kT}{e}\ln\left(\frac{J_s}{J_0} + 1\right) \simeq \frac{kT}{e}\ln\frac{J_s}{J_0} \quad (17.62a)$$

The power output of the device is

$$P = JV = \left\{J_s - J_0\left[\exp\left(\frac{eV}{kT}\right) - 1\right]\right\}V \quad (17.63)$$

This power has an optimum value if $dP/dV = 0$. This yields

$$J_s - J_0\left[\exp\left(\frac{eV}{kT}\right) - 1\right] - VJ_0\left(\frac{e}{kT}\right)\exp\left(\frac{eV}{kT}\right) = 0 \quad (17.64)$$

or

$$\exp\left(\frac{eV}{kT}\right)\left(1 + \frac{eV}{kT}\right) = 1 + \frac{J_s}{J_0} = \exp\left(\frac{eV_{oc}}{kT}\right) \quad (17.64a)$$

We now demonstrate how this equation can be solved by a rapidly converging method of successive approximations. We substitute $V = V_{oc}$ in the expression $1 + eV/kT$ and solve for V. This yields

$$V = V_1 = V_{oc} - \frac{kT}{e}\ln\left(1 + \frac{eV_{oc}}{kT}\right) \quad (17.65)$$

Next we substitute $V = V_1$ in the expression $1 + eV/kT$ and solve again for V. This yields

$$V = V_2 = V_{oc} - \frac{kT}{e}\ln\left(1 + \frac{eV_1}{kT}\right) \quad (17.66)$$

and so on. In the particular case when $V_{oc} = 0.600$ V and $kT/e = 25.8$ mV, we have

$$V_1 = 0.600 - 0.082 = 0.518 \text{ V}$$
$$V_2 = 0.600 - 0.078 = 0.522 \text{ V}$$

so that a two-step approximation is amply sufficient.

Suppose that the implicit equation (17.64a) has been solved and the solution is found to be $V = V_{mp}$. We now substitute $V = V_{mp}$ into J and find the corresponding current J_{mp}:

$$J_{mp} = J_s + J_0 - J_0 \exp\left(\frac{eV_{mp}}{kT}\right)$$

$$= (J_s + J_0)\frac{eV_{mp}/kT}{eV_{mp}/kT + 1} \tag{17.67}$$

$$\simeq J_s \frac{eV_{mp}/kT}{eV_{mp}/kT + 1}$$

since $J_s \gg J_0$. Hence

$$P_{\max} = J_{mp}V_{mp} = \frac{eV_{mp}/kT}{eV_{mp}/kT + 1}V_{mp}J_s \tag{17.68}$$

The optimum load impedance is

$$R_{mp} = \frac{V_{mp}}{J_{mp}} = \frac{1 + eV_{mp}/kT}{eJ_s/kT} \tag{17.69}$$

Since the power density input is E, the optimum efficiency of the energy converter is

$$\eta_{\max} = \frac{P_{\max}}{E} = \frac{eV_{mp}/kT}{eV_{mp}/kT + 1}\frac{V_{mp}J_s}{E} \tag{17.70}$$

V_{mp} increases with increasing gap width of the semiconductor and J_s decreases with increasing gap width. Consequently $\eta_{\max}$ first increases with increasing gap width, goes through a maximum at a gap width of about 1.5 eV, and decreases for higher gap widths. For the same reason R_{mp} increases steadily with increasing gap width.

Up to here we have neglected the series resistance r_s of the junction. If the series resistance r_s is taken into account, J_s and V_{oc} do not change, but the (J, V) characteristic for $r_s > 0$ lies below the characteristic for $r_s = 0$. Figure 17.8 shows these two characteristics and the operating points for maximum power (A for $r_s = 0$ and B for $r_s > 0$). The series resistance thus reduces the efficiency of the device, and for that reason it should be kept as small as possible.

The best results up to now have been obtained with silicon solar-energy converters. Theoretically, a somewhat larger gap width would give a higher efficiency, but for a number of reasons the actual efficiency is lower than for silicon converters. There is hope, however, that this situation may be reversed if more development work is performed.

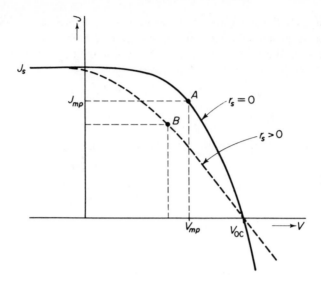

Fig. 17.8. (J, V) characteristic of a photovoltaic cell for $r_s = 0$ and for $r_s > 0$. In the first case the optimum power is obtained at the operating point A, in the second case at the operating point B.

17.2c. Avalanche Multiplication in Photodiodes

It often happens that the signal coming out of a photodiode drowns in the noise background of the associated receivers or amplifiers. In that case it is convenient to use avalanche multiplication in the photodiode. In the earlier days of avalanche multiplication, breakdown spots, called *microplasmas*, occurred in the junction. These microplasmas switched off and on intermittently and produced a large amount of noise. It has been possible, however, to design avalanche diodes that show uniform breakdown, and these diodes can be used for avalanche multiplication in the prebreakdown region of operation.

We shall give here a simplified version of the operation of the avalanche multiplier and of the noise associated with it. If

$$I_s(t) = I_{s0} + I_{s1} \cos \omega_p t \tag{17.71}$$

is the photocurrent, and I_0 the dark current of the device without multiplication, and if p is the number of hole–electron pairs produced when they traverse the space-charge region once, then the net signal current is

$$I_{sc} \cos \omega_p t = I_{s1} \cos \omega_p t [1 + p + p^2 + \ldots] = M I_{s1} \cos \omega_p t \tag{17.72}$$

where
$$M = \frac{1}{1 - p} \tag{17.73}$$

For the noise we start with the shot noise of $I_{s0} + I_0$, and this is multiplied. But since the generation of hole–electron pairs is also a noisy process, new noise is generated each time hole–electron pairs traverse the space-charge region. We shall assume that the secondary current generated each time shows full shot noise†. In that case

$$S_I(f) = 2e[(I_{s0} + I_0) + p(I_{s0} + I_0) + p^2(I_{s0} + I_0) + \ldots]M^2$$

$$= 2e(I_{s0} + I_0)\frac{M^2}{1 - p} = 2e(I_{s0} + I_0)M^3 \tag{17.74}$$

The signal-to-noise power ratio for a bandwidth B is therefore

$$\frac{S}{N} = \frac{M^2 I_{s1}^2}{2e[M^3(I_{s0} + I_0) + I_{eq}]B} = \frac{I_{s1}^2}{2eB[M(I_{s0} + I_0) + I_{eq}/M^2]} \tag{17.75}$$

where the noise of a receiver of bandwidth B is represented by a current generator $\sqrt{2eI_{eq}B}$ in parallel with the photodiode. Considered as a function of M, this has a maximum value

$$\frac{S}{N} = \frac{I_{s1}^2}{2e(I_{s0} + I_0)B} \cdot \frac{2}{3}\left(\frac{I_{s0} + I_0}{2I_{eq}}\right)^{1/3} \tag{17.76}$$

if

$$M = \left(\frac{2I_{eq}}{I_{s0} + I_0}\right)^{1/3} \tag{17.77}$$

Without amplification the S/N ratio would have been

$$\frac{S}{N} = \frac{I_{s1}^2}{2e(I_{s0} + I_0 + I_{eq})B} \tag{17.78}$$

A considerable improvement in signal-to-noise ratio is thus possible if $I_{eq} \gg I_{s0} + I_0$.

The first part of Eq. (17.76) represents the signal-to-noise ratio of the photodiode without amplification. Since the second part of (17.76) has a value smaller than unity, the avalanche multiplication plus the associated amplifier gives some deterioration of the S/N ratio, but it is much less than it would be if no avalanche multiplication were present.

The derivation of Eq. (17.74) is based on the presupposition that electrons and holes have the same ionizing power. The case in which electrons and holes have different ionizing power has been evaluated by McIntyre.‡ Let α and β be the ionization coefficients for electrons and holes; then the ratio $k = \beta/\alpha$ is a slow function of the field strength. Expressed in terms of k, McIntyre finds for $I_{S0} \ll I_0$

$$S_I(f) = 2eI_0M^3\left[1 + \frac{1 - k}{k}\left(\frac{M - 1}{M}\right)^2\right] \tag{17.79}$$

---

† In secondary emission noise we had

$$S_{\text{sec}}(f) = 2eI_{pr}\delta(\kappa - \delta)$$

and at relatively low voltages $\kappa - \delta \simeq 1$. The secondary emission noise then corresponds to full shot noise of the collected current $I_{pr}\delta$. We use the same idea in this derivation.

‡ R. J. McIntyre, *IEEE Trans. Electron. Devices*, **ED-13**, 164 (1966).

if the injected current I_0 consists of holes and

$$S_I(f) = 2eI_0M^3\left[1 - (1-k)\left(\frac{M-1}{M}\right)^2\right] \qquad (17.80)$$

if the injected current I_0 consists of electrons. The first expression is quite large for $k \ll 1$, equal to $2eI_0M^3$ for $k = 1$, and quite small for $k > 1$, so that the best results are obtained if the holes have the largest ionizing power. The second expression is quite large for $k \gg 1$, equal to $2eI_0M^3$ for $k = 1$, and quite small for $k \ll 1$, so that the best results are obtained if the electrons have the largest ionizing power. By proper design of the diodes a considerable improvement in $S_I(f)$, and hence in S/N, is thus possible. These predictions have been well verified by experiment.

17.2d. Schottky-Barrier Photodiode

A Schottky-barrier diode can be used as a photodiode in two ways:

1. If $E_b < hf < E_g$, where E_b is the barrier height seen from the metal side and E_g the gap width of the semiconductor, then photoelectrons produced in the metal can pass the barrier and be collected by the semiconductor. This process has a relatively low efficiency, since the photoelectrons rapidly lose energy to the electrons in the metal (problem 17.13).

2. If $hf > E_g$, then hole–electron pairs are produced in the semiconductor, and the holes can be collected by the metal. This is a diffusion process; the holes generated within a diffusion length L_p of the contact have a good chance of being collected (problem 17.14). Therefore, the quantum efficiency of this detector can be appreciable.

17.2e. Phototransistor

A phototransistor is a transistor with floating base irradiated with low-level light. It can be understood as a photovoltaic cell (Sec. 17.2a) with built-in gain.

Due to the light, a photocurrent I_{ph} flows to the base. But since the base is floating, for a *pnp* transistor with collector saturated current I_{C0}, we have

$$I_B = -(1 - \alpha_F)I_E + I_{C0} + I_{ph} = 0$$

or

$$I_C = I_E = \frac{I_{C0} + I_{ph}}{1 - \alpha_F} = (1 + \beta_f)(I_{C0} + I_{ph}) \qquad (17.81)$$

where β_F is the direct-current amplification factor so that the photocurrent is multiplied by a factor $(1 + \beta_F)$.

What is the noise at low-level light? Then $I_{ph} \ll I_{C0}$, and, since $(1 - \alpha_F)I_E$ and I_{C0} fluctuate independently, and each gives full shot noise,

$$\overline{i_b^2} = 4eI_{C0}\,\Delta f \qquad (17.82)$$

This noise is now multiplied by the alternating-current amplification factor β_0^2. In addition the current $I_{CE0} = (1 + \beta_f)I_{C0}$ gives full shot noise. Hence the collector noise at low frequencies is

$$\overline{i_c^2} = 2e(1 + \beta_F)I_{C0}\,\Delta f + \beta_0^2 4eI_{C0}\,\Delta f \tag{17.83}$$

Since $2\beta_0^2 \gg 1 + \beta_F$, the second term predominates.

For high-frequency modulated signals, the multiplication becomes frequency dependent and β_0 must be replaced by $\beta_0/(1 + jf/f_\beta)$. Typically, f_β is of the order of 1–10 kHz because of the low current levels involved. The signal at the collector thus becomes

$$i_s = \left(1 + \frac{\beta_0}{1 + jf/f_\beta}\right)I_{\text{ph}} \simeq \frac{\beta_0}{1 + jf/f_\beta}I_{\text{ph}} \tag{17.84}$$

and the collector noise becomes

$$i_c^2 = 2e(1 + \beta_F)I_{C0}\,\Delta f + \frac{\beta_0^2}{1 + f^2/f_\beta^2}4eI_{C0}\,\Delta f \tag{17.85}$$

Hence the signal-to-noise power ratio per unit bandwidth is

$$\frac{S}{N} = \frac{\overline{i_s^2}}{\overline{i_c^2}} = \frac{\beta_0^2 I_{\text{ph}}^2}{2e(1 + \beta_F)I_{C0}(1 + f^2/f_\beta^2) + \beta_0^2 \cdot 4eI_{C0}} \tag{17.86}$$

At low frequencies, since $1 + \beta_F \ll 2\beta_0^2$,

$$\frac{S}{N} \simeq \frac{I_{\text{ph}}^2}{4eI_{C0}} \tag{17.87}$$

which would be the signal-to-noise power ratio of the photovoltaic cell made by the base and the other electrodes. The multiplication process thus adds no noise at low frequencies.

At higher frequencies the signal-to-noise ratio decreases. It has decreased by a factor of 2 at a frequency f_1 such that

$$(1 + \beta_F)I_{C0}\frac{f_1^2}{f_\beta^2} = 2\beta_0^2 I_{C0} \quad \text{or} \quad f_1 = f_\beta\left(\frac{2\beta_0^2}{1 + \beta_F}\right)^{1/2} \tag{17.88}$$

This frequency usually lies between 10 and 100 kHz.

17.3. DIODE APPLICATIONS

17.3a. PIN Diode as a Microwave Switch

The *pin* diode has a low series resistance and a small capacitance when biased in the back direction and a low impedance when biased sufficiently far in the forward direction. Because of this, it can be used as an off–on switch in microwave circuits.

Figure 17.9 shows such a circuit. The diode is represented by an impedance $R_d + jX_d$, which incorporates the diode proper plus the leads. Let

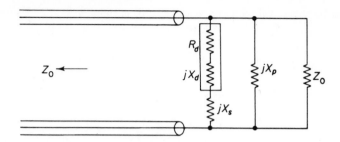

Fig. 17.9. The *p-i-n* diode as a microwave switch. The tuning of the switch is explained in the text.

$R_d + jX_d$ be equal to $R_f + jX_f$ for forward bias and equal to $R_r + jX_r$ for back bias. If we first add a series reactive component $X_s = -X_r$, the total impedance for back bias is R_r, which can be very small, since it consists only of the series resistance of the device. This state of bias can thus be used as the "off" state where the diode reflects nearly 100 per cent of the incoming power. For the other bias, the impedance is $R_f + j(X_f - X_r)$. If we add a parallel reactance $X_p = -(X_f - X_r)$, the net impedance is high, and nearly the full power is transmitted into the terminating impedance Z_0, which is the characteristic impedance of the line.

Pin diodes are very useful for this purpose since they combine a low back-bias capacitance with a low forward impedance.

17.3b. Tunnel Diode as an Amplifier and as an Oscillator

Let the tunnel diode have a negative conductance $-g_d$, and let a signal source consisting of a current generator i_s and an internal conductance g_s be connected to it, together with a load conductance g_L. The power fed into the load conductance is then increased by the presence of the negative conductance $-g_d$ (Fig. 17.10a). We assume first $g_s > g_d$.

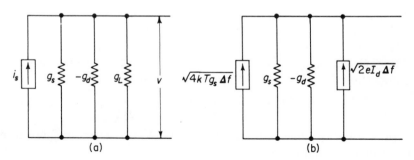

Fig. 17.10. (a) The tunnel diode as an amplifier. (b) Equivalent noise circuit of the tunnel diode.

The voltage v developed across the circuit is then

$$v = \frac{i_s}{g_s - g_d + g_L} \tag{17.89}$$

and the power fed into the load conductance g_L is

$$P_L = \frac{1}{2}|v|^2 g_L = \frac{1}{2}\frac{|i_s|^2 g_s}{2(g_s - g_d + g_L)^2}$$

$$= \frac{1}{8}\frac{|i_s|^2}{g_s}\frac{4g_s g_L}{(g_s - g_d + g_L)^2} \tag{17.90}$$

The power obtained when the source is *matched* to a passive load is called the *available power P_{av} of the source:*

$$P_{av} = \frac{1}{8}\frac{|i_s|^2}{g_s} \tag{17.91}$$

We now take $g_d \neq 0$, and $g_s > g_d$ and define the power gain of the circuit as the ratio of the power fed into the load to the power available at the source:

$$G = \frac{P_L}{P_{av}} = \frac{4g_s g_L}{(g_s - g_d + g_L)^2} \tag{17.92}$$

We now keep g_s and g_d fixed and vary g_L until the maximum power gain is obtained. This we find by differentiating Eq. (17.92) with respect to g_L and equating $dG/dg_L = 0$. The result is

$$G = G_{av} = \frac{g_s}{g_s - g_d}, \quad \text{for } g_L = g_s - g_d \tag{17.93}$$

This result is not surprising, for the signal source plus the negative conductance can be considered as a new signal source consisting of a current generator i_s and an internal conductance $g_s - g_d$.

If $g_s \rightarrow g_d$, the maximum power gain goes to infinity. What happens if $g_s < g_d$? As long as $g_s + g_L \geq g_d$, the preceding theory is still valid and the definition of the power gain still applies. But now an infinite power gain is obtained if $g_L = g_d - g_s$, and for $g_L < g_d - g_s$ oscillations occur.

Tunnel diode amplifiers operate well into the microwave range.

We now turn to the noise. The equivalent noise circuit is shown in Fig. 17.10b. Since it is common practice to count the noise of the load conductance g_L as belonging to the next stage, we have omitted the load. As we see by inspection, the noise figure of the tunnel diode circuit is

$$F = \frac{4kTg_s \Delta f + 2eI_d \Delta f}{4kTg_s \Delta f} = 1 + \frac{e}{2kT}\frac{I_d}{g_s} \tag{17.94}$$

To obtain a reasonable gain, g_s must approach g_d. The noise figure then approaches the limiting value

$$F_\infty = 1 + \frac{e}{2kT}\frac{I_d}{g_d} \tag{17.95}$$

In practice noise figures of the order of 2 ($= 3$ dB) can be achieved at microwave frequencies. This is quite acceptable, and hence tunnel diode microwave amplifiers have considerable appeal.

One might conclude from Eq. (17.94) that we can obtain a better noise figure by making g_s larger. In fact, we cannot, since the available gain G_{av} goes to unity. It may then be shown that the noise figure F_n of a large number n of identical stages connected in cascade may be written as

$$F_n = 1 + \frac{F-1}{1 - 1/G_{av}} = 1 + \frac{e}{2kT} \frac{I_d}{g_d} \qquad (17.96)$$

and this corresponds to (17.95).

If the minimum in the characteristic were rigorously zero, that is, if there were no excess current in the minimum, then the combination of (15.119) and (15.120) gives that the "voltage" I_d/g_d would attain its minimum value at $V = V_1$, and this minimum would be zero. Actually, because of the excess current, the minimum occurs at a lower value of V and is not zero.

We have seen that tunnel diodes can also be used as oscillators. Tunnel-diode oscillators have been designed that operate above 100 GHz. They afford the possibility of relatively cheap signal sources at microwave frequencies.

In the early days of tunnel-diode development it was hoped that tunnel diodes would find important applications in switching circuits. Tunnel-diode switching circuits indeed offer the possibility of ultrafast switching, but the fact that the tunnel diode is a two-terminal device makes it less desirable than the transistor. In special applications it may have distinct advantages.

17.3c. Varactor Diode; Step-Recovery Diode

A *pn* diode biased in the back direction has a voltage-dependent capacitance. It can therefore be used for capacitance mixing, also known as parametric mixing, parametric amplification, and frequency multiplication in the manner already discussed in Secs. 14.3c and d; we need not go into details here. *Pn* diodes suitable for this mode of operation are called *varactor diodes.*

Although varactor diodes can generate harmonic power with good efficiency, higher efficiencies can be obtained with *step-recovery diodes*. Typical results are frequency multiplication from 200 to 2000 MHz with 70 per cent efficiency. Also, much higher harmonics can be generated in single-stage operation than with varactor diodes.

The step-recovery diode is a *pin* diode. When the device has forward bias, it has practically zero forward impedance and it can store almost unlimited amounts of charge in the *i* region. When the device is back biased to the breakdown voltage in the *i* region, it can recover the original situation in an extremely short time. Its operation is such that it is forward biased during most of the cycle and back biased to breakdown voltage for the very small

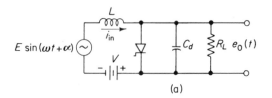

(a)

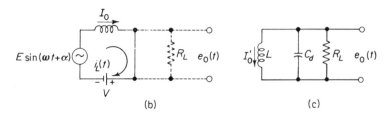

(b)   (c)

Fig. 17.11. Operation of the step recovery diode circuit. (a) Impulse generator circuit. (b) Equivalent circuit during conduction interval. (c) Equivalent circuit during impulse interval. (From S. A. Hamilton and R. D. Hall, *The Microwave Journal*, pp. 69–78, April 1967.)

remainder of the cycle. As a consequence, the voltage across the device shows extremely sharp pulses of high harmonic content, and this is why it can generate high harmonics.

We follow here the discussion by Hamilton and Hall.[†] Figure 17.11a shows the circuit used in the harmonic generation process. During the conduction interval the *pin* diode is a short-circuit and the current $i_L(t)$ through the inductance changes from I_0 to I_0' (Fig. 17.11b). During the recovery interval or impulse interval the current through the inductance switches back from I_0' to I_0 and the process repeats itself (Fig. 17.11c).

Suppose that the dc bias is $+V$, the forward bias for the conduction mode of operation is ϕ, and the ac driving signal is $E \sin(\omega t + \alpha)$. If the conduction mode starts at $t = 0$ when the current through L has the value I_0, then the current $i_{in}(t)$ through L is

$$i_{in}(t) = I_0 + \frac{E}{\omega L}[\cos \alpha - \cos(\omega t + \alpha)] - \frac{V + \phi}{\omega L}\omega t \qquad (17.97)$$

The charge stored in the diode is found by integration of $i_{in}(t)$. This gives

$$q(t) = \left(I_0 + \frac{E}{\omega L}\cos \alpha\right)t - (V + \phi)\frac{t^2}{2L} - \frac{E}{\omega^2 L}[\sin(\omega t + \alpha) - \sin \alpha]$$
$$(17.98)$$

We now require that $q(t) = 0$ at $t = t_1$, the end of the conduction cycle. By adjusting the bias voltage V, we can reduce the stored charge to zero at the

---

† S. A. Hamilton and R. D. Hall, *The Microwave Journal*, 69–78 (April 1967).

time when the current has its maximum negative value; that is, $di_{in}/dt = 0$. The voltages across diode and inductor are then both zero. This, in turn, means that at the beginning of the nonconducting (or depletion) interval, the instantaneous value of the generator voltage is equal to $-V$. Because the nonconducting interval is so short, we assume that $E \sin(\omega t + \alpha) - V$ is practically zero during that interval. The initial current I_0' is the significant excitation.

During the nonconducting interval the voltage $e_0(t)$ across R_L is found to be

$$e_0(t) = \frac{-I_0'\sqrt{L/C}}{\sqrt{1 - \zeta^2}} \exp\left[-\frac{\zeta}{\sqrt{1 - \zeta^2}} \omega_N(t - t_1)\right] \sin \omega_N(t - t_1) \quad (17.99)$$

where

$$\omega_N = \sqrt{\frac{1 - \zeta^2}{LC}}, \qquad \zeta = \frac{1}{2R_L}\sqrt{\frac{L}{C}} \quad (17.99a)$$

The current i_L in the inductor L during the interval is determined by integration of $e_0(t)$, since $e_0(t) = di_L/dt$. This yields

$$i_L(t) = I_0' \exp\left[-\frac{\zeta\omega_N(t - t_1)}{\sqrt{1 - \zeta^2}}\right]\left[\cos \omega_N(t - t_1) + \frac{\zeta \sin \omega_N(t - t_1)}{\sqrt{1 - \zeta^2}}\right]$$

$$(17.100)$$

Equation (17.99) represents a ringing pulse. Since diode conduction begins when the diode voltage goes positive, only the first half-cycle appears across R_L. The voltage across R_L during the nonconducting part of the cycle is therefore a half-sine pulse. The height of the pulse is limited to the breakdown voltage, and the width t_0 is

$$t_0 = \frac{\pi}{\omega_N} = \pi\sqrt{\frac{LC}{1 - \zeta^2}} \quad (17.101)$$

For high harmonic generation one should make $t_0 \ll t_1$.

But since $\omega_N(t - t_1) = \pi$ at the end of the nonconducting part of the cycle, we have

$$I_0 = -I_0' \exp\left(-\frac{\pi\zeta}{\sqrt{1 - \zeta^2}}\right) \quad (17.102)$$

Putting $i(t_1) = I_0'$, $q(t_1) = 0$, requiring that I_0' be a maximum at t_1, for maximum energy storage at the time of switching, and requiring finally that $L\, di_{in}/dt = 0$ for $t = t_1$ yields four equations in E, V, I_0, and α. It is thus possible to determine α for maximum energy storage.

One can then determine the input impedance seen by the generator under that condition, and hence the power delivered to the circuit by the generator of frequency ω. Finally, one can determine the power of frequency $\omega_N = n\omega$ developed into the load resistance R_L. The details are beyond the scope of this book; we refer to Hamilton and Hall's paper, which also gives practical circuits.

The high efficiencies obtainable in the circuit can be attributed to the small losses and to the extremely short transition times required for the device to change from a low impedance level to a very high impedance level. This problem is not yet fully understood. The thicker the *i* layer, the higher the breakdown voltage E_b and the longer the transition time. Empirically, Hamilton and Hall find the relationship

$$E_b < \frac{140}{\sqrt{f_n}} \quad V$$

where f_n is in gigahertz. At 10 GHz, this leads to $E_b < 45$ V and at 2 GHz, $E_b < 100$ V.

17.4. INJECTION LUMINESCENCE

When *pn* junctions are biased in the forward direction, minority carriers are injected and these minority carriers recombine with majority carriers. In some devices radiative recombination predominates, and almost every injected minority carrier gives rise to emission of a quantum with an energy approximately equal to the gap width of the material. This process is called *injection luminescence*. In good diodes the efficiency of the process is nearly 100 per cent.

In diodes in which the *p* and *n* regions are degenerate, a population inversion can occur when the applied forward voltage is larger than the gap width, and in that case lasing action can occur. We shall here devote our attention to injection luminescence; the junction laser is discussed in Sec. 17.5.

The effect of injection luminescence is most pronounced in materials in which direct band-to-band transitions are allowed. This is the case when the maximum energy in the valence band and the minimum energy in the conduction band occur at the same value of the wave vector **k** (usually at **k** = 0, as in GaAs and in some III–V and II–VI compounds.

17.4a. Theory of Injection Luminescence

If η_r is the quantum efficiency of the radiative process (that is, the average number of quanta emitted per injected electron), and if β is the probability of escape of the quantum, then the radiated power is

$$P_{\rm rad} = \frac{I}{e} hf \eta_r \beta = IV_{\rm ph}\eta_r\beta \tag{17.103}$$

where $V_{\rm ph}$ is the quantum energy in electron volts.

This power is radiated in random direction, and, since most materials used for *pn* injection luminescence have a relatively large refractive index *n* (for example, $n \simeq 3$–4), a major part of the emitted radiation is trapped by total reflection. If the *pn* junction is given a domelike shape, this difficulty can be overcome. Highly efficient lamps based on injection luminescence can thus be made.

If V is the applied voltage, then the power supplied by the battery is

$$P_b = IV \qquad (17.104)$$

and the efficiency of the device η_c is

$$\eta_c = \frac{P_{\text{rad}}}{P_b} = \eta_r \beta \frac{V_{\text{ph}}}{V} \qquad (17.105)$$

When η_r and β are close to unity, then η_c can be larger than unity if $V < V_{\text{ph}}$ (current starts to flow, and the junction starts to emit, long before the applied voltage V equals the gap width E_g).

It thus seems at first sight that the energy law is violated in the process. But in fact the remaining energy comes from cooling the junction. We can see this from Fig. 17.12, which shows the energy diagram of a pn junction with ohmic contacts. If an electron enters the n region from metal 1, it must overcome an energy barrier eE_1; therefore, the flow of current extracts the power IE_1 from metal 1. If an electron passes from the n region to the p region, it must overcome an energy barrier $e(V_{\text{dif}} - V)$, where V_{dif} is the diffusion potential and V the applied voltage; therefore, the flow of current extracts the power $I(V_{\text{dif}} - V)$ from the n region. If a hole passes from metal 2 to the p region it must overcome a barrier eE_2; therefore, the flow of current extracts the power IE_2 from metal 2. Since the battery supplied a power IV, the total amount of power available is

$$I(E_1 + V_{\text{dif}} - V + E_2 + V) = IE_g$$

where E_g is the gap width in electron volts. Since $V_{\text{ph}} \simeq E_g$, the energy balance is thus maintained, by virtue of the cooling effects at the contacts and at the junction.

Actually, the energy situation is even more favorable. The electrons do not arrive in the n region with zero energy, but with an average energy

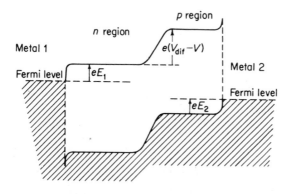

Fig. 17.12. Energy level diagram in a p-n junction to which a voltage V is applied.

$2kT/e$.† In the same way the holes arrive in the p region with an average energy $2kT/e$. The total energy available per hole–electron pair is therefore $(E_g + 4kT/e)$ electron volts, and this is amply sufficient to produce a photon of V_{ph} electron volts energy.

The shape of the spectral line is easily obtained. From the principle of detailed balance, the rate of radiative recombination at thermal equilibrium for a frequency interval dv at v is equal to the corresponding rate of generation of hole–electron pairs by thermal generation. This rate is $P(v)\rho(v)\,dv$ per unit volume, where $\rho(v)\,dv$ is the photon density in the range dv and $P(v)$ is the rate at which photons of frequency v are absorbed. The frequency distribution of the emitted radiation is thus $P(v)\rho(v)\,dv$. The problem is now to calculate $\rho(v)$ and $P(v)$.

Since $1/\lambda$ is the wave number of the quanta, the density of standing wave patterns is $4\pi(1/\lambda)^2\,d(1/\lambda)$ (see Sec. 3.3a). But each wave has two possible directions of polarization, and $1/\lambda = nv/c$, where n is the refractive index of the material. Hence the mode density is

$$8\pi\left(\frac{1}{\lambda}\right)^2 d\frac{1}{\lambda} = \frac{8\pi n^2 v^2\,dv}{c^3}\frac{d(nv)}{dv}$$

and the energy density, according to Planck's law, is

$$\rho(v)\,dv = \frac{8\pi n^2 v^2\,dv}{c^3}\frac{d(nv)}{dv}\frac{hv}{\exp{(hv/kT)} - 1} \qquad (17.106)$$

If $\alpha(v)$ is the absorption coefficient of the radiation at the frequency v, then the rate $P(v) = \alpha(v)v_g$, where v_g is the group velocity of the wave. But by definition

$$v_g = \frac{dv}{d(1/\lambda)} = c\,\frac{dv}{d(nv)}$$

since $1/\lambda = nv/c$. Consequently,

$$\begin{aligned} P(v)\rho(v)\,dv &= \frac{8\pi n^2 v^2\,dv}{c^2}\frac{hv\alpha(v)}{\exp{(hv/kT)} - 1} \\ &= \frac{8\pi n^2}{h^3 c^2}\frac{\alpha(E)E^3\,dE}{\exp{(E/kT)} - 1} \end{aligned} \qquad (17.107)$$

Since E/kT is large, $\exp{(E/kT)}$ increases very fast with increasing energy. The absorption coefficient $\alpha(E)$ increases very fast with increasing energy and reaches a plateau for quantum energies slightly larger than the gap width. As a consequence the emission line has a maximum at the energy $E = E_m$, where $\alpha(E)$ and $\exp{(E/kT)}$ vary equally fast with energy.

---

† The argument here is the same as for the electrons emitted by a thermionic cathode; we saw in Chapter 7 that an emitted electron carries an average energy $2kT_c/e$, where T_c is the cathode temperature.

At lower energies $\alpha(E)$ decreases much faster with decreasing energy than $\exp(E/kT)$, and at higher energies $\alpha(E)$ increases much slower with increasing energy than $\exp(E/kT)$. As a consequence, the emission line can be relatively sharp. Since both $\alpha(E)$ and $\exp(E/kT)$ increase faster with increasing energy for lower temperatures, the emission line becomes sharper at lower temperatures. Because the gap width E_g increases with decreasing temperature, the emission line shifts to higher energies for decreasing temperature. Figure 17.13 shows the line shape of GaAs luminescence at room temperature.

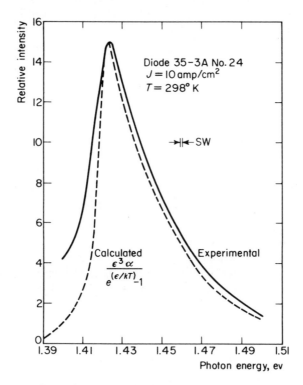

Fig. 17.13. Comparison of the calculated and observed emission bands of GaAs at 298°K (J. C. Sarace, R. H. Kaiser, J. M. Whelan, and R. C. C. Leite, *Phys. Rev.*, **137**, No. 2A, A 623, 1965).

17.4b. Practical Aspects of Injection Luminescence

The best-known light emitting diode is the GaAs diode, which emits around 9000 Å. It has a high external efficiency, especially if domelike structures are used.

For display purposes one wants light sources that emit visible radiation. This can be achieved by using Ga(As, P) diodes. Red emitters are $GaAs_{0.6}P_{0.4}$ (peak wavelength, 6490 Å), nitrogen-doped $GaAs_{0.35}P_{0.65}$ (peak wavelength, 6350 Å), and zinc plus oxygen-doped GaP (peak wavelength, 6925 Å). Yellow emitters are nitrogen-doped $GaAs_{0.15}P_{0.85}$ (peak wavelength, 5890 Å) and heavily nitrogen-doped ($N \approx 10^{20}/cm^3$) GaP (peak wavelength, 5900 Å). Nitrogen-doped GaP (peak wavelength, 5700 Å) emits in the green. At current densities of about 10 A/cm^2, the efficiency is of the order of a few tenths of 1 per cent. Higher efficiencies can be obtained under pulsed conditions.

For $GaAs_xP_{1-x}$ up to a certain value of x, the transitions are direct band-to-band transitions due to electron–hole recombination. Beyond that value of x, the transitions are indirect and involve phonons. For doped samples the transition will be from the conduction band to the trap level caused by the dopant.

The lifetime τ follows from the relationship

$$\frac{1}{\tau} = \frac{1}{\tau_R} + \frac{1}{\tau_{NR}} \tag{17.108}$$

where τ_R and τ_{NR} are the lifetimes for radiative and nonradiative transitions, respectively. The internal quantum efficiency can thus be written

$$\eta_{int} = \frac{1/\tau_R}{1/\tau} = \frac{1}{1 + \tau_R/\tau_{NR}} \tag{17.108a}$$

When traps participate in the recombination light, η_{int} is a function of the trap capture rates as well as of τ_R and τ_{NR}.

The diodes are grown either by vapor phase epitaxial (VPE) or liquid phase epitaxial (LPE) techniques. The first method is most common. For details we refer the reader to a review paper by Craford and Groves.†

17.5. JUNCTION LASER

In any distributed oscillator where there is a power gain coefficient g per unit length and a power loss coefficient γ per unit length, the condition for oscillation is

$$g - \gamma > 0 \tag{17.109}$$

We must thus evaluate g and γ for the junction laser, since it is nothing but a distributed oscillator at optical frequencies.

Let the junction laser have two energy levels E_1 and E_2, so that the emitted quantum $h\nu = E_2 - E_1$, with population densities N_1 and N_2 per unit length. Absorption is due to transitions $1 \longrightarrow 2$ and stimulated emission

---

† M. G. Craford and W. O. Groves, *Proc. IEEE*, **61**, 862 (1973).

to transitions $2 \longrightarrow 1$; the absorption and emission per unit length are proportional to N_1 and N_2, respectively, and the proportionality factors are identical. Hence g will be proportional to $N_2 - N_1$, so that $g > 0$ for $N_2 > N_1$ (population inversion).

As an example of the junction laser we discuss the GaAs junction laser. One of the chief requirements for materials that exhibit junction-laser action is that the absorbed frequency can also be reemitted. Materials in which the maximum energy in the valence band and the minimum energy in the conduction band lie at the same **k** value satisfy this condition, but materials in which this maximum and minimum occur at *different* **k** values do not.

The population inversion is here achieved by injection. To understand how this can be accomplished, consider a *pn* GaAs junction with degenerate *p* and *n* regions. The diffusion potential V_{dif} is then larger than the gap width E_g. We shall now see that for a population inversion the applied voltage V must be so chosen that

$$eV > h\nu \tag{17.110}$$

where ν is the frequency of the emitted light.

The proof is as follows. Let E_c be the energy of electrons in the conduction band and E_v the energy of holes in the valence band. The probabilities f_c and f_v that the energy states E_c and E_v are occupied are

$$f_c = \left[1 + \exp\left(\frac{E_c - \bar{\mu}_n}{kT}\right) \right]^{-1}$$

$$f_v = \left[1 + \exp\left(\frac{E_v - \bar{\mu}_p}{kT}\right) \right]^{-1}$$

respectively, where $\bar{\mu}_n$ and $\bar{\mu}_p$ are the quasi-Fermi levels for electrons and holes. Let P_ν be the radiation density of light of the frequency ν; then the number of photons absorbed per unit time is

$$\frac{dN_a}{dt} = CW_{cv}f_v(1 - f_c)P_\nu$$

where $1 - f_c$ is the probability that the energy level in the conduction band is empty and W_{cv} is the transition probability. In the same way the rate of stimulated emission of photons is

$$\frac{dN_c}{dt} = CW_{vc}f_c(1 - f_v)P_\nu$$

The constants are the same and $W_{cv} = W_{vc}$. Hence

$$\frac{dN_c}{dt} > \frac{dN_a}{dt}, \quad \text{if} \quad f_c(1 - f_v) > f_v(1 - f_c) \quad \text{or} \quad f_c > f_v$$

or $\bar{\mu}_n - \bar{\mu}_p > E_c - E_v = h\nu$. Since $\bar{\mu}_n - \bar{\mu}_p = eV$, we have hereby proved Eq. (17.110).

We now give a derivation for the minimum current density needed for starting laser action. Let J be the current density; then the number of carriers injected per square centimeter per second is J/e. If η is the quantum efficiency of spontaneous luminescence, then the number of quanta emitted per square centimeter per second is $\eta J/e$. If d is the width of the light-emission region, then the number of quanta emitted per unit volume per second is $\eta J/ed$. If $\Delta\nu$ is the effective bandwidth of the spontaneous-emission line, defined by the relation

$$\int_0^\infty G(\nu)\,d\nu = G_{\max}\,\Delta\nu$$

then the rate of spontaneous emission per unit frequency interval per second is $J\eta/(ed\,\Delta\nu)$. But the number of modes per unit frequency interval per unit volume, neglecting dispersion, is

$$n_\nu = \frac{8\pi n^3}{\lambda^2 c}$$

where n is the refractive index of the material. Hence the rate of spontaneous emission per mode is

$$\alpha = \frac{J\eta/(ed\,\Delta\nu)}{n_\nu} = \frac{J\eta}{ed\,\Delta\nu}\frac{\lambda^2 c}{8\pi n^3} \tag{17.111}$$

Up to now we have discussed only spontaneous emission. But the rate of stimulated emission per quantum per mode is equal to α; hence the rate of stimulated emission per mode is $N\alpha$, where N is the number of quanta per mode at the instant t. Hence

$$\frac{dN}{dt} = \alpha N \quad \text{or} \quad \frac{dN}{dx} = \frac{\alpha N}{dx/dt} = \frac{n\alpha N}{c} \tag{17.112}$$

The solution of this equation is

$$\ln\frac{N}{N_0} = \frac{n\alpha}{c}x \tag{17.113}$$

Hence the gain per unit length g is

$$g = \frac{n\alpha}{c} = \frac{J\eta\lambda^2}{8\pi n^2 ed\,\Delta\nu} \tag{17.114}$$

The onset of stimulated emission occurs if this gain per unit length is just offset by the losses per unit length. These losses are as follows:

1. The absorption coefficient α_0 of the light-emitting layer.
2. The losses due to transmission of the reflecting surfaces at the end of the laser. If R is the power reflection coefficient, then the transmission loss

per unit length is $-(\ln R)/L$, where L is the length of the active region. If R is close to unity, then $-\ln R \simeq 1 - R$.†

3. The diffraction losses of the active region. In the active region there is a negative absorption coefficient $-g$, and outside the active region there is a positive absorption coefficient α_s. Since the wave traveling in the active region spills over into the absorbing region by diffraction, there is a diffraction loss. This loss will be denoted by the loss factor α_{diff}.

The condition for oscillation is therefore

$$g > \alpha_0 - \frac{\ln R}{L} + \alpha_{\text{diff}} \qquad (17.115)$$

from which the minimum current density needed for oscillations can be determined. At liquid-helium temperatures, threshold current densities as low as 80 A/cm² have been measured. Beyond a certain temperature, however, the threshold current rises rapidly with temperature because the loss factors α_0 and α_{diff} increase with increasing temperature. Continuous-wave operation has been achieved at low temperature (liquid-nitrogen temperature and lower), but at room temperature the device is best operated under pulsed conditions. Often several frequencies are generated simultaneously, because several modes lie within the relatively broad spontaneous emission line.

The emitted radiation is not caused by a direct band-to-band transition, when $h\nu < E_g$. This means that either filled levels below the bottom of the conduction band or empty levels above the top of the filled band, or both, must be involved. These may either be donor, exciton‡, or acceptor levels.

Practical junction lasers are GaAs (0.85–0.90 μm), $\text{GaAs}_{1-x}\text{P}_x$ (0.64–0.90 μm), and $\text{Al}_x\text{Ga}_{1-x}\text{As}$ (0.628–0.90 μm). A transition from a direct to an indirect band gap occurs at 1.92 eV in (AlGa) As and at 1.96 eV in Ga(AsP). Continuous-wave operation at room temperature is possible in these lasers, but the emitted power increases with decreasing temperature because the device losses decrease. At 77°K, continuous-wave powers of a few watts are possible, but at those powers the lasers have a relatively short life. Examples of infrared lasers are InAs (3.1 μm), PbS (4.3 μm), InSb (5.4 μm), PbTe (6.5 μm), and PbSe (8.5 μm) diodes; these diodes must be cooled in order to lase. Continuous-wave powers of up to a few hundred milliwatts are possible, again with a short life.

---

† Usually no special steps are taken to increase the reflection coefficient of the end walls of the junction laser. If ϵ is the dielectric constant, then the refractive index $n = \sqrt{\epsilon}$ and the reflection coefficient $R = (n-1)^2/(n+1)^2$. Gallium arsenide has $n \simeq 3.6$, hence $R \simeq 0.32$, so that the relation $-\ln R \simeq 1 - R$ is a poor approximation.

‡ An exciton consists of an electron and a hole bound together in a medium of dielectric constant ϵ, forming a hydrogenlike atom.

The threshold current density of the GaAs, Ga(AsP), and (AlGa)As lasers is of the order of a few thousand amperes per square centimeter. It can be reduced by using single and double heterojunction devices, in which the material outside the active region has a larger gap width than the active region; this reduces absorption outside the active region. Also, partial counterdoping can reduce the threshold current. For further details, see Kressel's review paper.†

REFERENCES

CHANG, K. K., *Parametric and Tunnel Diodes*. Prentice-Hall, Inc., Englewood Cliffs, N.J., 1964.

GIBBONS, J. F., *Semiconductor Electronics*. McGraw-Hill Book Company, New York, 1966. Gives a simple introduction to lumped circuit models.

LE CAN, C., K. HART, and C. de RUYTER, *The Junction Transistor as a Switching Device*. Van Nostrand Reinhold Company, New York, 1962.

SZE, S. M., *Physics of Semiconductor Devices*. John Wiley & Sons, Inc. (Interscience Division), New York, 1969.

WATSON, H. A., ed., *Microwave Semiconductor Devices and Their Circuit Applications*. McGraw-Hill Book Company, New York, 1968.

PROBLEMS

1. Compare the results of Eqs. (17.18a) and (17.33c) for the cases $V_1 = V_2 = 5$ V, $V_{D0} = 0.70$ V, $\tau_p = 10^{-6}$ s.

Answer: (a) 0.28 μs; (b) 0.16 μs.

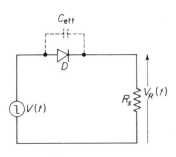

Fig. 17.14

2. In the diode circuit of Fig. 17.14, $V(t) = 0$ for $t < 0$ and $V(t) = V_1$ for $t > 0$. Express the output transient $V_R(t)$ in terms of the turn-on transient for $V_D(t)$ given in Sec. 17.1c.

Answer: $V_R(t) = V_1 - V_D(t)$.

3. In the diode circuit of Fig. 17.14, $V(t) = V_1$ for $t < 0$ and $V(t) = 0$ for

---

† H. Kressel, "Semiconductor Lasers," in A. K. Levine and A. J. DeMaria, eds., *Lasers*, vol. 3, Marcel Dekker, Inc., New York, 1971, p. 1.

$t > 0$. Express the output transient $V_R(t)$ in terms of the turn-off transient $V_D(t)$ given in Sec. 17.1c.

Answer: $V_R(t) = -V_D(t)$.

4. (a) A transistor has an alpha cutoff frequency of 1000 MHz and $\beta_F = 100$. If $C_{jc} = 0.5 \ \mu\mu F$, and $R_c = 2000 \ \Omega$, calculate τ_{BF} and τ_T.
(b) How would the result change if β_F were reduced to 10?

Answer: (a) $\tau_{BF} = 1.6 \times 10^{-8}$ s, $\tau_T = 11.6 \times 10^{-8}$ s.
(b) $\tau_{BF} = 1.6 \times 10^{-9}$ s, $\tau_T = 11.6 \times 10^{-9}$ s.

5. A transistor base is switched from -5 V to $+5$ V and back. Calculate the turn-on delay, turn-on time, turn-off delay, and turn-off time. $R = 3000 \ \Omega$, $R_c = 1000 \ \Omega$, $V_{CC} = 6$ V, $C_{Te} = 1.5 \ \mu\mu F$, $C_{jc} = 0.5 \ \mu\mu F$, $\tau_e = \frac{1}{2} \times 10^{-8}$ s, $\beta_F = 50$, $f_\alpha = 1000$ MHz, and $V_{BE0} = 0.70$ V.

Answer: $I_{Bs} = 0.12$ mA, $I_{B1} = 1.43$ mA, $I_{B2} = 1.90$ mA, $\tau_T = 33 \times 10^{-9}$ s, $\tau_0 = 5.1 \times 10^{-9}$ s, $\tau_1' = 2.8 \times 10^{-9}$ s, $\tau_2 = 2.5 \times 10^{-9}$ s, $\tau_3 = 2.1 \times 10^{-9}$ s. Total turn-on time 7.9×10^{-9} s; total turn-off time 4.6×10^{-9} s.

6. Repeat problem 5 if the base is switched from 0 V to $+5$ V and back.

Answer: $\tau_0 = 0.90 \times 10^{-9}$ s, $\tau_1 = 2.8 \times 10^{-9}$ s, $\tau_2 = 7.8 + 10^{-9}$ s, $\tau_3' = 13.7 \times 10^{-9}$ s. Total turn-on time 3.7×10^{-9} s, total turn-off time 21.5×10^{-9} s.

7. A silicon photodiode is irradiated with 2 mW of radiation with a quantum energy of 1.50 eV. The power reflection coefficient R is 25 per cent; the quantum efficiency is 100 per cent.
(a) Find the photocurrent.
(b) If the saturation current is 10^{-8} A, calculate the open-circuit voltage when the device is used as a photovoltaic cell. $T = 300°$K.

Answer: $I_s = 1$ mA; $V_{oc} = 0.30$ V.

8. For the diode of problem 7, find the optimum voltage V_{mp}, the optimum current I_{mp}, the optimum load resistor, the optimum power delivered to the load, and the efficiency. *Hint:* V_{mp} must be calculated by the iterative method discussed in Sec. 17.2b.

Answer: $eV_{mp}/kT = 9.2$, $V_{mp} = 0.237$ V, $I_{mp} = 0.90$ mA, $R_{mp} = 263 \ \Omega$, $P_{out} = 0.214 \times 10^{-3}$ W, $\eta_{max} = 10.7$ per cent.

9. An avalanche multiplier detector diode has $I_{s0} = I_0 = 10^{-8}$ A and the receiver has $I_{eq} = 10^{-5}$ A. What is the optimum multiplication factor and what is the optimum improvement in signal-to-noise ratio?

Answer: $M_{opt} = 10$; improvement $33\frac{1}{3}x$.

10. A tunnel diode has a characteristic

$$I_d = A[V(V_1 - V)^2 + B], \qquad \text{for} \quad 0 \leq V \leq V_1$$

with $V_1 = 0.15$ V.

(a) Determine B so that the maximum current at $V = \frac{1}{3}V_1$ is 10 times the minimum current at $V = V_1$.

(b) Determine $g_d = -dI_d/dV_d$ for $\frac{1}{3}V_1 \leq V \leq V_1$.

(c) Find graphically where I_d/g_d has its minimum value and determine this value.

Answer: (a) $B = \frac{4}{243}V_1^3 = 5.55 \times 10^{-5}$; (b) $g_d = A(V_1 - V)(3V - V_1)$; (c) $V = 0.133$ V, $(I_d/g_d)_{\min} = 2.22 \times 10^{-2}$ V.

11. What minimum noise figure F_∞ would one expect at that operating point?

Answer: $F_\infty = 1.43$.

12. Prove Eqs. (17.97) and (17.99).

13. A metal–n-type semiconductor photodiode is irradiated by quanta $h\nu$ so that $E_b < h\nu < E_g$, where E_g is the gap width of the semiconductor and E_b the barrier height of the metal–semiconductor contact. The photoelectrons are generated in the metal and escape into the semiconductor.

The power absorption law of light in the metal is $P(x) = P(0) \exp(-\alpha x)$. A photoelectron generated at x has a probability $\exp[-\beta(L - x)]$ of escaping into the semiconductor, if L is the thickness of the metal layer. The quantum efficiency of the production process of photoelectrons is η_p. Assume that $\beta \gg \alpha$.

(a) Show that the quantum efficiency η of the photodiode is

$$\eta = \eta_p \frac{\alpha}{\beta - \alpha}[\exp(-\alpha L) - \exp(-\beta L)]$$

(b) Show that the efficiency has a maximum $\eta_{\max}$ for

$$L = L_{\text{opt}} = \frac{\ln(\beta/\alpha)}{\beta - \alpha}, \qquad \eta_{\max} = \eta_p\left(\frac{\alpha}{\beta}\right)^{\beta/(\beta-\alpha)}$$

(c) Discuss this for the case $\beta = 10\alpha$ and $\alpha = 10^5/\text{cm}$. Find in particular $\eta_{\max}$ and L_{opt}.

14. A metal–n-type semiconductor photodiode is irradiated by quanta $h\nu$ so that $h\nu > E_g$, where E_g is the gap width. The power absorption coefficient for light is α_m in the metal and α_s in the semiconductor. The semiconductor has a diffusion length L_p for holes, so that the probability that a hole–electron pair, generated at a depth x in the semiconductor, is collected is $\exp(-x/L_p)$. The thickness of the metal layer is L. You may neglect the photoelectron production in the metal and you may assume that the quantum efficiency for pair production in the semiconductor is unity.

(a) Show that the quantum efficiency of the photodiode is

$$\eta = \exp(-\alpha_m L)\frac{\alpha_s}{\alpha_s + 1/L_p}$$

(b) Substitute $\alpha_m = \alpha_s = 10^5/\text{cm}$, and find how much effect the second term has for $D_p = 12.5$ cm/s and $\tau_p = 2 \times 10^{-8}$ s.

(c) What is the required thickness of the metal layer if $\eta = 0.70$?

Answer: (b) 2 per cent; (c) $L = 360$ Å.

18

Field Effect
Transistors

18.1. FIELD-EFFECT TRANSISTOR

18.1a. Field-Effect Devices; Equivalent
Circuits

Figure 18.1a shows an *n*-channel junction gate field-effect transistor. It consists of a conducting *n*-type channel between two ohmic contacts, called *source* and *drain*, respectively; the width of the channel is controlled by two back-biased *pn* junctions. The rectifying contact biased in the back direction is called the *gate;* the gate voltage controls the current flowing between source and drain.

Figure 18.1b shows an *n*-type metal-oxide-semiconductor (MOS) field-effect transistor. Here the source and drain are two *n*-type regions diffused into a *p*-type substrate. The part between source and drain is covered by an oxide layer, on top of which a metal gate is applied. When the appropriate voltage is applied to the gate, an *n*-type channel is induced at the surface of the substrate. If a voltage is applied between drain and source, current will flow from source to drain, and the magnitude of this current is controlled by the gate voltage.

There are two types of metal-oxide-semiconductor (MOS) field-effect transistors. It may happen that a channel is already developed at zero gate bias; this channel can then be cut off by negative biasing of the gate. This is called *depletion-type* operation. Or it may happen that the channel is formed

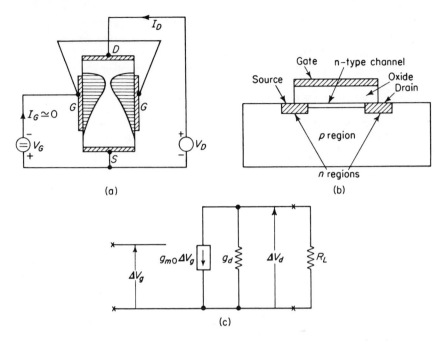

Fig. 18.1. (a) Schematic diagram of the field-effect transistor, showing the space-charge region and the channel. (b) Schematic diagram of an MOS field-effect transistor with an *n*-type channel. (c) Low-frequency equivalent circuit of the field-effect transistor. (Figs. 18.1–18.3 are taken from A. van der Ziel, *Electronics*, Allyn and Bacon Co., Boston, 1966.)

only for a sufficiently large positive bias of the gate; this is called *enhancement-mode* operation. Depletion-type devices can operate either in the depletion mode or in the enhancement mode, whereas enhancement-mode devices with *n*-type channel operate only for positive gate bias.

In both devices the drain current depends strongly on the drain voltage for small values of the drain voltage but only weakly for larger values (saturation). We shall see that this is important for the use of the device as an amplifier. In both devices the gate current is almost zero, and the drain current is a function of the gate and drain bias. That is,

$$I_g = 0, \qquad I_d = f(V_g, V_d) \qquad (18.1)$$

The response of the device to small increments ΔV_g and ΔV_d in the gate and drain voltages is therefore given by the relations

$$\Delta I_g = 0$$

$$\Delta I_d = \frac{\partial I_d}{\partial V_g} \Delta V_g + \frac{\partial I_d}{\partial V_d} \Delta V_d = g_{m0} \, \Delta V_g + g_d \, \Delta V_d \qquad (18.2)$$

where $\qquad\qquad g_{m0} = \dfrac{\partial I_d}{\partial V_g}, \qquad g_d = \dfrac{\partial I_d}{\partial V_d} \qquad (18.2a)$

The low-frequency equivalent circuit of the device can thus be represented by an open input and by an output consisting of a current generator g_{m0} ΔV_g in parallel with the *output conductance* g_d of the device (Fig. 18.1c). The parameter g_{m0} is called the *transconductance* of the device.

18.1b. Junction Field-Effect Transistor

We first consider the case in which the drain is unbiased and the gate has a bias V_g. The bias between channel and gate is then $W_0 = -V_g + V_{dif}$, where V_{dif} is the diffusion potential of the junction. We shall first show for an n-type channel that (Fig. 18.2a)

$$W_0 = \frac{eN_d a^2}{2\epsilon\epsilon_0}\left(1 - \frac{b}{a}\right)^2 = W_{00}\left(1 - \frac{b}{a}\right)^2 \tag{18.3}$$

Here N_d is the donor concentration of the n region, $2a$ is the width of the fully open channel, $2b$ is the width of the channel for the given bias, and

$$W_{00} = \frac{eN_d a^2}{2\epsilon\epsilon_0} \tag{18.3a}$$

The proof is as follows. In the space-charge region

$$\frac{d^2\psi}{dy^2} = -\frac{eN_d}{\epsilon\epsilon_0} \tag{18.4}$$

with $\psi = 0$ and $d\psi/dy = 0$ at $y = b$ and $\psi = -W_0$ at $y = a$. Integrating once gives

$$\frac{d\psi}{dy} = -\frac{eN_d}{\epsilon\epsilon_0}(y - b) \tag{18.5}$$

Integrating once more yields

$$\psi = -\frac{eN_d}{2\epsilon\epsilon_0}(y - b)^2 \tag{18.6}$$

Since $\psi = -W_0$ for $y = a$, we have hereby proved (18.3).

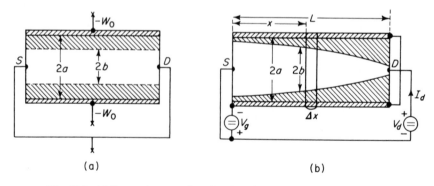

(a)   (b)

Fig. 18.2. (a) Space-charge regions in a junction gate field-effect transistor if a bias is applied between the channel and the gate. (b) Distribution of the space-charge regions in our model if the proper voltages are applied.

We now apply the drain voltage V_d. Then W_0 varies along the channel, but the relationship between W_0 and b is still approximately given by (18.3). The conductance for unit length of the channel will be shown to be

$$g(W_0) = g_0\left[1 - \left(\frac{W_0}{W_{00}}\right)^{1/2}\right], \qquad g_0 = 2\sigma_0 wa \qquad (18.7)$$

where σ_0 is the conductivity of the n-type material and w is the width of the channel. The reason is very simple. The conductance $g = 2\sigma_0 wb$, and

$$\frac{b}{a} = 1 - \left(\frac{W_0}{W_{00}}\right)^{1/2} \qquad (18.8)$$

so that

$$g = 2\sigma_0 wa\frac{b}{a} = 2\sigma_0 wa\left[1 - \left(\frac{W_0}{W_{00}}\right)^{1/2}\right] \qquad (18.8a)$$

If $V(x)$ is the channel potential at x, then $W_0 = V(x) - V_g + V_{\text{dif}}$.

The current I_d can be written as

$$I_d = -g(W_0)F_x = g(W_0)\frac{dW_0}{dx} \qquad (18.9)$$

flowing from drain to source; the reason is that $F_x = -dW_0/dx$. By integrating over the length of the channel and bearing in mind that $W_0 = W_s = -V_g + V_{\text{dif}}$ at $x = 0$ (source), and $W_0 = W_d = -V_g + V_{\text{dif}} + V_d$ at $x = L$ (drain), we obtain

$$\int_0^L I_d\,dx = I_d L = \int_{W_s}^{W_d} g(W_0)\,dW_0 = g_0\left(W_d - W_s - \frac{2}{3}\frac{W_d^{3/2} - W_s^{3/2}}{W_{00}^{1/2}}\right)$$

so that

$$I_d = \frac{g_0}{L}\left(W_d - W_s - \frac{2}{3}\frac{W_d^{3/2} - W_s^{3/2}}{W_{00}^{1/2}}\right) \qquad (18.10)$$

This expression holds for $W_d \leq W_{00}$. For $W_d = W_{00}$ the current reaches the saturated value

$$I_d = \frac{g_0}{L}\left[\frac{1}{3}W_{00} - W_s + \frac{2}{3}\frac{W_s^{3/2}}{W_{00}^{1/2}}\right] \qquad (18.10a)$$

For $W_d > W_{00}$ the expression is incorrect, but I_d is found to be almost independent of W_d experimentally, so that Eq. (18.10a) holds in the saturated region.

The output conductance is therefore

$$g_d = \frac{\partial I_d}{\partial V_d} = \frac{g_0}{L}\left[1 - \left(\frac{W_d}{W_{00}}\right)^{1/2}\right] \qquad (18.11)$$

which has zero value at saturation ($W_d = W_{00}$), and

$$g_d = g_{d0} = \frac{g_0}{L}\left[1 - \left(\frac{W_s}{W_{00}}\right)^{1/2}\right] \qquad (18.11a)$$

for zero drain bias ($V_d = 0$, $W_d = W_s$). The transconductance of the device is given by

$$g_{m0} = \frac{\partial I_d}{\partial V_g} = \frac{g_0}{L}\left[\left(\frac{W_d}{W_{00}}\right)^{1/2} - \left(\frac{W_s}{W_{00}}\right)^{1/2}\right] \quad (18.12)$$

which has zero value for zero drain bias and a maximum value

$$g_{max} = \frac{g_0}{L}\left[1 - \left(\frac{W_s}{W_{00}}\right)^{1/2}\right] = g_{d0} \quad (18.13)$$

in saturation ($W_d = W_{00}$).

The approximation used so far is called the *gradual channel* approximation; in this approximation it is assumed that Eq. (18.8a) is rigorously valid. If one solves the two-dimensional potential problem in the space-charge region accurately, one finds that this is not fully correct; we cannot go into details, however.

18.1c. MOS Field-Effect Transistor (MOSFET)

We give here the derivation of the characteristic of an MOS field-effect transistor on a p-type substrate, in which an n-type channel is induced when the gate voltage V_g is larger than a threshold voltage V_T. The device is made by diffusing an n^+-type source and an n^+-type drain into the substrate having a distance L. An oxide is grown or deposited on the region in between and a metal electrode is deposited on top of the oxide. The two n^+ regions act as source and drain, respectively, whereas the electrode on top of the oxide acts as a gate (Fig. 18.3).

If first no voltage is applied between source and drain and a voltage V_g is applied to the gate, the induced charge $Q(x)$ per unit length along the channel is proportional to V_g and may be written

$$Q(x) = -C_{ox}w(V_g - V_T) \quad (18.14)$$

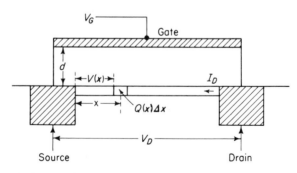

Fig. 18.3. Cross-section of an MOS field-effect transistor. V_G is the gate voltage, V_D the drain voltage, $V(x)$ the channel voltage at a distance x from the source, and $Q(x)\,\Delta x$ the charge induced in the section Δx of the channel.

where C_{ox} is the capacitance of the oxide layer per unit area and w is the width of the channel. Here

$$C_{ox} = \frac{\epsilon \epsilon_0}{d} \tag{18.14a}$$

where ϵ is the dielectric constant of the oxide and d the thickness of the oxide layer. The voltage V_T is called the *turn-on voltage;* it is determined by the amount of charge stored at the interface between source and oxide, the amount of charge stored in the oxide, and the difference in contact potential between channel and gate. We shall come back to this parameter later.

If now a voltage V_d is applied between drain and source, a voltage $V(x)$ will develop along the channel. It is assumed that Eq. (18.14) may be rewritten as

$$Q(x) = -C_{ox}w[V_g - V_T - V(x)] \tag{18.14b}$$

This is called the *gradual channel approximation;* it simply means that the charge per unit length $Q(x)$ is determined by the local voltage $V_g - V(x)$ between gate and drain at x. The channel conductance for unit length is therefore

$$g(x) = g(V) = -\mu Q(x) = \mu C_{ox}w[V_g - V_T - V(x)] \tag{18.15}$$

where $V_g > V_T$. Here μ is the surface mobility of the carriers; it is somewhat smaller than the mobility of bulk material, because the surface interferes with the electron motion. It should be noted that Eq. (18.15) only holds if the channel is nowhere pinched off; that is, $g(x) \neq 0$ for $0 < x < L$, where L is the channel length; this means that $V_d \leq V_g - V_T$.

If $F(x) = -dV(x)/dx$ is the field strength, then the current I_d flowing from source to drain is

$$I_d = -g(x)F(x) = g(V)\frac{dV}{dx} \tag{18.16}$$

or

$$I_d L = \int_0^L I_d \, dx = \int_0^{V_d} g(V) \, dV$$

or

$$I_d = \frac{1}{L}\int_0^{V_d} g(V) \, dV = \mu C_{ox}\frac{w}{L}\left[(V_g - V_T)V_d - \frac{1}{2}V_d^2\right]$$

$$= K[2(V_g - V_T)V_d - V_d^2] \tag{18.17}$$

for $V_D \leqq V_g - V_T$, where

$$K = \frac{1}{2}\mu C_{ox}\frac{w}{L} \tag{18.17a}$$

Note that K is proportional to w/L, so that I_d can be designed for a wide range of values by proper choice of the channel width w and the channel length L.

Experimentally, one finds that I_d is practically independent of V_d for

$V_d \geq V_g - V_T$. We may thus write

$$I_d = K(V_g - V_T)^2, \qquad \text{for } V_d \geq V_g - V_T \qquad (18.17b)$$

This is called the *pinch-off or saturated regime* of the characteristic, and I_d is called the *saturation* current.

For $0 < V_d \ll V_g - V_T$ the characteristic is linear, since the term V_d^2 in Eq. (18.17) can be neglected, or

$$I_d = 2K(V_g - V_T)V_d, \qquad \text{for } 0 < V_d \ll V_g - V_T \qquad (18.17c)$$

This is called the *ohmic regime* of the characteristic.

The output conductance g_d and the transconductance g_m are

$$g_d = \frac{\partial I_d}{\partial V_d} = 2K(V_g - V_T - V_d), \qquad g_m = \frac{\partial I_d}{\partial V_g} = 2KV_d \qquad (18.17d)$$

for $V_d < V_g - V_T$. For $V_d \geq V_g - V_T$, $g_d = 0$ and

$$g_m = 2K(V_g - V_T) \qquad (18.17e)$$

A similar calculation holds for the MOSFET with a p-type channel on an n-type substrate, provided one takes into account that the current and the applied voltages have opposite polarity. We thus have that current flows if $V_g < V_T$, and

$$I_d = -K[2(V_g - V_T)V_d - V_d^2], \qquad \text{for } 0 > V_d \geq V_g - V_T \qquad (18.18)$$

$$I_d = I_{ds} = -K(V_g - V_T)^2, \qquad \text{for } V_d \leq V_g - V_T \qquad (18.18a)$$

Something should be said about the parameter V_T. In many n-channel MOSFETs $V_T < 0$; if $V_T \leq V_g \leq 0$ the device is then said to be operating in the *depletion mode*. By proper design of the MOSFET, however, one can make $V_T > 0$. In that case current flows if $V_g > V_T$; this is called the *enhancement mode*. All present p-channel MOSFETs have $V_T < 0$; since the device operates only for $V_g < V_T$, it can be operated in the *enhancement mode only*.

Enhancement mode devices find widespread application in MOSFET logic circuits, because they allow for cascaded logic operation.

When one measures the output conductance g_d of a MOSFET in the pinch-off regime of the characteristic, one finds that g_d, although small, is not exactly zero. The reason is that the channel length L', which is equal to L at pinch off, decreases slowly with increasing V_d for an n-channel device. Therefore, the preceding theory is valid for $0 < x < L'$, whereas the current then flows by another mode for $L' < x < L$. Equation (18.17b) remains valid, provided that $K = \frac{1}{2}\mu C_{ox}w/L'$. We thus see that K increases slowly with increasing V_d. Hence

$$g_d = \frac{\partial I_d}{\partial V_d} = K(V_g - V_T)^2 \frac{1}{K}\frac{\partial K}{\partial V_d} = I_d\left(-\frac{1}{L'}\frac{\partial L'}{\partial V_d}\right) \qquad (18.19)$$

which is nonzero for $V_d > V_g - V_T$. A similar, but somewhat smaller, effect occurs for a JFET, and the explanation can be taken over directly.

18.1d. Effect of the Substrate on MOSFET Operation

Up to now we have neglected the space-charge region between channel and substrate. This is allowed if the substrate has negligible conductivity, for in that case there is hardly any charge stored in the space-charge region. Usually, however, the effect of this fixed space charge in the substrate must be taken into account. We do so for a p-type substrate, that is, for an n-channel device. Let Q_g be the induced charge by the gate and let Q_s be the fixed charge due to ionized acceptors; then the mobile charge is

$$Q = Q_g - Q_s \tag{18.20}$$

According to Sec. 18.1c,

$$Q_g = -C_{ox}[V_g - V_T - V(x)] \tag{18.21}$$

For Q_s the diode theory holds; hence, in analogy with Chapter 15, we have

$$Q_s = -\{2e\epsilon\epsilon_0 N_a[V_{dif} + V(x) - V_b]\}^{1/2} \tag{18.22}$$

where V_{dif} is the diffusion potential and V_b the substrate voltage. Hence

$$Q = -C_{ox}[V_g - V_T - V(x)] + \{2e\epsilon\epsilon_0 N_a[V_{dif} + V(x) - V_b]\}^{1/2} \tag{18.23}$$

so that the conductance $g(V)$ for unit length may be written

$$g(V) = -\mu w Q = \mu w C_{ox}[V'_g - V - 2\varphi^{1/2}(V + V'_b)^{1/2}] \tag{18.24}$$

where w is the channel width, and

$$V'_g = V_g - V_T, \qquad V'_b = V_{dif} - V_b, \qquad \varphi = \frac{1}{4}\left(\frac{2e\epsilon\epsilon_0 N_a}{C_{ox}^2}\right) \tag{18.24a}$$

We see that φ has the dimension of a *voltage*. Hence

$$I_d L = \int_0^L I_d\,dx = \int_0^{V_d} g(V)\,dV = \mu w C_{ox}\int_0^{V_d}[V'_g - V - 2\varphi^{1/2}(V + V'_b)^{1/2}]\,dV$$

Carrying out the integration, we obtain

$$I_d = \frac{\mu w C_{ox}}{L}\left\{V'_g V_d - \frac{1}{2}V_d^2 - \frac{4}{3}\varphi^{1/2}[(V_d + V'_b)^{3/2} - V_b'^{3/2}]\right\} \tag{18.25}$$

This gives our previous result back if N_a, and hence φ, goes to zero. Our previous theory is therefore correct in the limit $\varphi \to 0$. Equation (18.25) holds when the channel is nowhere pinched off.

Under that condition we have for the transconductance

$$g_m = \frac{\partial I_d}{\partial V_g} = \frac{\partial I_d}{\partial V'_g} = \frac{\mu w C_{ox}}{L}V_d \tag{18.26}$$

as in Sec. 18.1c whereas the drain conductance is

$$g_d = \frac{\partial I_d}{\partial V_d} = \frac{\mu w C_{ox}}{L}[V'_g - V_d - 2\varphi^{1/2}(V_d + V'_b)^{1/2}] = \frac{g(L)}{L} \qquad (18.27)$$

The zero-bias drain conductance is ($V_d = 0$)

$$g_d = \frac{\mu w C_{ox}}{L}[V'_g - 2\varphi^{1/2}V_b'^{1/2}] \qquad (18.27a)$$

which is zero if

$$V'_g = 2\varphi^{1/2}V_b'^{1/2} \quad \text{or} \quad V_g = V'_T = V_T + 2\varphi^{1/2}V_b'^{1/2} \qquad (18.27b)$$

Even if $V_T < 0$, it may thus happen that $V'_T > 0$ if N_a is sufficiently large. The device then operates in the *enhancement mode* only, but the characteristic is not fully quadratic because the last term in Eq. (18.25) depends on V_d.

When is the characteristic saturated? This is the case if $g_d = 0$ or $g(L) = 0$, or

$$V'_g - V_d - 2\varphi^{1/2}(V_d + V'_b)^{1/2} = 0 \qquad (18.28)$$

Solving for V_d yields

$$V_d = V_{ds} = V'_g + 2\varphi - [4\varphi^2 + 4\varphi(V'_g + V'_b)]^{1/2} \qquad (18.28a)$$

Hence the transconductance in saturation is

$$g_m = g_{max} = \frac{\mu w C_{ox}}{L}V_{ds}$$

$$= \frac{\mu w C_{ox}}{L}\{V'_g + 2\varphi - [4\varphi^2 + 4\varphi(V'_g + V'_b)]^{1/2}\} \qquad (18.29)$$

so that g_{max} is smaller than for $\varphi = 0$. The effect of the substrate is therefore to decrease g_{max}.

It should be noted that the device can also be driven in the enhancement mode by making V'_b larger, that is, by back biasing the substrate.

18.1e. Effect of a Field-Dependent Mobility in the Channel

For very short channel lengths, the field in the channel becomes so large that the mobility μ becomes field dependent. The results of Sec. 6.6 can be approximated by Shockley's equation

$$\mu = \frac{\mu_0}{[\frac{1}{2} + (\frac{1}{4} + F^2/F_0^2)^{1/2}]^{1/2}} \qquad (18.30)$$

For $F \ll F_0$, $\mu = \mu_0$, whereas $\mu = \mu_0(F_0/F)^{1/2}$ for $F \gg F_0$.

The drain current is now expressed as

$$I_d = \frac{g_0(V_0)F}{[\frac{1}{2} + (\frac{1}{4} + F^2/F_0^2)^{1/2}]^{1/2}} \qquad (18.31)$$

where $g_0(V_0) = \mu_0 C_{ox} w(V_g - V_T - V_0)$ is the low-field conductance of the channel for $V_d \leq V_g - V_T$; C_{ox} is the oxide capacitance per unit area, V_T the turn-on voltage, and V_0 the channel voltage at x. Solving for F, we obtain

$$F^2 = \left(\frac{dV_0}{dx}\right)^2 = \frac{I_d^2}{g_0^2(V_0)} + \frac{I_d^4}{g_0^4(V_0)F_0^2} \tag{18.31a}$$

Separating variables, integrating under the condition of saturation (that is, $V_d \to V_g - V_T$), and substituting $u = V_g - V_T - V_0$ yields

$$
\begin{aligned}
L &= \frac{\mu_0 C_{ox} w}{I_d} \int_0^{u_0} \frac{u^2\, du}{(u^2 + a^2)^{1/2}} \\
&= \frac{\mu_0 C_{ox} w}{I_d}\left\{\frac{1}{2}u_0(u_0^2 + a^2)^{1/2} - \frac{a^2}{2}\ln\left[\frac{u_0 + (u_0^2 + a^2)^{1/2}}{a}\right]\right\}
\end{aligned}
\tag{18.31b}
$$

where $a = I_d/\mu_0 C_{ox} w F_0$ and $u_0 = V_g - V_T$. Consequently,

$$\frac{2I_d L}{\mu_0 C_{ox} w u_0^2} = \left[1 + \left(\frac{a}{u_0}\right)^2\right]^{1/2} - \left(\frac{a}{u_0}\right)^2 \ln\left\{\frac{u_0}{a} + \left[\left(\frac{u_0}{a}\right)^2 + 1\right]^{1/2}\right\} \tag{18.32}$$

or in normalized, parametric form,

$$I_0 = z\left(1 + \frac{1}{z^2}\right)^{1/2} - z^2 \ln\left[\frac{1}{z} + \left(1 + \frac{1}{z^2}\right)^{1/2}\right], \qquad v_0 = \frac{2z}{I_0} \tag{18.33}$$

where

$$z = \frac{a}{u_0}, \qquad I_0 = \frac{2I_d L}{\mu_0 C_{ox} w(V_g - V_T)^2}, \qquad v_0 = \frac{V_g - V_T}{F_0 L} \tag{18.33a}$$

Here I_0 is nothing but the ratio of the actual drain current over the current expected for constant mobility. Making a Taylor expansion of I_0 with respect to $1/z$ yields, for large v_0,

$$I_0 \simeq \frac{2}{(3v_0)^{1/2}} \tag{18.34}$$

This asymptotic result would have been obtained directly by putting $\mu = \mu_0(F_0/F)^{1/2}$.

Defining the transconductance g_m as $\partial I_d/\partial V_g$ and defining the normalized transconductance G_m as the ratio of g_m over the value expected for constant mobility yields

$$G_m = \frac{g_m L}{\mu_0 C_{ox} w(V_g - V_T)} = I_0 + \frac{1}{2}\frac{dI_0}{dv_0} \tag{18.35}$$

Substituting the expression (18.34) yields, for large v_0,

$$G_m = \frac{3/2}{(3v_0)^{1/2}} = \frac{3}{4}I_0 \tag{18.35a}$$

The asymptotic expressions for I_0 and G_m are reasonably accurate for $z > 2$ and acceptable for $z > 1$.[†]

18.2. CALCULATION OF THE HIGH-FREQUENCY RESPONSE OF FETS

At higher frequencies one must take into account the capacitances C_{gs}, C_{gd}, and C_{ds} of the device. Here C_{gs} is the capacitance between gate and source, C_{gd} the capacitance between gate and drain, and C_{ds} the capacitance between drain and source.

Actually the gate and channel form a nonuniform distributed active *RC* network. We shall derive the wave equations of this *RC* network both for the junction gate and for the MOS field-effect transistors. Having solved this equation, we can write the ac gate current ΔI_g and the ac drain current ΔI_d in terms of the ac gate voltage ΔV_g and the ac drain voltage ΔV_d as

$$\begin{aligned} \Delta I_g &= (Y_{gs} + Y_{gd})\,\Delta Y_g - Y_{gd}\,\Delta V_d \\ \Delta I_d &= (Y_{dg} - Y_{gd})\,\Delta V_g + (Y_{ds} + Y_{gd})\,\Delta Y_d \end{aligned} \tag{18.36}$$

Here ΔI_g flows *into* the gate and ΔI_d *into* the drain; Y_{gs} is the gate-source admittance, Y_{gd} the feedback admittance, and Y_{ds} the drain-source admittance, whereas Y_{dg} replaces the low-frequency transconductance g_{m0}.

18.2a. Wave Equations of the Two Types of FETs

We first derive the wave equation of the junction FET. If $\Delta W(x)$ is the ac bias of the channel, then this corresponds to a fluctuating height $2\,\Delta b(x)$ of the channel. According to (18.8),

$$\Delta b(x) = -\frac{a\,\Delta W(x)}{2W_{00}^{1/2}W_0^{1/2}(x)} \tag{18.37}$$

This corresponds to a fluctuating charge $d\Delta Q/dx$ per unit length at the gate of

$$\frac{d\,\Delta Q}{dx} = -2\rho_0 w\,\Delta b(x) = \frac{wa\rho_0\,\Delta W(x)}{W_{00}^{1/2}W_0^{1/2}(x)} \tag{18.38}$$

where $\rho_0 = eN_d$ is the charge density of the space-charge region, and to an ac gate current per unit length $d\,\Delta i_g/dx$ of

$$\frac{d\,\Delta i_g}{dx} = j\omega\frac{d\,\Delta Q}{dx} = \frac{j\omega\rho_0 wa\,\Delta W(x)}{W_{00}^{1/2}W_0^{1/2}(x)} \tag{18.39}$$

---

† A. van der Ziel and A. Nussbaum, *Solid State Electronics*, **17**, 413 (1974).

As indicated in Fig. 18.4, if $\Delta I(x)$ is the ac current in the channel at a distance x from the source, then

$$\frac{d\,\Delta I}{dx} = \frac{d\,\Delta i_g}{dx} = \frac{j\omega\rho_0 wa\,\Delta W(x)}{W_{00}^{1/2}W_0^{1/2}(x)} \tag{18.40}$$

Because of the capacitive term on the right side, $\Delta I(x)$ depends on x and $\Delta I(0) \neq \Delta I(L)$. This is the basis for the ac behavior of the FET.

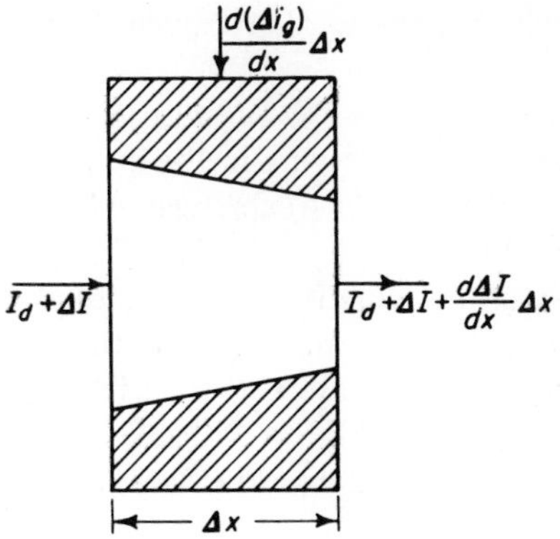

Fig. 18.4. Current flow in the section Δx of the junction FET.

Finally, ΔI must be calculated. Since

$$\Delta I = I - I_d$$
$$= g(W)\frac{dW}{dx} - g(W_0)\frac{dW_0}{dx} \tag{18.41}$$

we find, after manipulations, that

$$\Delta I(x) = \frac{d}{dx}[g(W_0)\,\Delta W(x)] \tag{18.42}$$

Substituting into (18.40), we obtain the *wave equation*

$$\frac{d^2}{dx^2}[g(W_0)\,\Delta W(x)] = \frac{j\omega\rho_0 wa\,\Delta W(x)}{W_{00}^{1/2}W_0^{1/2}(x)} \tag{18.43}$$

which has to be solved for $\Delta W(x)$ under the initial conditions

$$\begin{aligned} \Delta W(x) &= \Delta W(0), \qquad \text{at } x = 0 \\ \Delta W(x) &= \Delta W(L), \qquad \text{at } x = L \end{aligned} \tag{18.44}$$

If $\Delta V_d = 0$, the initial conditions are

$$\Delta W(0) = \Delta W(L) = -\Delta V_g \tag{18.44a}$$

If $\Delta V_g = 0$, the initial conditions are

$$\Delta W(0) = 0, \qquad \Delta W(L) = +\Delta V_d \tag{18.44b}$$

Ultimately, the interest is not so much in $\Delta W(x)$ as in the currents

$$\begin{aligned} \Delta I(0) &= \frac{d}{dx}[g(W_0)\,\Delta W]\big|_{x=0} \\ \Delta I(L) &= \frac{d}{dx}[g(W_0)\,\Delta W]\big|_{x=L} \end{aligned} \tag{18.45}$$

Of special interest are the currents

$$\Delta I_g = \int_0^L \frac{d(\Delta i_g)}{dx}dx = \Delta I(L) - \Delta I(0) \tag{18.46}$$

which is the total ac current flowing to the gate, and

$$\Delta I_d = \Delta I(L) \tag{18.47}$$

the current flowing in the output lead.

Having thus expressed ΔI_g and ΔI_d in terms of ΔV_g and ΔV_d, we can evaluate Y_{gs}, Y_{gd}, Y_{dg}, and Y_{ds}. Since the device is generally used in the saturated condition, we are especially interested in the saturated values of Y_{gs}, Y_{gd}, Y_{dg}, and Y_{ds}. To that end we evaluate the parameters for the unsaturated case and then let $W_d = -V_g + V_{\text{dif}} + V_d$ approach the saturation bias W_{00}.

We now derive the wave equation of the MOS field-effect transistor. To do so, we slightly change the notation of Sec. 18.1c. We take the direction of positive current flow from source to drain and put $V_g - V_T - V(x) = V_0(x)$. In that case Eq. (18.16) may be written

$$I(x) = \mu C_{\text{ox}} w V_0(x) \frac{dV_0}{dx} = \frac{1}{2} \mu C_{\text{ox}} w \frac{d}{dx}[V_0(x)]^2 \tag{18.48}$$

Introducing small variations $\Delta V_0(x)$, we have

$$\Delta I(x) = \frac{1}{2} \mu C_{\text{ox}} w \frac{d}{dx}\{[V_0(x) + \Delta V_0(x)]^2 - [V_0(x)^2\}$$

$$= \mu C_{\text{ox}} w \frac{d}{dx}[V_0(x)\, \Delta V_0(x)] \tag{18.49}$$

But in analogy with Fig. 18.4,

$$\frac{d}{dx}[\Delta I(x)] = j\omega C_{\text{ox}} w\, \Delta V_0(x) \tag{18.50}$$

Substituting (18.49) into (18.50) yields the wave equation

$$\frac{d^2}{dx^2}[V_0(x)\, \Delta V_0(x)] = \frac{j\omega}{\mu}\, \Delta V_0(x) \tag{18.51}$$

which shows similarity to Eq. (18.43).

This equation must now be solved under the initial conditions

$$\Delta V_0(x) = \Delta V_0(0) \quad \text{at } x = 0, \qquad \Delta V_0(x) = \Delta V_0(L) \quad \text{at } x = L$$

If $\Delta V_d = 0$, $\Delta V_0(0) = \Delta V_0(L) = \Delta V_g$, and if $\Delta V_g = 0$, $\Delta V_0(0) = 0$ and $\Delta V_0(L) = -\Delta V_d$. Again, $\Delta I_g = \Delta I(L) - \Delta I(0)$ and $\Delta I_d = \Delta I(L)$.

If Y_{gs}, Y_{gd}, Y_{dg}, and Y_{ds} are calculated by solving the two wave equations, one finds $Y_{gd} = Y_{ds} = 0$ at saturation. Furthermore, in first approximation,

$$Y_{gs} = j\omega C_{gs} + g_{gs} \tag{18.52}$$

where C_{gs} is independent of frequency and g_{gs} varies as ω^2 over a wide frequency range. This means that over a wide frequency range the input can be represented by a frequency-independent resistance r_{gs} in series with the capacitance C_{gs} such that

$$g_{gs} = (\omega C_{gs})^2 r_{gs} \quad \text{or} \quad r_{gs} = \frac{g_{gs}}{\omega^2 C_{gs}^2} \tag{18.52a}$$

Finally, it is found that $|Y_{dg}| \simeq g_{m0}$ over a wide frequency range and that $Y_{dg} = |Y_{dg}| \exp(-j\varphi_m)$, where φ_m increases with ω. Only at very high frequencies does $|Y_{dg}|$ decrease with increasing frequency (see next Section).

18.2b. Calculation of Y_{gs} and Y_{dg} for MOSFETs†

We start with Eq. (18.51) and substitute

$$V_0(x)\,\Delta V_0(x) = v(x)(V_{g0} - V_T)^2 \tag{18.53}$$

where V_{g0} is the dc gate bias. This yields

$$\frac{d^2v}{dx^2} = \frac{j\omega}{\mu}\frac{v(x)}{V_0(x)} \tag{18.54}$$

We next introduce

$$y = \frac{V_0(x)}{V_{g0} - V_T} \tag{18.55}$$

as a new dimensionless variable. We make this transformation as follows. According to Eq. (18.48),

$$I_d = \mu C_{ox} w V_0(x)\frac{dV_0(x)}{dx} = \mu C_{ox}(V_{g0} - V_T)^2 y\frac{dy}{dx} = 2I_{ds}Ly\frac{dy}{dx} \tag{18.56}$$

where I_{ds} is the saturation current. It is now easily shown that

$$\frac{d^2}{dx^2} = \left(\frac{I_d}{2I_{ds}L}\right)^2 \frac{1}{y}\frac{d}{dy}\left(\frac{1}{y}\frac{d}{dy}\right) \tag{18.57}$$

$$\frac{j\omega}{\mu}\frac{v(x)}{V_0(x)} = \frac{j\hat{\omega}}{L^2}\frac{1}{y}v(y) \tag{18.58}$$

where

$$\hat{\omega} = \omega\frac{L^2}{\mu(V_{g0} - V_T)} \tag{18.58a}$$

is a dimensionless quantity; $L^2/\mu(V_{g0} - V_T)$ is proportional to the transit time of the carriers in the channel. Substituting into Eq. (18.54) yields

$$\frac{d}{dy}\left(\frac{1}{y}\frac{dv}{dy}\right) = j\hat{\omega}\left(\frac{2I_{ds}}{I_d}\right)^2 v(y) \tag{18.59}$$

If we now make the final substitution

$$w = jy\left[\left(\frac{2I_{ds}}{I_d}\right)^2 \hat{\omega}\right]^{1/3} \tag{18.60}$$

we obtain

$$\frac{d}{dw}\left(\frac{1}{w}\frac{dv}{dw}\right) + v = 0 \tag{18.61}$$

---

† J. A. Geurst, *Solid State Electronics*, **8**, 88 (1965).

We shall now relate this equation to Stokes's equation

$$\frac{d^2u}{dw^2} + wu = 0 \tag{18.62}$$

by showing that if $u(w)$ is a solution of (18.62), then $v = du/dw$ is a solution of (18.61). This is done as follows:

$$\frac{1}{w}\frac{d^2u}{dw^2} + u = 0$$

for $w \neq 0$. Differentiation with respect to w yields

$$\frac{d}{dw}\left(\frac{1}{w}\frac{d^2u}{dw^2}\right) + \frac{du}{dw} = 0 \quad \text{or} \quad \frac{d}{dw}\left(\frac{1}{w}\frac{dv}{dw}\right) + v = 0$$

as had to be proved.

Now let $h_1(w)$ and $h_2(w)$ be two independent solutions of Stokes's equation; then the general solution is

$$u = Ah_1(w) + Bh_2(w) \tag{18.63}$$

and the solution of Eq. (18.61) is

$$v = \frac{du}{dw} = Ah'_1(w) + Bh'_2(w) \tag{18.63a}$$

where the prime indicates a derivative with respect to w. The constants A and B must now be determined from the boundary conditions at the source $(x = 0)$ and at the drain $(x = L)$. We do so for saturation $(I_d = I_{ds})$, and write

$$\begin{aligned} v(0) = v_s, & \quad y(0) = y_s, & \quad w(0) = w_s \\ v(L) = v_d, & \quad y(L) = y_d, & \quad w(L) = w_d \end{aligned} \tag{18.64}$$

and obtain, if ΔV_g is the ac amplitude of the gate voltage,

$$v_s = \frac{\Delta V_g}{V_{g0} - V_T}, \quad y_s = \frac{V_0(0)}{V_{g0} - V_T} = 1, \quad w_s = j(4\hat{\omega})^{1/3}$$

$$v_d = 0, \quad y_d = 0, \quad w_d = jy_d(4\hat{\omega})^{1/3} = 0 \tag{18.64a}$$

since $V_0(L) = 0$ and $\Delta V_0(L) = \Delta V_g$ at saturation. We thus have

$$\begin{aligned} v_s &= Ah'_1(w_s) + Bh'_2(w_s) \\ 0 &= Ah'_1(w_d) + Bh'_2(w_d) \end{aligned} \tag{18.65}$$

so that

$$A = \frac{h'_2(w_d)v_s}{\Delta(w_s, w_d)}, \quad B = -\frac{h'_1(w_d)v_s}{\Delta(w_s, w_d)},$$

$$\Delta(w_s, w_d) = \begin{vmatrix} h'_1(w_s) & h'_2(w_s) \\ h'_1(w_d) & h'_2(w_d) \end{vmatrix} \tag{18.65a}$$

Substitution into u yields

$$u = \frac{W(w, w_d)}{\Delta(w_s, w_d)}, \qquad W(w, w_d) = \begin{vmatrix} h_1(w) & h_2(w) \\ h_1'(w_d) & h_2'(w_d) \end{vmatrix} \qquad (18.65b)$$

Now, according to Eq. (18.49), after some manipulations

$$\Delta I_{d0}(x) = 2I_{ds}L\frac{dv}{dx} = I_{ds}\frac{1}{y}\frac{dv}{dy} = j^2(4\hat{\omega})^{2/3} I_{ds}\frac{1}{w}\frac{dv}{dw} \qquad (18.66)$$

$$= I_{ds}(4\hat{\omega})^{2/3}u$$

since

$$\frac{1}{w}\frac{dv}{dw} = \frac{1}{w}\frac{du^2}{dw^2} = -u \quad \text{and} \quad w = jy(4\hat{\omega})^{1/3}$$

Therefore, after substitution,

$$\Delta I_{d0}(x) = \frac{1}{2}g_{m0}(4\hat{\omega})^{2/3}\frac{W(w, w_d)}{\Delta(w_s, w_d)}\Delta V_g \qquad (18.66a)$$

where $g_{m0} = 2I_{ds}/(V_{g0} - V_T)$ is the low-frequency transconductance at saturation. Evaluating $\Delta I_{d0}(0)$ and $\Delta I_{d0}(L)$, we obtain

$$Y_{dg} = +\frac{\Delta I_{d0}(L)}{\Delta V_g} = g_{m0}(4\hat{\omega})^{2/3}\frac{W(w_d, w_d)}{2\Delta(w_s, w_d)} \qquad (18.67)$$

$$Y_{gs} = \frac{\Delta I_{d0}(0) - \Delta I_{d0}(L)}{\Delta V_g} = g_{m0}(4\hat{\omega})^{2/3}\frac{W(w_s, w_d) - W(w_d, w_d)}{2\Delta(w_s, w_d)} \qquad (18.68)$$

This solves the high-frequency response at saturation in principle.

We now find by substitution that the following series expansions are solutions of Stokes's equation:

$$h_1(w) = 1 - \frac{1}{6}w^3 + \frac{1}{30 \times 6}w^6 - \frac{1}{72 \times 30 \times 6}w^9$$

$$+ \frac{1}{132 \times 72 \times 30 \times 6}w^{12} + \cdots \qquad (18.69)$$

$$h_2(w) = w - \frac{1}{12}w^4 + \frac{1}{42 \times 12}w^7 - \frac{1}{90 \times 42 \times 12}w^{10}$$

$$+ \frac{1}{156 \times 90 \times 42 \times 12}w^{13} + \cdots \qquad (18.70)$$

Therefore, since $w_d = 0$, $h_1(0) = 1$, $h_1'(0) = 0$, $h_2(0) = 0$, and $h_2'(0) = 1$, we have

$$W(w_s, w_d) = h_1(w_s), \qquad W(w_d, w_d) = 1, \qquad \Delta(w_s, w_d) = h_1'(w_s) \qquad (18.71)$$

so that

$$Y_{dg} = g_{m0}\frac{(4\hat{\omega})^{2/3}}{2h_1'(w_s)}, \qquad Y_{gs} = g_{m0}(4\hat{\omega})^{2/3}\frac{h_1(w_s) - 1}{2h_1'(w_s)} \qquad (18.72)$$

Using the Taylor expansions (18.69) and (18.70), we obtain

$$Y_{dg} = \frac{g_{m0}}{1 + \frac{4}{15}j\hat{\omega} + \frac{1}{45}(j\hat{\omega})^2 + \frac{4}{4455}(j\hat{\omega})^3 + \cdots} \qquad (18.72a)$$

$$Y_{gs} = g_{m0}\frac{\frac{2}{3}j\hat{\omega} + \frac{4}{45}(j\hat{\omega})^2 + \frac{2}{405}(j\hat{\omega})^3 + \frac{2}{13,365}(j\hat{\omega})^4 + \cdots}{1 + \frac{4}{15}j\hat{\omega} + \frac{1}{45}(j\hat{\omega})^2 + \frac{4}{4455}(j\hat{\omega})^3 + \cdots} \qquad (18.72b)$$

If one writes $Y_{gs} = j\omega C_{gs} + g_{gs}$ and makes a Taylor expansion of (18.72b) in terms of $j\hat{\omega}$, one obtains

$$Y_{gs} = j\omega C_{gs} + g_{gs} = g_{m0}[\tfrac{2}{3}j\hat{\omega} + \tfrac{4}{45}\hat{\omega}^2] \qquad (18.73)$$

We thus see that g_{gs} varies as ω^2 over a wide frequency range and that $g_{gs}/g_{m0} = 1$ for $\hat{\omega} = (\frac{45}{4})^{1/2} \simeq 3.4$. A more accurate calculation based on Eq. (18.72b) yields $g_{gs}/g_{m0} = 1$ for $\hat{\omega} \simeq 3.9$.

According to Sec. 18.2e, the limit of usefulness of the noise performance of an FET is given by the condition $g_{gs}/g_{m0} = 1$. The series expansions (18.72a and b) are thus applicable for the whole useful range of MOSFET operation.

If one evaluates $|Y_{dg}|$, one obtains $|Y_{dg}|/g_{m0} = 0.836$ for $\hat{\omega} = 4.0$. The absolute value of $|Y_{dg}|$ is therefore practically constant for the whole useful range of MOSFET operation.

If one substitutes for g_{m0} and $\hat{\omega}$ in Eq. (18.73) and solves for C_{gs}, one obtains

$$C_{gs} = \frac{2}{3}g_{m0}\frac{\hat{\omega}}{\omega} = \frac{2}{3}C_{ox}wL = \frac{2}{3}C_0 \qquad (18.73a)$$

where C_0 is the oxide capacitance for a device area wL.

We now represent g_{gs} by a resistance r_{gs} in series with C_{gs}. According to Eqs. (18.52a) and (18.73), this yields

$$r_{gs} = \frac{g_{gs}}{\omega^2 C_{gs}^2} = \frac{1}{5g_{m0}} \qquad (18.73b)$$

This representation is useful for $\omega C_{gs}r_{gs} < 1$. Substituting for C_{gs} and r_{gs}, we obtain

$$\omega C_{gs}r_{gs} = 1 \quad \text{or} \quad \tfrac{2}{15}\hat{\omega} = 1 \quad \text{or} \quad \hat{\omega} = 7.5 \qquad (18.73c)$$

This is about twice as large as the frequency for which $g_{gs}/g_{m0} = 1$. A small calculation shows that the series representation C_{gs}, r_{gs} is valid for the whole useful range of the MOSFET; even at $\hat{\omega} = 4.0$ the error is only a few per cent.

Figure 18.5 shows that the condition $g_{gs} = g_{m0}$ corresponds approximately to the condition of unit voltage gain $|Y_{dg}| = g_{gs}$ of a cascaded tuned amplifier.

Figure 18.6 shows the equivalent circuit of a MOSFET at high frequencies for the case that the substrate is connected to the source. We have added here series resistances r_s and r_d in the source and drain leads, respectively.

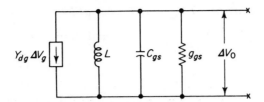

Fig. 18.5. Cut-off frequency of the cascaded tuned amplifier; the gain is unity if $|Y_{dg}| = g_{gs}$.

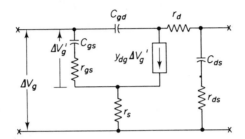

Fig. 18.6. Equivalent circuit of an FET at high frequencies for the case that the substrate is connected to the source.

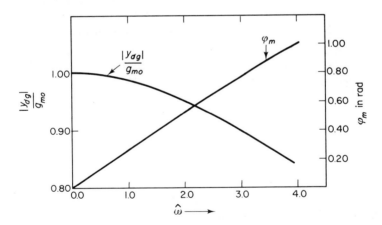

Fig. 18.7. $|Y_{dg}|/g_{m0}$ and φ_m versus ω.

Moreover, since most of the capacitance C_{ds} is between drain and substrate, and the substrate can have an appreciable resistance, we have added the resistance r_{ds} in series with the capacitance C_{ds}.

We now write $Y_{dg} = |Y_{dg}| \exp(-j\varphi_m)$. Figure 18.7 shows $|Y_{dg}|/g_{m0}$ and φ_m as a function $\hat{\omega}$.

18.2c. Calculation of Y_{gs} and Y_{dg} for JFETs

The calculation of $Y_{gs} = j\omega C_{gs} + g_{gs}$ and of Y_{dg} for JFETs proceeds along similar lines as for MOSFETs and the results are quite similar. Geurst has performed the calculation,† but has not given numerical values. It is to be expected that $|Y_{dg}|$ decreases only very slowly for the useful range of operation of the JFET and that in that range Y_{gs} can be represented by a capacitance C_{gs} and a resistance r_{gs} in series. The useful range of operation is again determined by the low-noise condition $g_{gs}/g_{m0} \lesssim 1$ (see Sec. 18.2e).

18.2d. Intrinsic Pulse Response of MOSFETs

The intrinsic pulse response of a MOSFET is determined by the time it takes to generate a conducting channel in a device that was first off. V. P. Singh‡ has calculated the effect numerically and has found that the time needed is of the order of the transit time of the carriers from source to drain. This time is given as

$$\tau = \frac{4L^2}{3\mu(V_g - V_T)} \tag{18.74}$$

A typical MOSFET has a channel length L of 10^{-3} cm. Assuming that $\mu = 1000$ cm^2/V s for an n-type channel and $V_g - V_T = 4$ V, yields $\tau = 0.33 \times 10^{-9}$ s. This is very short in comparison with the RC time of external circuits. For the load resistance of a MOSFET is of the order of several thousand ohms and the device capacitance is several picofarads, so the RC product is of the order of many nanoseconds. In comparison with this the intrinsic switching time of the MOSFET is negligible.

We finally prove Eq. (18.74) for a MOSFET that is just pinched off; that is, $V_d = V_g - V_T$. Then

$$\tau = \int_0^L \frac{dx}{\mu(dV_0/dx)} = \frac{1}{\mu I_{ds}^2} \int_0^{V_g - V_T} g^2(V_0)\, dV_0$$

$$= \frac{(\mu C_{ox} w)^2}{\mu I_{ds}^2} \int_0^{V_g - V_T} (V_g - V_T - V_0)^2\, dV_0 = \frac{4}{3} \frac{L^2}{\mu(V_g - V_T)}$$

since $dV_0/dx = I_{ds}/g(V_0)$, $dx = g(V_0)dV_0/I_{ds}$, $g(V_0) = \mu C_{ox} w(V_g - V_T - V_0)$, and $I_{ds} = \frac{1}{2}(\mu C_{ox} w/L)(V_g - V_T)^2$.

18.2e. Noise

Since the field-effect transistors are distributed RC networks, each element of which shows thermal noise, one would in first approximation expect full thermal noise associated with the resistances r_{gs}, r_s, and r_d and with the

---

† J. A. Geurst, *Solid State Electronics*, **8**, 563 (1965).

‡ V. P. Singh, University of Minnesota (unpublished).

transconductance g_{m0}. That is, if the source and drain are short-circuited, and the short-circuit currents are i_d and i_g, then

$$\overline{i_d^2} \simeq 4kTg_{m0}\,\Delta f, \qquad \overline{i_g^2} \simeq 4kTg_{gs}\,\Delta f \qquad (18.75)$$

if the noise of r_s and r_d is neglected. Actually, a more detailed calculation shows that these noise parameters must be written as

$$\overline{i_d^2} = n_d \cdot 4kTg_{m0}\,\Delta f, \qquad \overline{i_g^2} = n_g \cdot 4kTg_{gs}\,\Delta f \qquad (18.76)$$

Here n_d varies between $\frac{1}{2}$ and $\frac{2}{3}$ for a junction gate FET and is equal to $\frac{2}{3}$ for a MOSFET; n_g is slightly larger than unity for a junction gate FET and equal to $\frac{4}{3}$ for a MOSFET. These predictions are well satisfied for junction gate FET's, but in some MOSFETs n_d is from three to five times as large as predicted theoretically.

The currents i_g and i_d come from the same thermal noise sources; hence one would expect them to be correlated. Experimentally, however, one finds only a relatively small correlation that can be neglected for all practical purposes.

We now calculate the noise figure of the device under the assumption (18.75), neglecting the frequency dependence of $|Y_{dg}|$. We have from Fig. 18.8, neglecting the correlation between source and drain noise and assuming that the capacitance C_{gd} has been neutralized,

$$\overline{i^2} = \frac{4kTg_s\,\Delta f}{(g_s + g_{gs})^2}g_{m0}^2 + \frac{4kTg_{gs}\,\Delta f}{(g_s + g_{gs})^2}g_{m0}^2 + 4kTg_{m0}\,\Delta f \qquad (18.77)$$

and hence we find the noise figure F by dividing $\overline{i^2}$ by the first term. This yields

$$F = 1 + \frac{g_{gs}}{g_s} + \frac{(g_s + g_{gs})^2}{g_{m0}g_s} \qquad (18.78)$$

Its minimum value occurs at $g_s = (g_{gs}^2 + g_{m0}g_{gs})^{1/2}$, and is

$$F_{\min} = 1 + \frac{2g_{gs}}{g_{m0}} + 2\left[\frac{g_{gs}}{g_{m0}} + \left(\frac{g_{gs}}{g_{m0}}\right)^2\right]^{1/2} \qquad (18.78a)$$

which has a value of about 6 for $g_{gs} = g_{m0}$. We thus see that the useful range of operation of the device does not extend very much beyond the frequency f_t at which $g_{gs} \simeq g_{m0}$.

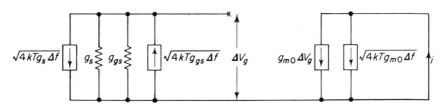

Fig. 18.8. Equivalent noise circuit of the FET amplifier at high frequencies.

At low frequencies the noise can be considerably larger than predicted here. The causes are various, and the details are beyond the scope of this book.

18.3. SURFACE PROPERTIES OF THE MOSFET

In this section we discuss the surface conditions in MOSFETs that determine the magnitude of the turn-on voltage V_T, defined in Sec. 18.1c, and of the diffusion potential V_{dif}, defined in Sec. 18.1d. Then we investigate the effect of a layer of opposite conductivity type buried in the substrate close to the interface. We finally apply these considerations to charge control devices.

18.3a. Calculation of V_T and V_{dif} for an N-Type Channel

Figure 18.9a shows the band structure of a metal-oxide–p-type semiconductor device for zero applied potential. There is a certain amount of band bending near the surface that can result in an n-type surface inversion layer. Since no voltage is applied, the Fermi level is everywhere at equal height. Also shown in the figure is the bottom of the conduction band of the SiO_2 layer.

If there are no charges stored at the interface or in the oxide layer, and the work functions of the metal and the p-type semiconductor are denoted by ϕ_m and ϕ_{sp}, respectively, then a voltage $V_{FB} = \phi_m - \phi_{sp}$ must be applied to the metal to straighten out the band structure. This voltage is called the *flatband* voltage (Fig. 18.9b).

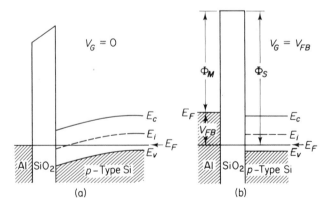

Fig. 18.9. The effect of metal-semiconductor work function difference on the potential distribution in an MOS structure. (a) Conditions for $V_g = 0$. (b) Flat-band condition.
(From A.S. Grove, *Physics and Technology of Semiconductor Devices*, John Wiley and Sons, New York, 1967, by permission.)

We now investigate the effect of interface energy states. These states store a charge Q_{ss} per unit area, and this gives rise to an induced charge $-Q_{ss}$ in the substrate. To overcome the effect of this charge, a voltage $-Q_{ss}/C_{ox}$ must be applied to the metal gate to restore the flatband condition; here C_{ox} is the capacitance per unit area of the metal-oxide-semiconductor capacitor.

In the early days of MOS technology there was also a charge distribution in the oxide due to Na^+ ions. Let there be a charge Q per square centimeter at a distance y from the metal; then this induces a charge $-Qy/d$ in the substrate, where d is the oxide thickness; this relation follows from solving Poisson's equation. If there were a *distributed* charge density $\rho(y)$ present in the oxide, then the induced charge in the channel would be

$$Q_{ind} = -\int_0^d \frac{y}{d} \rho(y)\, dy$$

Consequently, the flatband situation would be restored if a voltage

$$\frac{Q_{ind}}{C_{ox}} = -\frac{1}{C_{ox}}\int_0^d \frac{y}{d} \rho(y)\, dy$$

were applied to the metal. In modern device technology this effect is negligible. If the surface charge effect is taken into account, however, the flatband voltage is

$$V_{FB} = \phi_m - \phi_{sp} - \frac{Q_{ss}}{C_{ox}} \tag{18.79}$$

In MOS capacitors on a p-type semiconductor, Q_{ss} is positive and V_{FB} is usually negative. One can raise V_{FB} by replacing the metal gate electrode by a p^+-type polycrystalline silicon layer. Then ϕ_m must be replaced by the work function ϕ_{p+} of that layer, and $\phi_{p+} - \phi_{sp}$ is positive.

To evaluate V_T, we return to Fig. 18.9. We assume that no voltage is applied between drain and source. For $V_g = V_{FB}$, we had the flatband condition. How much band bending has to be produced to make the surface n type (inversion)? If we measure the Fermi level in the p region by the distance ϕ_{F_p} to the intrinsic level at midband, then a good criterion for the existence of an n-type channel on a p-type substrate is that the surface electron density n_s equal the acceptor concentration N_a of the substrate. Under this condition the Fermi level at the surface will be *above* the intrinsic level at midband by the same amount as it is *below* the intrinsic level in the bulk region. Therefore, the surface potential $\varphi_s(\text{inv})$ under that condition is

$$\varphi_s(\text{inv}) = 2\phi_{F_p} \tag{18.80}$$

Now the diffusion potential is defined as the potential difference between the surface of the channel and the substrate when inversion just sets in. Therefore,

$$V_{dif} = 2\phi_{F_p} \tag{18.80a}$$

Moreover, if V_g is the gate voltage, then the voltage developed across the oxide at the onset of inversion is

$$V_g - V_T = V_g - V_{FB} - \varphi_s(\text{inv}) = V_g - V_{FB} - 2\phi_{F_p}$$

so that

$$V_T = V_{FB} + 2\phi_{F_p} \tag{18.80b}$$

The basic equations (18.21) and (18.22) of Sec. 18.1d can now be derived.

For n-type channels, V_T is usually negative. By using a p^+-type silicon gate and a properly designed MOS capacitor, V_T can be positive.

For p-type channels on an n-type substrate, one can apply similar considerations. In Eq. (18.79) one must replace ϕ_{sp} by ϕ_{sn}. To achieve a p-type inversion layer, one must now have band bending in the opposite direction. If ϕ_{F_n} is the distance between the Fermi level in the substrate and the midband level, then the surface potential at inversion is

$$\varphi_s(\text{inv}) = -2\phi_{Fn}$$

so that, if V_g is the gate voltage, then the voltage developed at the onset of inversion is

$$V_g - V_T = V_g - V_{FB} - \phi_s(\text{inv}) = V_g - V_{FB} + 2\phi_{Fn}$$

so that

$$V_T = V_{FB} - 2\phi_{Fn} \tag{18.80c}$$

Presently, this voltage is negative for all p-type channels on n-type substrates.

18.3b. Change in Turn-on Voltage with a Buried Layer

By introducing a buried layer of opposite conductivity type near the surface in a p-type substrate, one can change the turn-on voltage of a MOSFET over several volts. We shall now evaluate the effect. To that end we assume that there is a buried donor layer of concentration N_d per square centimeter at $x = -x_0$, which gives rise to a space-charge region of width $L > x_0$ (Fig 18.10). We assume first for the sake of simplicity that there is no surface charge (at $x = 0$); in that case the flatband voltage is $V_{FB} = \phi_m - \phi_s$, as before. If the bulk region has an acceptor concentration N_a per cubic centimeter, then Poisson's equation for $-L < x < 0$ may written

$$\frac{d^2\psi}{dx^2} = \frac{eN_a}{\epsilon\epsilon_0} - \frac{eN_d\delta(x + x_0)}{\epsilon\epsilon_0} \tag{18.81}$$

where $\delta(x + x_0)$ is the Dirac delta function. The initial conditions are

$$\psi = 0, \quad \frac{d\psi}{dx} = 0 \quad \text{at } x = -L, \quad \psi = \varphi_s \quad \text{at } x = 0 \tag{18.81a}$$

where φ_s is the surface potential. Since there is no surface charge, $d\varphi/dx$ is

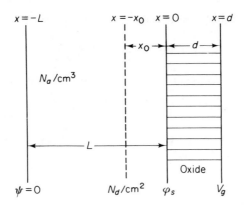

Fig. 18.10. *P*-type substrate with buried *n*-type layer at $x = -x_0$. There is also an oxide layer of thickness d the space charge region has a length L.

continuous at $x = 0$, or

$$\left(\frac{d\psi}{dx}\right)_0 = \frac{V_g - V_{FB} - \varphi_s}{d} \tag{18.81b}$$

where d is the oxide thickness and V_g the gate potential. Integrating once yields

$$\frac{d\psi}{dx} = \frac{eN_a}{\epsilon\epsilon_0}(x + L), \qquad \text{for } -L < x < -x_0$$

$$\frac{d\psi}{dx} = \frac{eN_a}{\epsilon\epsilon_0}(x + L) - \frac{eN_d}{\epsilon\epsilon_0}, \qquad \text{for } -x_0 < x < 0 \tag{18.82}$$

Now applying the condition (18.81b) at $x = 0$ yields

$$\frac{eN_aL}{\epsilon\epsilon_0} - \frac{eN_d}{\epsilon\epsilon_0} = \frac{V_g - V_{FB} - \varphi_s}{d} \quad \text{or} \quad V_g = V_{FB} + \varphi_s + \frac{eN_a}{\epsilon\epsilon_0}Ld - \frac{eN_d}{\epsilon\epsilon_0}d \tag{18.82a}$$

Integrating once more yields

$$\psi(x) = \frac{eN_a}{2\epsilon\epsilon_0}(x + L)^2, \qquad \text{for } -L < x < -x_0 \tag{18.83}$$

$$\psi(x) = \frac{eN_a}{2\epsilon\epsilon_0}(x + L)^2 - \frac{eN_d}{\epsilon\epsilon_0}(x + x_0), \qquad \text{for } -x_0 < x < 0 \tag{18.83a}$$

which satisfies the condition of continuity at $x = -x_0$.

A channel just begins at $x = -x_0$, if the Fermi level is just as much above the midband level at $x = -x_0$ as it is below it in the bulk. This is the case if

$$\psi(-x_0) = \frac{eN_a}{2\epsilon\epsilon_0}(L - x_0)^2 = 2\phi_{Fp}, \quad \text{or} \quad L = x_0 + \left(2\phi_{Fp} \cdot \frac{2\epsilon\epsilon_0}{eN_a}\right)^{1/2} \tag{18.83b}$$

Substituting $x = 0$ in (18.83a) yields

$$\varphi_s = \psi(0) = \frac{eN_a}{2\epsilon\epsilon_0}L^2 - \frac{eN_d}{\epsilon\epsilon_0}x_0 \qquad (18.83c)$$

Substituting into the expression for V_g, we thus obtain for the turn-on voltage

$$V_g = V_T = V_{FB} + \frac{eN_a}{2\epsilon\epsilon_0}L^2 - \frac{eN_d}{\epsilon\epsilon_0}x_0 + \frac{eN_a}{\epsilon\epsilon_0}Ld - \frac{eN_d}{\epsilon\epsilon_0}d \qquad (18.84)$$

The only effect of a charge Q_{ss} in the surface states is that it changes V_{FB} by an amount $-Q_{ss}/C_{ox}$, as discussed before.

As an example we take $\phi_{Fp} = 0.35$ V, $\epsilon = 12$, $N_a = 10^{15}/\text{cm}^3$, $N_d = 10^{12}/\text{cm}^2$, $d = 1000$ Å, and $x_0 = 2000$ Å. Then

$$V_T = V_{FB} + 2\phi_{Fp} = V_{FB} + 0.70 \text{ V}$$

without buried layer, whereas, according to (18.84) and (18.83b),

$$V_T = V_{FB} - 3.20 \text{ V}$$

for the buried-layer case. The turn-on voltage has thus changed by approximately 4 V. It is therefore clear that the turn-on voltage can change over several volts by proper choice of N_d and x_0.

The method also works for an n-type substrate with a p-type buried layer, but here the shift in V_T is toward positive voltages, of course.

18.3c. Charge-Coupled Devices[†]

Holes can be stored under properly biased electrodes made on MOS structures with an n-type substrate. If these structures are arranged in an array, and the electrodes are properly pulsed, one can transfer charges from the area under the one electrode to the area under the next one, and so on. These charge storage devices can thus be used as shift registers; they are known as charge-coupled devices (CCD's).

To understand the principle of operation, we turn to Fig. 18.11 which shows the electron energy versus distance through an MOS structure right after a voltage V_1 ($V_1 < 0$) has been applied to the metal electrode ($t = 0$). If now by some means holes are introduced into the area under the electrode and an equilibrium situation is reached, then the steady-state situation is such that the valence band at the interface is at about the same energy as the Fermi level in the bulk ($t = \infty$).

Next consider an array of electrodes 1–9 as shown in Fig. 18.12; voltages $-V_2$ are applied to electrodes 1, 4, and 7. The other electrodes are at the potential $-V_1$, with $V_1 < V_2$. Also shown are holes stored by some means under electrodes 1 and 7. The voltage to electrodes 2, 5, and 7 is lowered from

---

[†] W. S. Boyle and G. E. Smith, *Bell System Tech. J.*, **49**, 587 (1970).

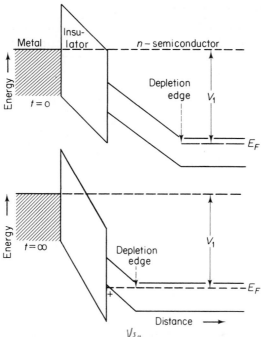

Fig. 18.11. A plot of electron energy is distance through an MIS structure both with (at time $t = \infty$) and without (at time $t = 0$) charge stored at the surface.
(From W.S. Boyle and G.E. Smith, *Bell Syst. Tech. J.*, **49**, 587, 1970, by permission.)

$-V_1$ to $-V_3$, with $V_3 > V_2$. As a consequence, the depletion region is now so shaped that holes move from under electrode 1 to under electrode 2 and from under electrode 7 to under electrode 8. If the voltage $-V_3$ is changed back to $-V_2$ and the voltage on electrodes 1, 4, and 7 is changed from $-V_2$ to $-V_1$, we have the situation shown in the figure. By now pulsing the third line, we can shift the charge one electrode further.

The case of zero stored charge is not a stable situation, however. A thermal current I_d flows because of generation–recombination centers in the depletion region under the contact. If Q is the equilibrium charge under the contact, then the time constant $\tau = Q/I_d$ is the *storage time* of the zero stored charge. These storage times are of the order of seconds.

If first no charge is stored under an electrode and then holes are moved from an adjacent electrode, part of the holes are stored in the interface and may not be taken out when the charge moves on. The stored charge thus decreases with time. This can be overcome by storing a small amount of positive charge, instead of storing zero charge; this is called a *fat zero*.

Generation of charge can be accomplished by a forward-biased *pn* diode

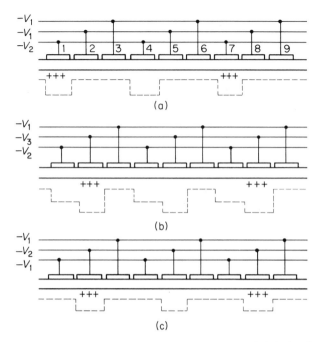

Fig. 18.12. Schematic of a three phase MIS charge coupled device (from W.S. Boyle and G.E. Smith, l.c., by permission.)

or by radiation-induced pair creation. Detection of the stored charge may be accomplished by current detection with a reverse-biased *pn* junction or Schottky barrier.

REFERENCES

Cobbold, R. S. C., *Theory and Application of Field Effect Transistors.* John Wiley & Sons, Inc. (Interscience Division), New York, 1970.

Frankl, D. R., *Electrical Properties of Semiconductor Surfaces.* Pergamon Press, Inc., Elmsford, N.Y., 1967.

Grove, A. S., *Physics and Technology of Semiconductor Devices.* John Wiley & Sons, Inc., New York, 1967.

Mayer, J. W., L. Erickson, and J. A. Davies, *Ion Implantation in Semiconductors, Silicon and Germanium.* Academic Press, Inc., New York, 1970.

Sze, S. M., *Physics of Semiconductor Devices.* John Wiley & Sons, Inc. (Interscience Division), New York, 1969.

Walmark, J. T., and H. Johnson, eds., *Field Effect Transistors.* Prentice-Hall, Inc., Englewood Cliffs, N.J., 1966.

PROBLEMS

1. Calculate the transit time τ of the carriers in the channel of a junction FET at saturation.

Answer:

$$\tau = \frac{3L^2}{2\mu W_{00}} \frac{1 + 3z^{1/2}}{(1 + 2z^{1/2})^2(1 - z^{1/2})} \qquad \text{where } z = W_s/W_{00}$$

2. Calculate the gate capacitance of a junction FET at saturation.

Answer:

$$C_{gs} = \frac{3eN_d awL}{W_{00}} \frac{1 + z^{1/2}}{(1 + 2z^{1/2})^2}$$

3. Express $f_0 = g_{m0}/2\pi C_{gs}$ at saturation in terms of τ.

Answer:

$$f_0 = \frac{1}{2\pi\tau} \frac{1 + 3z^{1/2}}{1 + z^{1/2}}$$

4. Calculate the transit time τ of the carriers in the channel of a MOSFET at saturation.

Answer: $\tau = 4L^2/[3\mu(V_g - V_T)]$.

5. Express for a MOSFET the frequency $f_0 = g_{m0}/2\pi C_{gs}$ at saturation in terms of τ.

Answer:

$$f_0 = \frac{4}{3} \frac{1}{2\pi\tau}$$

6. Calculate I_d and g_m from the asymptotic expressions (18.34) and (18.35a).

Answer:

$$I_d = \frac{1}{3^{1/2}} \frac{\mu_0 C_{ox}(V_g - V_T)^{3/2}(F_0 L)^{1/2}}{L};$$

$$g_m = \frac{3^{1/2}}{2} \frac{\mu_0 C_{ox}w(V_g - V_T)^{1/2}(F_0 L)^{1/2}}{L}$$

19

Miscellaneous
Semiconductor Devices

19.1. SPACE-CHARGE-LIMITED SOLID-STATE DIODES

Space-charge-limited (SCL) solid-state diodes occur in two forms. In the first form, the *single-injection SCL diode*, one contact injects carriers into the semiconductor and the other contact extracts carriers, whereas the current flow in the semiconductor itself is governed by the space charge of the injected carriers. In the second form, the *double-injection SCL diode*, one contact injects electrons that are collected by the other contact, whereas the other contact injects holes that are collected by the first contact.

19.1a. Single-Injection SCL Diodes

We assume that the semiconductor is nearly intrinsic so that the equilibrium carrier concentration can be neglected. We also neglect diffusion. The equations of the system for a device of length L are

Current equation: $J_n = e\mu_n nF$ (19.1)

Continuity equation: $e\dfrac{\partial n}{\partial t} = \dfrac{\partial J_n}{\partial x}$ (19.2)

Gauss's law: $\dfrac{\partial F}{\partial x} = -\dfrac{en}{\epsilon\epsilon_0}$,   or $n = -\dfrac{\epsilon\epsilon_0}{e}\dfrac{\partial F}{\partial x}$ (19.3)

464

Here J_n is the convection current density, n the carrier density, and F the field strength. Substituting n into (19.2) yields that the total current density

$$J(t) = e\mu_n nF + \epsilon\epsilon_0 \frac{\partial F}{\partial t} \tag{19.4}$$

is independent of x. Substituting for n from (19.3) yields

$$J(t) = -\frac{1}{2}\epsilon\epsilon_0\mu_n\frac{\partial(F^2)}{\partial x} + \epsilon\epsilon_0\frac{\partial F}{\partial t} \tag{19.5}$$

We now introduce dc terms (subscript 0) and small-signal ac terms (subscript 1):

$$J(t) = J_0 + J_1\exp(j\omega t), \qquad F = F_0 + F_1\exp(j\omega t),$$
$$V = V_0 + V_1\exp(j\omega t)$$

where $V(x, t)$ is the voltage at x and $F = -\partial V/\partial x$. This yields the following:

Dc equation: $J_0 = -\dfrac{1}{2}\mu_n\epsilon\epsilon_0\dfrac{d(F_0^2)}{dx}$ \hfill (19.6)

with $F_0 = 0$ and $V_0 = 0$ at $x = 0$ and $V_0(L) = V_a$.

Ac equation: $J_1 = -\mu_n\epsilon\epsilon_0\dfrac{d(F_0 F_1)}{dx} + j\omega\epsilon\epsilon_0 F_1$ \hfill (19.7)

with $F_1 = 0$ and $V_1 = 0$ at $x = 0$ and $V_1(L) = v_{a1}$.

Integrating (19.6) yields

$$-J_0 x = \frac{1}{2}\mu_n\epsilon\epsilon_0 F_0^2, \qquad \frac{dV_0}{dx} = \left(-\frac{2J_0}{\mu_n\epsilon\epsilon_0}\right)^{1/2}x^{1/2}$$
$$V_0(x) = \frac{2}{3}\left(-\frac{2J_0}{\mu_n\epsilon\epsilon_0}\right)^{1/2}x^{3/2} = V_a\left(\frac{x}{L}\right)^{3/2} \tag{19.8}$$

Hence substituting for $x = L$ and solving for J_0 yields

$$-J_0 = \frac{9}{8}\epsilon\epsilon_0\mu_n\frac{V_a^2}{L^3} \tag{19.9}$$

To integrate (19.7) we substitute $u = (x/L)^{1/2}$. Since $V_0(u) = V_a u^3$, we can express F_0 in terms of u and F_1 in terms of dV_1/du. This yields

$$\frac{d^2V_1}{dx^2} + j\omega\tau\frac{dV_1}{du} = -\frac{6J_1}{g_0}u \tag{19.10}$$

where \qquad $g_0 = -\dfrac{dJ_0}{dV_a} = \dfrac{9}{4}\mu_n\epsilon\epsilon_0\dfrac{V_a}{L^3}, \qquad \tau = \dfrac{4}{3}\dfrac{L^2}{\mu_n V_a}$ \hfill (19.10a)

Here g_0 is the low-frequency ac conductance and τ the transit time. Now $V_1 = 0$ and $dV_1/du = 0$ at $u = 0$ ($x = 0$), and $V_1 = v_{a1}$ at $u = 1$ ($x = L$). It is easily seen by substitution that the solution

$$V_1 = -\left[1 - j\omega\tau u + \frac{1}{2}(j\omega\tau u)^2 - \exp(-j\omega\tau u)\right]\frac{6J_1}{g_0(j\omega\tau)^3} \tag{19.11}$$

satisfies both the equation and the boundary conditions. Putting $u = 1$ yields v_{a1}; hence the device admittance is

$$Y = -\frac{J_1}{v_{a1}} = g + j\omega C = g_0 \frac{\frac{1}{6}(j\omega\tau)^3}{1 - j\omega\tau + \frac{1}{2}(j\omega\tau)^2 - \exp(-j\omega\tau)} \quad (19.12)$$

The conductance g is found to be equal to g_0 at low frequencies, as expected, goes through a minimum of $0.58g_0$ at $\omega\tau \simeq 7$, and oscillates around $\frac{2}{3}g_0$ for large $\omega\tau$. The capacitance changes from $\frac{1}{4}g_0\tau$ for $\omega\tau \ll 1$ to the high-frequency value $\frac{1}{3}g_0\tau$ for $\omega\tau \gg 1$; the latter corresponds to the electrode capacitance per unit area $C_d = \epsilon\epsilon_0/L$.

For small L the frequency response is very good so that the device is useful up to microwave frequencies.

Space-charge-limited triodes have also been reported,† but they have not lived up to expectations.

19.1b. Double-Injection NIP Diodes

We give here only some results obtained by Lampert and Rose.‡ They distinguish three regions of the J, V_a characteristic:

1. The ohmic regime, where

$$-J = e(n_0\mu_n + p_0\mu_n)\frac{V_a}{L} \quad (19.13)$$

It occurs at the lowest anode voltages V_a.

2. The semiconductor regime, where

$$-J = \frac{9}{8}e\mu_n\mu_p\tau \, |n_0 - p_0| \frac{V_a^2}{L^3} \quad (19.14)$$

It occurs at intermediate anode voltages; it may not be present if $n_0 = p_0$.

3. The insulator regime, where

$$-J = \frac{125}{18}\epsilon\epsilon_0\mu_n\mu_p\tau \frac{V_a^3}{L^3} \quad (19.15)$$

It occurs at the highest voltages.

In these equations n_0 and p_0 are the equilibrium carrier concentrations and τ is the predominant lifetime. The transition from one regime to the other occurs approximately at the points where the characteristics cross. The currents are larger than in the single-injection diode because the hole and electron space charge cancel each other to quite an extent.

There are no practical applications for double-injection diodes.

† R. Zuleeg, *Solid State Electronics*, **10**, 449 (1966); R. Zuleeg and P. Knoll, *Proc. IEEE*, **54**, 1196 (1966).

‡ M. Lampert and A. Rose, *Phys. Rev.*, **121**, 26 (1960).

19.2. SWITCHING DEVICES

In this section we discuss the *pnpn* switch, the silicon controlled rectifier, and the double-base diode or unijunction transistor. The first two operate along similar principles, whereas the third is based on conductivity modulation.

19.2a. The PNPN Switch

This is a four-layer device with ohmic contacts to the end *p* and *n* regions. There may also be an ohmic contact to the middle region; when current is flowing through that contact, the device operates as a *controlled rectifier*.

The device has two emitters e_1 and e_2 and two bases b_1 and b_2, and both bases are floating (Fig. 19.1). If a current I flows through the device, then a voltage V_{E1} is developed across the $e_1 b_1$ junction, a voltage V_{E2} is developed across the $b_2 e_2$ junction, and a voltage V_C is developed across the $b_1 b_2$ junction. A hole current I_{pE1} is injected from e_1 to b_1, a hole current I_{pC1} is injected from b_2 to b_1, an electron current I_{nE2} is injected from e_2 to b_2, and an electron current I_{nC2} is injected from b_1 into b_2. The corresponding current amplification factors are α_{f1}, α_{r1}, α_{f2}, and α_{r2}, respectively. Consequently,

$$I = I_{pE1} - \alpha_{r1}I_{pC1} = \alpha_{f1}I_{pE1} - I_{pC1} + \alpha_{f2}I_{nE2} - I_{nC2}$$
$$= I_{nE2} - \alpha_{r2}I_{nC2} \tag{19.16}$$

where
$$I_{pE1} = \frac{ep_{n1}D_pA}{w_1}\left[\exp\left(\frac{eV_{E1}}{kT}\right) - 1\right]$$

$$I_{pC1} = \frac{ep_{n1}D_pA}{w_1}\left[\exp\left(\frac{eV_C}{kT}\right) - 1\right]$$

$$I_{nE2} = \frac{en_{p2}D_nA}{w_2}\left[\exp\left(\frac{eV_{E2}}{kT}\right) - 1\right] \tag{19.16a}$$

$$I_{nC2} = \frac{en_{p2}D_nA}{w_2}\left[\exp\left(\frac{eV_C}{kT}\right) - 1\right]$$

Here the various parameters have their usual meaning, w_1 and w_2 are the widths of the two base regions, and p_{n1} and n_{p2} are the minority carrier concentrations in the two base regions, respectively. Furthermore

$$V = V_{E1} - V_C + V_{E2} \tag{19.17}$$

Eliminating I_{pE1} and I_{pE2} by putting

$$I_{pE1} = I + \alpha_{r1}I_{pC1}, \qquad I_{nE2} = I + \alpha_{r2}I_{nC2} \tag{19.18}$$

and substituting into the middle part of Eq. (19.16), we get an equation between I, I_{pC1}, and I_{nC2}. Solving for I yields

$$I = \frac{1 - \alpha_{r1}\alpha_{f1}}{\alpha_{f1} + \alpha_{f2} - 1}I_{pC1} + \frac{1 - \alpha_{r2}\alpha_{f2}}{\alpha_{f1} + \alpha_{f2} - 1}I_{nC2} \tag{19.19}$$

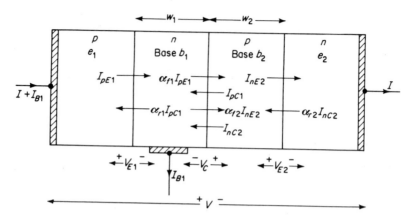

Fig. 19.1. Diagram of a *pnpn* device with the various junction voltages and currents. (Figs. 19.1-19.4 are taken from A. van der Ziel, *Electronics*, Allyn and Bacon, Co., Boston, 1966.)

There are now two possibilities:

1. $\alpha_{f1} + \alpha_{f2} < 1$. Then I_{pC1} and I_{nC2} must be negative if I is positive. This means that V_c must be negative, or that the junction $b_1 b_2$ is back biased. In the case V will be large and will rapidly increase with increasing I (non-conducting region).

2. $\alpha_{f1} + \alpha_{f2} > 1$. Then I_{pC1} and I_{nC2} must be positive for positive I, so that V_c is positive and relatively small; the junction $b_1 b_2$ is then forward biased. Hence V is quite small and increases only slowly with increasing I (conducting region).

For small forward currents in the $e_1 b_1$ and $e_2 b_2$ junctions, $\alpha_{f1} + \alpha_{f2} < 1$ because of recombination in the emitter space-charge regions. In that case I_{pC1} and I_{nC2} must be negative; the $b_1 b_2$ junction is thus back biased, and this back bias increases rapidly with increasing value of I. As a consequence, w_1 and w_2 decrease rapidly, and hence the injected minority currents increase more rapidly than the recombination current; in other words, $\alpha_{f1} + \alpha_{f2}$ increases. If the voltage V_c becomes sufficiently large, this development is materially aided by the onset of avalanche multiplication in the $b_1 b_2$ space-charge region. Since this is equivalent to a rapid increase in α_{f1} and α_{f2}, $\alpha_{f1} + \alpha_{f2}$ will pass the value unity. Then I_{pC1} and I_{nC2} must become positive, which means that the $b_1 b_2$ junction switches from back bias to forward bias. As a consequence, the voltage V drops drastically and the device switches from the nonconducting to the conducting mode of operation.

We shall now show that the device has a very low internal resistance dV/dI in the conducting mode of operation. To simplify matters, we put

$$\alpha_{f1} = \alpha_{f2} = \alpha_f, \qquad \alpha_{r1} = \alpha_{r2} = \alpha_r, \qquad \frac{ep_{n1}D_pA}{w_1} = \frac{en_{p2}D_nA}{w_2} = I_1$$

and we assume that $eV_C/kT \gg 1$, $eV_{E1}/kT \gg 1$, and $eV_{E2}/kT \gg 1$. In that case $I_{nC2} = I_{pC1}$ and hence $I_{pE1} = I_{nE2}$ and $V_{E1} = V_{E2}$, so that

$$I_{pE1} = I_{nE2} = I_1 \exp\left(\frac{eV_{E1}}{kT}\right), \qquad I_{nC2} = I_{pC1} = I_1 \exp\left(\frac{eV_C}{kT}\right) \quad (19.20)$$

Consequently, from (19.19) and (19.18),

$$I_{pC1} = I_{nC2} = \frac{2\alpha_f - 1}{2(1 - \alpha_r\alpha_f)} I$$

$$I_{pE1} = I_{nE2} = \frac{2 - \alpha_r}{2(1 - \alpha_r\alpha_f)} I \qquad (19.21)$$

Applying (19.20) yields

$$V = 2V_{E1} - V_C = \frac{kT}{e} \ln\left[\frac{(2 - \alpha_r)^2}{2(1 - \alpha_r\alpha_f)(2\alpha_f - 1)} \frac{I}{I_1}\right] \qquad (19.22)$$

For very large currents, for which α_r and α_f are almost independent of current, the internal resistance is

$$r = \frac{dV}{dI} \simeq \frac{kT}{eI} \qquad (19.23)$$

At intermediate currents there is a point where $dV/dI \simeq 0$. At lower currents dV/dI is negative; at higher currents dV/dI is positive and finally approaches the value given by (19.23) at large currents (Fig. 19.2).

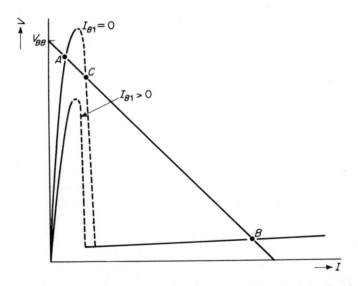

Fig. 19.2. (I, V) characteristic of the device with load line, for $I_{B1} = 0$ and for $I_{B1} > 0$. If the operating point was at A for $I_{B1} = 0$, then it must switch to B for $I_{B1} > 0$.

19.2b. Controlled Rectifier

If the current I_{B1} flows *out of* the base 1 contact, Eq. (19.16) may be written (see Fig. 19.1)

$$I + I_{B1} = I_{pE1} - \alpha_{r1}I_{pC1}$$
$$I = \alpha_{f1}I_{pE1} - I_{pC1} + \alpha_{f2}I_{nE2} - I_{nC2} \qquad (19.24)$$
$$= I_{nE2} - \alpha_{r2}I_{nC2}$$

Solving for I_{pE1} and I_{nE2} yields

$$I_{pE1} = I + I_{B1} + \alpha_{r1}I_{pC1}, \qquad I_{nE2} = I + \alpha_{r2}I_{nC2} \qquad (19.25)$$

Substituting into the second half of Eq. (19.24) yields, after manipulation,

$$-I(1 - \alpha_{f1} - \alpha_{f2}) + \alpha_{f1}I_{B1} = I_{pC1}(1 - \alpha_{f1}\alpha_{r1}) + I_{nC2}(1 - \alpha_{f2}\alpha_{r2}) \quad (19.26)$$

The $b_1 b_2$ junction is now back biased if

$$-I(1 - \alpha_{f1} - \alpha_{f2}) + \alpha_{f1}I_{B1} < 0$$

and forward biased if

$$-I(1 - \alpha_{f1} - \alpha_{f2}) + \alpha_{f1}I_{B1} > 0$$

so that switching occurs if

$$-I(1 - \alpha_{f1} - \alpha_{f2}) + \alpha_{f1}I_{B1} = 0$$

$$(19.27)$$

or

$$\alpha_{f1} + \alpha_{f2} + \alpha_{f1}\frac{I_{B1}}{I} = 1$$

Therefore the switching is easier if $I_{B1} > 0$. Hence it may happen that there are *three* points of intersection (A, C, and B) between the load line and the characteristic for $I_{B1} = 0$, and one point of intersection (B) if $I_{B1} > 0$ (Fig. 19.2). It is easily seen that point C is an unstable operating point. Hence if the operating point is at A for $I_{B1} = 0$, then the operating point will switch to B if I_{B1} is switched on and is sufficiently large. If I_{B1} is then turned off, the operating point stays at point B.

19.2c. Unijunction Transistor or Double-Base Diode

Figure 19.3 shows an n-type semiconducting bar of relatively high resistivity with two ohmic contacts B_1 and B_2 at each end. A p region E is alloyed in on the side, relatively close to the B_1 contact. If the junction thus formed is biased in the forward direction, it will inject holes into the n region; on the n side of the junction a hole density p will thus be maintained that is proportional to the emitter current I_E. If a positive voltage $V_{B_1 B_2}$ is applied between B_1 and B_2, the injected holes are swept down to the B_2 contact. Since space-charge neutrality is maintained at all times, the injected holes density p' changes the electron density from the value n_0 without injection to the new value $n_0 + p'$ in the region between the junction and the B_2 contact.

We first assume the emitter current I_E to be zero. Let the resistance between the B_1 contact and the n side of the junction and the resistance between the n side of the junction and the B_2 contact be R_{10} and R_{20}, respectively. Then the voltage between the emitter E and the contact B_2 is

$$V_{EB_2} = V_{EB_20} = V_{B_1B_2}\frac{R_{20}}{R_{10} + R_{20}}$$

$$\text{(19.28)}$$

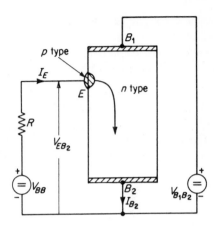

Fig. 19.3. Diagram of a unijunction transistor with bias voltages and current flow.

If the current I_E is made negative, the pn junction will be back biased and V_{EB_2} will decrease rapidly with decreasing I_E. If I_E is now increased to some positive value, the resistance between the B_1 contact and the n side of the junction will not change, but the resistance between the n side of the junction and the B_2 contact will decrease to a much lower value R_2 because of the increased conductivity of that region. Consequently, the voltage V_{EB_2} is now

$$V_{EB_2} = V_{B_1B_2}\frac{R_2}{R_{10} + R_2} + I_E\frac{R_{10}R_2}{R_{10} + R_2} + V_E \qquad \text{(19.29)}$$

where V_E is the relatively small voltage across the pn junction itself. Since R_2 decreases with increasing value of I_E, V_{EB_2} decreases with increasing value of I_E so that the device has a V_{EB_2}, I_E characteristic with a negative resistance region. Thus the device can be used in switching circuits.

We now discuss this problem in more detail, starting from the basic equations for the EB_2 region of the device, which is assumed to have a one-dimensional geometry. These equations are

$$J_p = e\mu_p pF - eD_p\frac{dp}{dx} \qquad \text{(19.30)}$$

$$J_n = e\mu_n nF + eD_n\frac{dn}{dx} \qquad \text{(19.31)}$$

$$0 = g - R - \frac{1}{e}\frac{dJ_p}{dx} \qquad \text{(19.32)}$$

$$0 = g - R + \frac{1}{e}\frac{dJ_n}{dx} \qquad \text{(19.33)}$$

$$\frac{dF}{dx} = -\frac{e}{\epsilon\epsilon_0}(p - p_0 - n + n_0) \qquad \text{(19.34)}$$

where g is the generation rate, R the recombination rate, $N_d = n_0 - p_0$ the donor concentration, and p_0 and n_0 the equilibrium hole and electron concentrations, respectively.

Combining (19.32) and (19.33), we find that

$$J = J_p + J_n \qquad (19.35)$$

is independent of x, where J is the total current density.

To simplify matters we shall assume that there is approximate space-charge neutrality in the EB_2 region. That is, $\partial n/\partial x = \partial p/\partial x$ and $n = p + N_d = p + n_0 - p_0$. Adding (19.30) and (19.31) and solving for F yields

$$F = \frac{J - eD_p(b - 1)\, dp/dx}{e\mu_p[p(b + 1) + b(n_0 - p_0)]} \qquad (19.36)$$

where $b = \mu_n/\mu_p$. This equation must be combined with Eqs. (19.30) and (19.32).

If we now write $J = I/A$ and $J_p = I_p/A$, these equations may be written

$$\frac{dI_p}{dx} = e(g - R)A \qquad (19.37)$$

Solving (19.30) for F and equating the result to (19.36), we get

$$\frac{dp}{dx} = \frac{pI - I_p[p(b + 1) + b(n_0 - p_0)]}{ebD_pA(2p + n_0 - p_0)} \qquad (19.38)$$

Since $g - R$ is a known function of p, this set of two simultaneous differential equations can be solved numerically on a computer if the initial conditions are known. Since the hole current is negligible in the EB_1 region, if the material is not too close to being intrinsic, we have $I_p(0) = I_E$. If the contact B_2 is a perfect sink for holes, we have $p(L_2) = p_0$, where L_2 is the distance between the emitter and the contact B_2.

To calculate the V_{EB}, I_E characteristic, we must know R_2 and V_E. Now $V_E = (kT/e)\ln[p(0)/p_0]$, and $p(0)$ is known. Moreover,

$$R_2 = \int_0^{L_2} \frac{dx}{e\mu_p A[p(b + 1) + (n_0 - p_0)b]} \qquad (19.39)$$

can be calculated. The schematic characteristic is shown in Fig. 19.4.

19.3. IMPATT DIODE OR READ OSCILLATOR

19.3a. Theory

A diode biased in the back direction so that it operates in the avalanche region can generate microwave frequencies. The diode in this mode of operation is called a Read oscillator, named after its inventor, W. T. Read.†

The operation of the device can be seen from Fig. 19.5a. It shows the

---

† W. T. Read, Jr., *Bell System Tech. J.*, **37**, 401 (1958).

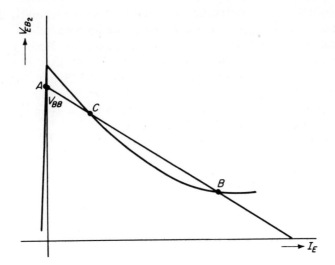

Fig. 19.4. (V_{EB2}, I_E) characteristic of a unijunction transistor with load line. The stable operating points are at A and B.

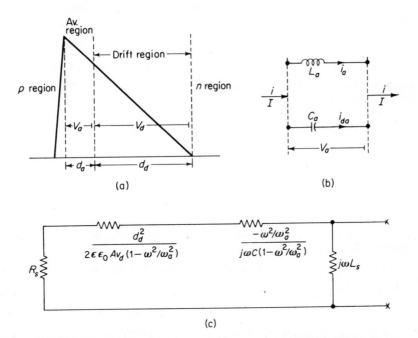

Fig. 19.5. (a) Space charge region of an avalanche diode with the avalanche and drift regions. (b) Current flow in the avalanche region. (c) Equivalent circuit for small θ.

field distribution in an avalanching p^+n diode. There is a relatively thin part of the space-charge region, on the n side of the junction, where avalanche multiplication occurs. In the remainder of the space-charge region the carriers drift but do not multiply. Because of the high fields involved, this drift will occur with the limiting drift velocity v_d for most of the space-charge region. To simplify our discussion we shall assume that the electrons and holes have the same ionizing power and that they move with the same limiting drift velocity v_d. We follow here Gilden and Hines.[†]

If p and n are the hole and electron densities, then the corresponding current densities are $J_p = -ev_d p$ and $J_n = -ev_d n$. Hole–electron pairs are being generated at the rate $\alpha v_d(n + p)$; we assume that the thermal generation is negligible. The equations of continuity are

$$\frac{\partial p}{\partial t} = -\frac{1}{e}\frac{\partial J_p}{\partial x} + \alpha v_d(n + p) \tag{19.40}$$

$$\frac{\partial n}{\partial t} = \frac{1}{e}\frac{\partial J_n}{\partial x} + \alpha v_d(n + p) \tag{19.41}$$

where α is the ionization rate.

We now add these two equations, put $J_p + J_n = J_a(t)$, eliminate p and n with the help of the definitions for J_p and J_n, and integrate with respect to x over the avalanche region. This yields, if $\tau_a = d_a/v_d$ is the transit time,

$$\tau_a \frac{dI_a}{dt} = +(I_p - I_n)\Big|_0^{d_a} + 2I_a \int_0^{d_a} \alpha \, dx \tag{19.42}$$

where we have switched from current densities to currents. If we now take into account that we are deep into the avalanche region, so that the junction saturation currents can be neglected, we have $I_p = I_a$ and $I_n = 0$ at $x = 0$, and $I_p = 0$ and $I_n = I_a$ at $x = d_a$. Hence Eq. (19.42) may be written

$$\frac{dI_a}{dt} = \frac{2I_a}{\tau_a}(\bar{\alpha}d_a - 1) \tag{19.43}$$

where

$$\bar{\alpha} = \frac{1}{d_a} \int_0^{d_a} \alpha dx \tag{19.43a}$$

is the average value of α over the avalanche region. For steady-state conditions, $\bar{\alpha}d_a$ is exactly unity.

We now introduce small-signal deviations. That is, we put

$$F = F_0 + F_a, \qquad I_a = I_0 + i_a \tag{19.44}$$

$$\bar{\alpha} = \bar{\alpha}_a + \bar{\alpha}_a' F_a, \qquad \bar{\alpha}d_a = 1 + d_a\bar{\alpha}_a' F_a \tag{19.45}$$

where $\bar{\alpha}_a'$ is the derivative of $\bar{\alpha}$ with respect to F. Taking into account that

---

† M. Gilden and M. E. Hines, *Trans. IEEE*, **ED-13**, 169 (1966).

F_a and i_a are small-signal quantities, substituting into (19.43), and neglecting second-order terms, we get for the avalanche conduction current, since $d/dt = j\omega$ if the signals vary as $\exp(j\omega t)$,

$$i_a = \frac{2\bar{\alpha}'_a \, d_a I_0 F_a}{j\omega\tau_a} = \frac{2\bar{\alpha}'_a I_0}{j\omega\tau_a} v_a \tag{19.46}$$

The displacement current in the avalanche zone is

$$i_{da} = j\omega\epsilon\epsilon_0 A F_a = j\omega\frac{\epsilon\epsilon_0 A}{d_a} v_a \tag{19.47}$$

where $v_a = F_a d_a$ is the ac voltage across the avalanche region and A the junction area (Fig. 19.5b). Consequently, the total current i is

$$i = i_a + i_{da} = \left(\frac{1}{j\omega L_a} + j\omega C_a\right) v_a \tag{19.48}$$

where
$$L_a = \frac{d_a/v_d}{2\bar{\alpha}'_a I_0}, \qquad C_a = \frac{\epsilon\epsilon_0 A}{d_a} \tag{19.48a}$$

This is the equation of a parallel tuned circuit, tuned at the angular frequency ω_a:

$$\omega_a = \frac{1}{\sqrt{L_a C_a}} = \sqrt{\frac{2\bar{\alpha}'_a v_d I_0}{\epsilon\epsilon_0 A}} \tag{19.48b}$$

We see that this frequency is independent of the length d_a. This is not so surprising, for L_a and C_a are *distributed* parameters, and the resonant frequency is a property of each small section of the avalanche region. This, in turn, is very convenient, for the width of the avalanche region is not too well defined.

We now introduce the impedance Z_a of the avalanche region as

$$Z_a = \frac{v_a}{i} = \left(\frac{1}{j\omega L_a} + j\omega C_a\right)^{-1} = \frac{1}{j\omega C_a}\frac{-\omega^2/\omega_a^2}{1 - \omega^2/\omega_a^2} \tag{19.49}$$

and introduce a parameter M, defined as

$$M = \frac{i_a}{i} = (1 - \omega^2 L_a C_a)^{-1} = \left(1 - \frac{\omega^2}{\omega_a^2}\right)^{-1} \tag{19.50}$$

The convection current i_a enters the drift region and propagates as an unattenuated wave with drift velocity v_d. Hence in the drift region, since the convection current is continuous at the boundary,

$$i_c(x) = i_a \exp\left(-\frac{j\omega x}{v_d}\right) = iM \exp\left(-\frac{j\omega x}{v_d}\right) \tag{19.51}$$

But since the displacement current is

$$i_d(x) = j\omega\epsilon\epsilon_0 A F_a(x) \tag{19.52}$$

and $i = i_c(x) + i_d(x)$ is independent of x, we have for the ac field strength in the drift region

$$F_a(x) = i\frac{1 - M\exp{(-j\omega x/v_d)}}{j\omega\epsilon\epsilon_0 A} \tag{19.53}$$

Integrating this expression over the drift length d_d yields for the ac voltage v_{dr} across the drift region

$$v_{dr} = \frac{d_d i}{j\omega\epsilon\epsilon_0 A}\left\{1 - \frac{1}{1 - \omega^2/\omega_a^2}\left[\frac{1 - \exp{(-j\theta)}}{j\theta}\right]\right\} \tag{19.54}$$

where $\theta = \omega d_d/v_d = \omega\tau_d$ is the transit angle and τ_d is the transit time through the drift region.

Consequently, if we take into account that $C_d = \epsilon\epsilon_0 A/d_d$ is the capacitance of the drift region, the impedance of the drift region is

$$\begin{aligned}
Z_d = \frac{v_{dr}}{i} &= \frac{1}{j\omega C_d}\left[1 - \frac{(\sin\theta)/\theta}{1 - \omega^2/\omega_a^2}\right] + \frac{1}{\omega C_d}\frac{(1 - \cos\theta)/\theta}{1 - \omega^2/\omega_a^2} \\
&= \frac{1}{j\omega C_d}\left[1 - \frac{(\sin\theta)/\theta}{1 - \omega^2/\omega_a^2}\right] + \frac{d_d^2}{2\epsilon\epsilon_0 Av_d}\frac{2(1 - \cos\theta)/\theta^2}{1 - \omega^2/\omega_a^2}
\end{aligned} \tag{19.55}$$

If we now bear in mind that the remaining part of the diode has a series resistance R_s, the total impedance of the device may be written

$$\begin{aligned}
Z_{tot} &= R_s + Z_d + Z_a \\
&= R_s + \frac{d_d^2}{2\epsilon\epsilon_0 Av_d}\left[\frac{2(1 - \cos\theta)/\theta^2}{1 - \omega^2/\omega_a^2}\right] \\
&\quad + \frac{1}{j\omega C_d}\frac{1 - \sin\theta/\theta - (\omega^2/\omega_a^2)(1 + d_a/d_d)}{1 - \omega^2/\omega_a^2}
\end{aligned} \tag{19.56}$$

as we find after some manipulations.

We now assume θ to be relatively small. In that case $(\sin\theta)/\theta \simeq 1$ and $2(1 - \cos\theta)/\theta^2 \simeq 1$ (for $\theta < \pi/4$). Equation (19.56) may then be written

$$Z \simeq R_s + \frac{d_d^2}{2\epsilon\epsilon_0 Av_d}\left(\frac{1}{1 - \omega^2/\omega_a^2}\right) + \frac{1}{j\omega C}\frac{-\omega^2/\omega_a^2}{1 - \omega^2/\omega_a^2} \tag{19.56a}$$

where $C = \epsilon\epsilon_0 A/(d_d + d_a)$ is the net capacitance of the space-charge region if the diode is biased just below avalanche breakdown. The second term of (19.56a) is an active resistance that is negative for $\omega > \omega_a$. The third term is a parallel resonant circuit, which includes the diode capacitance C and a shunt inductor. To make the diode oscillate, we must add a parallel inductance L_s to the equivalent circuit to tune it (Fig. 19.5c).

It is interesting to note that strong negative-resistance effects occur for relatively small transit angles. These negative-resistance effects explain why the diode will oscillate for frequencies $\omega > \omega_a$.

From the preceding discussion it is understandable that avalanche diodes will oscillate at microwave frequencies and that the oscillator is current-tunable and tunable by external means.

Perhaps the most accurate theoretical data have been obtained by Scharfetter and Gummel† by computer techniques. Their curves, plotting the microwave admittance of impatt diodes in the complex plane, show a reasonable resemblance to what would be expected from Eq. (19.56a), but their results are obtained with less assumptions and are thus closer to the actual situation. They have also been able to evaluate efficiencies.

In earlier versions of the impatt diode oscillator the efficiency was of the order of 5–10 per cent. By introducing properly shaped doping profiles the efficiency can be raised to 25 per cent. Since the power output of the device is limited by the dc power dissipation, it is obvious that much higher output powers can be obtained under pulsed conditions than under continuous-wave conditions.

The output power under continuous-wave conditions is of the order of 5–10 W at the X band. At higher frequencies the power decreases as $1/f^2$, as with most microwave oscillators. Oscillations above 100 GHz have been obtained.

19.3b. Trapatt Oscillator

The trapatt oscillator is a form of avalanche diode oscillator operating in a particular mode, the trapped plasma mode; hence the name. The interesting feature of this oscillator is that the current is high when the voltage is low, and vice versa, so that a very efficient conversion of direct current to radio-frequency energy, with up to 60 per cent efficiency, is possible.

The condition occurs if an impatt diode is operated in a high-Q circuit, where large radio-frequency voltages can develop. The large overdrive in voltage, up to about twice the breakdown voltage, causes the generation of enough charge to reduce the field to near zero at the boundary between the p and n regions. The field ahead of the carriers is still above the critical field, and avalanche generation begins to occur there. Thus an avalanche zone is created that travels through the diode, leaving the diode filled with a plasma of electrons and holes at a nearly zero electric field. The resulting low-field condition traps the plasma; because of the low field the voltage is small. The plasma is extracted at the boundary by space-charge limited-flow. A large terminal current can be maintained; this removes the plasma from the diode, and the diode voltage oscillates around the breakdown value and reduces the current to a relatively low value. After this the cycle repeats itself.

The rise in voltage, far in excess of the breakdown voltage, at the beginning of the cycle is possible because of the inherent delay in the avalanche process. After the diode has collapsed to a near zero voltage state, the diode

---

† D. L. Scharfetter and H. K. Gummel, *Trans. IEEE* **ED-16**, 64, (1969).

recovers. The recovery time depends on the density of the plasma and the magnitude of the recovery current; typically, the time needed is several times the transit time of the carriers moving at the saturated drift velocity.

A simple circuit for the trapatt mode of oscillations consists of a transmission line with the diode mounted at the one end and a low-pass filter at the other end. The low-pass filter is placed approximately one half-wavelength away from the diode and presents a short-circuit to all the harmonics of the trapatt oscillation frequency. The effect of these harmonics is important for the operation of the trapatt mode.

19.4. GUNN-EFFECT OSCILLATORS

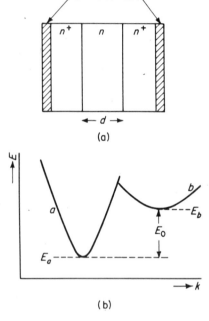

Fig. 19.6. (a) Structure used in Gunn effect devices. (b) Energy band structure in materials suitable for Gunn effect devices.

The Gunn-effect oscillator is based on negative conductivity effects in bulk semiconductors that have two conduction-band minima separated by an energy gap. Since it is a bulk semiconductor effect, a two-terminal negative conductance is present at microwave frequencies in *subcritical* devices. In *supercritical* devices, however, high-field domains can be formed that propagate with the drift velocity of the electrons. As a consequence a periodic waveform of a frequency corresponding to the drift time of the electrons through the material is generated. The device usually consists of an n^+nn^+ structure to which ohmic contacts are made. The active region is the n region between the two n^+ regions (Fig. 19.6a). One of the most suitable materials for this effect is GaAs.

The effect was demonstrated by Gunn,[†] hence its name; but it was predicted earlier by Ridley and Watkins. Ridley also predicted the domain structure.[‡]

[†] J. B. Gunn, *Solid State Communications*, **1**, 88 (1963).

[‡] B. K. Ridley and T. B. Watkins, *Proc. Phys. Soc.*, **78**, 291 (1961); B. K. Ridley, *Proc. Phys. Soc.*, **82**, 954 (1963).

19.4a. Negative Conductivity Effect

Consider a conduction band consisting of two subbands as pictured in the (E, k) diagram of Fig. 19.6b. The lower band is denoted by "a" and the upper by "b." The energies of the minima are E_a and E_b, and their difference will be denoted by E_0. Their effective density states are N_a and N_b, their effective masses m_a^* and m_b^*, their mobilities μ_a and μ_b, and their electron densities n_a and n_b, respectively. We shall assume that $\mu_a \gg \mu_b$. In GaAs, for example, $\mu_a/\mu_b \simeq 60$.

At low fields almost all electrons are in the lower subband a, but at high fields the electrons become "hotter" and the upper subband b becomes more and more populated. As a consequence the conductivity g decreases with increasing field strength, because $\mu_b \ll \mu_a$. The current density

$$J = gF \tag{19.57}$$

and

$$\frac{dJ}{dF} = g + F\frac{dg}{dF} \tag{19.58}$$

so that the condition for a negative conductivity is

$$-F\frac{dg}{dF} > g \tag{19.58a}$$

Since g decreases with increasing field strength at high fields, there is a distinct possibility that this will happen.

To find out whether it happens indeed, we calculate dg/dF. Since

$$g = e(\mu_a n_a + \mu_b n_b) \tag{19.59}$$

$$\frac{dg}{dF} = e\left(\mu_a \frac{dn_a}{dF} + \mu_b \frac{dn_b}{dF}\right) + e\left(n_a \frac{d\mu_a}{dF} + n_b \frac{d\mu_b}{dF}\right) \tag{19.60}$$

or, putting $n_a + n_b = n$, where n is a constant, and assuming that μ_a and μ_b are proportional to F^{-p}, we have

$$\frac{dg}{dF} = e(\mu_a - \mu_b)\frac{dn_a}{dF} - e(\mu_a n_a + \mu_b n_b)\frac{p}{F}$$

$$\simeq e\mu_a \frac{dn_a}{dF} - e\mu_a n_a \frac{p}{F} \tag{19.61}$$

if $\mu_a \gg \mu_b$ and $n_a\mu_a \gg n_b\mu_b$.

Substituting into (19.58a) yields

$$-\frac{F}{n_a}\frac{dn_a}{dF} + p > 1 \tag{19.62}$$

Since dn_a/dF is negative, this condition can be easily satisfied, especially if $p \neq 0$, which is true at high fields. There is often a region where $p \simeq \frac{1}{2}$ or $\mu_a = \mu_0(F_0/F)^{1/2}$, and a region where $p \simeq 1$ or $\mu_a = u_c/F$, where u_c is the limiting velocity of the carriers. Especially in the latter case the slightest negative value of dn_a/dF will satisfy the inequality (19.62). Ridley and Watkins have shown in a more detailed calculation that this is indeed the case.

We conclude therefore that a band structure of the type shown in Fig. 19.6b can indeed give negative conductance effects at high field strengths. For a suitable operation the band separation should be a few tenths of a volt. If it is much larger, the b band remains empty; if it is much lower, the b band is already partly populated at low field strengths.

19.4b. Domain Formation

It will now be shown that in the negative conductivity regime the uniform field distribution is not stable. Space-charge accumulation layers will be formed, which tend to lower the field on the cathode side and raise the field on the anode side of the disturbance (domain formation); these charge accumulation layers tend to grow and decay, and they propagate with the drift velocity of the carriers.

To understand this growth-and-decay process, we have the following equations. If ρ_k is the particle density,

$$\frac{\partial F}{\partial x} = -\frac{e}{\epsilon\epsilon_0} \sum_k \rho_k \qquad (19.63)$$

where $k = 1$ refers to the fast (= high mobility) electrons and $k = 2$ to the slow (= low mobility) electrons. Furthermore, there is continuity for each of the components. That is, if I_k is the *particle* current density,

$$\frac{d\rho_k}{dt} = -\frac{\partial I_k}{\partial x} + \frac{\partial \rho_k}{\partial t} \qquad (19.64)$$

Moreover, since no new electrons are created but fast electrons are promoted to slow electrons, and vice versa, we have

$$\sum_k \frac{\partial \rho_k}{\partial t} = 0 \qquad (19.65)$$

Finally, the current density is

$$J = -\sum_k eI_k \qquad (19.66)$$

Differentiating (19.63) with respect to time (total differential!) gives

$$\frac{\partial}{\partial x}\dot{F} = -\frac{e}{\epsilon\epsilon_0} \sum_k \frac{d\rho_k}{dt} = \frac{e}{\epsilon\epsilon_0} \sum_k \frac{\partial I_k}{\partial x} - \frac{e}{\epsilon\epsilon_0} \sum_k \frac{\partial \rho_k}{\partial t}$$
$$= -\frac{1}{\epsilon\epsilon_0} \frac{\partial J}{\partial x} \qquad (19.67)$$

because of Eqs. (19.64)–(19.66).

We now assume that we start with a small space-charge accumulation of width Δx and we investigate what happens to this space-charge layer. To that end we integrate Eq. (19.67) over the layer. This yields

$$\Delta\dot{F} = \dot{F}_2 - \dot{F}_1 = -\frac{1}{\epsilon\epsilon_0}(J_2 - J_1) = -\frac{1}{\epsilon\epsilon_0}\frac{dJ}{dF}\Delta F = -\frac{\Delta F}{\tau} \quad (19.68)$$

which has the solution

$$\Delta F = \Delta F_0 \exp\left(-\frac{t}{\tau}\right) \quad (19.69)$$

Here τ, the differential relaxation time, is given by

$$\frac{1}{\tau} = \frac{1}{\epsilon\epsilon_0}\frac{dJ}{dF} \quad (19.69a)$$

It is positive when dJ/dF is positive, and hence under positive conductivity conditions the dipole-layer disturbances will tend to die out exponentially. It is negative when dJ/dF is negative, and hence under negative conductivity conditions the accumulation-layer disturbances will tend to grow exponentially.

This is as far as the "small-signal" theory can bring us. But it is sufficient to show that in the negative conductivity mode of operation a uniform field distribution is unstable and that space-charge accumulation layers will be formed at the cathode that can grow and propagate. The growth does not go unchecked, however, for the total voltage across the sample remains constant. Any increase in field strength at the anode side of the disturbance is accompanied by a decrease in field strength on the cathode side. Finally, the field strength at the cathode becomes so small that no new excitations of high-mobility electrons to low-mobility electrons can take place. The disturbance then begins to die out until the equilibrium position is restored. This happens at the moment that the accumulation layer reaches the anode. The process then starts over again, a new disturbance starts at the cathode, grows and propagates, and the cycle repeats itself (Fig. 19.7).

Actually, most samples contain inhomogeneities. At such inhomogeneities, dipole layers may start, which then grow in the same manner as the space-charge layers discussed earlier. In some samples a space-charge layer starts at the cathode and then changes into a dipole layer at an inhomogeneity; in other samples the dipole layer starts directly at the inhomogeneity.[†]

Because of the existence of these "running domains" the terminal dc (I, V) characteristic of the device does not show a negative conductance region. It looks more like a current increasing with increasing V, saturating, and then breaking into oscillations.

Since the period T of the periodic waveform is approximately equal to the transit time $\tau = d/v$ of the carriers, where d is the length of the active region

---

[†]The growth time of the domain is usually short in comparison with the transit time $\tau = d/v$ of these domains across the sample.

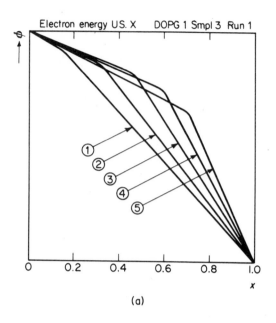

(a)

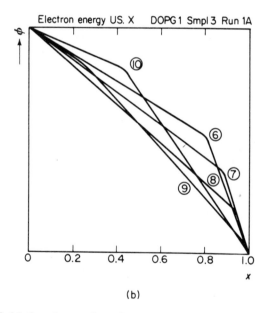

(b)

Fig. 19.7. Motion of space-charge layers in Gunn diodes. (a) Initial formation of a pure accumulation layer. (b) Disappearance and renucleation of pure accumulation layer.
(From H. Kroemer, *IEEE Transactions, ED-13*, 27, 1966.) Circled numbers denote increments in time.

and v the drift velocity of the carriers, the frequency of the oscillation would be expected to be

$$f \simeq \frac{1}{\tau} = \frac{v}{d} \qquad (19.70)$$

For devices with an active region of 100–10 μm, this frequency lies in the 1–10 GHz range. Gunn-effect oscillators thus appear to be attractive sources of microwave power. Continuous-wave operation with powers above 100 μW and pulsed operation with powers higher than 100 W are easily obtainable with efficiencies as high as several per cent.

19.4c. Negative Conductance Oscillations[†]

There is another mode of operation of Gunn diodes in which the frequency is determined by the external tuned circuit rather than by the transit time. In this mode of operation the device is connected in parallel with a tuned circuit. If the circuit is oscillating and the voltage swing across the tuned circuit is large enough to reduce the voltage across the device below the threshold value for negative conductance during part of each cycle, and the ratio of frequency to carrier concentration is within a certain range (see below), then the oscillations are maintained. This mode of operation is called the limited-space-charge accumulation (LSA) mode.

When the sample is oscillating in the LSA mode, the electric field across the diode rises from below the threshold value to a value more than twice the threshold value so quickly that the space-charge distribution associated with a high field domain has no time to form. If the period of the oscillation frequency is shorter than, or not more than several times greater than, the negative dielectric relaxation time, additional space-charge distribution will not build up appreciably before the field across the diode decreases again into the positive conductance range. This may be written as

$$\frac{1}{f} \lesssim \frac{3\epsilon\epsilon_0}{en|\mu_n|} \qquad (19.71)$$

where ϵ is the dielectric constant, μ_n the average negative mobility,[‡] and e the electron charge.

In order for the accumulation layer to disappear during each cycle, the dielectric relaxation time when the field is below threshold must be short compared to the fraction of a period when the field in the diode is below threshold. This may be written as

$$\frac{1}{f} \gg \frac{\epsilon\epsilon_0}{en\mu_0} \qquad (19.72)$$

[†] J. A. Copeland, *Proc. IEEE*, **54** 1479 (1966).

[‡] The negative conductivity can be described by a negative average mobility; $|\mu_n|$ is much smaller than the low-field mobility μ_0.

Substituting $\epsilon = 12.4$, $\mu_0 = 5000 \text{ cm}^2/\text{V s}$, and $|\mu_n| = 100 \text{ cm}^2/\text{V s}$ gives

$$2 \times 10^5 \gtrsim \frac{n}{f} \gg 1.4 \times 10^3 \qquad \text{s/cm}^3$$

In practice one finds the LSA mode of oscillation for

$$2 \times 10^5 > \frac{n}{f} > 10^4$$

Frequencies as high as 88 GHz have been obtained, with efficiencies up to about 2 per cent for continuous-wave operation and higher efficiencies for pulsed operation.

REFERENCES

BOTT, I. B., and W. FAWCETT, "The Gunn Effect in Gallium Arsenide, in L. Yung, ed., *Advances in Microwaves*, vol. 3. Academic Press, Inc., 1968, pp. 223–300.

GENTRY, F. E., F. W. GUTZWILLER, N. HOLONYAK, Jr., and E. E. VON ZASTROW, *Semiconductor Controlled Rectifiers*. Prentice-Hall, Inc., Englewood Cliffs, N.J., 1964.

LAMPERT, M., and P. MARK, *Current Injection in Solids*. Academic Press, Inc., New York, 1970.

SZE, S. M., *Physics of Semiconductor Devices*. John Wiley & Sons, Inc. (Interscience Division), New York, 1969.

PROBLEMS

1. Evaluate the characteristic of a space-charge-limited diode under the condition that $\mu = \mu_0 (F_0/F)^{1/2}$.

Answer: $J_a = \frac{10}{9}(\frac{5}{3})^{1/2} \epsilon\epsilon_0 \mu_0 F_0^{1/2} V_a^{3/2}/d^{5/2}$.

2. Evaluate the characteristic of a space-charge-limited diode under the condition that $\mu' = u_c/F$ or $u_d = \mu F = u_c$, where u_c is the limiting velocity of carriers at high fields.

Answer: $J_a = 2\epsilon\epsilon_0 u_c V_a/d^2$.

3. Plot the real part g/g_0 of Y/g_0, as given in Eq. (19.12), as a function of $\omega\tau$.

20

Miscellaneous
Semiconductor Problems

20.1. HALL EFFECT AND MAGNETORESISTANCE

20.1a. Hall Effect

In a semiconducting bar of length L and a width w_y in the Y direction, let an electric field $\mathbf{i}F_x$ be applied in the X direction and a static magnetic field $\mathbf{k}B_z$ in the Z direction; then a field $\mathbf{j}F_y$ is found to be developed in the Y direction. This effect is called the *Hall effect*. We now calculate the magnitude of the effect.

For an n-type semiconductor the current density in the X direction is

$$J_{nx} = -enu_{dn} = en\mu_n F_x \tag{20.1}$$

where $u_{dn} = -\mu_n F_x$ is the drift velocity of the electrons. The magnetic field gives an average force $-e\,\mathbf{v} \times \mathbf{B} = \mathbf{j}eu_{dn}B_z$ in the Y direction. Hence the net force in the Y direction is

$$-eF_y + eu_{dn}B_z = -e(F_y + \mu_n B_z F_x) \tag{20.2}$$

and the current density in the Y direction is

$$J_{ny} = e\mu_n n(F_y + \mu_n B_z F_x) \tag{20.3}$$

Since no net current is flowing in the Y direction, $J_{ny} = 0$, or

$$F_y = -\mu_n B_z F_x = RJ_{nx}B_z \quad \text{or} \quad R = -\frac{1}{en} \tag{20.4}$$

because of (20.1). The factor R is called the Hall constant or the *Hall coefficient.*

For a p-type semiconductor the current density in the X direction is

$$J_{px} = epu_{dp} = ep\mu_p F_x \qquad (20.1a)$$

where $u_{dp} = \mu_p F_x$ is the drift velocity of the holes. The magnetic field gives an average force $e\mathbf{v} \times \mathbf{B} = -j e u_{dp} B_z$ in the Y direction. The total net force in the Y direction is

$$eF_y - eu_{dp}B_z = e(F_y - \mu_p B_z F_x) \qquad (20.2a)$$

and hence the current density in the Y direction is

$$J_{py} = e\mu_p p(F_y - \mu_p B_z F_x) \qquad (20.3a)$$

Since no net current is flowing in the Y direction, $J_{py} = 0$, or

$$F_y = \mu_p B_z F_x = RJ_{px}B_z \quad \text{or} \quad R = \frac{1}{ep} \qquad (20.4a)$$

because of (20.1a).

If both electrons and holes are present, the current density in the X direction is

$$J_x = J_{nx} + J_{px} = e(n\mu_n + p\mu_p)F_x \qquad (20.1b)$$

and the current density in the Y direction is

$$J_y = J_{ny} + J_{py} = e\mu_n n F_y + e\mu_p p F_y + e\mu_n^2 n B_z F_x - e\mu_p^2 p B_z F_x \qquad (20.3b)$$

Since no net current is flowing, $J_y = 0$, or

$$F_y = \frac{\mu_p^2 p - \mu_n^2 n}{\mu_n n + \mu_p p} B_z F_x = RB_z J_x \quad \text{or} \quad R = \frac{1}{e}\frac{\mu_p^2 p - \mu_n^2 n}{(\mu_n n + \mu_p p)^2} \qquad (20.4b)$$

because of (20.1b). This expression reduces to (20.4) for $p = 0$ (n-type semiconductor) and to (20.4a) for $n = 0$ (p-type semiconductor). For intrinsic material $n = p = n_i$ and

$$R = R_i = \frac{1}{en_i}\frac{\mu_p - \mu_n}{\mu_p + \mu_n} \qquad (20.4c)$$

We can easily measure the field F_y by determining the potential difference

$$V_y = -F_y w_y \qquad (20.5)$$

between two points on the opposite faces perpendicular to the Y direction; these points must be so chosen that they have zero potential difference without a magnetic field.

Equations (20.4), (20.4a), and (20.4b), although correct as far as order of magnitude is concerned, are somewhat inaccurate. To find more accurate expressions, we need a more careful averaging of the magnetic force. *We shall see that for semiconductors Eqs.* (20.4), (20.4a), (20.4b), and (20.4c) *must be multiplied by the factor* $3\pi/8$ *if lattice scattering predominates.* To derive this factor, we must solve the Boltzmann transport equation.

20.1b. Derivation of the Basic Equations from the Boltzmann Transport Equation

For n-type material the Boltzmann transport equation is

$$-\mathbf{V}_p f \cdot \left(-e\mathbf{F} - \frac{e}{m^*}\mathbf{p} \times \mathbf{B} \right) - \frac{f - f_0}{\tau} = 0 \qquad (20.6)$$

or, if $\omega_0 = eB_z/m^*$,

$$\frac{f - f_0}{\tau} = (eF_x + \omega_0 p_y)\frac{\partial f}{\partial p_x} + (eF_y - \omega_0 p_x)\frac{\partial f}{\partial p_y} \qquad (20.7)$$

We now try as a solution

$$f = f_0 + A(p)p_x + B(p)p_y \qquad (20.8)$$

where $A(p)$ and $B(p)$, which are functions of $p = (p_x^2 + p_y^2 + p_z^2)^{1/2}$ only, are small, first-order terms in F_x and F_y. Since

$$\frac{\partial f_0}{\partial p_x} = -f_0\frac{p_x}{m^*kT}, \qquad \frac{\partial f_0}{\partial p_y} = -f_0\frac{p_y}{m^*kT}$$

$$\frac{\partial}{\partial p_x} = \frac{p_x}{p}\frac{d}{dp}, \qquad \frac{\partial}{\partial p_y} = \frac{p_y}{p}\frac{d}{dp} \qquad (20.9)$$

by substituting (20.8) into (20.7) and after manipulation, we have

$$\frac{A}{\tau}p_x + \frac{B}{\tau}p_y = \left(-\frac{eF_x}{m^*kT}f_0 - \omega_0 B \right)p_x + \left(-\frac{eF_y}{m^*kT}f_0 + \omega_0 A \right)p_y \qquad (20.10)$$

where the second-order terms

$$eF_x\left(A + \frac{p_x^2}{p}\frac{dA}{dp} + \frac{p_x p_y}{p}\frac{dB}{dp} \right) \quad \text{and} \quad eF_y\left(B + \frac{p_x p_y}{p}\frac{dA}{dp} + \frac{p_y^2}{p}\frac{dB}{dp} \right)$$

have been neglected. Since Eq. (20.10) must be true for all p_x and p_y,

$$\frac{A}{\tau} = \frac{eF_x}{m^*kT}f_0 - \omega_0 B \quad \text{or} \quad A + \omega_0\tau B = -\frac{eF_x\tau}{m^*kT}f_0$$

$$\frac{B}{\tau} = -\frac{eF_y}{m^*kT}f_0 + \omega_0 A \quad \text{or} \quad -\omega_0\tau A + B = -\frac{eF_y\tau}{m^*kT}f_0 \qquad (20.11)$$

Solving for A and B yields

$$A = -\frac{eF_x\tau - eF_y\omega_0\tau^2}{1 + \omega_0^2\tau^2}\frac{f_0}{m^*kT}$$

$$B = -\frac{eF_x\omega_0\tau^2 + eF_y\tau}{1 + \omega_0^2\tau^2}\frac{f_0}{m^*kT} \qquad (20.11a)$$

so that A and B are indeed first-order terms in F_x and F_y.

For the current densities we thus have

$$J_x = \frac{-e}{m^*} \int_{-\infty}^{\infty} \int_{-\infty}^{\infty} \int_{-\infty}^{\infty} p_x f \, dp_x \, dp_y \, dp_z,$$

$$J_y = \frac{-e}{m^*} \int_{-\infty}^{\infty} \int_{-\infty}^{\infty} \int_{-\infty}^{\infty} p_y f \, dp_x \, dp_y \, dp_z \tag{20.12}$$

where f is given by (20.8) and A and B are given by (20.11a). To evaluate these expressions, we bear in mind that the averages $\langle p_x \rangle = \langle p_y \rangle = 0$ and that

$$\langle p_x p_y \rangle = \int_{-\infty}^{\infty} \int_{-\infty}^{\infty} \int_{-\infty}^{\infty} \frac{1}{n} p_x p_y f_0 \, dp_x \, dp_y \, dp_z = 0$$

Consequently, the expressions are simplified greatly and become

$$J_x = \frac{e}{m^{*2}kT} \int_{-\infty}^{\infty} \int_{-\infty}^{\infty} \int_{-\infty}^{\infty} \left(\frac{eF_x\tau - eF_y\omega_0\tau^2}{1 + \omega_0^2\tau^2} \right) p_x^2 f_0 \, dp_x \, dp_y \, dp_z \tag{20.13}$$

$$J_y = \frac{e}{m^{*2}kT} \int_{-\infty}^{\infty} \int_{-\infty}^{\infty} \int_{-\infty}^{\infty} \left(\frac{eF_x\omega_0\tau^2 + eF_y\tau}{1 + \omega_0^2\tau^2} \right) p_y^2 f_0 \, dp_x \, dp_y \, dp_z \tag{20.14}$$

We now introduce polar coordinates (p, ϑ, φ) in the (p_x, p_y, p_z) space. Then $p_x = p \sin \vartheta \cos \varphi$, $p_y = p \sin \vartheta \sin \varphi$, and $dp_x \, dp_y \, dp_z$ is replaced by $p^2 \, dp \sin \vartheta \, d\vartheta \, d\varphi$. Bearing in mind that

$$\int_0^\pi \sin^3 \vartheta \, d\vartheta = \tfrac{4}{3}, \qquad \int_0^{2\pi} \cos^2 \varphi \, d\varphi = \int_0^{2\pi} \sin^2 \varphi \, d\varphi = \pi$$

we obtain,† if $\langle \ \rangle$ denotes an average value,

$$J_x = \frac{e^2}{3kT} \left[F_x \int_0^\infty \frac{\tau v^2}{1 + \omega_0^2\tau^2} f_0 \cdot 4\pi p^2 \, dp - \omega_0 F_y \int_0^\infty \frac{\tau^2 v^2}{1 + \omega_0^2\tau^2} f_0 \cdot 4\pi p^2 \, dp \right]$$

$$= \frac{e^2 n}{3kT} \left[F_x \left\langle \frac{v^2\tau}{1 + \omega_0^2\tau^2} \right\rangle - \omega_0 F_y \left\langle \frac{v^2\tau^2}{1 + \omega_0^2\tau^2} \right\rangle \right] \tag{20.15}$$

and in the same way

$$J_y = \frac{e^2 n}{3kT} \left[\omega_0 F_x \left\langle \frac{v^2\tau^2}{1 + \omega_0^2\tau^2} \right\rangle + F_y \left\langle \frac{v^2\tau}{1 + \omega_0^2\tau^2} \right\rangle \right] \tag{20.16}$$

Since $J_y = 0$, for the Hall effect, we have

$$F_y = -\omega_0 F_x \frac{\langle v^2\tau^2/(1 + \omega_0^2\tau^2) \rangle}{\langle v^2\tau/(1 + \omega_0^2\tau^2) \rangle} \tag{20.17}$$

---

† According to the definition of averages

$$\langle g(p) \rangle = \frac{1}{n} \int_0^\infty g(p) f_0 \cdot 4\pi p^2 \, dp$$

where n is the carrier density.

and, consequently,

$$J_x = \sigma_x F_x = \frac{e^2 n}{3kT}\left[\left\langle\frac{v^2\tau}{1+\omega_0^2\tau^2}\right\rangle + \omega_0^2\frac{\langle v^2\tau^2/(1+\omega_0^2\tau^2)\rangle^2}{\langle v^2\tau/(1+\omega_0^2\tau^2)\rangle}\right]F_x \quad (20.18)$$

from which the conductivity σ_x can be determined. It is seen that σ_x is a function of ω_0^2, so that σ_x is an even function of B_z. Evaluation of the averages shows that σ_x decreases monotonically with increasing value of B_z. This is known as the *transverse magnetoresistance effect*.

20.1c. Evaluation of the Hall Effect and the Magnetoresistance Effect

For $\omega_0\tau \ll 1$, Eqs. (20.17) and (20.18) may be written

$$F_y = -\omega_0 F_x\frac{\langle v^2\tau^2\rangle}{\langle v^2\tau\rangle}, \qquad J_x = \frac{e^2 n}{3kT}\langle v^2\tau\rangle F_x = en\mu F_x \quad (20.19)$$

Consequently, the angle θ between the resultant electric field and the X axis is

$$\tan\theta = \frac{F_y}{F_x} = -\omega_0\frac{\langle v^2\tau^2\rangle}{\langle v^2\tau\rangle} = -\mu_H B_z \quad (20.20)$$

where μ_H is called the *Hall mobility*.

$$\mu_H = \frac{e}{m^*}\frac{\langle v^2\tau^2\rangle}{\langle v^2\tau\rangle} = \mu\frac{\langle v^2\tau^2\rangle\langle v^2\rangle}{\langle v^2\tau\rangle^2} \quad (20.20a)$$

because of the definition (20.19) of μ, and since $\langle v^2\rangle = 3kT/m^*$. Consequently, the Hall constant R follows from

$$F_y = -\mu_H B_z F_x = -\frac{\mu_H}{\mu}\frac{B_z J_x}{en} = RB_z J_x \quad (20.21)$$

so that

$$R = -\frac{\mu_H}{\mu}\frac{1}{en} \quad (20.21a)$$

Our initial result was thus off by a factor μ_H/μ. We now calculate this factor for semiconductors in which lattice scattering predominates. In that case $\tau = l/v$, and hence

$$\langle v^2\rangle = \frac{3kT}{m^*}, \qquad \langle v^2\tau^2\rangle = l^2, \qquad \langle v^2\tau\rangle = l\langle v\rangle = l\left(\frac{8kT}{\pi m^*}\right)^{1/2}$$

so that

$$\frac{\mu_H}{\mu} = \frac{l^2(3kT/m^*)}{l^2(8kT/\pi m^*)} = \frac{3\pi}{8} \quad (20.22)$$

as had to be proved. The same factor is found for p-type semiconductors and for semiconductors in which both electrons and holes are present.

For very large fields, $\omega_0\tau \gg 1$; hence, replacing $1+\omega_0^2\tau^2$ by $\omega_0^2\tau^2$, and neglecting small terms, we get, from (20.17) and (20.18),

$$F_y = -\omega_0 F_x \frac{\langle v^2 \rangle}{\langle v^2/\tau \rangle}$$

$$J_x = \frac{e^2 n}{3kT} \frac{\langle v^2 \rangle^2}{\langle v^2/\tau \rangle} F_x = \frac{e^2 n}{m^*} \frac{\langle v^2 \rangle}{\langle v^2/\tau \rangle} F_x \qquad (20.23)$$

Hence

$$F_y = -\frac{e}{m^*} B_z \cdot \frac{m^*}{e^2 n} J_x = -\frac{1}{en} B_z J_x = RB_z J_x \qquad (20.24)$$

so that the Hall constant now becomes

$$R = -\frac{1}{en} \qquad (20.24a)$$

A more detailed evaluation shows that $|R|$ decreases steadily from the low-field value $(3\pi/8)(1/en)$ to the high-field value $1/en$.

We now turn to the magnetoresistive effect. According to Eq. (20.18),

$$\sigma_x = \frac{e^2 n}{3kT} \left[\left\langle \frac{v^2 \tau}{1 + \omega_0^2 \tau^2} \right\rangle + \omega_0^2 \frac{\langle v^2 \tau^2/(1 + \omega_0^2 \tau^2) \rangle^2}{\langle v^2 \tau/(1 + \omega_0^2 \tau^2) \rangle} \right] \qquad (20.25)$$

As long as $\omega_0^2 \tau^2 < 1$, we may try a Taylor expansion in $\omega_0 \tau$; if terms higher than ω_0^2 are neglected, this yields

$$\sigma_x = \frac{e^2 n}{3kT} \left[\langle v^2 \tau \rangle - \omega_0^2 \langle v^2 \tau^3 \rangle + \omega_0^2 \frac{\langle v^2 \tau^2 \rangle^2}{\langle v^2 \tau \rangle} \right] \qquad (20.25a)$$

If the averages are evaluated, it is found that the second term predominates over the third, so that σ_x decreases with increasing B_z. If $\omega_0 \tau \gg 1$, we may try a Taylor expansion in $1/\omega_0 \tau$. Retaining only the lowest-order term yields

$$\sigma_x = \sigma_\infty = \frac{e^2 n}{3kT} \frac{\langle v^2 \rangle^2}{\langle v^2/\tau \rangle} = \frac{e^2 n}{3kT} \langle v^2 \tau \rangle \cdot \frac{\langle v^2 \rangle^2}{\langle v^2/\tau \rangle \langle v^2 \tau \rangle} \qquad (20.25b)$$

Since the low-field conductivity is

$$\sigma_0 = \frac{e^2 n}{3kT} \langle v^2 \tau \rangle \qquad (20.26)$$

we see that the resistance ratio is

$$\frac{R_\infty}{R_0} = \frac{\sigma_0}{\sigma_\infty} = \frac{\langle v^2/\tau \rangle \langle v^2 \tau \rangle}{\langle v^2 \rangle^2} \qquad (20.27)$$

Evaluating the averages for lattice scattering $(\tau = l/v)$ yields

$$\frac{R_\infty}{R_0} = \frac{32}{9\pi} = 1.132 \qquad (20.28)$$

since

$$\left\langle \frac{v^2}{\tau} \right\rangle = \frac{4}{l} \left(\frac{8}{\pi} \right)^{1/2} \left(\frac{kT}{m} \right)^{3/2}$$

This result was derived for spherical energy surfaces and when lattice scattering predominates. Much larger effects occur in materials with strong anisotropy; in that case there can also be a *longitudinal effect*.

20.1d. Applications

One of the main applications of the Hall effect is in the measurement of the carrier concentration of semiconductors. As a corollary, after having calibrated a given semiconductor sample, one can use it for the measurement of magnetic field strength.

Magnetoresistance effects give information about energy surfaces in semiconductors. For that reason a study of those effects helps us understand these materials.

Magnetoresistance effects can also be used for magnetic field measurements. Since the change in resistance

$$\Delta R = \gamma B^2 \tag{20.29}$$

over a wide range of field strengths, one can determine γ by putting the sample in a known magnetic field. Unknown magnetic fields can then be measured by simple resistance measurements with a Wheatstone bridge arrangement. Bismuth, which has a large magnetoresistance effect, is very useful for that purpose.

20.2. THERMOELECTRIC EFFECT

20.2a. Peltier Effect; Seebeck Effect; Thomson Effect

When two conductors a and b are joined together and a current I flows through the junction, then heat is generated or absorbed at the junction at a constant rate (Fig. 20.1a). The rate Q of heat generation is directly proportional to the current and changes sign if the current changes sign. We may thus write for the heat generated

$$Q = \Pi_{ab} I \tag{20.30}$$

where the notation Π_{ab} indicates that the current flows from conductor a to conductor b. Obviously

$$\Pi_{ba} = -\Pi_{ab} \tag{20.30a}$$

The effect is called *Peltier effect*, the rate of heat generation Q is known as *Peltier heat*, and the coefficient Π_{ab} is known as the *Peltier coefficient*. At the end of this section we shall give a simple illustration of this effect with the help of an ohmic metal–semiconductor contact.

If two conductors a and b are joined at two points 1 and 2 and a temperature difference ΔT is maintained between the two junctions, then an open-

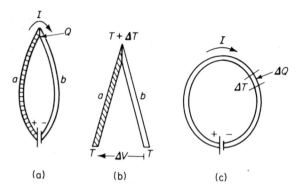

Fig. 20.1. (a) A junction *ab* carrying a current *I* shows Peltier effect. (b) A junction *ab* kept at the temperature difference ΔT shows Seebeck effect. (c) A uniform material carrying a current *I* and having a temperature gradient $\Delta T/\Delta x$ shows Thomson effect.

circuit potential difference ΔV is developed (Fig. 20.1b). This effect is called the *Seebeck effect*. The differential *Seebeck coefficient* α_{ab} is defined by

$$\alpha_{ab} = \lim_{\Delta T \to 0} \frac{\Delta V}{\Delta T} \qquad (20.31)$$

Finally, Fig. 20.1c shows a current *I* passed through a portion of a single homogeneous conductor over which there is a temperature difference ΔT. It is found that heat is emitted or absorbed at a rate ΔQ that is proportional to the current *I* and to the temperature difference ΔT. This effect is called *Thomson effect*. The Thomson coefficient γ is defined by

$$\gamma = \lim_{\Delta T \to 0} \frac{\Delta Q}{I \Delta T} \qquad (20.32)$$

The Thomson coefficient is taken to be positive if heat is evolved when a positive current passes from a higher to a lower temperature.

The three effects are thermodynamically related by means of the so-called *Kelvin relations*. We shall derive these relations in the next section.

Another way of expressing thermoelectric effects consists in introducing the *absolute thermoelectric force per degree* $\alpha = d\theta/dT$ of the material, where

$$\alpha_{ab} = \alpha_a - \alpha_b \qquad (20.33)$$

α_{ab} then becomes simply the difference between two material coefficients. We shall see that α is related to the Thomson coefficient γ.

We now illustrate the Peltier effect for ohmic metal–semiconductor contacts, and we shall use it to derive expressions for the Peltier coefficient of such contacts.

First, we consider an ohmic metal–*n*-type contact (Fig. 20.2a), and we

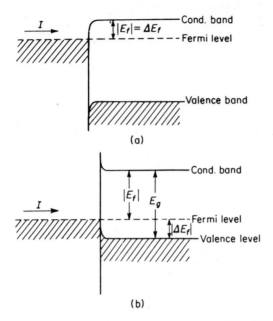

Fig. 20.2. (a) Diagram for Peltier effect in *n*-type material. (b) Diagram for Peltier effect in *p*-type material.

assume that positive current is flowing from the metal into the semiconductor. Electrons thus flow from the semiconductor into the metal. The electrons enter the metal with a kinetic energy $2kT$, $\frac{1}{2}kT$ for two perpendicular directions of motion parallel to the contact and kT for the direction of motion perpendicular to the contact (as in the thermionic case). Their average energy is thus $|E_f| + 2kT$ above the Fermi level of the metal, and so an electron gives an average energy $|E_f| + 2kT$ to the metal. If a current I is flowing, the rate at which energy is developed at the junction is

$$Q = \frac{I}{e}(|E_f| + 2kT) \tag{20.34}$$

Hence

$$\Pi_{mn} = \frac{|E_f| + 2kT}{e} \tag{20.34a}$$

The symbol Π_{mn} indicates that current flows from the metal to the *n*-type semiconductor, and the positive sign indicates that heat is extracted from the semiconductor and supplied to the metal.

If the direction of current flow is changed, heat is extracted from the metal. The electrons now enter the semiconductor with an energy $2kT$ and have to climb a potential barrier $|E_f|$. Hence the average energy supplied to an electron by the metal is $|E_f| + 2kT$, and the rate at which energy is supplied to the junction is

$$Q = -\frac{I}{e}(|E_f| + 2kT) \qquad (20.35)$$

so that the effect has indeed changed sign and

$$\Pi_{nm} = -\frac{|E_f| + 2kT}{e} = -\Pi_{mn} \qquad (20.35a)$$

Next we turn to an ohmic metal–p-type contact (Fig. 20.2b), and we assume that positive current is flowing from the metal into the semiconductor. Holes are now emitted into the p-type semiconductor with an average energy $2kT$, and holes generated at the Fermi level have to climb a potential barrier $E_g - |E_f|$. Consequently, the average energy supplied by the metal to a hole crossing the contact is $E_g - |E_f| + 2kT$. The rate Q at which heat is *supplied* to the junction if a current I is flowing is therefore

$$Q = -\frac{I}{e}(E_g - |E_f| + 2kT) \qquad (20.36)$$

and hence

$$\Pi_{mp} = -\frac{E_g - |E_f| + 2kT}{e} \qquad (20.36a)$$

We can rewrite Eqs. (20.34a) and (20.36a) by introducing the distance ΔE_f between the Fermi level and the band edge. For an n-type semiconductor, $\Delta E_f = |E_f|$, and for a p-type semiconductor, $\Delta E_f = E_g - |E_f|$. We then have

$$\Pi_{ms} = \mp \frac{\Delta E_f + 2kT}{e} \qquad (20.37)$$

where the minus sign holds for p-type material and the plus sign for n-type material. The direction of current flow is from the metal into the semiconductor in each case.

20.2b. Kelvin Relations

We shall here derive the Kelvin relations for bulk material. In a certain wire, with the X axis in the direction of the axis of the wire, let a potential gradient $\nabla \psi$ and a temperature gradient ∇T be present. Let I be the electric current and K the heat current, so that K/T is the entropy current. As long as the system is linear, I and K/T should depend linearly upon $\nabla \psi$ and ∇T. That is,

$$I = -L_{11}\nabla\psi - L_{12}\nabla T$$
$$\frac{K}{T} = -L_{21}\nabla\psi - L_{22}\nabla T \qquad (20.38)$$

This is the equivalent of a passive, reciprocal, linear electrical network; thus

$$L_{12} = L_{21} \qquad (20.38a)$$

This is known as the *Onsager relation*, and it can be proved with the help of irreversible thermodynamics.

If we now express $-\nabla\psi$ and K in terms of I and ∇T, we obtain

$$-\nabla\psi = \frac{I}{L_{11}} + \frac{L_{12}}{L_{11}}\nabla T = \rho I + \alpha\nabla T$$

$$K = T\frac{L_{21}}{L_{11}}I + T\left(\frac{L_{12}L_{21}}{L_{11}} - L_{22}\right)\nabla T = \Pi I - \lambda\nabla T$$

(20.38b)

where ρ is the resistivity of the wire, α the Seebeck coefficient, Π the Peltier coefficient, and λ the heat conductivity. Because of (20.38a), we have

$$\alpha = \frac{\Pi}{T} \qquad (20.39)$$

This is called the first *Kelvin relation*.

We now take a section Δx of the wire (Fig. 20.3) and adjust the current I such that $\nabla\psi = 0$. This means that

$$I = -\frac{\alpha}{\rho}\nabla T \qquad (20.40)$$

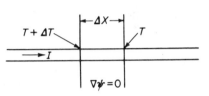

Fig. 20.3. Circuit for the derivation of the Kelvin relations.

Then $I\nabla\psi = 0$, and the entropy production in the section Δx must be zero. The entropy entering into the section Δx is $\Pi(T + \Delta T)I/(T + \Delta T)$, the entropy lost by the Thomson effect is $\Delta Q/T = \gamma I\,\Delta T/T$, and the entropy leaving section Δx is $\Pi(T)I/T$. Hence

$$\frac{\Pi(T + \Delta T)I}{T + \Delta T} - \frac{\Pi(T)I}{T} - \frac{\gamma I\,\Delta T}{T} = 0$$

Dividing by $I\,\Delta T$ yields

$$\frac{\Pi(T + \Delta T)/(T + \Delta T) - \Pi(T)/T}{\Delta T} - \frac{\gamma}{T} = 0$$

or
$$\frac{d(\Pi/T)}{dT} = \frac{d\alpha}{dT} = \frac{\gamma}{T} \qquad (20.41)$$

This is called the *second Kelvin relation*. It may also be written

$$\alpha = \int_0^T \frac{\gamma(T')}{T'}\,dT' \qquad (20.41\text{a})$$

Having defined Π and α for a single material, we have for junctions

$$\Pi_{ab} = \Pi_a - \Pi_b, \qquad \alpha_{ab} = \alpha_a - \alpha_b \qquad (20.42)$$

As a consequence, for a metal–semiconductor junction we have

$$\alpha_{ms} = \alpha_m - \alpha_s = \frac{\Pi_{ms}}{T} = \mp\left(\frac{\Delta E_f}{eT} + \frac{2k}{e}\right) \qquad (20.43)$$

where the minus sign holds for p-type material and the plus sign for n-type material. Since $\alpha_m \ll \alpha_s$, this corresponds to

$$\alpha_s = \pm\left(\frac{\Delta E_f}{eT} + \frac{2k}{e}\right) \tag{20.44}$$

Consequently, the Thomson coefficient γ_s of the semiconductor is, according to (20.41),

$$\gamma_s = T\frac{d\alpha_s}{dT} = \pm T\frac{d}{dT}\left(\frac{\Delta E_f}{eT} + \frac{2k}{e}\right) \tag{20.45}$$

In the last two equations the plus sign holds for p-type materials and the minus sign for n-type materials.

We now bear in mind that for n-type materials in which practically all donors are ionized

$$\Delta E_f = kT \ln \frac{P_1}{N_d}, \qquad P_1 = 2\left(\frac{2\pi mkT}{h^2}\right)^{3/2} \tag{20.46}$$

where N_d is the donor concentration. Consequently, near room temperature,

$$\alpha_s = -\frac{k}{e}\left(\ln\frac{P_1}{N_d} + 2\right) \tag{20.44a}$$

$$\gamma_s = -\frac{3}{2}\frac{k}{e} \tag{20.45a}$$

Corresponding equations hold for p-type materials. This is left as a problem for the reader.

The preceding equations were derived for nondegenerate semiconductors in which almost all impurity centers are ionized. The materials used in practical thermoelectric devices, such as thermoelectric power generators and thermoelectric coolers, have carrier concentrations of the order of $10^{19}/cm^3$ and are degenerate semiconductors. Although it is possible to derive expressions for the thermoelectric force per degree for these materials, it is often more convenient to *measure* the thermoelectric force per degree and use the *measured* values in the design of thermoelectric devices.

20.2c. Thermoelectric Power Generators

In the simplest case a thermoelectric power generator consists of two rods made of two materials a and b with two junctions 1 and 2 kept at a temperature difference ΔT (Fig. 20.4). The device can then be represented by an emf $\alpha_{ab} \Delta T$ with an internal resistance R, which is the total series resistance of the two parts a and b of the thermoelectric circuit. If an external load resistance R_0 is applied, the current is $\alpha_{ab} \Delta T/(R + R_0)$, and the power fed into the external load is

$$P = \left(\frac{\alpha_{ab} \Delta T}{R + R_0}\right)^2 R_0 \tag{20.47}$$

which has a maximum value

$$P_{\max} = \frac{(\alpha_{ab}\,\Delta T)^2}{4R} \quad (20.47a)$$

for $R_0 = R$. The factor α_{ab}^2/R is
called the figure of merit for this
application.

The conditions for a large maximum power are therefore large ΔT
and a large value of α_{ab}^2/R. The first
is easily accomplished, although for
large values of ΔT the expression α_{ab}
ΔT must be replaced by $\bar{\alpha}_{ab}\,\Delta T$,
where $\bar{\alpha}_{ab}$ is the average value of α_{ab}
over the temperature interval ΔT:

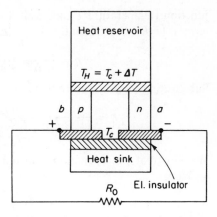

Fig. 20.4. Diagram of a thermoelectric generator.

$$\bar{\alpha}_{ab}\,\Delta T = \int_{T}^{T+\Delta T} \alpha_{ab}\,dT \qquad (20.48)$$

The second condition requires more careful consideration. First, one
should make the lengths of the rods small and the diameters of the rods
large, since that decreases the resistance. One cannot go too far in that direction, however, because the heat reservoir then loses a large amount of power
by heat conduction, diminishing the overall efficiency of the system. Next
one increases the impurity concentration of the rods. This tends to decrease
α_{ab} and decrease R, but since at first R decreases faster than α_{ab}, the optimum
design lies at high impurity concentrations (around $10^{19}/\text{cm}^3$). Finally, it is
important to give rods a and b opposite thermoelectric forces per degree.
If the two rods have equal resistance and equal and opposite thermoelectric
forces per degree, α_{ab}, R, and α_{ab}^2/R each increase by a factor of 2.

Rather than optimizing the power, it is often more meaningful to optimize
the efficiency η, defined as the ratio of the power P_0 developed into the load
resistance R_0 to the heat flow q_H from the source at the hot junction:

$$\eta = \frac{P_0}{q_H} \qquad (20.49)$$

Now

$$P_0 = I^2 R_0, \qquad I = \frac{\alpha_{ab}\,\Delta T}{R + R_0} \qquad (20.49a)$$

$$q_H = K\,\Delta T + (T + \Delta T)\alpha_{ab}I - \tfrac{1}{2}I^2 R \qquad (20.49b)$$

Here I is the current flowing through the circuit, R the resistance of the
thermocouple, K the thermal conductance, $T + \Delta T$ the temperature of the
hot junction, $(T + \Delta T)\dot{\alpha}_{ab}$ the Peltier coefficient, and $I^2 R$ the heat developed
in the thermocouple. Of this, half the heat flows to the heat sink at the cold

junction (temperature T) and half flows back to the heat source (hence the term $-\frac{1}{2}I^2R$). Consequently, if $T + \Delta T = T_H$,

$$\eta = \frac{I^2 R_0}{K \Delta T + \alpha_{ab} T_H I - \frac{1}{2} I^2 R} \tag{20.50}$$

Substituting for I and introducing $m = R_0/R$ as a new variable, one obtains

$$\eta = \frac{m \, \Delta T/T_H}{\dfrac{(1 + m)^2}{T_H} \cdot \dfrac{RK}{\alpha_{ab}^2} + 1 + m - \dfrac{1}{2} \dfrac{\Delta T}{T_H}} \tag{20.50a}$$

We now treat η as a function of m and RK, and must optimize RK for a given m. Now

$$K = \frac{\lambda_a A_a}{L_a} + \frac{\lambda_b A_b}{L_b} = \lambda_a \delta_a + \lambda_b \delta_b \tag{20.51}$$

$$R = \frac{\rho_a}{\delta_a} + \frac{\rho_b}{\delta_b} \tag{20.52}$$

where A is the cross-sectional area of each bar, L its length, and $\delta = A/L$; λ_a and λ_b are the heat conductivities and ρ_a and ρ_b are the electrical resistivities. Consequently,

$$RK = \lambda_a \rho_a + \lambda_a \rho_b \left(\frac{\delta_a}{\delta_b} \right) + \lambda_b \rho_a \left(\frac{\delta_b}{\delta_a} \right) + \lambda_b \rho_b \tag{20.53}$$

which has a minimum value

$$(RK)_{\min} = [(\rho_a \lambda_a)^{1/2} + (\rho_b \lambda_b)^{1/2}]^2 \tag{20.53a}$$

when

$$\frac{\delta_a}{\delta_b} = \left(\frac{\rho_a \lambda_b}{\rho_b \lambda_a} \right)^{1/2} \tag{20.53b}$$

If we now introduce the figure of merit of the junction,

$$Z = \frac{\alpha_{ab}^2}{(RK)_{\min}} = \frac{(|\alpha_a| + |\alpha_b|)^2}{[(\rho_a \lambda_a)^{1/2} + (\rho_b \lambda_b)^{1/2}]^2} \tag{20.54}$$

the expression for η may be written

$$\eta = \frac{m \Delta T/T_H}{[(1 + m)^2/Z T_H] + 1 + m - \frac{1}{2}(\Delta T/T_H)} \tag{20.55}$$

which has an optimum value

$$\eta = \eta_{\max} = \frac{(m_{\mathrm{opt}} - 1) \, \Delta T/T_H}{m_{\mathrm{opt}} + T_c/T_H} \tag{20.55a}$$

when

$$m = m_{\mathrm{opt}} = (1 + Z T_{av})^{1/2} \tag{20.55b}$$

Here $\Delta T = T_H - T_c$, T_c is the temperature of the cold junction, T_H the temperature of the hot junction, and $T_{av} = \frac{1}{2}(T_H + T_c)$ the average value.

It thus makes sense to call

$$M = \frac{\alpha^2}{\lambda \rho} \tag{20.56}$$

the figure of merit of the material. Preferably, the figures of merit of the bars a and b should be comparable.

We must now investigate how M depends on the impurity concentration N of the material. The thermoelectric force per degree α decreases slowly with increasing N. The electrical conductivity $1/\rho$ increases linearly with N as long as lattice scattering predominates, and much slower than linearly when impurity scattering begins to predominate. The heat conductivity λ consists of a term due to lattice conduction and a term due to heat conduction by the carriers; the latter increases with increasing N. As a consequence, M attains a maximum at a relatively large impurity concentration, around $10^{19}/cm^3$. At those concentrations the material is degenerate, and the simple expressions for α derived in the previous section cannot be used.

The most promising thermoelements use Sb_2Te_3, Bi_2Te_3, and some of their alloys. The details are beyond the scope of this book.

To increase the efficiency of the thermoelectric generators, one should make ΔT large. Unfortunately, one cannot go too far in that direction, since α depends upon T, and $\Delta T/T_H$ increases slower than $\Delta T/T_c$ when ΔT becomes comparable to T_c. It may then be convenient to use two thermoelectric bars in series, one more suitable for higher temperature and the other for lower temperature. By proper design of the two bars, the temperature of the point where the two bars meet can be about halfway between T_c and T_H.

The thermoelectric bars should be provided with low-resistance contacts, so that the contact resistance is small in comparison with the resistance of the bars. If two bars are used in series, as suggested in the preceding paragraph, good ohmic contacts should be made to the adjoining faces of the bars.

20.2d. Thermoelectric Refrigerators

Thermoelectric refrigerators are based on Peltier cooling. Figure 20.5 shows such an arrangement.

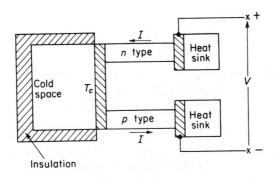

Fig. 20.5. Diagram of a thermoelectric refrigerator.

The rate at which heat is removed from the cold reservoir can be written as

$$q_c = \alpha_{ab} T_c I - \tfrac{1}{2} I^2 R - K\,\Delta T \qquad (20.57)$$

similar to the previous case. Here $K\,\Delta T$ is the heat flowing back from the hot reservoir and $I^2 R$ is the Joule heat developed in the elements; half of this heat returns to the cold reservoir. The heat pumping rate has an optimum value

$$(q_c)_{\text{opt}} = \frac{\tfrac{1}{2}\alpha_{ab}^2 T_c^2}{R} - K\,\Delta T, \qquad \text{when} \quad I = \frac{\alpha_{ab} T_c}{R} \qquad (20.57a)$$

If we set the heat removed from the cold reservoir equal to zero, the temperature difference that the refrigerator maintains becomes a maximum. Hence

$$(\Delta T)_{\max} = \frac{\tfrac{1}{2}\alpha_{ab}^2 T_c^2}{RK} \qquad (20.57b)$$

We can now design the refrigerator such that the product RK is minimized. This yields

$$(\Delta T)_{\max} = \tfrac{1}{2} T_c^2 Z \qquad (20.58)$$

where Z is the figure of merit introduced for the thermoelectric generator.

This is not necessarily the most economical operation of the device, however. Often it is better to optimize the coefficient of performance.

$$\beta = \frac{q_c}{P} \qquad (20.59)$$

where P is the electrical power input.

Since the voltage that must be applied must supply the IR volt drop across the thermocouple and overcome the Seebeck emf is

$$V = \alpha_{ab}\Delta T + IR \qquad (20.60)$$

the electrical power input is

$$P = VI = \alpha_{ab} I\,\Delta T + I^2 R \qquad (20.61)$$

Substituting into β, one obtains

$$\beta = \frac{\alpha_{ab} T_c I - \tfrac{1}{2} I^2 R - K\,\Delta T}{\alpha_{ab} I\,\Delta T + I^2 R} \qquad (20.62)$$

To maximize this we introduce $f = IR/\alpha_{ab}$ as a new variable. This gives

$$\beta = \frac{f T_c - \tfrac{1}{2} f^2 - RK\,\Delta T/\alpha_{ab}^2}{f\,\Delta T + f^2} \qquad (20.62a)$$

We optimize this in two steps. First, for a given f, we minimize RK; this introduces again the figure of merit Z encountered earlier. Next β is maximized as a function of f by proper choice of I. This is left as a problem for the reader (problem 2).

20.3. AMORPHOUS SEMICONDUCTORS

20.3a. *Properties of Amorphous Semiconductors*

Amorphous semiconductors are semiconductors with a more or less random structure. One can arrive at the energy band structure by comparing it with a crystalline material.

In a crystalline material one has the valence band and the conduction band. Both consist of extended energy states; that is, the wave functions of these states extend all through the crystal. Any impurity or dangling (that is, nonconnected) bond produces a localized energy state in the forbidden gap for which the electron wave function extends over a small region only. The more disturbances there are, the more localized energy states there will be. In an amorphous semiconductor there are so many localized states, located all over the energy region between the valence band and the conduction band, that the whole energy gap disappears (Fig. 20.6).

Even though there is no region without energy states, there is something equivalent to a gap. The localized energy states allow conduction by hopping from the one localized energy state to the next, but this results in a very small mobility. The extended energy states, however, allow conduction with a much larger mobility. There is thus a *mobility gap* with boundaries E_c' and E_v', so that the mobility for electrons with $E_c' > E > E_v'$ is very small, and the mobility outside that region is much larger. The Fermi level E_f lies somewhere between E_c' and E_v', but not necessarily at midway.

If one measures the light absorption coefficient α_a, one finds for the region of large absorption ($\alpha_a \simeq 10^3$–10^4 cm^{-1})

$$\alpha_a = \text{const} \frac{(h\nu - E_{g0})^n}{h\nu} \tag{20.63}$$

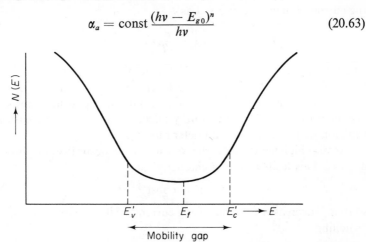

Fig. 20.6. Density of states N(E) versus energy E in amorphous semiconductors, showing the mobility gap and the Fermi level.

with n between 0.5 and 3; the parameter E_{g0} is called the *optical gap*. It does not coincide with $E'_c - E'_v$, but is generally larger. If one measures the photo-conductivity as a function of photon energy, one finds a definite threshold just as in the crystalline case. However, this threshold energy is not equal to $E'_c - E'_v$ either.

At low temperatures the current flow is by carrier hopping. According to Mott[†], the conductivity σ associated with this process is given by

$$\sigma = \text{const} \exp\left[-\left(\frac{\alpha^3}{\rho_0 kT}\right)^{1/4}\right] \tag{20.64}$$

Here ρ_0 is the density of localized states at the Fermi level, and α^{-1} is the length of range of the localized wave function. This temperature dependence comes about because of a continuum of possible activation energies for hopping between localized states.

At higher temperatures there can be excitation into the conduction band. The activation energy is $E'_c - E'_v$; hence the conductivity can be given as

$$\sigma = \sigma_0(T) \exp\left(-\frac{E'_c - E'_v}{kT}\right) \tag{20.65}$$

where $\sigma_0(T)$ is a slow function of T; one can thus determine the width of the mobility gap from the conductivity data. Often this mechanism predominates at room temperature.

At high fields ($E > 10^4$ V/cm) the current increases strongly with increasing field strength E. This could be caused by a Schottky effect at a contact, but it is usually due to the lowering of the activation energy of a trapped carrier by the applied field; this is called the *Poole–Frenkel* effect. In that case the current I may be expressed as

$$I = \text{const} \exp\left[-\frac{e}{kT}V_t + \frac{e}{kT}\left(\frac{eV}{\pi\epsilon\epsilon_0 d}\right)^{1/2}\right] \tag{20.66}$$

where V is the applied voltage, V_t the activation energy of the trap in electron volts, d the thickness of the layer, and ϵ its dielectric constant. The reader will recognize that the second term in the exponent corresponds to the exponent observed in the Schottky effect (Chapter 7). By measuring I as a function of V and T, one can determine V_t.

At very high fields avalanche ionization can occur in amorphous semiconductors. This leads to a V, J characteristic

$$V = \text{const} \, J^{-1/2} \tag{20.67}$$

so that V decreases with increasing current.[‡] This is important for explaining switching.

---

† N. F. Mott, *Phil. Mag.*, **19**, 835 (1969).
‡ N. K. Hindley, *J. Non-Crystalline Solids*, **8–10**, 557 (1972).

20.3b. Applications

One can use inorganic amorphous semiconductors in electrophotography. After charging of the material in question by a corona discharge, an electrostatic image is produced through a selective discharge by photoconductivity. This electrostatic image is used in turn to control the deposition of charged pigment particles. This is the basis of modern document copying procedures. The Xerox Corporation uses Se and As–Se compositions as the photoconductive element. These materials have high resistivities in the dark, combined with reasonable photoconductive processes.

Selective crystallization in amorphous films, through incident light, can be used for image handling, such as photography and electrophotography. The same selective crystallization can be induced by a laser beam, and can be used for writing digital information. The laser beam can also produce local melting and subsequent quenching and thus can restore the amorphous material to its original form, thereby erasing the written information.

Interesting applications of amorphous semiconductors are in electronic switches and memories. Figure 20.7a shows the characteristic of an electronic switch, and Fig. 20.7b shows the characteristic of a memory device.

In the first device the characteristic switches from a high-voltage, low-current to a low-voltage, high-current state if the applied voltage V exceeds a certain threshold voltage V_T. In the high-current state the voltage does not depend very much upon current. The high-current and low-current branch meet at a voltage V_H (holding voltage), at very low currents the high-current state switches back to the low current state. The switching path in the IV

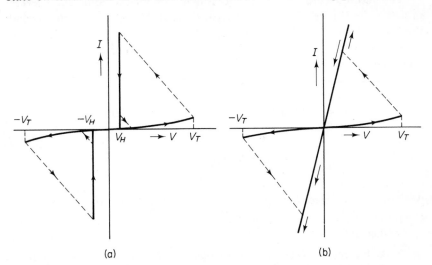

(a)   (b)

Fig. 20.7. (a) Characteristic of electronic switch. (b) Characteristic of memory device.

plane depends on the resistance in the external circuit. These switching processes are probably electronic in nature.

In the memory device there is also a high- and low-conductivity state, but the two parts of the characteristic pass through the origin. Depending on past history, the device is in either the low- or the high-conductivity state and can be changed from the one state to the other by applying a sufficiently large pulse. It is believed that a high-conducting path is formed thermally by switching from the low- to the high-conductivity state, and that this high-conducting path is destroyed by switching from the high- to the low- conductivity state.

These devices have the advantage that they can be made cheaply by evaporation or sputtering techniques. They are very insensitive to radiation damage; in view of the disturbed structure of the materials the latter is not surprising. The devices have a lifetime of 10^8–10^9 switchings, which is sufficient for many applications.

One disadvantage is that switching times are relatively slow, of the order of a few tenths of a microsecond or even longer. The switching time τ_d decreases with increasing applied voltage V; if V is above the threshold voltage V_T, the approximate relationship is

$$\tau_d = \tau_{d0} \exp\left(-\frac{V - V_T}{V_0}\right) \tag{20.68}$$

where τ_{d0} and V_0 are constants but show some fluctuation from one switching to the next.

The nature of the switching process is not fully clear; apparently several processes contribute. First, there is local Joule heating in the sample; second, there is a field-dependent current due to the Poole–Frenkel effect; and, third, there may be local avalanche ionization. The three effects together are certainly sufficient to produce switching.

There can be stretches of low-conductivity paths in the material separated by small regions of low conductivity. Switching then only has to occur in these relatively small regions, and not all through the sample.

REFERENCES

ANGRIST, S. W., *Direct Energy Conversion*. Allyn and Bacon, Inc., Boston, 1965.

GOLDSMITH, H. J., *Applications of Thermoelectricity*. Methuen & Company Ltd., London, 1960.

IOFFE, A. F., *Semiconductor Thermoelements and Thermoelectric Cooling*. Inforsearch Ltd., London, 1957.

Journal of Non-Crystalline Solids, vols. 8–10 (conference papers), 1972.

National Academy of Sciences, *Fundamentals of Amorphous Semiconductors.* The
Academy, Washington, D.C., 1972.

SEITZ, F., *The Modern Theory of Solids.* McGraw-Hill Book Company, New York,
1940.

PROBLEMS

1. Carry out the differentiation of Eq. (20.55) with respect to m and prove Eqs.
(20.55a) and (20.55b). *Hint:* Put $Z = (m_{opt}^2 - 1)/T_{av}$.

2. Find the value of f for which the expression (20.62a) has its optimum value.

Answer:
$$f_{opt} = \frac{\Delta T}{ZT_{av}}[1 + (1 + ZT_{av})^{1/2}]$$

Appendix **A**

A.1. APPROXIMATION METHODS IN WAVE MECHANICS

A.1a. Perturbation Theory for Nondegenerate States

Suppose that in the wave equation†

$$H\varphi = E\varphi \tag{A.1}$$

the Hamiltonian can be written as

$$H = H^0 + \lambda H' \tag{A.1a}$$

where λ is a parameter and $\lambda H'$ is a relatively small term. Suppose that the equation

$$H^0\varphi - E\varphi = 0 \tag{A.2}$$

has been solved already and the energies (or eigenvalues) are $E_1^0, \ldots, E_n^0$ with the corresponding wave functions $\varphi_1^0, \ldots, \varphi_n^0$. Equation (A.1) can then be solved as follows:

Develop φ_k and E_k in a Taylor series in λ,

$$\varphi_k = \varphi_k^0 + \lambda\varphi_k' + \lambda^2\varphi_k'' + \ldots \tag{A.3}$$

$$E_k = E_k^0 + \lambda E_k' + \lambda^2 E_k'' + \ldots \tag{A.4}$$

---

† We drop here the subscript *op* of H_{op}.

Now substitute into Eq. (A.1) and equate the terms of the same order in λ. This yields

$$\text{zero order:} \quad H^0\varphi_k^0 - E_k^0\varphi_k^0 = 0 \tag{A.5}$$

$$\text{first order:} \quad H^0\varphi_k' - E_k^0\varphi_k' = (E_k' - H')\varphi_k^0 \tag{A.6}$$

$$\text{second order:} \quad H^0\varphi_k'' - E_k^0\varphi_k'' = (E_k' - H')\varphi_k' + E_k''\varphi_k^0 \tag{A.7}$$

According to a well-known theorem in the theory of differential equations, the nonhomogeneous equation (A.6) has a solution only if the right side is orthogonal with respect to the solution of the homogeneous equation

$$H^0\varphi_k - E_k^0\varphi_k = 0 \tag{A.6a}$$

But we know that this has the solution φ_k^0. Hence

$$\int \varphi_k^{0*}(E_k' - H')\varphi_k^0 \, dV = 0 \quad \text{or} \quad E_k' \int \varphi_k^{0*}\varphi_k^0 \, dV - \int \varphi_k^{0*}H'\varphi_k^0 \, dV = 0$$

Applying the normalization rule yields

$$E_k' = \int \varphi_k^{0*}H'\varphi_k^0 \, dV = H_{kk}' \tag{A.8}$$

where H_{kk}' are the diagonal matrix elements of H' (Sec. 2.3).

We now develop φ_k' with respect to the eigenfunctions φ_l^0:

$$\varphi_k' = \sum_l a_{kl}\varphi_l^0 \tag{A.9}$$

To calculate a_{kl}, we substitute (A.9) into (A.6) and replace l by u:

$$\sum_u a_{ku}(H^0\varphi_u^0 - E_k^0\varphi_u^0) = \sum_u a_{ku}(E_u^0 - E_k^0)\varphi_u^0 = (E_k' - H')\varphi_k^0 \tag{A.10}$$

Now multiply both sides by φ_l^{0*} and integrate over the volume. This gives zero terms in the left side except when $u = l$. Hence

$$a_{kl}(E_l^0 - E_k^0) = \int \varphi_l^{0*}(E_k' - H')\varphi_k^0 \, dV$$

$$= E_k' \int \varphi_l^{0*}\varphi_k^0 \, dV - \int \varphi_l^{0*}H'\varphi_k^0 \, dV \tag{A.11}$$

Putting $l = k$ gives Eq. (A.8):

$$0 = E_k' - \int \varphi_k^{0*}H'\varphi_k^0 \, dV$$

Taking $l \neq k$, we have $E_l^0 \neq E_k^0$, and the term with E_k' is zero. Hence

$$a_{kl} = \frac{\int \varphi_l^{0*}H'\varphi_k^0 \, dV}{E_k^0 - E_l^0} = \frac{H_{lk}'}{E_k^0 - E_l^0} \tag{A.12}$$

so that

$$\varphi_k' = \sum_l{}' \frac{H_{lk}'}{E_k^0 - E_l^0}\varphi_l^0 \tag{A.13}$$

where $\sum_l'$ means that the term with $l = k$ must be omitted. Here H'_{lk} are the off-diagonal matrix elements of H':

$$H'_{lk} = \int \varphi_l^{0*} H' \varphi_k^0 \, dV \qquad \text{(A.12a)}$$

We now turn to Eq. (A.7). It has a solution only if the right side is orthogonal with respect to the solution of the homogeneous equation, or orthogonal with respect to φ_k^0. Hence

$$\int \varphi_k^{0*}[(E'_k - H')\varphi'_k + E''_k \varphi_k^0] \, dV = 0 \qquad \text{(A.14)}$$

which gives, by substitution of (A.13),

$$E''_k = \sum_l' a_{kl} \int \varphi_k^{0*} H' \varphi_l^0 \, dV = \sum_l' \frac{H'_{kl} H'_{lk}}{E_k^0 - E_l^0} \qquad \text{(A.15)}$$

The solution up to the second order thus gives for the energy

$$E_k = E_k^0 + \lambda H'_{kk} + \lambda^2 \sum_l' \frac{H'_{kl} H'_{lk}}{E_k^0 - E_l^0} \qquad \text{(A.16)}$$

The energy can thus be expressed in terms of the matrix elements H'_{kl} of the perturbation Hamiltonian H'.

Afterward we can, without loss of generality, put $\lambda = 1$ if we wish. For example, if the perturbation is a potential term V', the energy in second-order approximation is (see Section 4.1b)

$$E = E_k^0 + V'_{kk} + \sum_l' \frac{V'_{kl} V'_{lk}}{E_k^0 - E_l^0} \qquad \text{(A.16a)}$$

A.1b. Perturbation Theory for Degenerate States

Suppose that to the energy E_k^0 of the unperturbed system belong N wave functions φ_{km}^0 $(m = 1, \ldots, N)$, where the φ_{km}^0's are normalized and orthogonal. If the perturbation term $\lambda H'$ in the Hamiltonian is taken into account, the N-fold degeneracy may be removed.

The trouble is that we do not know what wave function to choose. To remedy the situation, we take a linear combination

$$\varphi_{km} = \sum_{n=1}^N \alpha_{mn} \varphi_{kn}^0 \qquad \text{(A.17)}$$

as the zero-order wave function. We now apply perturbation theory and try to determine the α_{mn}'s. By doing this, we "match" the wave functions φ_{km} to the perturbation.

We apply first-order perturbation theory by writing

$$H = H^0 + \lambda H', \qquad E_k = E_k^0 + \lambda E'_k$$
$$\varphi_{km} = \sum_{n=1}^N \alpha_{mn} \varphi_{kn}^0 + \lambda \varphi'_{km} \qquad \text{(A.18)}$$

We substitute into the equation

$$H\varphi_{km} = E_k\varphi_{km} \tag{A.19}$$

and bear in mind that

$$H^0\varphi_{kn}^0 - E_k^0\varphi_{kn}^0 = 0 \tag{A.20}$$

for $n = 1, \ldots, N$. Equating terms of the same order in λ yields, for the zero-order terms,

$$H_0 \sum_{n=1}^{N} \alpha_{mn}\varphi_{kn}^0 = E_k^0 \sum_{n=1}^{N} \alpha_{mn}\varphi_{kn}^0$$

which is already satisfied, and for the first-order terms (λ),

$$H_0\varphi_{km}' + H' \sum_{n=1}^{N} \alpha_{mn}\varphi_{kn}^0 = E_k^0\varphi_{km}' + E_k' \sum_{n=1}^{N} \alpha_{mn}\varphi_{kn}^0$$

which may be written

$$H^0\varphi_{km}' - E_k^0\varphi_{km}' = (E_k' - H') \sum_{n=1}^{N} \alpha_{mn}\varphi_{kn}^0 \tag{A.21}$$

Again Eq. (A.21) has a solution only if the right side is orthogonal to the solutions of the homogeneous equation (A.20). Thus it must be orthogonal to each function $\varphi_{kj}^0 (j = 1, \ldots, N)$. Hence

$$\int \varphi_{kj}^{0*}\left[(E_k' - H') \sum_{n=1}^{N} \alpha_{mn}\varphi_{kn}^0\right] dV = 0 \tag{A.22}$$

Putting

$$H_{kj,\,kn}' = \int \varphi_{kj}^{0*} H' \varphi_{kn}^0 \, dV \tag{A.22a}$$

in Eq. (A.22), and making use of the normalization conditions, yields the set of equations

$$(E_k' - H_{k1,\,k1}')\alpha_{m1} - H_{k1,\,k2}'\alpha_{m2} \ldots - H_{k1,\,kN}'\alpha_{mN} = 0$$
$$-H_{k2,\,k1}'\alpha_{m1} + (E_k' - H_{k2,\,k2}')\alpha_{m2} \ldots - H_{k2,\,kN}'\alpha_{mN} = 0$$
$$-H_{kN,\,k1}'\alpha_{m1} - H_{kN,\,k2}'\alpha_{m2} \ldots + (E_k' - H_{kN,\,kN}')\alpha_{mN} = 0$$

which has a solution only if its determinant is zero. Hence

$$\begin{vmatrix} E_k' - H_{k1,\,k1}' & -H_{k1,\,k2}' & \cdots & -H_{k1,\,kN}' \\ -H_{k2,\,k1}' & E_k' - H_{k2,\,k2}' & \cdots & -H_{k2,\,kN}' \\ \vdots & & & \\ -H_{kN,\,k1}' & -H_{kN,\,k2}' & \cdots & E_k' - H_{kN,\,kN}' \end{vmatrix} = 0 \tag{A.23}$$

which is an Nth-order equation with N solutions $E_{k1}', \ldots, E_{kN}'$. If the values E_{ki}' are all different, the perturbation has removed the degeneracy completely. If some of the values are equal, the perturbation has removed the degeneracy in parts. Equation (A.23) gives thus the N values of the energy up to first-order approximation:

$$E_{ki} = E_k^0 + \lambda E_{ki}', \qquad i = 1, \ldots, N \qquad (A.24)$$

Next we can take $\lambda = 1$ without loss of generality.

A.1c. WKB Method for Determining the Transparency of Potential Barriers

For the purpose of calculating the transparency of potential barriers to atomic particles an approximation method, known as the WKB method, is often used.

Let a particle wave be described by the wave equation

$$\frac{d^2\varphi}{dx^2} - f(x)\varphi = 0 \qquad (A.25)$$

where $f(x)$ is positive for $x_1 < x < x_2$ (Fig. A.1). For $x_1 < x < x_2$, we then try the solution

$$\varphi = \exp\left[\alpha(x)\right] \qquad (A.26)$$

Substituting into (A.25) gives the following relation for α:

$$\alpha''(x) + [\alpha'(x)]^2 - f(x) = 0 \qquad (A.26a)$$

It is now assumed that α is a slowly varying function of x, so that

$$\alpha''(x) \ll [\alpha'(x)]^2 \qquad (A.26b)$$

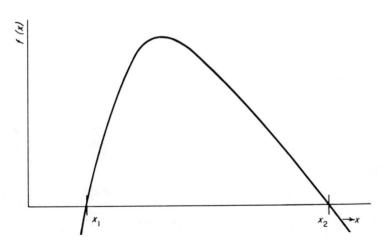

Fig. A.1.

Equation (A.26a) then has the approximate solution

$$\alpha(x) = \pm \int_{x_1}^{x_2} \sqrt{f(u)}\, du, \qquad \text{for } x_1 < x < x_2 \tag{A.27}$$

We need a negative sign, for we look for a wave traveling from left to right. Hence the transparency of the barrier is

$$T = \varphi\varphi^* \Big|_{x=x_2} = \exp\left[-2 \int_{x_1}^{x_2} \sqrt{f(x)}\, dx\right] \tag{A.28}$$

In the case of a potential barrier $V(x)$ (Section 7.3),

$$f(x) = \frac{2m}{\hbar^2}[V(x) - E] \tag{A.29}$$

where $V(x)$ is the potential energy and E the energy of the electron. Hence

$$T = \exp\left\{-2\left(\frac{2m}{\hbar^2}\right)^{1/2} \int_{x_1}^{x_2} [V(x) - E]^{1/2}\, dx\right\} \tag{A.30}$$

A.2. NOISE

A.2a. Spectral Intensity

A noise voltage or current $X(t)$ is a quantity that has zero average value $\overline{X(t)}$ and a nonzero mean-square value $\overline{X^2(t)}$. The noise voltages or currents that are normally encountered are of such a nature that averages taken over a short time interval are independent of time. Such noise processes are said to be *stationary*.

If the fluctuating quantity could be measured directly, it would be sufficient to determine its mean value $\overline{X^2}$. But often the fluctuating quantity is amplified and filtered, and then there may be no simple relationship between the mean square of the input and the mean square of the output of the electronic equipment.

To better characterize the noise in such cases, one introduces the *autocorrelation function*

$$\overline{X(t)X(t + s)} \tag{A.31}$$

of $X(t)$, which tells how long the fluctuation at t *persists* at later times (for $s > 0$) or how it *builds up* from earlier fluctuations (for $s < 0$). Generally, $\overline{X(t)X(t - s)} = \overline{X(t)X(t + s)}$.

Next the spectral intensity $S(f)$ of $X(t)$ is introduced by the definition

$$S(f) = 2 \int_{-\infty}^{\infty} \overline{X(t)X(t + s)} \cos 2\pi f s\, ds \tag{A.32}$$

That is, $S(f)$ is twice the Fourier transform of $\overline{X(t)X(t + s)}$. If $\overline{X(t)X(t + s)}$

has zero value for $s > \tau$, then $S(f)$ is constant for $2\pi f \tau \ll 1$; such a spectrum is called a *white* spectrum.

This spectral intensity has the following properties:

(a)
$$\overline{X^2} = \int_0^\infty S(f)\, df \qquad\qquad (A.33)$$

That is, if we know the spectral intensity, we can reconstruct $\overline{X^2}$ by a simple integration process. Hence we have not lost anything by introducing $S(f)$.

(b) If a fluctuating signal $X(t)$ of spectral intensity $S(f)$ is applied to an amplifier with a gain function $g(f)$, then the mean-square value $\overline{Y^2}$ of the output signal $Y(t)$ is

$$\overline{Y^2} = \int_0^\infty |g(f)|^2 S(f)\, df \qquad\qquad (A.34)$$

This solves the filter problem mentioned earlier. It is beyond the scope of this book to give proofs of these theorems.

We can formulate these results as follows. In any electronic equipment any noise-current generator $i(t)$ can be represented by an equivalent-current generator $(S_i \Delta f)^{1/2}$ in a small frequency interval, and any noise emf $e(t)$ can be represented by an equivalent emf $(S_e \Delta f)^{1/2}$ in a small frequency interval Δf. Then we can find the mean-square output signal of the equipment by applying steady-state circuit analysis and adding the effects of all the frequency intervals Δf quadratically. In practice this amounts to integrating over the passband of the equipment.

We can now see how the spectral intensity $S(f)$ can be measured. Let the signal $X(t)$ under investigation be fed into a narrow-band amplifier tuned at the center frequency f_0. Unless $S(f)$ changes very rapidly with frequency, the value of $S(f)$ will be nearly equal to $S(f_0)$ for all frequencies for which $g(f)$ has an appreciable value. Consequently, Eq. (A.34) may be written

$$\overline{Y^2} = S(f_0) \int_0^\infty |g(f)|^2\, df = S(f_0) g_0^2 B_{\text{eff}} \qquad\qquad (A.35)$$

where g_0 is the midband value of $|g(f)|$ and

$$B_{\text{eff}} = \frac{1}{g_0^2} \int_0^\infty |g(f)|^2\, df \qquad\qquad (A.35a)$$

is the effective bandwidth of the system. *Narrow band* means that $B_{\text{eff}} \ll f_0$. Measuring $\overline{Y^2}$, g_0, and B_{eff} yields $S(f_0)$.

A.2b. Thermal Noise and Shot Noise

The thermal noise of a resistance R at the temperature T is caused by the random motion of carriers. Its magnitude is given by *Nyquist's theorem:* The thermal noise of a resistance R kept at the temperature T can be represented by a noise emf $[S_v(f)\,\Delta f]^{1/2}$ in series with R or by a noise-current generator $[S_i(f)\,\Delta f]^{1/2}$ in parallel with $g = 1/R$, where

$$S_v(f) = 4kTR, \qquad S_i(f) = 4kTg \qquad (A.36)$$

and k is Boltzmann's constant. This is white noise.

As a consequence, one can characterize the noise of any two-terminal device with an open-circuit voltage spectral intensity $S_v(f)$ by a noise resistance R_n, and any two-terminal device with a short-circuit current spectral intensity $S_i(f)$ by a noise conductance g_n with the help of the equations

$$S_v(f) = 4kTR_n, \qquad S_i(f) = 4kTg_n \qquad (A.37)$$

Equation (A.36) is a direct consequence of the equipartition theorem, for if the resistance R is connected to a capacitor C, then the mean-square value of the voltage v developed across C is given as

$$\overline{v^2} = \int_0^\infty \frac{S_v(f)}{1 + \omega^2 C^2 R^2}\, df = \frac{kT}{C} \qquad (A.38)$$

so that $\frac{1}{2}C\overline{v^2} = \frac{1}{2}kT$, in agreement with the equipartition theorem.

Nyquist's theorem holds for any conductor at thermal equilibrium. It also holds for most conductors through which direct current is flowing, as long as the carrier density in the sample does not fluctuate. If the carrier density shows spontaneous fluctuations, however, because carriers are generated and recombine, then the resistance of the sample will fluctuate, and the direct current flowing through the sample transforms these fluctuations into a fluctuating emf at the terminals that shows up as noise (generation–recombination noise). The effect is absent in semiconductors in which the carrier density does not fluctuate, which is the case, for example, if all donors and acceptors are ionized.

Shot noise occurs as follows. Suppose that a series of events occur independently and at random at the average rate n. The actual number of events, n, occurring during a particular second will fluctuate around its average value $\bar{n}$. Then

$$\text{var } n = \overline{(n - \bar{n})^2} = \overline{n^2} - (\bar{n})^2 = \bar{n} \qquad (A.39)$$

and the spectral intensity $S_n(f)$ of this fluctuation $\Delta n = (n - \bar{n})$ is

$$S_n(f) = 2\bar{n} \qquad (A.40)$$

As an example, take the emission of electrons by the cathode of a saturated diode. Then $\bar{n} = I_d/e$ is the average number of electrons emitted per second, where I_d is the emission current. Consequently,

$$S_n(f) = 2\bar{n} = \frac{2I_d}{e} \qquad (A.41)$$

But if $\Delta n = n - \bar{n}$ is the fluctuation in number emitted per second, then the fluctuating current is $\Delta I_d = e \, \Delta n$; hence its spectral intensity is

$$S_i(f) = e^2 S_n(f) = 2eI_d \tag{A.42}$$

This called *Schottky's* theorem. It holds for any phenomenon consisting of a series of equal, independent, random events carrying a charge e.

As a consequence, one can characterize the noise of any two-terminal device with a short-circuit current spectral intensity $S_i(f)$ by its equivalent saturated diode current I_{eq} with the help of the equation

$$S_i(f) = 2eI_{eq} \tag{A.43}$$

If the events occur at random but are not independent, it is no longer true that var $n = \bar{n}$. Equation (A. 40) then becomes

$$S_n(f) = 2 \text{ var } n \tag{A.44}$$

of which (A.40) is a special case.

In semiconductors we introduced the generation rate $g(t)$ and the recombination rate $r(t)$ with equilibrium values $g_0 = r_0$. But $g(t)$ and $r(t)$ describe random processes to which Eq. (A.40) can be applied. The spectral intensities are therefore

$$S_g(f) = 2g_0, \qquad S_r(f) = 2r_0 \tag{A.41a}$$

The equations in this section have been stated without proof. The readers who are interested in proofs should consult textbooks on the subject.

A.2c. Noise Figure

It is often convenient to refer all noise sources in an amplifier back to the input. One can then represent the noise by an equivalent emf $(\overline{e_{eq}^2})^{1/2}$ in series with the input resistance R_s or by an equivalent-current generator $(\overline{i_{eq}^2})^{1/2}$ in parallel with the input conductance $g_s = 1/R_s$; $\overline{e_{eq}^2}$ and $\overline{i_{eq}^2}$ can now be compared with the thermal noise of the resistance R_s and the conductance g_s, respectively. The spot noise figure F of the circuit is defined by the equations

$$F \cdot 4kTR_s \, \Delta f = \overline{e_{eq}^2}, \qquad F = \frac{\overline{e_{eq}^2}}{4kTR_s \, \Delta f} \tag{A.45}$$

$$F \cdot 4kTg_s \, \Delta f = \overline{i_{eq}^2} \qquad F = \frac{\overline{i_{eq}^2}}{4kTg_s \, \Delta f} \tag{A.46}$$

The two definitions are equivalent, since $\overline{e_{eq}^2} = \overline{i_{eq}^2} R_s^2$. Obviously, $(F - 1) \cdot 4kTR_s \, \Delta f$ represents the noise of the stage minus the noise of the source resistance R_s. F is often expressed in decibels.

The noise figure of an amplifier is easily measured. To that end a saturated diode is connected in parallel with the source, and its current I_d is adjusted in such a way that the output noise power of the amplifier is doubled. Then

$$\overline{i_{eq}^2} = 2qI_d\,\Delta f \quad \text{and} \quad F = \frac{q}{2kT}I_dR_s \simeq 20I_dR_s \qquad (A.47)$$

We are thus able to define and measure the noise figure of amplifiers or of individual stages. In most cases only the first stage contributes to the noise figure of the amplifier.

REFERENCES

VAN DER ZIEL, A., *Noise: Sources, Characterization, Measurement.* Prentice-Hall, Inc., Englewood Cliffs, N.J., 1970.

Appendix *B*

Short Table of Physical Constants

Velocity of light:
$$c = 299{,}792.9 \pm 0.8 \text{ kilometers/second}$$

Avogadro's number:
$$N = (6.02472 \pm 0.00036) \times 10^{23} \text{ per gram molecule}$$

Electronic charge:
$$e = (1.60207 \pm 0.00007) \times 10^{-19} \text{ coulomb}$$

Electron rest mass:
$$m = (9.1085 \pm 0.0006) \times 10^{-31} \text{ kilogram}$$

Proton rest mass:
$$m_p = (1.67243 \pm 0.00010) \times 10^{-27} \text{ kilogram}$$

Planck's constant:
$$h = (6.6252 \pm 0.0005) \times 10^{-34} \text{ joule second}$$

Boltzmann's constant:
$$k = (1.38042 \pm 0.00010) \times 10^{-23} \text{ joule/degree}$$

Electric conversion factor:
$$\epsilon_0 = \frac{10^7}{4\pi c^2} = 8.854 \times 10^{-12} \text{ farad/meter}$$

Magnetic conversion factor:
$$\mu_0 = 4\pi \times 10^{-7} = 1.257 \times 10^{-6} \text{ henry/meter}$$

Index